The Falklands Factor

The Battle that Changed the Shape of Ships

The Royal Navy's first wartime experience of a subtle new threat. On May 4th 1982, HMS *Arrow* attempts to fight the fires raging in HMS *Sheffield*. Specifically designed to protect the Fleet from aircraft and missile attack, *Sheffield* was detected by an Argentine Super Etendard aircraft from 15 miles away. The plane descended to sea-level, then 'popped up' to 200 ft, risking entry into *Sheffield*'s radar horizon to lock its missile's radar onto its target. It then descended back to sea-level and at 9 miles fired the missile, leaving the area after a quick look at the damage to its target.

Sheffield was the first Royal Navy ship to be sunk in action since the Second World War.

'The Old and the New ...'

TOP: Pre-Falklands: Type 42 – Batch 3 Destroyer HMS *Edinburgh*, ordered 1979
BOTTOM: Post-Falklands: Type 45 Destroyer HMS *Daring*, ordered 1999
(Photos of ships are to the same scale)

The Falklands Factor

The Battle that Changed the Shape of Ships

David Laurent Giles

Foreword by the Late Admiral Lord Hill-Norton

Unicorn Press

First published in Great Britain in 2025 by Unicorn Press
60 Bracondale
Norwich
NR1 2BE

tradfordhugh@gmail.com
www.unicornpublishing.org

A CIP record of this book can be obtained from the British Library

ISBN 978 1 7391640 7 2

Design by newtonworks.uk
Printed in Malta by Gutenberg Press

Contents

Foreword

by The Late Admiral of the Fleet the Lord Hill-Norton, GCB, ADC (formerly First Sea Lord, Chief of the British Defence Staff and Chairman of the NATO Military Committee)

This is the story of a man who had a brilliant idea and, by his courage and determination in the face of every conceivable obstacle, drove it through to a successful conclusion.

His idea was for an entirely new form of ship design and propulsion, for both warships and merchantmen. Convinced that this mould-breaking and revolutionary innovation would bring with it a dramatic improvement in warship performance, he offered it to the Royal Navy. It was at once rejected, not by the professionals but by the Technical Department responsible for warship design – the Royal Corps of Naval Constructors – with contumely and contempt.

A vigorous controversy soon arose between those professionals who liked and supported David Giles's notion in principle, and those, like the RCNC, who did not, and the dispute became known as the Short/Fat versus Long/Thin argument. It was much discussed in the Wardrooms of the Fleet, where a clear majority of the younger officers were sure that Giles's ideas had great possibilities.

I became involved in the affair in about 1983, six years after I left my last effective appointment as Chairman of the Military Committee of NATO, when some supporters asked me, as a former First Sea Lord, to set up and run an informal Committee to examine the problem.

Because my personal inclinations were generally on their side, I agreed to do so. The Hill-Norton Committee, as it became known, was composed of half a dozen men well known in various disciplines and with impressive qualifications. We made it clear that we had no access to any official, much less confidential, documents in a report and submitted (unasked) to the Prime Minister. We were, of course, unpaid and minor expenses were paid for by Giles's supporters.

We reported that we had no doubt that the evidence we had examined made it clear that the apparent advantages of the Short/Fat hull were strong enough to demand a formal investigation by an Official Committee of Inquiry to be established under the Chairmanship of a learned Judge or Queen's Council. The Prime Minister was sufficiently impressed to order that such an official Inquiry be set up to examine the matter in depth. She appointed the Defence Secretary, George Younger, to set up an Inquiry and he appointed Lloyds Register of Shipping under Mr Roderick MacLeod, its Chairman.

The Warship Hull Design Inquiry lasted for over a year and is said to have cost £1 million. It reported in July 1988 and published its detailed findings in two volumes with a long and technical Report which made it clear that Mr MacLeod could find no support for the view that Giles's design had any merit for warships of the size of destroyers and frigates. Mr Younger could hardly do otherwise than accept the opinion of his proper advisers.

The Short/Fat lobby, to which I by this time belonged, felt unable to accept this verdict and held, for a wide variety of reasons, that the official Report was seriously flawed. I must at once reveal that The Report was not only flawed but tainted, as Mr Giles learned when he was sent, anonymously, a copy of a loose minute from the Chairman of the Hull Committee of the Defence Scientific Advisory Council of the Ministry of Defence, informing the Inquiry, even before it had begun, of what conclusions it would be wise to reach. I have never, in some seventy years on the Active List of the Royal Navy, previously encountered so disgraceful an act by a Government servant in a similar influential and senior position. It was, however, typical of the behaviour of such powerful vested interests before, during and after the Report was published.

The reader will learn of the remarkable campaign of lies, half-truths and dirty tricks mounted by these same vested interests in this splendid book. Suffice it for me to say here that this campaign made it clear to Giles that he would have to look elsewhere to find the support he needed if he were to bring his ideas into reality. So he packed his bags and went to the United States.

October 13th, 2003

Prologue

Lafayette Square, Washington DC: St John's Church lies behind the statue of General Jackson astride his rearing horse. The brick-fronted Federal Courts Building can be seen on the right, on the site of what was previously Commodore Rodgers's house.

The United States and, in particular, Washington, is a place where I had been living and working continuously or periodically, since my boyhood there during the Second World War. So it was entirely appropriate that October 16th 2016, the most important day of my working life, should occur in the heart of the USA, in Lafayette Square, just across Pennsylvania Avenue from the White House. It was also the most precarious: the culmination of a thirty-five year battle over my patented and controversial naval hull design with two of the world's greatest Navies.

It was half an hour's drive from my apartment in Alexandria, Virginia, on that brilliant autumn morning, past the gold-flecked maples fringing the George Washington Parkway and across Memorial Bridge to Lafayette Square. As I drove, I knew that my life was bound to change: for the better if the US Law upheld my patents; for the worse if it upheld the defence contrived by the US Department of Justice.

Walking across Lafayette Square to the US Federal Claims Court reminded me of scenes from the five years of my Washington childhood: the famous statue of General (later President) Jackson, flourishing his tricorn hat; also St John's Church, where we were taken most Sundays. We had been forced to leave our home on the South Coast of England, due to my father, a yacht designer, being transferred from his quaint Lymington office to work for the Royal Navy's Design Department in Bath. Thereafter, with my mother, sister and twin brother, we were sent to the USA, where we spent much of the War living with our widowed Great Aunt Freda Giles and her brother, the gruff old Admiral William Rodgers. As grandson of the great Commodore John Rodgers, he would entertain us with rousing tales of his driving the British out of Washington at the end of the War of 1812.

In 1943 my father, frustrated after three years working on naval vessels within the Admiralty bureaucracy in Bath, joined us in Washington. Instead he was employed by the British Air Commission selecting US-produced materials involved in the design and construction of wooden aircraft: the Mosquito bomber and the D-Day gliders. It was he who, from the first, led me into the world of planes, and ships and the sea. On Sundays, after Church, flying model aircraft below the Washington Monument; or sailing model yachts on the pond that is now the World War 2 Memorial. Or, later, sailing in an old crabbing skiff on the Chesapeake Bay.

*****

During the 1980s my company, Thornycroft, Giles & Associates Ltd, operating from a tin shed on the Isle of Wight and a garret in South London, had caused a major public controversy in Britain. Our so-called 'Short, Fat Ship' was the subject of secret, unauthorised tank-testing and design analysis by some of the most powerful defence interests in the country: the Royal Navy, the Ministry of Defence, and British Shipbuilders. With others, in 1987, we had won a High Court action as Osprey Ltd, defending our design copyrights. After seven years we were fully vindicated by a favourable settlement, causing great offence and several 'early retirements' within Government, Naval and industrial circles. Today, thirty-five years later, the same story had been repeated. Now it concerned the infringement of my US Patents by the US Navy, represented by the US Department of Justice.

Surely, I thought, on the evidence, they would not be allowed to get away with it. But going to court is an uncertain business. Particularly when up against the near-infinite resources of the US Government and Lockheed Martin, its largest defence contractor – and with both determined to discredit your claims. You have to wonder if taking on such an unequal contest is worthwhile. But we had all the evidence we needed and the best people in America to present it, all financed, not by Contingency Lawyers, but, personally, by our own Patent Agent. What more could one ask for to strengthen one's confidence?

*****

This is the story of a battle against the military-industrial complexes of two governments: in the UK and USA. Both, while publicly discrediting my ideas, were apparently covertly intent on using them. First, major UK Naval interests were caught secretly investigating an elaborate series of tank-tests of a number of my company's designs. Then an Official UK Inquiry, ordered by Prime Minister Thatcher for a similar public evaluation, was caught in a blatant act of subterfuge, as described by Lord Hill-Norton in his Foreword. Later, for a further sixteen years, my work in the USA, apart from a major commercial ship-design project, involved the US Navy's infringement of my 1990 US Patent. All of its claims were incorporated in a 'revolutionary' new class of US Navy Fast Frigate, the 47-knot Littoral Combat Ship, or LCS, the fastest-ever ocean-going warship. Hence my reason to enter the US Federal Courthouse on that October day in 2016.

Then, returning to Britain in 2021, it describes the clear evidence that the Ministry of Defence and British Aerospace Systems's newly-introduced Type 45 Destroyer incorporates my Company's key design principles shared with them in the 1980s. These allow a large hull to benefit from 'hydrodynamic lift' with increasing speed. Over a twenty-five year development, this has become the essential feature in resolving a key lesson of the 1982 Falklands Campaign, allowing an increase in the hull's beam, or width, without penalty. Contrary to the 'received wisdom' of the day it provides greater stability to install a much higher and greatly-improved air-defensive radar capability, without loss of speed or excessive power: 'An important contradiction to traditional theory', and a major solution to the Falklands Factor.

The Opening of the Falklands Campaign

On a calm April evening in 1982, with my family, I sailed our 36-foot ketch into the charming Puerto Andraitx in Majorca. The first thing we noticed was a huge Union Jack being flown by a German yacht. We couldn't understand why, until its owner hailed us: "Glückwunch! You're at war with Argentina."

My family immediately wanted to know what would happen to me? For, despite my on-going Court Case against British Shipbuilders, I was simultaneously in discussion with leading MOD Ministers, Admirals and senior executives of British Aerospace Systems. Surprisingly, these powerful sources also supported me in my attempt to introduce a new type of hull, more adapted to the future needs of air-defence than the traditional 'long, thin' designs that were to prove so vulnerable to air attack a few weeks later. I told my family: "A phone will probably ring."

In a few days a motley Task Force was assembled. It comprised a variety of frigates and destroyers, one new and one Second World War aircraft carrier, and a number of commercial ships and cruise liners, including *Queen Elizabeth 2*, for transporting helicopters, stores and troops. Two weeks after the Task Force sailed for the Falklands, my phone rang as expected. It was Geoffrey Pattie, the Minister of State for Defence Procurement. He wanted to start moving immediately on our 'Short, Fat' Sirius, or 'S90', an enlarged version of our Osprey patrol vessel in service with the Danish and other navies. It would act as an alternative to a traditional 'Long, Thin' frigate, the Type 23. We should agree a plan of action in co-operation with Admiral Sir Raymond Lygo, the Chief Executive of BAE Dynamics, and several other leading defence contactors. He would arrange for us to meet Vice-Admiral Sir Lindsay Bryson, Controller of the Navy and Third Sea Lord. He claimed that Bryson was not a seaman, but a career engineering officer, the first non-seaman ever to be appointed Controller. He had spent much of the previous years working in company with our chief opponents, the Royal Corps of Naval Constructors. The Naval Staff Requirement for the proposed Type 23 frigate had just been issued and I would receive it as a working document, together with security clearance, but it did not incorporate any lessons from the Falklands. "On yer bike!" said Pattie, with his customary brevity, and rang off.

Days later, on the afternoon of May 4th, I was driving down Primrose Hill, in North London, with a colleague when we heard the news flash on the car radio. The MOD spokesman, Ian McDonald, told us in his iconic staccato phrases: 'HMS *Sheffield* has been struck by an Exocet missile ...

It was fired from an Argentine aircraft ... This has caused a fire and some casualties ... The ship has been abandoned.' Later that evening we met Geoffrey Pattie at the House of Commons. It was clear that a traditional design had been found wanting, with tragic consequences. Therefore we must press for more agile, stable and capable air-defence ships with greater urgency. "Get the skids under your project", said Pattie. He would give us his full support.

Warship design now became a political football after a quarter of our surface fleet had been sunk or disabled in the Falklands in only two months. The politicians had to show a disgruntled Parliament and electorate that alternative designs were being considered, despite their commitment to Privatisation based on existing projects. Pattie gave us a copy of a letter from Admiral Bryson to Sir John Charnley, the MOD Controller of Research Establishments, with a copy sent to Sir Ronald Mason, the Chief Scientific Adviser to the MOD. Bryson's letter mentioned the possibility of a short research programme for our Sirius. It was uncomplimentary about the potential of our project: 'These contentions have been based on a Standard Methodical Series for resistance and sea-keeping trials on the Osprey in model and full scale.' The only 'standard methodical series' ever applied to Osprey was the elaborate unauthorised tanktests of British Shipbuilders and others, the subject of our current High Court Action. Likewise, the only full-scale Trials Reports were those of the Ospreys, *Havørnen* and *Indaw*, that we had passed to the MOD in confidence. Mason had replied that 'there would be much value in funding a modest programme ... *since it is important that the study be perceived from the outset as unbiased and independent.*[1] [my italics]. Mason's words gave us some hope that, in the light of the Falklands experience, the MOD might be seriously considering an alternative, the S90, a Shorter, Fatter frigate. However his remark about it 'being perceived as unbiased and independent' sounded ominous. Was that not to be assumed in such cases?

Yet we enjoyed good support from the Navy's seafaring branch. A friend who had attended the production of a TV programme on the Falklands Campaign told me how Captain Sam Salt, former Commanding Officer of the ill-fated HMS *Sheffield*, had pointed to a drawing of our sturdy, beamy S90, saying: "That's the sort of ship we need."

1 Memorandum, Sir Charles Mason to The Controller of the Navy, 7/6/82.

A sea-change was overdue in warship design, still encumbered by a century of 'received wisdom'. But all this did not suddenly spring into my life in the 1980s. It was something engendered in the mind: partly genetic, partly based on youthful experience, as in Wordsworth's 'voluntary power instinct'. Imparted to me by my father, an innovative yacht designer, it was my youthful inspiration leading me ever towards boats, ships and the sea.

A sea-change: Two recent classes of warship, each incorporating the author's patented feature of Dynamic Lift in large hull designs.

February 2006: Launching of the first Type 45 Destroyer, HMS *Daring*, incorporating the lessons of the Falklands. Claimed by the RN to be 'The most capable destroyer ever' and 'The world's best air-defence ship'.

September 2006: Launching of USS *Freedom*, a 3,450-ton, 45-knot 'Short, Fast Frigate', first of a new class of warship, based on the lessons of the Gulf Wars.

CHAPTER 1

'Jack of All Trades'

> Time present and time past
> Are both perhaps present in time future
> And time future contained in time past.
>
> TS Eliot: 'Burnt Norton'

Steaming Back to Britain

For us nine-year-old twins the war ended in Washington with the climax of VJ Day: parties, fire engine rides and waving flags. We left New York in September 1945 on the old Cunard liner *Aquitania*, the last of the great four-stackers, to return to our home in Lymington on the Solent. With clouds of smoke rising from her four tall funnels, we steamed past the Statue of Liberty, the New York skyline receding fast. We joined gangs of British kids as they raced around the former luxury liner with unlimited access, exploring the maze of passages, shouting at the top of our very American voices. To live with the pulsating heartbeat of the four propellers and the great steam turbines, to watch the seething whirlpools churned up astern and to hear the constant humming and singing of the wind in the rigging of the masts and funnels. It was a thrilling experience.

On our last evening aboard, I stood on deck with my father as we entered the English Channel. In the watery sunset there was a whole fleet of Breton Tunnymen under tanned sails becalmed on a pink, windless swell, their rods hoisted high into the air, like ghosts from a bygone age. Without any mention of his war work designing decoy Tunny boats to transfer agents across the same waters, father drew on his pipe and said, "That's Europe for you – another world, far from America."

We manoeuvred slowly into the Southampton docks, or what was left of them. The skyline was a mass of broken buildings with great gaping spaces between them. The walls at the Cunard dock, close to the railway station, were riddled with shrapnel and bullet holes. Some had collapsed, with makeshift huts dotted among the rubble. Everything seemed very small, the few cars looking like toys.

The station itself was a ruin: roofless with some of the platforms blown away. Eventually a small, black tank-engine arrived, pushing four dirty green carriages. Our baggage was tossed heavily into the luggage van and Mother led us to one of the few compartments with an interconnecting toilet, a quaint feature unknown to us from our trips in America's corridor carriages. The guard blew his shrill whistle and waved a green flag. With much chuff-chuffing, clinking of chain couplings and banging of buffers, we clattered down the line. It reminded me of the clockwork Hornby trains I had left behind in Washington.

After a short ride through the heath and woodland of the New Forest, we arrived at Brockenhurst Station: "Braakn'urst ... This is Braakn'urst," rang out the loudspeaker, in its harmonious Hampshire accent. "Change 'ere for Limminin' Tain, Limminin' Pier, Yaaarmth and the Oile o' Woight ...!"

We were home!

'The Myth'

In October 1946, there was a diphtheria outbreak at my boarding school, and we were sent home for three weeks.

One morning during our short holiday, we were introduced to a balding man with a cheerful round face and sparkling eyes. This was the legendary ocean-racing skipper Captain John Illingworth, who had established my father's reputation as a designer of ocean racers before the war with his *Maid of Malham*. His name was usually mentioned with reverence at home and I was surprised to meet such a mild-mannered person when so many of my father's clients seemed quite ferocious. The two of them disappeared into the sitting room, where my father had set up his drawing board. There they pored over lines and sail plans for most of the day.

There are many views on who did what in the design of this 'revolutionary' and eccentric craft, which was also known by my father as the 'Ugly Duckling'. However much these two men had battled over the plans in the sitting room, the outcome was supremely successful and changed forever the nature of the ocean racing yacht. My father did the hull lines and the structure: light, but of extraordinary strength, owing much to his wartime work with wooden aircraft. But there was much of Illingworth's inventive genius in the sail-plan, subtleties of the interior design and gadgets for trimming the sails.

There followed long and bitter arguments with the Royal Ocean Racing Club's Official Measurer about whether innovations and improvements should be penalised in the interest of maintaining fair competition between yachts of widely differing age and design. There were groans and grunts as my father loudly argued his case for innovation over the hall telephone while, in the kitchen, my mother would bang and crash the pans in noisy protest at the endless negotiations. However little I understood of the detail, I was fascinated by the passion of his sense of innovation: how he reacted to his beloved *Myth* being insulted as 'one of those monstrous square shapeless boxes', as she was described in the pages of the official history of the Royal Ocean Racing Club (RORC). In some ways foretelling the future battle with his erstwhile wartime Naval Constructor colleagues and the contempt with which one of their kind had coined the phrase 'The Short, Fat Ship'.

The summer of 1947

A freezing winter of snow, strikes and power cuts gave way, surprisingly, to the sunlit days of a glorious summer. They seemed endlessly perfect, like the Chesapeake summers, thanks partly to the retention of the wartime measure of 'double summer time' that gave us sunlight almost to midnight. I shall relate my first sight of the *Myth*, exactly as I remember it. It was off Cowes; the yachts were circling in the balmy Solent morning, awaiting the start which was due at noon. With their dipping sheers and deck-lines, graceful long, pointed bows, sterns and bowsprits, they rise and fall on the gentle swell as they await a breath of fickle wind: a fashion parade of classic yachts.

Slowly creeping round the breakwater was the ugliest little boat I had ever seen. What was worse, she was being towed. She had sawn-off ends, quite vertical at bow and stern; her deck-line seemed hogged, unlike the sweeping hollow sheer of other boats. Her mast was tall and almost amidships; she had a long low deckhouse like a casing; she resembled a submarine in some ways. And she was going to win the Fastnet? I was not impressed. Yet ... because she looked so modest and unexciting, I rather hoped she'd pull it off. It would be a real feather in Dad's cap: none of his boats had ever won the Fastnet.

Myth of Malham: Twice winner of the Fastnet Race in consecutive years. Her owner, the great ocean-racing skipper John Illingworth, claimed 'She altered the face of ocean racing.' And a Myth she did become.

A few days later we sat listening to the BBC 6.30 West of England News. The measured voice of the announcer went through a few items of West Country interest and paused:

> "This year's Fastnet Race has been won by a radical new type of ocean racer, the *Myth of Malham*, which despite being one of the smallest boats in the race, crossed the line well ahead of many much larger yachts."

I felt a great wave of pride. So things that don't look as everyone expects are not necessarily wrong. And things may look the way they do for a good reason – in the *Myth*'s case it was to save weight in her hull, using the principles of lightweight wooden aircraft structures that my father had developed during the war. He thus could transfer the weight saved down to the keel, 'where it would earn its keep', as he said ... just like our model. So she sailed fast even in bad weather, when many yachts chose to heave-to. She was more efficient than those yachts that looked beautiful, some for beauty's sake. I had learned the first essential of good design: in pursuit of efficiency you cannot please everyone: hence 'Jack's Ugly Duckling'. Yet she did employ a number of the lessons learned on those Sundays spent messing about with models beneath the Washington Monument. In particular, her vertical centre of gravity: hence her stability.

With my mother's approval, father maintained a drawing board in the sitting room in order to continue work after returning from his office on Lymington's cobbled Quay Hill. As we boys listened to Dick Barton on the radio, he would be hunched over his designs muttering to himself, humming tuneless ditties as he pushed his weights and splines around the board; or scratched intricate sketches of minor structural details onto linen-backed paper, with carefully composed notes on their execution in Indian ink. Or, puffing on an empty pipe, he would sigh as he gazed across the garden towards the distant salt marshes with the placid waters of the Solent glinting beyond. There was always some exciting new idea taking shape on my father's drawing board, where he worked late into the evening designing not only boats, but also sheet winches, runner levers and other yachting paraphernalia.

With my brother, I would visit small boatyards with him during school holidays, savouring the aromas of mahogany, oak, western red cedar, ash or, rarely in those times, teak; with the more pungent smells of Stockholm tar, anti-fouling, fresh paint and varnish all adding to the sensual experience. There was also the music of boat building – the clumping of mallets upon chisels and adzes upon wood, the shriek of band-saws, the constant hammering to clench the copper rivets and the soft hollow sounds which echoed back through the wooden hulls. As my father sighted the sheer

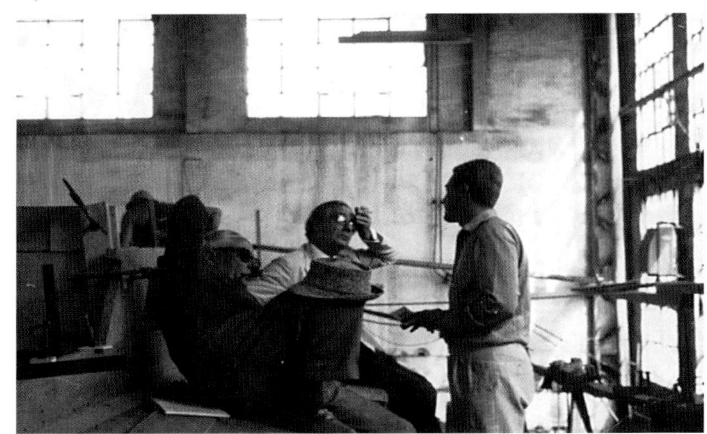

A passion for boats: Father (centre) building *Susanna* in Italy with Vittorio Beltrami, the great yacht builder (left), and Snr. Brainovich, her owner.

strake of a small Vertue class sloop, he would beckon me to come and look with him. I would lay my head on the gunwale and gaze along the length of her deck to discern the flow of that saucy 'nip' in the sheer towards the stern, so dear to his eye and his style.

For me that same passion stirred one afternoon on the waters of the Solent. For fifty pounds, father had bought us a twelve-foot sailing dinghy, which had been retired from service at the Dartmouth Naval College: a Dart One design. With a huge 'standing lug' sail, her hull was varnished elm and mahogany but, with a heavy steel centreboard, she leaked like a sieve. Father's saying was: "One boy sails; the other bales."

Dart was on a broad reach in the Hurst Narrows, sailing fast in the strong wind that often blows in from the west as the sun goes down. In a sudden gust, as the wind hit her huge sail, she shot forward; my father and I had to sit far out on the windward gunwale to keep her level, as she sliced through the waves in a cloud of spray. "We're planing!" I cried out.

"Move aft, get your weight back in the boat, otherwise she'll bury her bow," my father shouted above the roar of wind and wake. The beamy little *Dart* had suddenly accelerated into a new dimension of speed. Instead of pitching up and down across the short, steep Solent chop, the boat seemed to lift and race through the waves, almost level, with a sheet of spray thrown up on either side. The rudder sent a trembling pulse through the lightly built mahogany hull and the water flew cleanly off her stern as she cut a smooth, flattened wake through the water. Father explained that we weren't 'planing', but surfing, or 'semi-planing'. Rather than skimming across the surface of the water like a speedboat, *Dart* was being lifted, very slightly, by the crest of what he called the 'captive wave' generated beneath her hull's slightly hollowed stern. "It's like an aircraft wing lifting gently to the onrush of air at take-off," he said ... "That's why we had to move our weight towards the stern: to compensate for the lift beneath it – otherwise *Dart* might bury her bow in an oncoming wave."

Dart.

"Could a big ship be made to behave like *Dart*?" I asked, remembering my trips to and from America in the old *Samaria* and *Aquitania*.

It was all a matter of proportion, he replied. No marine engine could ever be as efficient as *Dart*'s sail, which dissipates none of its thrust through exhausts, gearboxes, shafts or thrashing propellers. For a ship of such size to semi-plane, some new type of engine was needed: much lighter, more powerful and more efficient than the heavy engines and the boilers and steam turbines – and the propellers – used by ships in those days. Something like the new use of jet propulsion in aircraft. But, in the future, given stronger, lighter material for hulls, sails and masts – and sufficient wind – he thought good-sized ocean racing yachts might also semi-plane: "Like big dinghies," he said, as we sailed back to our mooring in Oxey Lake. At home he showed me how the rounded bottom of a teaspoon, held beside the water flowing from a tap, was sucked into the water, while the hollowed, or scooped, side was pushed out: "An exaggertated lesson of what was happening beneath little *Dart*," he said, "as she rose to the lift of the captive wave generated beneath her slightly hollowed stern ... A traditional rounded or straight stern is sucked down into the water just like that teaspoon, creating a big increase in drag and, therefore, power."

I would sneak into the sitting room when he was away from home, to pull back the dustsheet that covered his drawings, feasting my eyes on designs for the latest ocean racer or cruising boat. I would dream of sailing across oceans, exploring in the wake of the great seafaring authors like Claud Worth or Alain Gerbault, whose books were firmly locked up in my father's bookcase until I found where he kept the key. Thus he failed to discourage my desire to enter the hazardous and unprofitable world of yacht design: "Something a family could only afford once in ten generations." His income from 'chocolate boats', as he called them, being augmented by his mother hailing from the Fry's Chocolate family.

*****

Having signed up with the Royal Naval Volunteer Reserve in 1953, my last year at school, I was assured a place in the Navy for my National Service as an Ordinary Seaman. During a day-trip in HM Submarine *Scorcher*, one of her 'VR' officers described the uniquely informal society of the Submarine Service. In 1954, promoted from Upper Yardsman to Midshipman, I took a Submarine Officer's Qualifying Course at HMS *Dolphin* in Gosport.

Drafted to join HM Submarine *Trenchant* in Malta, I was the most junior officer in the First Submarine Squadron, commanded by the redoubtable wartime submariner, Captain 'Jackie' Slaughter. He, together with many of the wartime 'boats' – and a few of their submariners – were still serving with me in Malta.

Being classified as a 'Submersible Boat', *Trenchant* did not bear the exalted title of Her Majesty's Ship. About the oldest submarine in operation, she'd survived the entire war with a fine record of sinking twenty-one enemy ships, including, in twenty-five minutes, the 14,750-ton Japanese cruiser *Ashigara* – by five torpedoes from a salvo of eight. Life in an ancient submarine was rather like a small sailing yacht: dark and damp – but congenial. The crew was a motley collection of 'roughy-toughies' who might put the Pirates of the Caribbean to shame. 'An illigitimate, piratical fringe' as a well-known submariner has termed it. A daily record of their lives at sea and ashore was captured in a scurrilous periodical known as *The Oily Bilge*, named after a particularly oleaginous part of a submarine's internal anatomy. Edited by the charming, literary, but incorrigibly mischievous Able Seaman Albert Plumeridge, his on-board library included several classics of the French avant-garde Olympia Press: "'Ere y'are, Middy – this'll teach you a thing or two!" he said, passing me a copy of *The Story of O*, a renowned erotic classic of the day.

The editor of *The Oily Bilge*: Able Seaman Plumeridge, the quintessential submariner of those days.

During eighteen months in *Trenchant* we visited many Mediterranean ports, to be royally entertained by the local British Consul or other dignitaries. My father disparagingly referred to it as a 'Med cruise in a submersible yacht'. We joined NATO exercises with the US Sixth Fleet and others. We also carried out a few long-distance sonar trials in the mid-Atlantic with one of our Home Fleet 'boats', *Thule*, equipped with a secret high-performance listening or 'side-scanning' sonar. Out from Gibraltar, in mid-Atlantic, she was reputed to hear the *Queen Mary* entering New York about an hour after the event, the speed of sound under water travelling at about three times its speed in the air. Alas, every time we joined up with *Thule* – maybe 200 miles away from our original rendezvous – there was that same damned Soviet 'Whisky' class submarine trailing us. All our orders were in special code via the Portland Admiralty Underwater Weapons Establishment. It later transpired that our position, but little data of use, was communicated to Moscow via a Ruislip bungalow by the moles of the infamous Portland Spy Ring. We abandoned the trials, with the Whisky trying to maintain 'silent routine' – but loudly present on *Thule*'s amazing sonar – about fifty miles away, we did however learn how inferior the Soviet equipment must have been. I left my time with a great affection for the Royal Navy, in particular the Submarine Service and its people. Not so for my ancient 'T' class submarine with its afflictions of old age and some inexpert design features – such as a wayward Induction Hull Valve that, but for a 'loo-chain' and flap-valve, sometimes threatened our survival. Yet, for all its faults, it did create a wonderful sense of bonhomie amongst its crew.

During the two 'gap years' of my late departure occasioned by the Suez Crisis and the end of National Service, before university I worked in a factory as a lathe setter-operator making parts for cars and aircraft. In my spare time, in our driveway, I converted *Minion*, an open-cockpit day-boat of my father's design, into a miniature cruising boat with three bunks, galley with a gas stove, chart table and 'heads'. Later, being Quartermaster of the Lymington to Isle of Wight ferry was about the most entertaining job I ever had. It included steering the ancient coal-fired paddle-steamer *Freshwater* between Lymington and Yarmouth, 'Oile o' Woight'. Like most old paddle

Our 'Mini-*Myth*' *Minion* at Ile de Tatihou. My twin sits atop the 'doghouse' and, lying astern, our 4-foot Prout Coracle folding dinghy, called 'Sauvetage pliant' by the local fishermen.

steamers, she was not a manoeuvrable vessel. In a strong south-westerly wind, due to her high varnished mahogany superstructure aft of the bridge, she needed a jib to blow her bow around the Cocked-Hat mark at a tight bend in the Lymington River. My namesake, Captain 'Jinxer' Giles, would lower the sash window of the bridge with a crash, shouting to Willy Harper on his steam winch: "Hoist the jib!" The passengers cheered loudly as the ancient tanned sail was hoisted slowly up the forestay.

Culture and 'Airnarchy'

Between 1958 and 1961 I spent three years at Oxford, mostly under the tutelage of John Bayley and Christopher Tolkien: practical academics, one of whom had driven an army lorry and the other a Spitfire during World War II. It was an intellectual hothouse, full of temptations and pleasure. But there were consolations away from Oxford too, with the 'high thinking and low living' of reading parties in an Alpine chalet or sailing around the St Kilda and Muckle Flugga, the northern extremeties of the British Isles, in a small boat of my father's design. It gave much-needed increment from the old *News Chronicle*, which sponsored the trip.

Charming, unforgettable: Oxford was an experience of enchantment, but an intellectual hothouse which I, for one, couldn't endure for more than my allotted three years. As an intense cultural adventure after the years in the Navy, the engineering factory, the ferry and boat-building, it was intellectually stimulating; but by the end, aged twenty-six, I was glad to 'buckle to' and go forth into the real world.

My immediate choice was to follow another teenage dream. Inspired by the Comet jet airliner and Britain's 'white heat of technology' in other 1950s aeronautical projects, I joined the de Havilland Aircraft Company. While they were seeking a graduate in engineering or a related scientific discipline, although an arts graduate with some practical engineering experience, I applied. I was surprised to be accepted on two conditions: one was that I attend the local 'Tech' for evening classes in aeronautical engineering; and the other that I should apply my training in the Humanities to a special task – I was to write a critique on the design and failure of the de Havilland Comet 1, which, with three disastrous crashes in a few months, had shattered the dream of Britain's aviation supremacy.

As a member of a sales engineering team in the Aerodynamics Department, working on comparative performance of competing aircraft and the development of new designs, more from the economic than aerodynamic viewpoint. It was exciting work. Involved in the great 1960s decade of jet aircraft development, I worked with many original designs, from the DH 125, the first business jet, the HS 146, to the Trident and HS 132, an early progenitor of the Airbus. We were in an age of pioneering technology, including a 'super-critical' wing to reduce subsonic drag for the Airbus – and beyond to supersonic speeds.

At de Havilland's, I learned a subtle trick of aerodynamics that was to dominate much of my working life: the Supercritical wing, or 'Roof-top Rear-loaded Aerofoil'. This reduced wing drag-rise at transonic speeds between subsonic pressure, or 'wave-drag' (as it's called in both aero- and hydrodynamics) and supersonic or frictional drag caused by the increase in air density at transonic speed; also raising skin-friction and heating, at supersonic speeds. Father called it 'A semi-planing wing', as his teaspoon test had shown how Dart's concave stern employed dynamic lift, conversely, to reduce the transition from friction to wave-drag. (Also see **Appendix 1**, Figs. 2 and 3.)

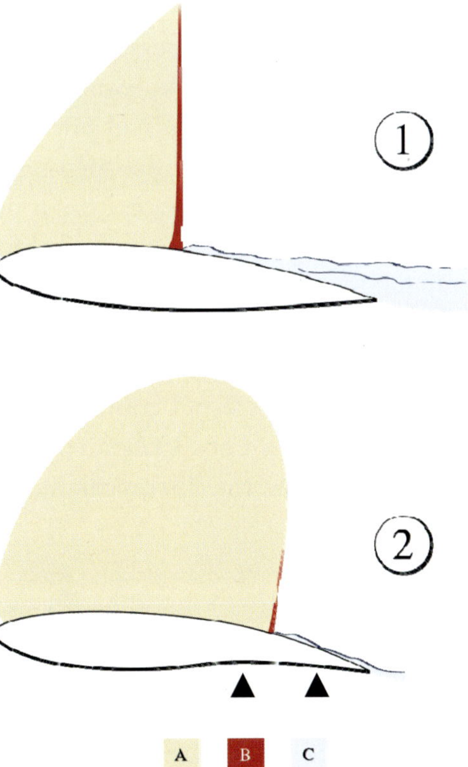

Cross-section (1) of a normal wing. Cross-section (2) of the de Havilland 'Roof-Top Rear-Loaded Aerofoil', with the concave 'Hook' lifting the trailing edge, indicated by the arrows. The wing thus assumes a more level attitude, to reduce lift, passing from the dominance of lifting, or wave-drag, to frictional drag as it enters the increasing air-density at Mach 1. This also reduces the shock-wave that occurs at the transition from sub- to supersonic speeds. Conversely, a hull at increasing speed above its limiting frictional-drag speed, or Froude Number, can obtain increased lift as it transitions from the dominance of frictional drag into the domain of pressure, or wave drag. (See **Appendix 1**, Figs. 2 and 3.)

So began an experience that led me through the most exciting period in the evolution of the modern jet airliner. It offered great challenges, a steady income and a close affinity with the elements, in this case the air. When I joined de Havilland's in 1961, the world was still in love with the style and beauty of the new jets in all their variety of design and layout. There were arrangements of two, three and four engines: at the back of the fuselage, within, under and over the wing, high tail, mid tail and low tail designs. However, by 1972, design improvements had become incremental, based on an economic formula to carry the greatest number of passengers with the least number of engines: 'hot bottoms in cold seats', as they say in the airline business. The inspiring vision of a Comet being replaced by the commercial realities of an Airbus.

During my time at de Havilland's I met Lord Cohen, formerly Chairman of the Official Committee of Inquiry into the catastrophic losses of three of the early Comet 1s. He was convinced that the critical cracking and failure of the fuselage skin was *not* 'beyond the knowledge of current metallurgy', as his Inquiry had officially concluded. More likely, it was due to the structural penalty of burying the bulky DH 'Ghost' centrifugal-flow jet engines in the most highly-stressed part of the airframe – the wing-root – thus realizing the aerodynamicist's dream of a 'clean wing'. The American Civil Aeronautics Authority had previously refused certification of the Comet 1 for operation by Capital Airways in the USA on those very grounds. To carry sufficient passengers for a profit, risks were taken to save weight in other parts of the structure. Most US prop-driven airliners had been flying for several years at lower altitudes, with lower cabin pressure and frequency of service – but with thicker pressurized fuselage skins. This was endorsed by many of my de Havilland engineering colleagues. It was all due to the eternal conflict between economics, engineering and aerodynamics. As Lord Cohen concluded gloomily, he was 'only the Judge; the Jury were the Technical Experts' – and they were defending their

The Comet 1, showing the clean wing with engines buried in the wing-root.

backsides. So much for Public Inquiries, as I was to learn to my cost many years later.

'*Vale!*': Farewell to Father

Father had an uncanny genius for dreaming up names for his boats, such as *Myth of Malham*. Towards the end of his life he designed the last of four yachts for a redoubtable Yorkshireman, Colonel Norman Birch, Commodore of the Humber Yawl Club. His earlier boats had been called *Maid*, *Rose* and *Rosette* – all '*of York*'. Now he invited Father to design his last boat. In a moment of inspiration, Father suggested *Vale of York*: a single word embracing both the beauty of the Yorkshire vales and the Latin for farewell, '*vale*'.

Father had first tried working at Camper & Nicholson's under the great designer 'Old Charlie' Nicholson for a few months. But he always wanted to do things differently. He had learned a lot: chiefly how to do it Old Charlie's way, which was usually very good ... but father had other ideas. There must have been a better way to approach some of the problems than just 'pretty yachts' with their long overhangs and heavy hulls: expensive to build and short on usable interior volume. It was the antithesis of what was needed for popular family cruising: boats like his famous 'Vertue' class five-tonners, which he thought should be the real future for sailing boats. Which it was – and there are still many sailing today. Later, one of his last designs became the most numerous class of family cruising yachts in the world: the Westerly Centaur, of which 2,500 were built – alas, long after selling his shares to his partners for a pittance.

"If you think of all the painstaking detail and passion which went into so many of these early proposals that were never built, you'll understand why there's not much left over for you boys and Tam," was the gist of the words he intoned so gloomily. Finally, near to death, when I asked him what it was that had kept him going, I only obtained one short response: "Passion and battles, old boy. Passion for the force of wind and sea. Battles with people."

CHAPTER 2

1973: Back to Boats

During 1971–72 I spent a year in Southern California working on a Pan American Airways project, which involved the use of Dassault Falcon business-jets operating as commercial aircraft. Alongside me at Pan Am, a brilliant young entrepreneur, Fred Smith, was contemplating use of the same planes on another startling innovation: his fledgling Federal Express. For my wife Vanessa and me, with three small boys between two and six, our time in California was a great experience – although she hardly shared my enthusiasm for America. The climate was wonderful and entertainment for the boys easy: the choice of beaches, mountains or desert made for a relaxed existence. But I had always intended to return to the sea and boats once my father was no longer there to remind me of the financial perils involved. The idea of a 'lifting' hull, with propellers for smaller ships – and an equivalent to jet propulsion for much larger ones – which my father had light-heartedly raised during our afternoon in the dinghy *Dart*, twenty-five years before, was beginning to take hold.

If the design of commercial aircraft had become largely routine by 1973, innovation seemed possible in the naval field. Faster ships were needed to match the increasing speeds and operating depths of Soviet nuclear submarines and torpedoes. For faster, larger ships a better form of propulsion was needed, in view of the poor efficiency – and loud noise – of propellers at speeds above thirty-five knots. Lightweight marine gas turbines, derived from aircraft jet engines, were starting to replace the heavy boilers and steam turbines in smaller warships. A marine gas turbine uses jet power to turn a shaft – just like a prop-jet engine in an aircraft, while sucking in and exhasting air through ports in the funnels. It can drive a propeller or the newly emergent water jet, the water jet being more efficient at higher speeds in larger hulls. Gas turbines were still very thirsty, however, and when fuel was low, sea-water had to be flooded into the fuel tanks to maintain stability of those 'long, thin' hulls. An unhappy arrangement since, in heavy weather, fuel and water can mix into an emulsion that cannot be burned efficiently. Above the waterline, guns were being replaced by missile launchers. Missiles were lighter in weight and fewer in number, with no need for magazines full of shells and warheads which, like the weight of boilers and steam turbines, were carried in the bottom of the ship, increasing its stability. Radar aerials had to be mounted as high as possible above the waterline to extend the ship's offensive and defensive horizons. Thus weight was moving ever upwards, eroding the limited stability of the long sleek destroyer hulls, still based on the universally accepted principles of naval architecture refined by Brunel's railway engineer-cum-naval architect, William Froude, in the 1860s.

A job in the marine business would give me a good opportunity to introduce my ideas, inspired by the simplicity of the de Havilland wing, into the military or commercial nautical fields. But, whatever cropped up, it was clear that I could never survive as a 'big company' man. I had to follow my own judgement and ideas, rather than the corporate policies of either the large aerospace or shipbuilding companies. I was looking for the big vision; it was there, somewhere out at sea. It was no longer in the air. It was unlikely to come from the ship owners, institutional naval architects or shipyards; I would have to find it on my own. Micawber-like, I waited for something to turn up.

During my time with Pan Am I had introduced the designer Jon Bannenberg to Avions Marcel Dassault to style the interior of Monsieur Dassault's personal Falcon jet. In 1973, when Jon heard that my venture with Pan Am had ended, thanks to the OPEC fuel crisis, the Yom Kippur War and the collapse of our chief sponsors, Clarkson's Tours, he invited me to work as Project Manager for a new series of 70-foot fast motor-yachts being built in Italian yards, partly to his design. These fast, lightweight motor-yachts had much in common with an aircraft. 'Shoe-boxes', he called them, their hard-chine hulls being formed from four flat surfaces of plywood. Weight-saving was at a premium in order for the hull to perform efficiently – but their flat surfaces pounded severely at any speed in only moderate waves. Jon's designs were revolutionary, hence his disregard for these boxy 'ply-flyers'. He changed the whole style and appearance of large motor yachts to their present rounded, sculpted contours. He also had a keen appreciation of my father's early contribution to this process. We would sit sipping espresso in Portofino or Rapallo looking out over the yachts, some of them my father's creations, talking boats into the early hours. Jon would sketch out dream ships and I would quiz him on the emerging technology of gas turbines and water jets. When we visited the yards, the naval architects would show us sketches of motor-yacht hulls, while Jon would bring them life, with a sudden flourish of his ball point – slanting a mast or deckhouse here, or altering an aft-deck there – leading them towards a new generation of the fulsome, rounded mega-yachts of today.

In the yards of the Ligurian coast, my father's name was often greeted with kind words and sometimes a glass was raised to honour the memory of 'Il Maestro'. However much I wished to make my own way, it was always a delight for me to hear of the great regard that still existed for my father amongst the Italian yards where some of his finest creations took shape. But I was looking for a new type of hull for a further series of motor-yachts: one that might combine greater space with superior seaworthiness at speed.

Jon introduced me to Commander Peter Thornycroft, who was famed for his design of the semi-planing Nelson launches. 'Semi-displacement', as they are sometimes called, seemed a misnomer, as their displacement is not reduced, compared with traditional fast so-called 'displacement' hulls. They were broad in the beam but, with their subtly hollowed 'lifting' underwater hull lines, are able to maintain higher speeds comfortably in heavy weather and without excessive power. For this reason they had been adopted as the standard design for the Trinity House Pilot Boats (see **Appendix 2**, Fig. 1) and as tenders for the Royal Yacht *Britannia*. Hundreds were in worldwide service. Thornycroft had recently designed a larger version as a Patrol Craft for the Mexican Navy, three times the size of the Nelson 40: the 'Azteca' class. Twenty-one of these were being built in Clyde shipyards and a further ten in Mexico. Jon arranged for me to attend the sea trials of the first on the Clyde in September 1973. I had once driven a Nelson 40 through wild weather in the Channel and relished that same exhilarating speed and smooth passage through choppy seas that I had first experienced in little *Dart* so many years ago.

Enter Peter Thornycroft

"Wonderful morning for sea trials!" A jolly voice rang out across the dining room of the Marine Hotel at Troon. Looking up, I saw a raffish nautical type tacking between the tables like a boat seeking its mooring in a crowded anchorage. A red spotted handkerchief flew from the breast pocket of his reefer jacket like a burgee, while the blue and white Docksides, frayed navy-blue shirt and scarlet tie all complemented the 'shabby-nautical' flavour of his attire. He brandished a battered yachting cap with a tarnished Royal Yacht Squadron badge which, removed with a flourish, revealed a distinguished head of silver hair.

It was 'The Commander' joining us for breakfast after his trip from London by overnight sleeper: "Ah now, porridge and Spithead Pheasant ... that's why I came here for breakfast," cried Peter. "The English hotel kipper, these days, is no more than a boil-in-the-bag plastic invertebrate." But this was the authentic Loch Fyne variety with backbone, head and fins intact. We watched in awe as he dismembered the glistening pelagian with the skill of a surgeon. I was immediately relieved. Here was a kindred spirit – not a stuffy withdrawn type like some other naval architects I had met. With his charming turn of phrase and cheerful outgoing manner, he even reminded me of my father – and, like him, he had successfully introduced innovation into the obscure mysteries of hull design.

"Today we'll confound all those skeptics at Vosper's who said we'd be lucky to get over twenty knots ... I call it the Nelson Factor – You'll see!"

Breakfast over, we headed for the shipyard. As we drove past the famous Troon golf course, wind rocked the car and the rain lashed across the greens and bunkers, large waves breaking on the beach. Ducking through the driving rain, we ran across the quay to the *Andrés Quintana Roo*, the lead ship of the Azteca class, to be greeted by John Halbert, the businessman who had masterminded the deal. He was accompanied by members of the Mexican Navy, a Lloyd's Register Surveyor and a skipper from the Ailsa Shipyard.

"Blowing hard from the southwest," Peter shouted above wind and rain. "The ebb tide coming down the Clyde will have stacked up a pretty good sea – just the weather we need. We could run south towards Ailsa Craig, see how she takes the weather coming in from the Irish Sea and return to do some speed runs on the calmer waters of the Paisley measured mile."

The Trials were a huge success. The boat surged ahead to 26 knots and, as she picked up speed, moving through the highest waves without the usual pitching or slamming expected in these conditions. "Steady as a rock," said the Lloyd's surveyor. Unlike many fast boats in my experience, at full speed her wake was extraordinarily flat, apparently allowing her an untroubled passage through the water without the usual fountain-like cascade of water dissipating the propellers' thrust.

Conversely, she had a high bow-wave, of which Peter quoted the words of 'Old Charlie' Nicholson, the great yacht designer: "It's easier to push water than to pull it."

On the Paisley trials course she exceeded the speed estimates of the Vickers' Dumbarton test-tank by over three knots. Immediately the Scottish 'Champagne' was produced – the best Islay vintage. Knocking it back, Peter asked for a telex to be sent to his former colleagues at Vosper Thornycroft, boasting that his wide-beam design had far exceeded their own speed estimates, or the 'Nelson Factor' or (1–x), as he called it rather than (1+x) as usual for conventional hulls: "From Mr. 26 to Mr. 22 knots". In such circumstances, irresistible; but it created an enduring impression in the Halls of Naval Architecture – like the first Torpedo Boats of his grandfather.

An 'Azteca' patrol craft of the Mexican Navy at sea, en-route to Mexico.

Over dinner in the sleeper back to London, Peter described his family history: how his grandfather, the great Victorian engineer Sir John Isaac Thornycroft, had changed the shape of ships. In his Chiswick boatyard in the 1860s, he built his 'quick steam launches' for Thames river parties, the only craft allowing the ladies to keep up with their gentlemen in the new Racing Eights. On hearing that Thornycroft's launches made 16 knots, the designer of the *Great Eastern*, John Scott-Russell, exclaimed: "Either Thornycroft's a liar or he can't read a stop-watch." John Thornycroft's first semi-planing hull, the 30-foot *Gyrinus*, later won two gold medals for powered craft in the 1908 Olympics, being the only boat to finish over a stormy course off the Isle of Wight. Later his 40-knot 'skimming', or planing, 35-knot Coastal Motor Boats were used in the Royal Navy's successful attack against the Bolshevik fleet at Kronstadt in 1919: the forerunners of the Motor Torpedo Boats, or MTBs, of World War II.

As a small boy, Peter had watched his 80-year-old grandfather investigating a series of extraordinary hull-shapes in his private testing tank on the Isle of Wight. I felt certain that Peter had learned much from this experience, scaling up the lessons of tiny models and small craft to the size of the 112-foot Azteca. After leaving the Royal Corps of Naval Constructors in 1956, Peter established quite a cult for his semi-planing Nelson launches among the smart Cowes set: 'the seagoing version of the Range Rover,' as some described them. His Keith Nelson Yard was a cluster of tin sheds on the side of Bembridge Harbour. Across the road he built a snug hideaway by setting piles in concrete poured into the hulk of an old Thames barge, the *Blackwater*, lying sunk in the mud. There he was surrounded by Thornycroft memorabilia – his favourite being a postcard of

An inventive genius at work on breaking the 'Laws': Sir John Thornycroft with models at his private testing tank, April 1916. The black model on the right is the revolutionary semi-planing hull *Gyrinus*. The three pale models to its left are 'skimming' hulls used for coastal motor boats in World War I. To the left are 'ideas' with lifting foils beneath their sterns.

one of his grandfather's steam lorries, dated 1898, from Khartoum. The scrawled message read, "Dear Thornycroft: We have several steam lorries in the Soudan. Yours are the best ... Kitchener." Little did I realize that old Sir John had also made lorries and even elegant luxury cars. But the *Blackwater* was the perfect place to dream up new boat designs.

Peter recounted how, in 1962, he had won the first order from Trinity House for his 40-foot fast pilot boat to be built in the new material, GRP, or fibreglass. On trials, the first Nelson 40 went about two knots faster than expected from the tank-tests – and she seemed to have just a little sparkle in her performance that the earlier boats had lacked. This may have been due to a slight hollow or 'tuck in the buttocks' which, above its traditional hull-speed, stimulated a beneficial high-pressure captive wave – slight but significant – very like the effect of the hollow beneath the trailing edge of the de Havilland wing. This gave slightly more lift for a little less drag. The Nelsons all had it as they became larger. They were highly successful and sold in their hundreds all over the world. They are still being built today. Later, Peter had briefly worked for the residue of his old family company, Vosper Thornycroft, after selling them the building rights to some of his Nelson designs. However, he found the bureaucracy and 'traditional wisdom' of some ex-MOD naval architects at Vosper's hard to accept. In 1970, having towed his caravan-cum-office out of Vosper's car park, he established his own firm, now called TT Boat Designs, in the same old tin sheds on the side of Bembridge Harbour. He then set up his drawing office in the *Blackwater*, with a brilliant young naval architect and draughtsman, Arthur Mursell, to assist him.

Peter Thornycroft and Arthur Mursell at work in the *Blackwater*.

As we finished our dinner, I asked Peter the question I had put to my father so many years before: could a semi-planing hull be made much bigger, perhaps the size of a destroyer or even larger? Could there even be economies of scale in ships such as I had experienced in jet airliners, which grew to over seven times their size over fifteen years – from the Comet 1 to the Boeing 747 and beyond, into a new drag regime beyond the Sound Barrier? Could a much larger vessel benefit from the lift generated by its own 'captive wave' to overcome the restraints of prohibitive wave-drag?

"Yes," he replied, "I believe it could."

However, he insisted, this would be true only if you maintained the same proportions of length, volume and power. The speed should then also rise proportionately with a constant – rather than exponential – increase in power above the hull's 'Threshold Speed', or traditional limiting 'Froude Number' based on the traditional view of 'wave-making resistance'. That is the speed above which power increases exponentially, or 'prohibitively' with higher speed in a classic 'long, thin' destroyer hull. Thus a shorter, wider hull – like the Azteca – can reach a higher speed without prohibitive increase in power, as shown in today's Trials.

Alas, he emphasised, to increase the size of the 112-foot Azteca to only 150 feet, one was up against the so-called 'laws' of naval architecture and – more importantly – those who believed in them with almost religious intensity. A wide hull would require greater power than a longer, thinner hull of the same displacement, or volume of water displaced by the hull, even if that would be outweighed by significant advantages of stability, sea-worthiness and reduced construction and operating costs. The suggestion that such a hull, enlarged, would 'lift' gently above a certain speed, thus reducing the effects of 'wave-making resistance', was mere 'heresy', among certain High Priests of Naval Architecture within the MOD – as we were to learn, to our cost and theirs, in the future.

CHAPTER 3

The Wave

After the sea-ship, after the whistling winds,
After the white-gray sails taut to their spars and ropes,
Below, a myriad, myriad waves hastening, lifting up their necks,
Tending in ceaseless flow towards the track of the ship
Waves of the ocean bubbling and gurgling, blithely prying,
Waves, undulating waves, liquid, uneven, emulous waves ...

Walt Whitman: 'After the Sea-Ship'

An aircraft is totally immersed in one fluid: air, whose density diminishes with altitude – but increases with speed. A submarine, when dived, is totally immersed in another: water. But a ship moves at the confluence of both. Air is compressible, while water is incompressible and about 820 times the density of air. Thus a tsunami has far greater destructive power than a hurricane-force wind. Think of the destruction of Bandar Aceh in 2005: a whole city flattened by an incompressible wall of water moving at a mere 30 mph. A wind of equivalent force would have to blow at about 17,250 mph due to the different density of wind and water. As previously mentioned, there is a similarity between aircraft and ships in overcoming the effects of pressure – or 'wave-drag', as it's known in both Aero- and Hydrodynamics. There are similarities between Mach and Froude Number, as between the transonic 'boom' and a hull's captive wave, with critical speed in each case affected by the different density of air and water.

The elements of drag that slow the movement of a ship are more complex than those that apply to a plane or a submarine. That is why few people, save naval architects, understand the precepts that determine the shape and performance of ships and boats. Even today, naval architecture remains a complex discipline, combining elements of science, mathematics and art. Its so-called 'traditional wisdom' still remains at the mercy of the unpredictable forces of Nature and the infinite interpretations of mankind.

Newton's Third Law of Motion tells us that, for every action, there is an equal and opposite reaction. A breeze blowing across the water generates waves. The longer and harder the wind blows, the longer and higher the waves become. The ripples on a lake are only a foot apart, inches high and move at walking pace. But an ocean wave may have travelled for thousands of miles. Its crests can be over a thousand feet apart and thirty feet high: a huge mass of water moving at thirty knots across an expanse of open sea. A mini-tsunami.

A ship moving through water creates drag according to its shape, size, proportions, submerged volume (or displacement) and speed. This drag is visible in the waves and wake it generates. In the 1860s William Froude measured this phenomenon by comparing the drag, or 'resistance', of model hulls, of different length, beam and displacement as they were towed through his Torquay testing tank. By comparing the results he established his basic formula, or Froude Number, whereby a ship's maximum practical speed is defined by its length, beam and displacement. Longer, narrower ships, according to Froude, were able to go faster than shorter, wider ones, due to the increased length of their self-created, or captive wave – similar to ocean waves whose speed relates to the length between their crests. This remained the received wisdom among naval architects for the next thirty or so years. Making the hull longer and narrower would allow an increase in practical speed up to the point at which a traditional rounded hull begins to sink, creating a prohibitive increase in drag (or power), for any higher speed. Long narrow hulls were faster, but they were less stable, seaworthy, spacious and more costly to build. At the end of the 19th century when speed was the priority and higher, more efficient power became available, as in John Thornycroft's torpedo boats and destroyers, the extreme 'long, thin' hull was the only choice. But there remained another way to overcome Froude's Wave Barrier, the very opposite of aerodynamics: reducing a wing's lift to penetrate the increasing air-density beyond Mach's Sound Barrier. So, in the case of a hull, causing it to lift, to 'swim instead of sink', at that critical speed as it enters the domain of wave-drag. However, in certain Halls of UK Naval Architecture, that was only accepted for 'quite small craft', as explained by the MOD and the millions of pounds spent attempting to discredit such a simple principle during the 1980s. So, if tank-tests scale up for a larger sinking hull, why not for a lifting hull? Quite apart from the traditional shapes of hulls, it was impractical to increase the speed of large ocean-going ships, due to the limited efficiency of propellers at speeds above 30 knots, as in planes when approaching the sound barrier.

Peter and I both believed that this traditional wisdom could be overcome by subtle variations of hull shape, or 'hull form' of warships and in much larger ships, by more efficient lightweight high-powered gas turbines. In such commerical and military ships, at higher speeds, water jets would be more efficient than propellers. This might apply in much larger hulls than the Nelsons and the Azteca and other 'small craft' to which the principle had hitherto been limited, as we proved in later projects, such as the 32,500-ton/40-knot FastShip.

Like any fluid, water prefers to travel in a straight line. If it encounters an irregularity in a passing surface, it avoids Nature's abhorred vacuum by accelerating to catch up with itself. As demonstrated by my father's 'teaspoon test' years before, a slight bulge or straight upward profile from the keel, in the underwater surface of a hull will create low pressure, or 'sinkage', increasing its displacement of water as speed increases. A slight hollow will create upward pressure, or 'lift'. This can lift the hull, reducing the amount of water the vessel displaces at higher speeds and thus a significant portion of wave-drag. Given cunning design, the hull can be wider for its length, more stable and roomy, and thus able to carry a greater useful load at higher speeds, without excessive increase in power, sustained by its captive wave on the same principle as an aircraft's wing. To me this seemed a concept of great simplicity – even beauty – similar to the principle of flight. It had been proven in little *Dart*, in the Nelson power boats and now in the Azteca. It had been applied in principle to larger patrol boats of up to 370 tonnes – but without a significant increase in beam. So why not, with a wider beam, or wing area, in something larger? Thus the enlarged semi-planing hull seemed the simple way to overcome the restraints of Mr Froude's 'Laws'.

As my father had forcast thirty years before, the propulsion now existed for much larger ships to exceed the limiting speed of propellers, as the jet engine had moved to large aircraft into a new era of flight: the Jet Age.

Preconceptions

> 'Even if the propeller had the power of propelling a vessel it would be found altogether useless in practice, because of the power being applied in the stern it would be absolutely impossible to make the vessel steer'.
>
> (Sir William Symonds, Surveyor of the Navy 1837, on Brunel's idea for the Screw Propeller)

> 'The laws which govern the wave-making resistance of ships are not yet fully understood.'
>
> (Sir William White, Director of Naval Construction: *Naval Architecture*, 1910)

> 'Most of the teaching in naval architecture was based upon the premise that 'this is how it is done' and was much concerned with imposing what the wisdom of previous generations had developed.'
>
> (KJ Rawson, Deputy Director of Ship Department, MOD (Navy), *Ever the Apprentice*, 2006, p. 40)

> 'Elementary knowledge of the effect of Froude Number on a ship's wave-making resistance would have ruled out such a form for frigate size and speed.'
>
> (Prof. David Andrews, RCNC, FRINA, 2005, on the S90 Project)

> 'Even if the tank-tests showed the traditional theories were wrong, the Navy would not accept them.'
>
> (Admiral Sir Henry Leach, BBC *Newsnight*, Dec. 1982)

Preconceptions have always encumbered the progress of ship design. From the days of the Greek trireme and, later, the Viking longship, speed had always been associated with a 'longer, thinner' hull.

However, stability – the hull's ability to resist capsizing – has usually been of even greater importance. Increasing the beam of a hull provides a two-fold advantage: it increases both the useful volume of the hull and its stability, thus allowing a greater load to be carried, and higher in the ship. That is why the Viking merchant ships were 'short and fat': but, by comparison with the famed 'longships', they were slow and ponderous. Sailing against the wind, as they could, the more warlike longships obtained the necessary stability by their warriors sitting up to windward, in addition to stone or other ballast – thus transferring everything to the other side each time they tacked ... as the Shetland Scows do to this day.

After the Greek and later galleys with their human engine rooms, ships from the Vikings onwards followed similar principles until the arrival of the clippers in the nineteenth century. These changed the whole nature of ship design. Inspired innovation in hull form – based on observation and experience as much as mathematical calculation – gave them the ability to combine high speeds with a good commercial load.

Cutty Sark: For her size and tonnage, perhaps the fastest merchant ship in history.

Basil Lubbock's classic, *The Log of the 'Cutty Sark'*,[2] explains what a contradiction she, and other clipper ships, presented to Froude's preconceptions. Her displacement relative to her length was nearly three times that of the *Queen Mary* and her beam-for-length 40 per cent greater – about the same as an Azteca. Yet, in high winds and seas, she could maintain a speed of 17 knots, a Froude No. of 0.35, for several hours: a relative speed for her length that has yet to be exceeded by any conventionally-powered ocean-going cargo ship. Even under shortened sail she could generate greater power than any possible combination of engine and propeller in a ship of her size. Her fine, deep bow was inherited from the earlier clipper *Tweed* and her hull sections came originally from the Baltimore and Boston privateers that had out-sailed the Royal Navy in the war of 1812. Neither flat-bottomed nor generously rounded in the tradition of previous sailing merchantmen, her hull had cambered sections, known as 'dead-rise', angled upwards either side of the keel, reducing the downward suction of the hull at speed.

Cutty Sark also embodied a 'tuck in her buttocks': a pronounced hollow in her wide-arching 'counter', that part of the stern that was only immersed at her highest speeds in high wind and waves. 'Its power is very apparent ...

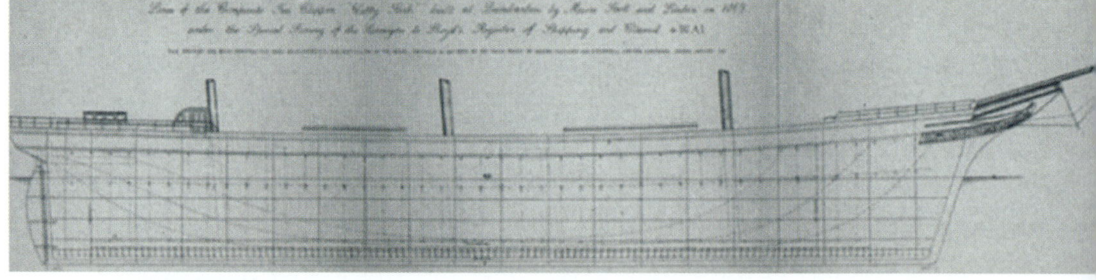

The secrets of the *Cutty Sark*: lines plan showing the concave lines of her stern, or 'tuck in her buttocks' (left side of the drawing) and her fine bow (right side).

2 Published by Chapman & Co., Glasgow, 1928

Cutty Sark: Showing the slight concave 'tuck in her buttocks', from the port side.

compared to the short bird-like counters of the earlier 'cracks' in the tea trade', wrote Basil Lubbock. This generated the beginning of a high pressure wave which, above about 15 knots, could sustain her stern from sinking, providing a degree of lift as she was driven at a steady 17 knots, the same relative speed as the SS *United States*, the world's fastest liner. Was this the genius of her designer, Hercules Linton?

Many longer, narrower clippers, like the ill-fated *Ariel*, were faster in modest winds, but in heavy weather the ship could become unmanageable with a high risk of broaching, rolling onto her beam-ends sideways to the huge seas, usually ending in disaster. Eric Newby describes this dramatically in 'The Last Grain Race'. Many believed that was how *Ariel* was lost without trace in the Southern Ocean in 1872. By contrast, the great Captain Woodget, in Lubbock, wrote of the extraordinary performance of his *Cutty Sark*: "She was a wonderful runner: she was never pooped and kept wonderfully dry aft. I never hove her to and always ran everything out." Was her stern uplifted by the power of her captive wave?

The clippers were doomed when the world of shipping changed radically in the 1870s. This was thanks to three factors: the opening of the Suez Canal, the development of effective steam power, and the model tests of William Froude in his experimental tank at Torquay, that allowed mathematical discipline to be applied to ship design. With the advent of the Suez Canal, the new all-steam ships were no longer dependent on a fair wind and the need to maintain high speeds in the huge seas of the Southern Ocean. Froude had demonstrated how they could steam on a minimum of power from their primitive engines, for a modest calm water speed. The new routes tended to follow coastlines, with intermediate ports being created for trade and coaling. With the opening of the Panama Canal in 1914, only a dwindling number of sailing ships, the 'Windjammers', continued to traverse the wild waters of the Southern Ocean.

Although Froude translated the mysteries of hull speed and stability into a mathematical formula depending on the length, displacement and proportions of a hull, he showed little concern for its underwater shape. The discipline of naval architecture became constrained by mathematical theory, which has persisted ever since. Only in the experiments of more liberal and questing minds, like those of Sir John Thornycroft and his 'quick launches' scaled-up to destroyers and, later his planing and semi-planing hulls – or the designers of racing yachts – have attempts been made to break away from the obstacles raised by 'traditional wisdom'. They still respect the fundamental tenets of Froude, but without being obstructed by them. Thus, in the right conditions, today's ocean racing yachts achieve speeds far in excess of Froude's 'Laws'.[3]

Someone who has worked with the mysteries of flight is more likely to look at the opportunities for comparable benefits from high pressure when an object moves through water. Similar contradictions of lift, sink and drag will occur, although due to a hull's buoyancy and the much greater density of water, at much lower speeds and with different effect. That is the difference: one that some naval architects seem unable to accept. Sir William White's 1910 pronouncement on the lack of understanding of 'wave-making resistance' is typical. But John Thornycroft had grasped its significance when he designed his first semi-planing boat, *Gyrinus*, in 1908 – as did his grandson, fifty years later.

3 Not all naval architects are so rigid in their thinking, freely acknowledging that their science (or art, as some claim) is still at the empirical stage, and therefore their 'laws' are subject to revision. My conversations with the leading yacht designer Olin Stephens, Professor RE Bishop of Brunel University, Ray Adams of Ocean Transport and Michael Ayres of Townsend-Thoresen were particularly informative in this respect. Nor has any leading UK or international testing tank ever given any hint of a scaling problem with our large semi-planing hulls of up to 37,500 tons.

CHAPTER 4

Designing a Hull

Like humans, most ships carry the genes – and some of the character – of their forebears. As John Thornycroft said, "a destroyer is *practically* an enlarged torpedo-boat." What he did not say was that the torpedo-boat was itself based on his experience with his fifteen-foot 'quick' Thames launches. Likewise, his first destroyer, the 260-ton HMS *Daring*, was enlarged, or 'scaled-up', using the same design principles over the course of 100 years, into thousands of 'Long, Thin' destroyers for all the world's navies. Likewise, my father used to say that his sailing boat designs, whether cruising or ocean-racing, all harked back to ideas first proven in earlier designs, often by other designers.

So one starts with a design that is known to work well and builds up from there. Peter Thornycroft took the 40-foot Nelson and scaled it up to 50-feet, then 75-feet and, finally, the 112-foot Azteca. The success of this encouraged further experiments. Through several stages, and over many years, we would subsequently scale this up to a ship of over 700 feet in length, and 42,500 tons and 45 knots, powered by water jets; as confirmed by several international testing tanks, Classification Societies and MIT's Department of Ocean Engineering. At each variant, or scaling up, the designer incorporates the lessons of the 'parent-form', adjusting the hull's underwater shape according to the relative speed requirement – or Froude Number. Even at such size, the power-for-speed remains 'efficient'. Assuming the same category of hull, one then comes up with a lines plan showing the three-dimensional contours of the chosen hull form. This is the point at which, based on the lines plan, the designer has to create a scale model of the hull, allowing for the weight and distribution of hull-structure, propulsion, machinery, accommodation, armament, stores, crew, passengers, etc., taking account of the resulting effect on the stability of the hull in calm and rough water – and at different speeds and heading to the waves. If there's a need to refine the most efficient hull variant for a certain application, it is necessary to test a 'standard methodical series' of hull variations based on the 'parent-form' to establish the best combination, or 'hybrid'. Then comes the Big Question: 'Will it work – and, if so, how well?'

What are Testing Tanks – And Why are They So Important?

> 'The short fat ship school insisted that William Froude in 1872 and every ship model towing tank superintendent around the world ever since, had somehow failed to see their contention that pushing a sponge in the bath broadside on, made smaller waves than pushing its narrow face forward.'
>
> (Peter Usher, CBE, PRINA, 2006)

The Testing Tank at SSPA, Sweden, with paddle-generated waves visible.

A testing tank is the marine equivalent of a wind-tunnel with more variables to consider, being the confluence of two fluids of widely different density – both unpredictable and opposing a ship's progress. It is a miniature ocean. Its sea can be flat calm or agitated by waves – from mere ripples to the billows of a hurricane – generated by giant flaps at one end, the waves breaking onto a wooden 'beach' at the other. It is as near to Nature as Man can aspire. The waves can be projected at 'regular' or 'irregular' intervals, but without the wind that creates them at sea – thus no 'white horses'. The complex effect of wind and waves combined, can only be tested by self-propelled large-scale models (one of known performance at full-scale) compared in the open sea, as we did with S90 and a Leander frigate.

The earliest testing tank was the French Navy's Basin d'Essais, at Brest in the late 18th century. The first English version was constructed at the Admiralty's expense by William Froude at Torquay, in 1872. Today the world's largest tank is the US Navy's David Taylor Model Basin in Maryland:

Froude's models, 'Swan' (above) and 'Raven', Swan being more efficient at higher speeds. Brilliant as Froude was, he took little account of the shape of a hull under water – more concerned with its beam and displacement (or volume) for its length. In this respect his work was superseded by later hydrodynamic studies, which have shown that the underwater shape of a hull can influence performance dramatically.

The large bridge-like towing-carriage of the NMI testing-tank working on our S90 design. Left to right: Peter Knowles and Mike Blea (BHC), Michael Ranken, Peter Thornycroft and my son, Dan.

7 metres deep, 15.5 metres wide, 575 metres long – about six times the length of a football pitch. Many tanks are much smaller. Of the nine we used, in the UK the smallest were at Southampton University (138m in length) and British Hovercraft Corporation (BHC) on the Isle of Wight; the largest was at the National Maritime Institute (NMI) in London (385m). Contrary to the MOD, they all accepted scaling-up. The ships and boats that are 'towed' along a testing tank are scale models – progenitors of their full-sized offspring in every detail of the shape of their hulls, above and

Our 770-foot, 32,500-tonne 'FastShip' of our 1994 design: a 1/40th-scale self-propelled model breaking through 7.5 metre/25-foot average waves at 40 knots in the Swedish SSPA test-tank.

below the waterline. They are weighted to represent their full-scale weight, or displacement of water, and Longitudinal Centre of Gravity (LCG) or Trim. They are first towed by a moving 'carriage' that spans the width of the tank like a bridge. This is initially to measure the drag, or resistance, of their hulls as they pass through calm water. These are known as Calm Water Resistance Tests.

Next they are towed through waves of different heights – and against the movement of the waves at different headings (Head Seas) or in the same direction (Following Seas) or any other angle. They can then be moved across the carriage as it runs along the tank, thus creating the equivalent of seas from different angles, or multi-directional seas (Bow, Beam and Quartering). In these cases the rolling and pitching motions of the models are recorded, with their Vertical Centre of Gravity (VCG) adjusted to represent the ships loaded in different conditions, taking account of the distribution of cargo, fuel, passengers and other factors affecting their pitching, rolling and stability.

Hardly a 'small craft', at 27,025 tonnes: the FastShip 1/40th self-propelled model, with scaled KaMeWa water jets, at 41 knots, in the Gothenburg University tank (SSPA), 1995.

Finally they are subject to Self-Propelled Tests, powered by electric motors representing the scaled horsepower of their engines measured by dynamometers, driving carefully scaled model propellers or water jets. These are run in various seas (Head, Following and Multi-Directional) to measure the hull's behaviour and the speed-reduction caused by the rolling and pitching motions of the ship.

From these tests the efficiency of the propulsion system is calculated as a 'Propulsion Factor'. This usually varies between 50 and 75 per cent efficiency, depending on the type of hull and propulsion employed. For instance, a tug-boat will employ propellers that are most efficient at low speed; while a destroyer will use those that are most efficient at high speed. Some of the latest fast ferries and naval vessels use water jets that become more efficient than propellers above 30 knots, although less efficient at lower speeds.

More recently Computational Fluid Dynamics (CFD) has allowed the accurate simulation of a hull's speed-induced underwater pressures, based upon computer-based data of all the above features. This has clarified an area of hydrodynamics that, for over a century since Froude, had been largely ignored by naval architects and testing tank engineers, who only thought in terms of resistance, or drag, measured with a towing-tank model,

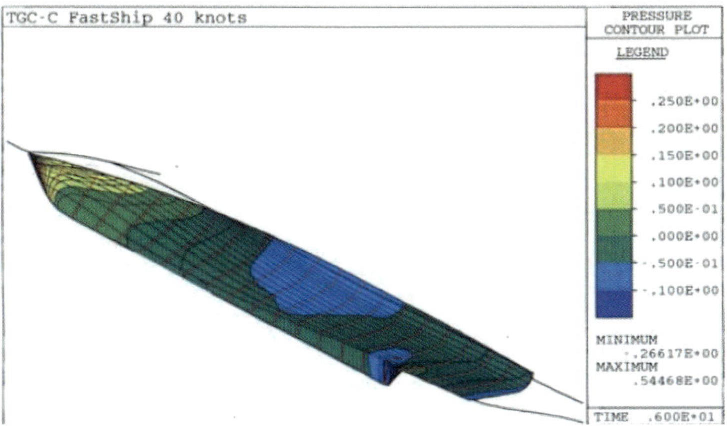

A CFD simulation of a semi-planing hull moving at 40 knots, viewed from below at an angle of 45°. The high-pressure 'lifting' area (green) covers a much greater portion of the underwater hull than the low-pressure 'suction' area (blue). The underwater area of Froude's 'Swan' and 'Raven' models would exhibit a uniformly blue area of very low pressure, or suction, due to their strongly convex rounded shape, at almost any speed. (Source: SSPA test-tank, Gothenburg)

rather than relating it to the variation of water pressure beneath the hull at different speeds and displacements. Testing tanks are generally manned by experts in Fluid Dynamics – in particular Hydroynamics and Naval Architecture. Also involved are experts in Computational Fluid Dynamics, who run simulations to supplement 'raw' tank data. Not all experts are equal, however; some have more experience in testing conventional warship hulls, others in testing yachts, speedboats or even seaplanes. The BHC tank on the Isle of Wight, for example, had a long and distinguished history in the testing of large flying boats, in which the critical effect of 'sink' or 'lift' was carefully measured to ensure harmony at take-off. This was also important for high-speed planing and semi-planing hulls. The staff at other tanks knew little of such work, and were unable to replicate it or even judge it in an informed way. In 1981, the Director of the long-respected Vickers St Albans tank described such measurements as "something of which we know very little".

Assuming the ship is finally built, its performance is measured in the closest possible equivalent sea conditions to the tank tests. Finally, a 'Correlation Factor' is calculated. This is based on the measured power required to drive the ship at a given speed (a) during tank tests, and (b) at sea trials. It is generally known as a factor of (1+x), since most hulls need marginally greater power for a given speed in full-scale, thanks to external effects such as roughness or inaccuracies in the finished hull, protrusions such as propeller brackets, rudders and anti-roll fins – even weeds or barnacles and, finally, the efficiency of the wake or 'wake-fraction'.

A normal correlation factor, or (1+x), scaling up from a model to a full-size hull, would be 1.11 to 1.125. In certain cases, however, a 'minus' value (1–x) is recorded, as explained in **Appendix 3**. This often applies to advanced or unusual hull forms, such as semi-planing hulls, which perform better in speed-trials at sea than in testing tanks. This was chiefly due to the lack of experience of testing and trials of larger variants of such designs, previously restricted to 'quite small craft'. Their superior propulsive efficiency, chiefly resulted from a flat wake flowing downwards off the high-pressure stern into the propellers' thrust-line, rather than dissipated in a high fountain-like 'rooster-tail'. When such variations occur, naval architects are puzzled, upset, even hostile, since the 'laws' of their profession appear to have been undermined or 'misinterpreted'. They tend to suspect a trick or faulty measurements. Hence the remark by the First Sea Lord,

Admiral Sir Henry Leach, on a December 1982 BBC *Newsnight* programme, concerning our later S90 frigate design: "Even if the tank tests showed the traditional theories were wrong, the Navy would not accept them." Such was the conviction of those who were determined to thwart us ... the die was cast.

CHAPTER 5

Joining Forces

Convinced of a new avenue to explore in ship design, I took the hydrofoil to Cowes on a chilly day in November 1975, a savage east wind rattling across the steep Solent seas. Peter met me at the Fountain Pier and we drove in his old Rover to the headquarters of the Royal Yacht Squadron, known as 'The Castle', for lunch. There, sitting before a cheerful coal fire, we agreed a basis for pooling our resources. It took up just one small sheet of Squadron notepaper. Our large vessel activities would be shared in the name of Thornycroft, Giles & Associates Ltd, or TGA; Peter would keep his work on smaller vessels in TT Boat Designs; and together we would sell second-hand Nelsons and boats of my father's design to provide the initial income. This proved a successful business model for our first five years, until the design work produced adequate income.

At the end of December 1975, TGA sold a second-hand motor yacht to the ruling Al-Sabah family of Kuwait. A few days later, on my fortieth birthday, January 9th 1976, I paid a deposit of £40,000 into the TGA account. My new life had begun.

Peter and Arthur Mursell immediately set to work on an enlarged version of the Azteca. The Kuwait Navy was looking for a new type of fast patrol craft, so Azteca grew into a 165-foot vessel with a speed of 36 knots (see **Appendix 2**, Fig. 2). It carried four Exocet missiles and had a wide aft deck, for a small helicopter. Westland Helicopters strongly approved and financed the initial tank tests at their BHC facility, the old Saunders-Roe tank in Cowes, built to test the hulls of flying-boats and seaplanes up to their point of take-off. Peter and my father had used it for many years for semi-planing hulls. We called our new design Osprey, its role being partly anti-submarine, to catch 'fish' with its Westland Lynx helicopter, thanks to its wide landing deck for such a small, fast hull.

There was also a world market emerging for Offshore Patrol Vessels, or OPVs, to police the newly-declared 200-mile offshore 'Exclusive Economic Zones' – first claimed by Mexico and Iceland, and soon to be universally recognised. The Royal Navy was looking for a fresh generation of vessels to augment their ex-trawlers – or frail and vulnerable frigates in 'Cod Wars' barging tactics – as formerly used for Fishery Protection. The newly declared British EEZ was a huge sea area, some 3.5 million square miles, with valuable new resources of offshore oil and gas as well as traditional fisheries. The Osprey seemed well suited to the task for the same reasons as the Azteca had appealed to the Mexican Navy with its vast sea areas to police, off two coastlines. Greater numbers of cheaper, smaller vessels, with a good turn of speed and reasonable seaworthiness, could cover the huge sea area at lower cost than traditional designs. An embarked helicopter could greatly extend their surveillance horizon. As all countries were by now committed to declaring their 200-mile EEZs, the export market would expand rapidly. For any design to enter the MOD's Naval procurement process, particularly from outside the usual suppliers, it needs powerful partners in the defence industry. Out of the blue, in February 1976, Peter's old family firm, Vosper Thornycroft, approached us. Their Sales Director raised the possibility of co-operating over a Kuwait Navy order for patrol craft, for which our yacht contract with the Ruler's family stood us in good stead. The Osprey was the ideal size for Vosper's to build it, so we called it 'Vosprey'. I was forthwith summoned to breakfast with Sir John Rix, the Chairman of Vosper's, at London's Savoy Hotel.

As we sat down to a table laden with coffee and pastries, he produced photos of the Azteca's wake at 26 knots. I had taken these myself during the Azteca trials and given them to one of his colleagues a few weeks before. I was surprised that he also had photos of the Osprey tank tests that we had just completed at BHC.

"The BHC tests were supposed to be confidential," I said.

"You shouldn't worry about that," he replied. "This sort of thing happens the whole time ... Once a model's in the tank it's more or less public knowledge these days. Some of our technical people are quite impressed. They've been trying to produce a wake as smooth as the Azteca's for a long time. Others claim there's nothing unusual."

We got down to business. Under the current Labour Government, Vosper Thornycroft was due to be nationalised, to become a part of British Shipbuilders, generally known by the unflattering acronym of 'BS'. If this happened, would we consider the possibility of Vosper having an exclusive agreement to build and sell the Osprey? From within a nationalised BS, Rix was authorised to set up a Patrol Craft Committee to consider construction of the Osprey. We would have the support of Vosper, the Navy and the MOD's Defence Sales Organisation. Financing would come from Government.

It was a tempting prospect, but it carried certain risks. Knowing the controversy that Peter's ideas for the Azteca had aroused within Vosper's technical fraternity, we suspected we would face opposition. An exclusive agreement could be the kiss of death: they could do as they wished with our design – 'use it, or lose it' – and the latter seemed more likely in the short term. Despite our misgivings, we decided to sign.

In early 1977, I was introduced to Commander Charles Adams, Deputy Head of the UK Coastguard, who was involved in Offshore Protection policy. Although he had spent much of the Second World War as a Commanding Officer (or CO) of corvettes and frigates escorting Arctic convoys, he was more of a gentle maritime philosopher than the hardened sea dog one would expect. Having played a part in resolving an earlier 'Cod War' with Iceland as CO of HM Coastguard Vessel *Miranda*, Adams had a keen appreciation of the essentials required. He had also been a member of the Cabinet Office under Harold Macmillan, and knew the ways of Whitehall and its bureaucrats. He was to become an invaluable source of wisdom and pithy humour, a trusted adviser, supporter and friend.

The Second 'Cod War' with Iceland had just been settled and Adams was intrigued by the possibilities of the Osprey. He had witnessed the

Commander Charles Adams, DSC, commanding HM Coastguard Vessel *Miranda*.

sturdy, small, beamy Icelandic gunboats running circles around our longer, narrower, lightly-built frigates. The media had been full of HM ships limping back into the dockyards bearing the scars of ramming and collision. Osprey would be quick and nimble, with a helicopter to extend its surveillance horizon, giving warning of possible intervention.

Adams was also convinced that a shorter, wider hull would be stronger and cheaper for the job. But his experience of the British Civil Service had taught him to be careful in his approach. When the BBC programme *Tomorrow's World* featured six of the twenty-one Clyde-built Aztecas crossing to Mexico, he warned us against publicly promoting our designs: "You will only be creating antibodies within Whitehall." It was good advice, as was his memorable parting shot: "In such circumstances, as a mere outsider, there's only one thing worse than being wrong ... It's being right."

'The Proud Man's Contumely ...'

At a Greenwich Forum on Maritime Affairs, Adams introduced me to Julian Taylor, Managing Director of the renowned old shipping company the Blue Funnel Line, now a part of Ocean Transport. Taylor also had much experience in government and the maritime world. During a visit to his Liverpool office in 1978, I met his Chief Naval Architect, Marshall Meek. Julian described Meek as an evangelical Christian, whose belief in the traditional Laws of Naval Architecture was as unshakable as his faith in the Ten Commandments. Meek did not hide his professional's distaste for the enthusiastic amateur. When he published his memoirs, *There Go The Ships*, in 2003,[4] he opened a Pandora's Box that revealed the murky background to a battle that was to dominate the rest of my life. In his book, from the very first, Meek is extremely uncomplimentary towards me and my ideas – often both malicious and false. Like others connected with the MOD, his memoirs give a telling and valuable retrospective account of matters as seen through an official's protective eyes.

... the Insolence of Office

To the considerable annoyance of Meek, Taylor took Thornycroft and Giles under his wing. He arranged a meeting with the Navy Minister Patrick Duffy, the Controller of the Navy Admiral Sir Richard Clayton, and his

4 Marshall Meek CBE, *There Go The Ships*, the Memoir Club, 2003, p. 192. Further references shown as 'Meek'. Also see **Appendix 5**, Fig. 5, Item 3.

technical adviser Jack Daniel, who was also the former Director General of the Ministry of Defence Ship Department in Bath. Daniel was another of the country's leading naval architects who was to play a leading role at British Shipbuilders in their illegal tank-tests of our designs.

When I explained some of the features of our Osprey design I was informed that there was 'nothing new' about the Osprey and that it would be too small for the operation of a helicopter – whatever the views of the experts at Westland, who had financed the initial Osprey tank-tests and specialized in providing ship-borne helicopters for the Navy. This struck me as odd. Despite the eminence of the MOD's team, their response was hasty and ill-informed. I had no way of knowing what lay behind it, or what a strange and ruthless twilight world I had just entered – the world of defence contracts and the military-industrial complex.

Over the next 18 months we tried to find a way of building the Osprey in Britain, but wherever we turned we were directed back to BS and John Rix's Patrol Boat Committee, which had yet to call its first meeting. The British shipyards, which seemed demoralised and short of work, were awaiting the lifeline of nationalisation promised by Mr Callaghan's Labour Government. They were generally over-manned and under-capitalised, using building methods that had scarcely changed since the Second World War.

We were getting nowhere. Or was there something we hadn't thought about – a commercial use for our ideas? Inspiration came from an unexpected source.

Laura Ashley: Logistical Problems of the Rag-Trade

Bernard Ashley had bought *Maid o' the Mist*, a Nelson 40, after a sea trial off Guernsey.

Meanwhile his wife, Laura, was expanding her dress-making business from a small shop in South Kensington to New York and beyond.

I rang her at their home in Carno, North Wales, about some equipment for the Nelson. To a background of barking dogs and kitchen noises, she told me what a great boat it was. To my surprise she then asked me if it could be enlarged to be a fast ocean-going cargo ship to take her dresses to the USA.

Her logistical problems were a nightmare, she said. The container ships seemed unable to work to any serious schedule, so they had to send everything weeks ahead of when it was needed, which was both expensive and inefficient. At times the demand for some new line was overwhelming so the stock couldn't be sent out immediately, by which time the market had disappeared with the end of a heat wave. Or a crane might have broken down in one port so the ship had to go to another, resulting in a stack of containers full of frilly cotton frocks sitting in the back of beyond waiting to be collected – or pilfered – by heaven knows who.

Sending it by air was a last resort. It was terribly expensive and not that reliable, even if major carriers with their dedicated air-freighters were used. When the planes were full, rates soared and there was no space. When they were empty services were cut back. "It was so much better when it all went by lorry from Carno down the M4 to London," she said.

I sympathised. There was a pause and then she let fly: "Why can't you take one of your hulls and make it much bigger? The way that Nelson went through the waves the other day was amazing – and she was going so fast. There's a huge business developing out there for time-sensitive products like ours."

"Once we've made the dresses we want to get our money back as quickly as possible. Within a week or two of them hitting the racks they'll go into discount – and head south from the USA, along with the prices. The money we lose on discount is far greater than the saving on cheap sea freight. Having a fast service that you can depend on, matters – and you'll get a much faster return on your investment in the ships ..."

"Go for it! What are you waiting for ...?" And, many years later, I did.

I was amazed. Here was one of the brightest stars in the market telling me I was right, when everyone else was saying that there was no need for speed in sea freight. They had said the same about the first ideas for a jet-powered airliner back in the mid-1940s when the idea of the Comet, the first jet airliner, was first mooted as the Type IV in the 1943 Second Brabazon Committee. "The cost would be prohibitive," had been its reaction. It was left to the lone passion and genius of Geoffrey de Havilland and his team to take the initiative over what they called 'the economics of speed', based on the faster return on investment outpacing their increased operating costs. This was exactly what Laura Ashley was proposing.

CHAPTER 6

'Our Brothers, the Danes'

In May 1977 I met a Danish pilot who ran a number of Nelson 40s from the Skagen Pilot Station in the wild waters of the Skagerrak. His local shipyard, Frederikshavn Vaerft in Jutland, had just finished building a new class of patrol boats – a change from their usual output of commercial vessels. He arranged an introductory visit to the yard.

The 'Vaerft' was the biggest employer in Frederikshavn – and unlike any shipyard we had seen in Britain. To extend the building hall while assuaging the local historians, a 400-year-old powder magazine had been moved a quarter of a mile to the northwest. Unlike British yards I had visited, some of which still had earth floors in their workshops, the whole establishment was spotless. The offices were like an IKEA showroom. Many sections of ship were under construction in covered bays, with very few people to be seen. Those visible wore smart blue overalls and white hard-hats and were going about their business in a determined and cheerful fashion. The yard even had a brass band that was proudly paraded at the christening of every ship. Launchings were a thing of the past. Ships were built under cover in 'blocks', or modules; finally joined together in the outside building dock, floated out for finishing touches and dock-trials, christened and taken out on sea-trials the same day.

In July 1977, having heard nothing from the BS Patrol Boat Committee, we signed a design contract with Frederikshavn Vaerft. At the request of the Danes, Osprey Ltd was formed in Guernsey to hold the design rights. The shareholders were Peter and I, plus David Browning, a Guernsey resident who owned a boatyard where the new Nelson 44 pilot boats were being built. In joining forces with the Danes we were taking a risk. The advice of John Rix was that the Osprey would be seen to have flown the nest and was now fair game for our detractors in the MOD and BS.

"Expect no mercy!" Rix cautioned. It sounded more like a threat. And soon it became one.

The Frederikshavn yard quickly received an order from the Danish Fisheries Ministry for the first Osprey, re-arranged as an Offshore Patrol

Vessel, or OPV. They signed contracts with us for further work on larger and smaller versions. Everything in Denmark seemed to move ahead more smoothly than we could have hoped. But we still kept the door open for the British yards to build military versions, such as the six new patrol craft for the Navy's Hong Kong Squadron that were being mooted in MOD circles.

A Change of Wind

In March 1979, a Conservative Government was elected; its first priority to de-nationalise. Free enterprise would flourish again – or so we were led to believe by our first-ever woman Prime Minister, Margaret Thatcher. All the energies devoted to the nationalisation of BS were now thrown into reverse. A new word was in the air: 'Privatisation'.

Peter phoned in a state of excitement. "David, I just got a letter from Lord Strathcona. He writes: 'Francis Pym (the new Defence Secretary) has asked me to be an Under-Secretary of State for Defence Procurement. My particular pigeon is the Navy. I'm to be known as Pym's Number One!'"

"Maybe that'll help us," Peter reflected. "We have a friend at Court: a belted Earl, no less!"

Euan Strathcona was an old friend of Peter's, a scion of the family that had founded the Canadian Pacific railway and shipping empire. A tall, impressively bearded seafaring man, he owned the Scottish west coast island of Colonsay, travelling to and from the mainland in his Nelson 34. He had been an RNVR skipper of wartime patrol craft and had a salty, distinctive seafaring style that appealed to politicians – perhaps less so to bureaucrats. We thought he might give us a fair shot at the Hong Kong Patrol Craft order.

By May 1979 we had a ship (see **Appendix 2**, Fig. 3). The christening of the first Osprey, named *Havørnen* (the Danish for 'Sea Eagle') took place on the precise day contracted eighteen months before. It was bright and breezy as we trooped across the road from the Jutlandia Hotel. Painted pale green with white superstructure and a bright orange deck, *Havørnen* was dressed overall, a rainbow of flags fluttering above her in the brisk wind. The Vaerft brass band marched smartly around the corner of the main construction hall. We went aboard for the christening ceremony, and the Danish Fisheries Minister formally took ownership of the vessel. The Vaerft flag was lowered and the swallow-tailed Danish Naval ensign was raised with a loud cheer. The acceptance trials took place immediately.

The Frederikshavn Vaerft band plays for the christening of *Havørnen*, May 1979.

Havørnen headed out into the Kattegat for speed trials, accompanied by the customary Soviet trawlers that attended such occasions in the Baltic. Peter discarded his Royal Yacht Squadron blazer in favour of a bright blue boiler suit and chromium torch.

"She'll turn 360 degrees in one-and-a-half ship's lengths at full power," he announced – and she did, remaining bolt upright – a welcome change from the outward lurch of slender naval ships. Now for the real test: the final speed trials. We stood on the bridge with our stopwatches as *Havørnen* made runs over the measured mile off the Swedish coast. We need not have worried. Though barely operating above the minimum semi-planing speed of her hull design, at 22 knots she was still about two and a half knots faster than tank-tests by the Danish Technical University had predicted, based on conventional hull designs. The (1–x) was in action again: the 'Nelson Factor', contrary to the traditional addition of a (1+x) for propeller loss.

Havørnen cost less than £1.5 million to build. By comparison, the Royal Navy's new OPV-2 patrol vessels cost £10 million each, with little more in

A pair of 'Ospreys': *Havørnen* seen from the Burmese *Indaw* in the North Sea.

the way of performance, internal volume or general utility. At a post-trials celebration, after seagull stew on the Isle of Læsø, out came the Aquavit. Peter, quoting Nelson's message before the Battle of Copenhagen, raised his glass to "Our brothers, the Danes".

A few weeks later, Frederikshavn Vaerft announced an order from the Burmese Navy for three more Ospreys. The first, to be delivered in 1980, would be built in just six months. In addition there was the continuing order for 21 Aztecas built on the Clyde and a further ten in Mexico. In all, 39 naval patrol vessels based on Peter's Azteca and Osprey semi-planing hull form were in service or on order.

'The Risk Business'

We were riding on a wave of euphoria: everything seemed to be going our way. The Danes were delighted with the performance of their new ship and they proudly offered to bring it to London to show off its paces to the Royal Navy and foreign naval attachés. After deck-landing trials at sea with a Trinity House helicopter, she would sail from Harwich, up the Thames past Greenwich and through Tower Bridge to berth alongside HMS *Belfast*. It sounded like good publicity – so why not?

Charles Adams drove me to Harwich, where we were to join *Havørnen*. After a tour of the ship with Captain Jensen, he was impressed. But he did utter a warning: "You've won a bit of a victory. Don't rub it in."

The next day, aboard *Havørnen* at sea on the way to London, a small Trinity House helicopter landed on the broad aft deck very easily, like a bee alighting on a flower. Then I had a trial flight, the pilot commenting on the size of deck for such a small ship: "Much easier than one of our lightships," he said as he gently touched down with a few deft movements of his controls.

However: "Doesn't prove a thing," said one young Naval type. "It's flat calm."

Next, the 21-foot inspection boat was launched from the stern slipway – a yellow 'RIB' that sped away after a two-minute launching; and returned to be hauled up its slipway beneath the helicopter deck in two minutes. Even the Naval types aboard were impressed: "Better than trying to launch a boat over the side by davits, rolling through forty-five degrees in a seaway," said one, referring to the RN's £10 million 'Castle' class OPV.

Later we executed various manoeuvres and high-speed runs and turns before entering the Thames. Tower Bridge was duly raised and, without tugs, Captain Jensen deftly berthed his ship alongside HMS *Belfast* – to the considerable annoyance of the attendant tug-skippers.

The great day, June 19th 1979, dawned full of hope: a beautiful morning with London at its midsummer best. *Havørnen* was dressed overall and the stiff breeze sent the bunting fluttering horizontally. A Royal Marines Band played on the helicopter deck, while the dignitaries – diplomats, foreign naval attachés, bureaucrats and others – strolled across the gangway from HMS *Belfast*. Admiral Sir William Pillar, appropriately entitled the Chief of Fleet Support, arrived from the MOD with another Admiral: senior men interested in seeing what all the fuss was about. They were welcomed aboard by Captain Jensen and almost immediately ambushed by a BBC camera team, with its attendant furry microphone and a reporter from *The Risk Business* TV series.

"Why aren't the Royal Navy buying this ship? Why does it have to be sold to the Danes so that they can bring it to London to show it off to you? And it's only a fraction of the cost of the new 20-knot 'Castle' class that's building for your Fishery Protection Squadron. What about all those fragile frigates battered in the last 'Cod War' by the small, nimble, Icelandic gunboats? Aren't you missing the boat?"

It was exactly as Charles Adams had foretold. The Admirals were appalled: no interviews had been arranged, and they were completely unprepared. Acutely embarrassed, they gave what lame diplomatic answers they could and left.

Pilloried in Parliament

On November 8th 1979, in a Parliamentary Adjournment Debate on fishery protection, the Navy Minister, Keith Speed, stated, "For reasons on which I shall expand later, the Royal Navy finds 'Osprey' to be completely unsuitable for the offshore patrol duties that would be asked of her. That view is shared by the Fishery Departments, whose policy dictates the way in which the fishery protection task is carried out. There is, therefore, no question at all of her being used for these operations." Osprey might also hazard the smooth process of issuing orders for the privatisation of the shipyards.

He listed the Osprey's supposed disadvantages: Too slow (16 or 17 knots; the 'Castles' claimed 20 – yet Osprey made 22 knots on trials); constructed without separate watertight compartments, which would make her vulnerable in a collision; unable to stay at sea for more than ten days; and, finally, an awkward arrangement for launching an RIB inspection boat. His remarks were clearly drawn from the adverse comments by Kenneth Rawson, Deputy Director of the MOD Ship Department, in a report on his *Havørnen* trial trip which I was shown a few days previously by Strathcona.

Speed's public criticisms were just as partisan as Rawson's, as shown by the following facts:

First, contrary to Speed's claim, the satisfaction of Peter Dereham, the Chief Inspector at the Ministry of Agriculture, Fisheries and Food, who attended the same trials. He was a professional mariner and considered *Havørnen* to be a perfectly good ship – fast, spacious, commodious, responsive and manoeuvrable ... She did have a quick roll, due to the heavy commercial Danish diesel engines required by the Government, but she was perfectly comfortable for most seafaring people and remarkably good value for money. "However," he said gloomily, "Rawson and the MOD are out for your blood: I feel sorry for you." Indeed they were ...

Second, the thirty Osprey variants that were later bought by the Danish Navy and other countries, with satisfaction expressed by many of their commanders. These included the captain of the Burmese *Indaw*, and the Danish Pilot who accompanied her, expressed his satisfaction to a rather surprised Lord Strathcona; also, the skipper of a Greek Navy Osprey, whom I met at MIT some years later. Today, forty-five years after delivery of the first Azteca to the Mexican Navy, twenty of these vessels are recorded as remaining on active service.

In defence of the Navy's laborious design procedures, Keith Speed quoted Peter Thornycroft's offhand remark in a previous TV series, *Tomorrow's World*, about the Aztecas. Jokingly, he declared that he "designed by the seat of his pants" – the "victory of practice over theory," as he put it. It was a classic piece of self-deprecating modesty, considering the hundreds of his Nelson launches in service worldwide and the 30 Aztecas under construction or on order for the Mexican Navy. The success of these vessels far exceeded the sales of any other British designs, a success made possible by Peter's free-thinking approach. The Minister took the remark out of context and chose to interpret it as a sign of amateurism. Having dismissed the Osprey as a Fishery Protection Vessel, Speed surprised us by mentioning that a version might be appropriate for the Navy's new Patrol Craft for the Hong Kong Squadron, "for which procurement will proceed on the basis of a design and build competition amongst UK shipyards." This was puzzling. The two roles were very similar, whether in the seas off Hong Kong or the British Isles. As Peter observed, with his customary quote from *Alice in Wonderland*, it sounded like "jam tomorrow": a tempting prospect held out to keep us quiet and obedient: "Curiouser and curiouser," said the Commander.

Meanwhile serious damage was done to our reputation. As a result of the Minister's statement, the Burmese Government immediately deferred their order for two further Ospreys until their first vessel, still under construction, had shown her paces. The Maritime Press, led by Lloyd's List, carried headlines such as 'The Osprey is unsuitable for fishery protection.' However, once *Indaw* had shown her paces on her delivery voyage from Denmark to Burma through a series of gales, the Burmese Navy immediately restored its order for two further Ospreys.

*****

Considering the widely-broadcast disapproval of Osprey within the MOD, we were surprised to be invited in May 1980 to enter a competitive tender for the Navy's new Hong Kong Patrol Craft. Our initial design was to be based on a faster, lighter Osprey (see **Appendix 2**, Fig. 4) but later enlarged to the 600-ton 'Condor'. Then a further bizarre surprise followed. Through the intercession of Julian Taylor, we were invited to submit the Osprey plans for assessment, at their expense, by the former Vickers Testing tank, a high security establishment run by British Shipbuilders (BS) at St Albans, normally used for warships and nuclear submarines. Their sister-tank, Vickers tank at Dumbarton in Scotland had already undertaken all the Azteca model-tests. Peter's TT Boat Designs was then hastily certified to the exacting standards for all designers of Royal Navy vessels (Defence Standard 05/21). On his payroll there were only Peter, Arthur Mursell and Jill Swallow, his secretary. As Peter said: "Perhaps we should have included Nelson, the office cat, as 'Chief Rodent Control Officer'." So why did they bother ...? The Osprey had been thrown out by the MOD, in both Parliament and the Press, months before. It all seemed very strange considering the contempt in which TGA and the Osprey were held within the MOD, and more publicly 'pilloried' as some said, by the Navy Minister, in the Parliamentary Debate.

During our May 16th visit to St Albans, with the encouragement of the BS and Vickers officials, we agreed to lend the Osprey plans and relevant tank-tests, having received signed undertakings covering our Copyright and non-copying or use of the plans without our written permission or supervision. We also offered Vickers the use of the large Osprey model stored at BHC. This was exactly the 1/10th size (16′ 6″ in length) that would be employed by Vickers, and our gesture was clearly appreciated by David Moor, the Superintendent. He claimed the Osprey design was a topic of

The Burmese Navy's First Osprey, *Indaw*.

much interest and controversy among the Naval Architectural elite and he wanted to settle the argument once and for all. We therefore removed the one-tenth scale Osprey model from BHC to Peter's shed for its possible despatch to Vickers.

Alas, two weeks later, the Osprey plans were returned to me by Moor with a peremptory note, as 'being of no immediate interest'. BS thereafter obstructed us from obtaining a bid from any nationalised British yard to tender with our design. "Curiouser and curiouser," repeated Peter again. With all the conflicting opinions it was becoming more than ever like the Mad Hatter's Tea-Party.

The following month I attended open-sea measurements of the ship-motion and sea-keeping qualities of the Osprey design, this time financed by the Department of Industry. Surprisingly perhaps, these were witnessed and approved by an MOD Ship Department representative working for Rawson. The trials took place in the North Sea aboard *Indaw*, the first of the Burmese Navy's three Ospreys. In 1979, BHC had previously tested a 1/10th scale self-propelled Osprey model in the Solent under a preliminary DOI grant, with excellent results, leading to the further DOI Grant for these full-scale *Indaw* trials.

These North Sea trials further endorsed the earlier 1/10th model tests and were equally encouraging, contradicting the subjective judgements of Rawson, Daniel and Navy Minister Speed in Parliament. Far from a 'potential source of disaster', during the *Indaw* trials we drove through a ten-foot head sea at full power, with hardly any pitching, while clouds of spray were thrown past the sides of the ship – just like the Azteca. All, even including the Man-from-the-MOD, were impressed.

In July 1980, *Indaw* berthed at Tilbury on her way from Denmark to Burma. At last we were able to persuade Lord Strathcona to visit our ship, of which he had heard such unflattering reports. With Julian Taylor, I joined him at the Ministry of Defence in Whitehall, and his driver took us to Tilbury Pier. As a former wartime CO of patrol boats, he cast a critical eye over *Indaw* and seemed surprised. She was rather a handsome ship, he suggested – better than he had been led to believe by his officials: "Were they trying to pull the wool over my eyes?" he asked. We then met the Danish Pilot assigned to accompany the vessel on its voyage to Burma. He was most complimentary. Knowing of the Nelson Pilot launches run by his colleagues in the stormy waters of the Skaw – the confluence of the

Skagerrak and Kattegatt – he was not at all surprised: "This ship is the shape only the English understand: a good sea-boat and fast as well – even in a big sea." As former Director General of the MOD Ship Department – now Board Member for Warship Construction of BS, Daniel could fix the whole thing in anyone's favour: he was 'gamekeeper turned poacher'. Strathcona then arranged for me to visit Appledore Shipbuilders in Devon, one of the BS yards competing for the Hong Kong Patrol Craft. He told me to speak to Joe Ball, the Managing Director, for discussions about a possible bid with Osprey, or her bigger sister, 'Condor'.

This was a complete change from his earlier attitude. Had Strathcona achieved something, we wondered? Was the tide really turning?

Our hopes were soon dashed. In September, having received the huge Design and Build Specification for the Hong Kong boat, Ball told me we were too late. He could hardly read the volumes of Requirements by the tender-date in late October – only six weeks away – let alone be sure of matching them. We were mortified.

An Admiral Breaks Ranks

In October 1980, Charles Adams arranged for me to attend a two-day seminar on Warship Procurement, held by the International Institute for Strategic Studies at Madingley Hall, outside Cambridge. I was introduced to Admiral Sir James Eberle (Commander-in-Chief NATO Northeast Atlantic Fleet), who showed a lively interest in the Osprey controversy. When I described the possibility of enlarging the Osprey to corvette or frigate size, he suggested that I send him our ideas. So was born a new design: a 246-foot, 1,750-ton, 30-knot semi-planing vessel, named after the brightest star, Sirius, my old friend of night-watches in the Navy.

This could serve both as a fast Offshore Protection Vessel and as a light frigate. With gas turbine engines it could combine sufficient speed and sea-worthiness for escort duties with the NATO Northern Fleet. Our extended programme of model and full-scale Osprey tests had shown that only a scaled-up Osprey could achieve a practical speed of 30 knots in ocean seas on so short a hull-length and heavy displacement, and at lower cost than the traditional 'long, thin' alternatives. Replacing the weight of traditional boilers or diesel engines with lightweight gas turbines, the wider beam of Sirius would also improve stability for a higher (thus longer-range) radar, useful for both fishery protection and escort duties.

Encouraged by Admiral Eberle, I wrote a paper: 'The Sirius Pocket Destroyer – More of a Beagle than a Greyhound', Eberle being Master of the Navy's Beagles. Afterwards, at Strathcona's suggestion, I began a correspondence with Kenneth Rawson, the Royal Navy's Chief Naval Architect. Surprisingly – considering his damning report on the *Havørnen* in October 1979 – he now seemed far more receptive to our ideas. At his request I provided him with the Osprey tank-test results, the 1/10th scale self-propelled model tests and the full-scale speed trials and measurements of the *Indaw*'s sea-keeping. He compared the Osprey with Royal Navy ships of similar size and agreed that its results for sea-keeping were generally superior.

Lord Strathcona remained keen that a variant of the Osprey design should be considered as a private industry alternative for the Navy's Hong Kong Patrol Craft (see **Appendix 2**, Fig. 4). The tender date was now delayed from October 1980, apparently because a Hong Kong shipyard was being invited to provide an outside bid.

That November, Strathcona arranged for me to meet Sir Robert Atkinson, Chairman of BS, to discuss lack of progress with our Hong Kong tender. In the event, I was confronted not only by the Chairman, but the entire top technical management of BS, who roundly criticised the Osprey – while requesting further details of our own BHC tank-tests and Osprey powering data.

After half an hour of being criticised for our ideas, I agreed to supply the data on the understanding that Vickers would re-consider undertaking tank-tests on the Osprey at their cost and under our supervision, as suggested the previous June. They said they would let me know their decision in due course. But, again, why were they doing this in the face of their resounding public opposition to the Osprey? There was one exception to the BS crowd's generally dismissive attitude: the towering presence of Jim Venus, former Chairman of Appledore Shipbuilders and Austin & Pickersgill – the only true shipbuilder among them. As they left, he whispered in my ear, "You're right young man – and they know it! Good luck." Did that explain everything ...? I wondered.

Was Strathcona at last beginning to flex his ministerial muscles: 'to learn the trade', as he put it? He told me he had summoned the 'Bath people' – Rawson, Daniel and others – to his office and informed them there would be swingeing financial penalties if the Hong Kong boat did not make its required speed; yet, by Naval standards, the contract price limit was extremely low. A Vosper design could make the speed, but not the price. Another bid, by Brooke Marine, could achieve the price but not the speed. Thus nothing so far matched the specified cost/speed requirement. It appeared that an Osprey-based design could achieve both price and speed. But we did not have a shipyard. Without it we could not apply for the Design and Build Specification.

"You *must* find a yard," Strathcona said. But where?

Mrs Thatcher's policy was to privatise the main warship yards, which were to be given naval or military contracts to reinforce their share value when sold on the Stock Market. A number of smaller yards were not included in this plan, so naval funding was (unofficially) not available for them. The only independent yard we approached was not interested. In the circumstances this was not surprising: what non-BS shipyard would dare bid against the major yards with solid Government support?

In November 1980 an 'Op Ed' appeared in the *Times*, entitled 'A Big Prize for the Right Little Ship'. Written by Julian Taylor, it was intended to avoid open MOD controversy. It emphasised the need to reduce the price of ships whilst increasing their capability and numbers to cope with the burgeoning activities of the Soviet Northern Fleet. It also argued that the demands of modern air-defensive weapons systems and ship-borne helicopter operations required greater stability and, therefore, increased beam.

By now, the editor of *Jane's Fighting Ships* and *Jane's Defence Review*, Captain John Moore, had become a supporter. Formerly Director of Naval Intelligence, he was an unofficial adviser to the Prime Minister on naval matters. I spent an evening with him and his wife in his old cottage on the South Downs, while a mighty gale roared outside. We chatted about the deficiencies of the current generation of frigates and destroyers designed to operate under constant air cover in a NATO scenario. He wanted to know what might be done to improve the defensive ability of smaller fighting ships against air attack in non-NATO engagements. He asked for a summary of my views that he could hand to the PM at a meeting due to take place a few days later. I sent him 'Sirius – The Pocket Destroyer', written earlier for Admiral Eberle.

In keeping with the unusual mood of goodwill, I was invited just before Christmas to meet Strathcona for a drink at White's Club in St James's. Before a roaring fire in the oak-panelled hall of this fine old London institution, he told me again how strongly he had insisted to the MOD people that the Hong Kong boat should make its speed. Basking in the transitory glory of Public Office, he must have felt every bit as contented and confident as he appeared.

"Keep alongside British Shipbuilders, David," he advised with a seasonal pat on the shoulder. "They know the odds. The Hong Kong boat *has* to make its speed and cost targets. They know you have the ability to do it with your design. If they don't, they'll be in real trouble. I've made that very clear to them!" I walked back to the Tube, up St James's through the Christmas shoppers, with a light heart. So long as BS were openly considering our ideas and our Belted Earl was enjoying Christmas on his Scottish island, then God was in his Heaven and all seemed well with his world. But not in ours. Despite sending Osprey propulsion data to BS, we received no news of any tank-tests for which they agreed our involvement would be required. So we presumed, as in the previous June, that they were taking no further action on Osprey.

CHAPTER 7

The Changing of the Guard

On New Year's Day 1981, at the unusually early hour of nine in the morning, I had a call from Peter Thornycroft.

"There's been a Cabinet purge," he said. "Strathcona's been sacked."

His replacement as Minister of State for Defence Procurement was Lord Trenchard. With a father who was Founder of the RAF, his own background in aviation and running an Air-Freight company, Treffield Aviation, he had neither experience nor interest in naval matters.

I arranged to meet Ian Gow, an old school-friend, at the House of Commons. As Mrs Thatcher's Parliamentary Private Secretary, he had the ears of the Highest in the Land. I shall never forget our short conversation.

"Why was Strathcona sacked?" I asked.

"He wasn't sacked, David ... He was *shot*!"

"But *why*?"

"All I can say is, he got across his officials. That's dynamite ... Beyond that I'm unable to comment."

I suggested it sounded like *Yes Minister*, the BBC's popular political-cum-civil service satire.

"It could be," Ian replied, with a marble-eyed expression behind horn-rimmed glasses and the usual charming smile. "Now tell me about your Reverend twin brother, John: I have something in mind for him. What about the living of Greenwich? It's in the PM's gift and he strikes me as being just the sort of chap for the job."

Was this the *quid pro quo* for the loss of Strathcona, our best supporter? Ian was clearly uncomfortable about his removal, but that wasn't going to upset the smooth flow of politics. He was already onto the next item of his agenda. I left with a feeling of deepest gloom. Our political and ministerial support seemed to have faded away.

This looked to me like the end of the road – at least in Britain. I began to think it was time to take our ideas across the Atlantic. On January 20th 1981 a new US President, Ronald Reagan, was sworn in. He had promised a 600-ship Navy to counter the threat of the Soviet Northern Fleet: the subject of that Cold War classic, *The Hunt for Red October*. Washington was alive with the optimism of the incoming Administration: new appointments, new defence money – and a six-hundred ship Navy. We thought they might just be interested in one of our designs.

Arrangements were made via our former Ambassador, Peter Jay, for me to meet a distinguished nuclear physicist and expert in fluid dynamics and defence technology, Dr Richard L Garwin. As Scientific Adviser to five US Presidents, he had also co-authored a book on naval policy with Paul Nitze, Kennedy's Navy Secretary. He had neither political nor military-industrial axes to grind – nor the Halls of Academe to fear. As a Fellow of IBM's Thomas J Watson Research Centre, his brief was simple: to consider new ideas – to examine, experiment and dismiss or recommend them. He seemed the ideal person to pass judgement on my theories.

On a freezing January afternoon, I trudged down a snow-piled Constitution Avenue. Passing the huge bronze bust of a brooding Einstein, I mounted the steps of the US National Academy of Sciences. Little did I realise how much my future would depend on the outcome of this meeting. My briefcase was crammed with tank-test and trials reports on the Osprey, and my report on the Sirius Project. At the reception desk I was told that Dr Garwin would be finishing his lecture shortly.

Waiting on the mahogany bench in the lofty marble atrium, I contemplated the words engraved around the inside of the cupola high above me: 'Ages and cycles of Nature in ceaseless sequence moving ...' Glittering mosaics in Hildreth Meière's 1924 dome, graphically illustrated this process. I too thought of my own struggles with the forces around me, both human and natural: people and the sea ...

Dr Richard L Garwin, renowned nuclear physicist, expert in fluid dynamics and defence technology. A former Chairman of the Naval Warfare Panel of the President's Science Advisory Committee, he quickly grasped the value of our designs and became our most important ally for the next forty years. In May 1981 US *Science* magazine devoted a four-page article to him with the headline: 'The physicist whom Fermi (his mentor at the University of Chicago Graduate School) called a genius, sees a failure of rational debate on defence issues.'

My thoughts were interrupted by a hubbub from the lecture theatre. The tall doors opened, releasing a mass of people who crowded about the man whose company I sought. With his steel-rimmed spectacles and domed forehead capped by a cloud of greying hair, he was the perfect vision of a nuclear physicist. He slowly descended the marble staircase, chatting amiably with members of his audience. Once they had dispersed, he headed purposefully towards me.

We sat together on the bench and glanced through the tank-test reports. He could spare precisely one hour in which to consider my ideas; he then had to catch the Eastern Airlines Shuttle back to New York. After some time spent scanning our data he announced his verdict with his renowned economy of words: "I have some understanding of fluid dynamics," he said modestly. "I presume your hull, at this point, begins to lift, thus shedding its displacement of water, rather than increasing it by sinking at the stern? If so, that is of some interest."

After a short pause for further consideration, Garwin agreed that, if our hull were scaled up, it might allow reduced power-for-speed in such a relatively short length and high volume of hull. Alternatively, it might allow unusually high speed for such a hull – but without the usual prohibitive increase in power: "It could be quite significant for increasing the speed and operational capabilities of future warships." In a moment he had grasped the principle that would not be accepted by much of the British naval architectural fraternity for half a lifetime: hydrodynamic lift, generated beneath the stern to reduce the wave-drag and allow higher practical speeds and stability, could apply in a ship of any size.

If this were the case, he suggested, it should be of interest to any navy. True, more power would be required for a much higher speed than was previously considered practical for a ship of the same proportions and size. But that should be offset by the reduction in hull cost, greater stability, and a greater useful load for its size. With the latest developments in Soviet submarine and torpedo technology, maintaining speed in higher ocean seas was also of great importance. That would be a factor to investigate.

He saw no fundamental flaw in our argument. We talked on. His prescribed hour was extended to include supper at the Georgetown Bistro. He caught the last Eastern Shuttle back to New York, and I returned to my hotel – and to London – a much happier man.

CHAPTER 8

The Bombshell

The phone call came on February 16th 1981 – just a fortnight after my return from Washington. It was from a furious Martin Stevens, Head of Hydrodynamics at the BHC test-tank.

"David, what the hell d'you think you're doing, running our 16-foot Osprey model in Vickers' tank at St Albans without telling me? They know nothing about semi-planing hulls – they're towing it like a bloody tanker! And last year you promised we'd be involved if Vickers went ahead!"

His boss, Bill Crago, Head of BHC, had been in St Albans the previous week. He had seen what he believed was BHC's 16-foot Osprey model in their high-security experimental tank. It was being tested with other models of Moor's design and Azteca models of different proportions, in a 'Standard Methodical Series' – a major programme examining variations in the shape and dimensions of a 'parent-form' – clearly in this case the Osprey and Azteca, both subject to our copyright. David Moor, the Superintendent we met the previous summer, had asked Crago for advice: he was apparently 'way out of his depth' in testing a semi-planing design.

To Martin, it must have appeared a shocking breach of trust on our part: as if we had removed the model from BHC, taken it to Vickers covertly, and gone ahead with a major series of tests – all behind his back.

Naturally Crago assumed he had seen the same Osprey model that we had removed from his BHC store, with assurances of their involvement after our visit to Vickers the previous summer: it was exactly the same size and of the same yellow colour.

Considering the years of experience Crago had in the testing of our designs with consistent accuracy, for us to test our designs at Vickers in such an elaborate programme of scaled sizes, speeds and proportions, without his advice, was highly offensive.

I phoned Peter Thornycroft in Bembridge. He confirmed that the model being tested at Vickers was not BHC's. Their model was still in his boat shed. It had lain there on trestles since we collected it from BHC nine months before, in anticipation of Vickers' needing to test it with BHC's involvement. But Moor had declined, saying BS had no further interest in the Osprey.

I called Martin back to reassure him that we were not involved. Whatever model was being tested at Vickers, it was not from BHC. Nor had we known anything about it, apart from such a possibility being mentioned at my meeting with BS in November, months after Vickers had rejected the idea and returned our lines plans – which they must have copied and retained for their own use, either then or in the future. Hence their model.

"Oh shit! Look, David, forget I ever spoke ... Please!" It was too late. My loss of innocence had begun.

*****

After hearing Martin Stevens' outburst, what should we do? Go to the Vickers high-security tank and accuse them of violating our written agreement? That would surely result in a denial and the immediate destruction of the model and probably all other evidence. Would it not be better to pop a few 'ferrets' into the MOD's Whitehall burrow: drop a word into the ear of an Admiral or some other intimate of the Ministry? BS might not risk destruction of evidence for fear of harbouring a whistleblower. Could we obtain the release of the model by a Court Order? Surely not: our only evidence was hearsay.

Little did I know I had thus initiated a process that would dominate the rest of my working life. To refresh my shocked brain, I wandered out into my secluded Wandsworth garden to think over the consequences of Martin's message. Its narrow paths and beds led to the bosky dell with its huge pear tree and other fruit trees – a welcome haven of peace behind the house. The big mahonia was still in flower, its strong scent of lilies-of-the-valley wafting across the lawn. The usual flock of starlings was rising and falling in twittering vortices around the hundred-year-old elms which stood, still un-diseased, in the garden of Seymour Lodge, a couple of houses away.

In a few minutes Vanessa would be returning with the boys from their schools. The three of them would come roaring into the kitchen, toss down their satchels and jump, fighting, onto the day-bed to watch John Craven's *Newsround* and *Blue Peter*. Then they'd devour quantities of tea and baked beans on toast – before settling down to Latin texts or practising on trumpet or clarinet. The noisiest, but one of the best times of the day.

But first, here were a few moments of peace in which to reflect on the extraordinary position in which I found myself: to try to formulate an effective strategy for coping with what looked like a major deception by some of the most powerful forces in the land ... Whilst publicly advising everyone, including politicians and the media, of their lack of interest in our ideas, the forces of reaction within the military-industrial establishment were, in fact, undertaking unauthorized tests of several models of our design family – and of varying scaled sizes, speeds and proportions.

"Take no hasty action," I cautioned Peter. "Let's think it through ... Play the fool."

*****

The following week, in London, I attended a NATO Discussion Day at the Royal United Services Institute, generally known as the 'RUSI'. Admiral Eberle was there. Striding up to me in an unusually hostile manner, he said he'd visited Ship Department at Bath the previous day. Expecting to hear of future plans for warship and submarine procurement by Ken Rawson and his team, he was surprised to find that all they could talk about was Thornycroft and Giles and their crazy ideas. He admitted to being punch-drunk by the end of their session, but had finally succumbed to their arguments. I would receive his official debunking letter shortly; the people at Bath were composing it for him right now. They had claimed it was all due to our inferior 'resistance co-efficient'. I would shortly be sent what he called the 'Bum's Rush'.

I told him that Bath was jumping the gun, since I was in the middle of a protracted correspondence with Rawson and Jack Daniel comparing the Osprey's performance with traditional hulls. They had become quite encouraging – even offering to do some more studies based on possible Osprey tank-tests and our sea-keeping trials aboard the Osprey *Indaw* the previous summer. Again it didn't add up. What was the true activity behind the veiled threats and smiling faces?

"Only a few weeks ago both were asking for the Osprey powering information. They claim they're doing a further re-assessment – so why such haste to slam the door in my face?"

"Beats me," he replied, bewildered. "But they do seem to have it in for you in a big way. Go to Bath and sort 'em out – that's all I can suggest." This was the moment to strike.

"So why have they been testing an elaborate series of models of our designs in the Vickers tank? Secretly – without our knowledge, advice or approval?"

He looked stunned ... I could almost hear the sound of my first ferret – its little feet scuttling down the MOD burrow.

In preparation for my meeting with Rawson at Bath as suggested by Eberle, I visited Dr Tony Morrall at the National Maritime Institute, part of the famed National Physical Laboratory in West London. He had been in charge of sea-keeping measurements of *Indaw* the previous summer and had close contacts with the MOD and BS. It was a chance to check on the Vickers tests – and, perhaps, to insert another ferret. I asked him point-blank about David Moor running an Osprey model at St Albans. "Oh yes ..." he responded. "You know all about it, don't you?" implying that I should.

"I'm not supposed to know a thing: that's why I asked the question." He appeared surprised; but seemed happy to enlarge upon the subject.

"Yes ... they're running it at the same time as their design for Hall Russell's bid for the Hong Kong Patrol Craft. David Moor's in charge and he's floundering around trying to make the contract speed."

Morrall told me that Moor's design, based on a standard Vickers hull, was too slow. The performance penalties were stiff, leading to outright rejection if it were one knot below the contract speed. Alongside his own design, Moor was apparently testing large Osprey and Azteca models in a major test programme. He cheerfully claimed that, if anyone outside BS or the MOD found out, David Moor could be in big trouble. Mrs Thatcher was determined to close down either the NMI or Vickers tank in her privatisation programme – so no wonder Morrall was keen to co-operate. "What are you going to do?" he asked.

"What *can* I do? It's a high-security tank. How could I get anywhere near the model?"

"Don't tell anyone what I've told you." he said darkly.

"Of course not," I replied.

My second ferret scuttled away down the Whitehall burrow. I knew that Morrall and NMI had close contacts with Bath and the MOD.

The next day I took the train to Bath to meet Kenneth Rawson. An old blue Navy bus delivered me to the depressing hutments comprising the MOD Ship Department at Foxhill. Formerly a military hospital, it stood on top of the Somerset hills – an unlikely setting for the headquarters of the

Navy's ship design elite: the Royal Corps of Naval Constructors, the very organisation from which my father had resigned during the War. Rawson had devoted a lifetime to ship design and had written a leading textbook on naval architecture. Based on our previous correspondence and the debate in Parliament, he was surprisingly complimentary about the Osprey. The recent sea-keeping measurements by NMI of *Indaw*, had confirmed it was in many respects better than other hulls of the same size at speed in high seas. Also, due to the added beam there were advantages in stability, cost, and accommodation in the middle of the ship – where the greatest volume was available and the motion was least. He conceded that 'there might be some merit' in our ideas and, surprisingly, he declared that it might be better for the Navy and for Britain to develop a new approach to warship design rather than just offering the same designs as other countries. Far from being antagonistic, he favoured continuing our correspondence.

It was indeed difficult to reconcile his conciliatory attitude with his official rejection given to Admiral Eberle a few days before, let alone his opinions reported so publicly in the House of Commons ... No wonder, as Tony Morrall told me at NMI, the Hong Kong Patrol Craft contract was being stretched out far beyond the original tender date awaiting their test results of the Osprey design to help them reach the contract speed.

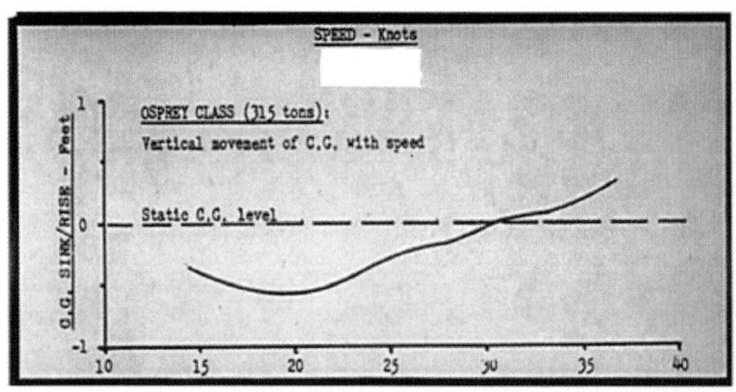

One of many tank-test measurements by BHC, of the sink and rise of the Osprey model centre of gravity (c.g.) above a 'threshold' speed of 20 knots, rising to its static waterline at 31 knots.

We parted on friendly terms: he even donned his iconic deerstalker hat and drove me to Bath station in his old green Morris Minor. In parting, I undertook to send him BHC's Osprey model tests – including comprehensive test results and measurements of dynamic lift above a 'threshold speed' of 20 knots, with increasing speed lifting the hull by seven inches, back to its static waterline at 30 knots. That was the maximum operational speed of the later Danish Navy's 'Flyvefisken' Corvettes, based on the Osprey design by Frederikshavn's sister-yard Aalborg. Compared with the Osprey *Havørnen*'s speed trials, the full-scale vessel was faster than predicted by the BHC tank-tests – an unusual result which he would certainly wish to investigate. I was not about to show my hand over my knowledge of the secret Osprey tests at Vickers; nor was he about to show his.

CHAPTER 9

The Ferrets Go to Work

It was time for some legal advice. I called Peter Martin of Frere Cholmeley, an aviation lawyer I had dealt with in the past. He arranged for his copyright expert, Frank Presland, to meet me. Used to representing pop idols like The Bee Gees, The Who and Elton John, he had a jovial, expansive presence, enveloping me with his enthusiasm for such an obvious case of 'skulduggery', as he called it.

"The fact is, though, I can't do anything until you have solid evidence that they're up to no good. And that could be difficult with the MOD working in such a respected high-security institution as Vickers."

At Presland's suggestion, I contacted Bill Crago at BHC to challenge Moor about the Vickers tests. He became evasive, claiming that in such profoundly disturbing circumstances he preferred not to get involved. He feared it might be a warning to 'the very powerful interests concerned of possible action on our part'. He expressed outrage over Moor's conduct – although I detected a slight note of satisfaction at the embarrassment of an arch-rival. Like Morrall he knew that, due to Mrs Thatcher's purge of British industry, at least one major testing tank was for the chop. I was careful to retain tape-recordings of all three phone conversations between Crago, Peter Thornycroft and myself. On 25th March Peter phoned. He, too, had just heard what he called 'Crago's confession', all recorded by his answering machine. Crago had 'come clean' and told Peter that he had met David Moor at St Albans on 19th February: "The model was in the tank," Bill said. "I saw it all happening. I had heard the rumours, but couldn't believe them. Moor told me that this was what he'd been told to do, but he didn't know much about what he was doing." It was a huge programme involving several models of variations of our designs, scaled from 150 to 7,000 tonnes and speeds to 40 knots. So much for being 'limited to quite small craft', as became the bedrock of the MOD's later ruthless public rejection of our ideas ...

Peter Thornycroft had just mentioned this to an old friend, John Coates, who, having recently retired from being Rawson's boss as Director of the

Forward Design Group of the Royal Corps, spoke in hushed tones of 'the bonfire at Bath'. This had followed the revelation within the MOD of our knowledge of the Vickers tests on the Osprey.

Now we had the true story, certainly enough to satisfy Frank Presland that 'they were up to no good'. It also explained Tony Morrall's ill-concealed glee when I told him that we knew about the Osprey model at St Albans. Like Crago, he was digging the dirt on David Moor, hoping this might keep NMI afloat in the threatened closure of testing facilities.

"We've got the evidence," said Peter with delight. "Let's use it!"

"No, just sit tight," I suggested. "A rabbit will pop up shortly. Then we can go to work."

Later that day I phoned Jack Daniel to check on how his 're-evaluation' of the Osprey was proceeding. He was too busy to answer my call.

A Jack Rabbit

That weekend, on the morning of Saturday 28th March 1981, I tore open an ivory envelope lying on my doormat, from the BS Knightsbridge HQ:

> From: RJ Daniel, OBE, F.Eng, R.C.N.C. Board Member for Warship Building.
>
> Dear Mr Giles,
>
> I am deeply concerned at the time that has passed since you left the data on the Osprey family with me for examination in order that we in British Shipbuilders, and myself in particular, could form a measured opinion of the characteristics and performance of the form. I found that I needed more, and more methodical data on the hydrodynamic characteristics in calm water and waves.
>
> You will recall that some time ago you called on David Moor at the British Shipbuilders Tank at St Albans last May and asked him to run a model of OSPREY and gave him a lines plan. You indicated that you would not be able to pay for the experiments. At the time it did not seem possible to do this work but you will now be pleased to know that in an effort to provide additional data David Moor has made a model and it has been run in the tank as a minor Research Item. The delay in getting results for discussion with you has of course been because commercial work has to take priority at St Albans, however they have now almost completed the runs ... and we will soon be able to discuss the results. Should you wish to see the model and have a look at the

rough results, please get in touch with me and we will arrange a visit on a convenient date. We will, of course, give you a copy of the final results.

Yours sincerely,

(signed) Jack Daniel.

P.S. I have just been told that you have been ringing!

A Jack Rabbit indeed – straight out of the BS/MOD burrow: and a former Director General of MOD Ship Department at Bath, preceding Rawson. So they were all 'thick as thieves'.

The Opening of 'Ospreygate'

On 30th March, I took the letter from Daniel to our lawyer, Frank Presland, and discussed our possible course of action. He suggested issuing an immediate Writ for Breach of Copyright and Confidentiality.

At the beginning, he warned, it would all move very slowly: we would have to brief a Junior Counsel and apply for a preliminary High Court Hearing. That's when the meter would start to tick: we would need about £60,000 just to get through the Preliminary Hearing. To brief Junior Counsel could be even more expensive, and BS would make it as difficult as possible for us. If the Ministry of Defence was involved, they were probably destroying whatever documents they could, especially at Bath: the 'Bonfire'.

By suing BS, a state-owned corporation, we were likely to be up against not only the Ministry of Defence and the Navy, but also The Crown. There would be Treasury Solicitors, the Official Secrets Act, and Whitehall obstruction. I told Frank that I doubted we could raise enough funding ourselves, but perhaps we might manage it with Osprey Ltd. Shareholders included the Danish Shipyard which owned the copyright in the design. There were several new overseas orders for Ospreys and more design work was coming to us from Frederikshavn. Of course Peter and the other shareholder, David Browning, would have to agree – as would Frederikshavn Vaerft, since they had a vested interest. "We might be able to raise a first payment," I said.

"If not, you'll have to kiss goodbye to any rights you have in the design and it will become public property," said Frank.

So that was their game! Steal our ideas, confident that we could not afford to pursue them in the High Court, restricted to important Civil cases requiring much higher funding than in the local courts. If we failed our Court Action, they could 'lose or use' our designs, depending on how they wished to proceed. However, if we *did* pursue them there could be bad publicity. The finger of blame in this case seemed to be pointing at the MOD officials at Bath, as well as Jack Daniel, their former DG Ships now at BS, with serious political and personal consequences. The damage to their reputations would make us unpopular; but how could we now trust anything we signed with BS or the MOD? Rawson's memoir suggests that his 'resignation' from the MOD in 1983 must have been forced upon him by higher powers, whose directions he was only obeying, but who were ruthless in their execution. Others clearly suffered the same fate. In Frank's view, we had a chance of frightening them to the point where they might come up with a settlement. But we did not just want a settlement, we wanted to retain the copyright in our design, since we were confident it could lead to bigger things in the future.

"You'll have to fight all the way," said Frank. "And that means you and anyone else likely to benefit from holding on to the rights, such as the Danes. They would surely be prepared to help out, wouldn't they?"

He agreed to start the legal process. We required written undertakings that BS and Vickers would:

(1) Within 24 hours return all copies of the lines plans and any documents containing information elicited from them;
(2) Allow us to inspect the model, and hand it over to us if we wished;
(3) Supply us with a list of all those involved in the copying and testing of the Osprey;
(4) Suspend all further testing, research, development and production concerning any naval design derived from the lines plan;
(5) Pay our legal costs.

The letter, together with our Writ for Breach of Copyright, was duly dispatched to BS to arrive on April 1st. But this was no April fool.

We fixed a meeting with a Junior Counsel, John Mummery QC, a well-known expert in copyright law. He sympathised with our frustration, but warned us of what lay ahead: there would be Injunctions, Orders in Discovery, and Affidavits, followed by a series of interlocutory hearings to force the pace on Discovery. BS would certainly obstruct us at every opportunity. Did we really want to go down such a long and costly road? Could we not obtain some accommodation from BS and the Navy, in return for allowing them to use certain of our design ideas? If respected establishments, of the calibre of Vickers, had proposed investigating the feasibility of our claims, why had we not handed over all our data and drawings unconditionally?

I recounted the story of Frank Whittle and the sorry treatment he had endured from the Air Ministry, the RAF and others during the initial development of his jet engine: 'Impracticable', they called it, as they refused to pay the £5 renewal fee for his patent, thus leaving it open to the German Hans von Ohain, whose jet engine flew years ahead of Whittle's ...

Mummery suggested that we were hardly in the same league as Frank Whittle, nor were our ideas likely to be as significant as the jet engine.

This was the first of many such warnings, no doubt well intended. But if Vickers, BS and the MOD were worthy of our trust, why had they acted in secrecy, poured scorn on our designs, and thus denied the validity of our own test results by two such highly-respected tanks as BHC and the Danish Technical University?

There followed the usual exchange of letters between lawyers. At first BS claimed that I had approached them and asked them to run tests so that they could arrive at a realistic assessment of our design. Jack Daniel, they said, had authorised the Vickers model tests 'on my behalf'.

Frank wrote a strong reply, claiming that, if a designer lends his plans to a testing tank, that does not mean that he assumes they will be used to construct a model and test it without his knowledge – particularly when he has obtained a written undertaking expressly to prevent this. He also reminded BS that Julian Taylor had passed on to me the invitation, from Moor to visit the St Albans tank. It was not my initiative. Nor was this a 'relatively minor difference', as BS claimed.

Viewing the Corpse

BS finally allowed us to inspect their Osprey model 2219: 'showing a bit of leg', as Frank put it. Frank, Peter Thornycroft, David Browning and I arrived at the Vickers tank on 8th April. After passing through the security formalities, we were admitted to the lobby of the building where the offending item was lying on a trestle beneath a shroud, like a corpse. Both Moor and his assistant appeared, dressed in black suits, white shirts and drab ties, accompanied by their lawyer.

"Do look, David ... The undertaker with his professional mourner!" Peter whispered. He was in his usual yachting kit, with a bright red spotted hanky in his blazer pocket, RYS cap in hand – hardly 'rig of the day' for such a solemn occasion. He was urged to lower his voice by an embarrassed Presland, who had assumed the solicitous smile he reserved for such occasions. The sharp edge of Peter's wit could prick any bubble of pomp or circumstance, sometimes giving offence.

With due reverence David Moor lifted the shroud for us to identify the body. It was certainly an Osprey, a yellow 16-foot fibreglass model: something we had never used ourselves, preferring wood because it could be altered more easily as testing proceeded. The model had its propeller shaft brackets and tunnels fitted, but the propellers, shafts and rudders had been removed. Normally models are first tested 'naked', for bare hull resistance and then for estimates of speed and power, with propellers and rudders.

"There's also something funny about her buttocks," Peter whispered quietly to me, as he stroked that crucial part of the model's anatomy. "There may be a bit of bad blood there."

The Vickers crowd looked pale.

"It's an Osprey all right," Peter agreed aloud, "but a fairly crude model, if I may say so."

Once in the car we dissolved into peals of laughter. "Most unseemly," said Frank. "You must restrain yourselves."

*****

Meanwhile the correspondence with Rawson continued. Eleven days after we had issued our writ against Vickers and BS, on April 10th, I received a letter from him that was in stark contrast to all that had gone before. He wrote in vengeful terms so strong that they hardly seemed to be from the pen of the friendly open-minded gentleman who had been so receptive and interested during our conversations at Bath.

"I have yet to complete the study promised to you in my letter of 25 March," he admitted, *"but I believe you will be further disappointed with the comparison of your claims for short fat ships with the established performance of long thin ones."* After further dismissive technical comments, he closed with this

unexpectedly hostile outburst: "*While understanding that your objective is to keep your proposals in the public eye at any cost, I believe that rape of this department's views, which have been consistently fair and objective, will not dispel the widespread public distrust of your claims among professional engineers and scientists.*"

It seemed there was even a note of panic about his denunciation. He had changed his position, from one of open-minded interest to outright rejection. By now of course he must have heard about our writ and our visit to St Albans and had been told to be on the offensive: he had no further need for my co-operation. His mood had turned from defence to attack, from Killing with Kindness to Death by Decree. Like a dutiful civil servant, he had done as instructed. He must surely have been informed of our writ against BS, contrary to his denial of any complicity, as claimed in his memoir.

CHAPTER 10

Rawson's Revenge

Only three days later, on Monday 13th April, Rawson sent his promised 'study'. Its implication was pretty clear: the BS and MOD were apparently investigating 'an important contradiction of traditional theory'. Why else would they have reacted in such a vengeful way so soon after we issued our writ and inspected their model? They had been caught red-handed in the midst of a secretive test series of our design family.

Famously, he wrote what was to become an Instrument of Policy for two major navies for the next forty years, although contradicted by their major test-tanks and technical institutions such as NPL/NMI and MIT.

> *You will remember, at our meeting on 27 February I said that if there were any significant dynamic lift on the OSPREY form, it really would represent an important contradiction of traditional theory. Unhappily, the information you now supply does not show such evidence and it must be presumed that conventional wisdom is not to be denied. Such dynamic advantage is confined to quite small craft and is lost at OSPREY displacements – not surprisingly because displacement varies as the cube of the dimension and dynamic lift as the square. We are safe in assuming that ships of SIRIUS size will certainly obey normal laws.*

Despite his confident tone, the two graphs accompanying the letter were clearly written in haste and in his own hand, in some places almost scribbled. In one place he had mis-spelt Osprey as 'Osrey'. His letter was circulated to everyone with any possible interest in the matter, within and without the Ministry of Defence. I wondered if any of them would bother to read it, let alone understand or question it. His graph of 'Osrey' power/speed did not come from any data I had given him. It compares the Osprey power/speed with a 'Ton' class minesweeper, with both at 373 Imperial Tons. The BS ex-Vickers tanks at St Albans and Dumbarton alone in the UK still used Imperial units. So whence came the 'Ton' and Osprey comparison? It must have been from one or other of those tanks using existing Imperial Ton data.

Reviewing the dates of his letters, it became clear that something significant must have happened over that weekend, between Friday 10th April (when he dispatched his 'rape' letter) and when he sent his 'Revenge' on 13th. He included in his Figure 1, estimates of power and speed which had come neither from TGA nor BHC. He had also insisted that "Estimates should come from tank tests, not extrapolation, if they are to be convincing." Since no BS documents had been disclosed by mid-April his graph could not be checked. The comparative tests being in Imperial units, they must have come from a Vickers tank.

But could he have had the Osprey model re-run at St Albans over the weekend of 11th–12th?

That seemed improbable. BS and Moor had assured us that Vickers had suspended all testing on the Osprey model as soon as our writ was issued. Frank thought it most unlikely they would risk it, considering the undertakings given by BS and their fear of a traitor in their midst. To have run a model after those proceedings started would have been a criminal act. Or was it another model that was used elsewhere? But that, too, would have been illegal. The 'Ton' model displacement was converted from Imperial to metric units. So Dumbarton it was, as disclosed in later evidence, by a series of frantic telexes between Moor and the Dumbarton tank carriage log, confirming the tests were carried out over the weekend 11th–12th April, no doubt in time for Rawson to write his letter on April 13th.

*****

A few days later, Dr Garwin called. He was in London on his way home from Moscow where, with other concerned scientists, he had been trying to obtain the release of André Sakharov from the Gulag. I showed him the 'Revenge' letter. He was intrigued by Rawson's blunt statement that the Osprey hull would not scale up to the larger size of Sirius. He promised to consider it and send a reply. Later his wife, Lois, who was with him in London, told me that her husband was 'extraordinarily angry' about what Rawson had written. I eagerly looked forward to his response.

The BS Solicitors wrote on April 13th – the same date as Rawson's 'Revenge' letter – changing their story:

> "Sometime after the November meeting we decided to accede to your clients' request and carry out a full evaluation. Our model, which was based on our best professional interpretation of all the data available

> to us including the design material furnished to us by your clients since November and other information available from Press publications, was intended to be as representative of the "Osprey" type and size of design as we could make it. Incidentally, the lines plan ... was not then available to BS and could not be used since it had been returned to your clients by Mr Moor under cover of a letter dated 27th June 1980."

"That," as our Queen's Council (QC), was to write six years later, in his Short Opening for the Final Trial ... "is something you should tell to the marines."

The Confessional

At David Moor's invitation, we agreed to visit St Albans again to discuss their latest version of the provenance of the Osprey model. BS would be fielding the 'A Team', led by Jack Daniel and David Moor, including their in-house QC, and they would give us their account of exactly what they were doing and how they did it: there had, they said, been some trivial misunderstanding, which they wished to explain.

"Copied the lines plan; then made a plug, a mould and a model, I presume," said Peter when I phoned him about the meeting. "Pretty straightforward, I'd say."

We were herded into the Vickers Boardroom where a cheerless collection of BS people had assembled. The walls were hung with models of great battleships and submarines, remnants of the arms sales of Sir Basil Zaharoff – the original 'Merchant of Death' who, at the turn of the century, was renowned for selling Vickers armaments to friend and foe alike.

As if it were news to us, Daniel started by talking about the nature of a testing tank, 'a place where no secrets are hidden: it has the atmosphere of the confessional'. He implied that, by merely entering the hallowed halls of the Vickers tank and speaking to David Moor, we were as good as giving away our ideas, that they were like sins confessed to a priest. David Moor was renowned for his photographic memory. Once they were in his head he couldn't forget them, except perhaps by lobotomy. But, again, as with the confessional, there would be no question of them being repeated outside.

After listening to a series of inconsistent, halting explanations concerning the model, Frank suggested that the two groups should separate for private discussions. We retired into an office, not daring to speak above a whisper for fear of hidden microphones. After a few minutes of muffled discussion we returned to Moor's office. The following quotations are from the meeting transcript:

"Now come on, gentlemen," said Frank cheerfully. "You're going to have to make up your minds. Which story do you wish us to believe? According to your letter of April 2nd, in response to our writ, it appears that you copied the lines plan that Giles gave you and that 'JM' did a version of it from which the model was built. But, according to your letter of April 13th, you made it up yourselves from published data, photographic memory and a touch of genius, as Mr Daniel has also suggested today. Of course, 'JM' could probably tell us much more about the process ..."

Their Counsel squirmed. He could not confirm the position either way: there was still much confusion as to how the model had been designed. He undertook to provide a complete description of how it is possible to produce an Osprey hull from the Azteca lines plans in the possession of the Vickers Dumbarton tank, at which its model was tested, and now a part of BS. But the Azteca copyright was still owned by Peter's TT Boat Designs.

"Very well," said Frank, "but you must make up your mind and provide a full written explanation containing dates and names involved in the design of the model. Both Daniel and Moor must agree with its contents and it must be delivered by May 8th."

The crucial question was why, when their model was exactly the same size as the one we had originally offered Moor – and removed from BHC for that purpose – had they gone to all the trouble, secrecy and risk of making one themselves?

Frank thought there was a rift between Daniel and Moor, though we would have to wait for the evidence to see who would take the blame for the tests. He cautioned against believing either of them at that moment – and against any social or professional contact with anyone from the other side.

This remark was aimed at Peter who had been approached by Daniel in a conciliatory fashion at the end of the session, suggesting dinner with him and his new young wife to sort out a 'small misunderstanding'.

"Don't even think about it," said Frank with unusual firmness.

Frank phoned the next day: he had a letter from BS. It contained this key statement: "Where the letter of April 13th says the original lines plan was not used, it is not correct. The original plan was copied in case it was needed for future reference. When in November instructions were given for a model to be constructed offsets were taken and were scaled up and re-plotted on the lines plan from which the model was constructed ..."

So they *had* lied ... and our plans *were* used for the model, just as we thought. All that nonsense about starting the model in January after my November 1980 meeting with the BS Board and Daniel (and about Moor's photographic memory) was now 'inoperative'. As admitted later, 'the model' was being driven from St Albans up and down to Dumbarton for testing over Christmas and Hogmanay.

At last we could get to work on obtaining the other documents concerning the design and testing of the model; and since they seemed reluctant to release anything, it would probably mean taking them to court. They were playing hardball, trying to increase our costs and grind us into submission while they, to judge by their procrastination and legal feints, must have had access to unlimited funds. The process had already cost us good money.

In the High Court on 19th June 1981, we obtained our first Court Order. BS were to stop any further testing and to disclose all relevant documents. The Court also called for a speedy trial and awarded costs to Osprey Ltd. That was a start. However, we remained concerned that Moor might still be using some of the information gleaned from the Osprey testing for the Hong Kong design that he was submitting to the MOD at that very moment.

Four days later, on 23rd June 1981, John Nott, the Defence Secretary, announced that Hall Russell had been awarded the Hong Kong Patrol Craft contract. He claimed that the order was placed 'after a hard-fought design and build competition' and mentioned the severe penalties for non-compliance of speed. Now that the order was announced, BS could get on with the privatisation of Hall Russell – a keystone of the new Government's industrial policy.

We thought it likely, however, that some of our design features had been copied in the final Hong Kong design. We returned to the High Court on 3rd July and sought an injunction restraining BS from using any information derived from their Osprey tests in any future design of whatever size; an order that extracted protests from them on the grounds that they could hardly 'unlearn' what they might have picked up from testing the Osprey – the 'lobotomy' Daniel had spoken of at our last Vickers meeting.

The Judge in this hearing was Mr Justice Martin Nourse. He had an amused, paternal smile throughout the proceedings that sometimes verged on laughter. John Mummery, our Junior Counsel, seldom employed sentences of more than about five words, interrupted by long periods of silence: perhaps a courtroom ploy to concentrate the attention of his audience. He obtained the injunction we sought. Thus the contract announced by John Nott a few days earlier was stymied. It was subject to a complete redesign by Vickers – officially without reference to anything gleaned from their Osprey model tests. However the speed trials confirmed the beneficial effect of dynamic lift that Moor still incorporated in his design, which Defence Secretary Heseltine unsuccessfully later tried to conceal with a Parliamentary secrecy order.

CHAPTER 11

The Battle of the Giants of Academe

In June 1981 another important envelope tumbled through my letterbox. This was from IBM's Thomas J Watson Research Centre at Yorktown Heights, outside New York. It was the response from Dick Garwin to Rawson's 'Revenge'. Garwin's stiletto-like logic thrust straight at the heart of Rawson's argument, challenging his claim that the 'square-cube law' prevented the Osprey hull from obtaining benefit from dynamic lift even if a much smaller version could do so at the same proportional speed.

Rawson had written, *"Such dynamic advantage is confined to quite small craft and is lost at OSPREY displacements – not surprisingly because displacement varies as the cube of the dimension and dynamic lift as the square. We are safe in assuming that ships of SIRIUS size will certainly obey normal laws."*

Garwin challenged him directly: "Since dynamic lift goes as the square of the speed, the lift per square meter for a larger ship ... increases linearly with dimension. Hence if a small ship avoids catastrophic drag rise by dynamic lift, a larger ship of similar form will do it also to the same degree."

Of course Garwin was right: the Osprey really *did* represent 'an important contradiction of traditional theory', to use Rawson's own words.

I circulated Garwin's letter to the people in the MOD, the Navy and Whitehall who had received copies of Rawson's 'Revenge', hoping to shake their unquestioning acceptance of Rawson's argument. I received no response. Presumably they closed ranks or, more likely, did not bother to read his reply – even in the unlikely event they could understand it.

Rawson tried the finest weapons in the bureaucrat's armoury – arrogance, obfuscation, splitting hairs – but Garwin stuck to his guns, insisting that "dynamic lift is just as much available (at speeds where it is needed) on the large ship as on the small ship".

This is exactly what I had suggested during my meeting with Rawson at Bath. And he had agreed. But his July 8th response to Garwin was dismissive: "The assumptions of square or cube variation are too crude to demonstrate more than a general trend of diminishing significance of dynamic lift with size."

This assertion was even less convincing than those he had made earlier. Anyone knows that the lift of a Boeing 707 wing did not diminish with speed as it was scaled up to the size of the 747. There was no 'general trend of diminishing significance of dynamic lift with size'. Did Rawson truly believe he could pull the wool over Garwin's eyes as he had clearly done with his naval, political and official superiors? Perhaps he 'forgot' about the importance of maintaining the same proportional speed, as Garwin had pointed out: 'at speeds where it is needed ...' Or perhaps he didn't: he was just 'blinding them with his personal science'. I could hardly wait to see Dr Garwin's reply.

Enter more Giants ...

For some time we had enjoyed increasingly strong support from Captain John Moore, RN, the respected Editor of *Jane's Fighting Ships*, Chief of Naval Intelligence – also a former Commanding Officer of my old submarine, *Trenchant*. Throughout his time at Jane's he had acquired a reputation for detailed analysis of advances in Soviet submarine and warship design. He was famous for speaking his mind, not only in *Fighting Ships*, but also in *Jane's Annual Defence Review* and other publications. He was known to be involved in several controversies that the MOD would have preferred to keep out of the public eye.

At John Moore's suggestion, I sent copies of the Rawson–Garwin correspondence to one of the scientific prodigies of the Second World War, Dr RV Jones, FRS: "the formidable Dr Jones; the man who bent the bloody beams," as Winston Churchill described him in his history of the Second World War. Among Jones's many achievements was his success in diverting German bombers away from their targets in Britain with false radio directional beams, rather than those transmitted by the Luftwaffe. He was now Professor of Natural Philosophy at Aberdeen University. I hoped he might cast new light on the argument between Rawson and Garwin.

Dr Jones replied courteously, saying he was sympathetic, citing the battle between advocates of the de Havilland Mosquito and supporters of the heavy bomber during the War. He suggested I should contact Professor Sir James Lighthill, FRS, Provost of University College, London, formerly Director of the Royal Aircraft Establishment at Farnborough and then Lucasian Professor of Mathematics at Cambridge (a post in which he was preceded by Isaac Newton and succeeded by Dr Stephen Hawking).

Jones further encouraged me by saying, "If you have Dr Garwin on your side you ought certainly to be able to give our officials a stirring run for their money." With Garwin's involvement, we were at last attracting the attention of qualified people who might give an unbiased view on whether or not our design represented 'an important contradiction of traditional theory'.

On July 13th, I received a reply from Sir James Lighthill supporting Rawson's position. This was a disappointment, but no surprise: Sir James was, after all, Provost of University College of which Rawson was his Visiting Professor of Naval Architecture. "I am afraid that I remain sceptical of the ideas put forward," he wrote ... "I see the disadvantages of 'large lifting area' and 'wider form' (disadvantages arising from increased frictional resistance and wave-making resistance) as likely to be substantially greater than any potential advantages of dynamic lift at these displacements. 'The square-cube law is at it again!'"

The light-hearted tone of this last remark, intended no doubt to make peace between the correspondents, could not disguise its bizarre implications. Was Sir James suggesting that a large plane wouldn't fly as well as a small one – reminiscent of the old de Havilland joke: 'sparrows can fly, but swans can't'? I could hardly believe this, coming from the former Director of the Royal Aircraft Establishment and such a universally acknowledged expert on fluid dynamics. Furthermore he was undermining the whole technique of tank testing. We were talking about Osprey measurements by the BHC tank showing the hull's centre of gravity lifting almost a foot between 20 and 30 knots at full scale.

I needed to escape from the desiccated world of Whitehall, the Ministry of Defence and BS: their terse little inaccuracies; their carefully-crafted phraseology intended to conceal, but with implication ... those knavish little words kept on file to deflate the innovator or enthusiast, whilst conveying an implied squeak of concealed glee: '*sadly*', or '*unhappily* ...!' 'Mandarins playing verbal billiards using elliptical balls with flexible cues', was how Charles Adams described it. All those mincing words of rejection to some of the great ideas of their day:

'*Sadly*, Mr Harrison, we have to tell you that your Chronometer is not sufficiently accurate for use in calculating Longitude ...'; '*Unfortunately*, Mr Brunel, even if the propeller had the power of propelling a vessel it would be found altogether useless in practice ...'; '*Reluctantly*, Mr Parsons, we have to report that we cannot justify the time or effort to further examine your ideas for a steam turbine ...'; '*Sadly*, Pilot Officer Whittle, we find that your ideas for jet propulsion are impracticable ...'. And, dare I quote Rawson's verdict on the presence of 'lift' in the Osprey: '*Unhappily*, Mr Giles, the information you now supply does not show such evidence and it must be presumed that conventional wisdom is not to be denied ...'

It was time to go sailing with my family and friends ... to escape from the stuffy and deceptive dialogue of the Mandarins and their servants. Captain Richard Sharpe, then Director of Naval Future Policy, has described his job in the *Times* as 'writing long-term papers which nobody had time to read, languishing in an over-manned, ineffective and defeatist bureaucracy'. He later succeeded John Moore as Editor of *Jane's Fighting Ships*.

Gentlemanly Giants

Back in London at the end of August, my sanity restored from sailing around the Balearic Islands, I received a copy of Garwin's reply to Rawson's latest letter. Its tone was more determined, with a hint of annoyance:

> "I have nothing to say about whether a particular design should exhibit a reduction of wave drag due to dynamic lift – only that the effect is as important for a big ship of given design as for a small one or a model."

I had been concerned during our holiday, not over the idea that our theory might be wrong, but that Garwin might follow the example of Lighthill and capitulate in order to 'accommodate' his scientific colleagues. But Garwin was no lightweight. He was sticking to his guns, as he would do, successfully and publicly, over Reagan's 1983 Star Wars ABM programme.

Now we had to wait for what must be a final response. Surely Rawson could not go on clutching at the straw of official insouciance – or deceit? In the unaccustomed context of physicists of the calibre of Garwin and RV Jones he might sink like any 'small craft'. He should 'put up or shut up', to use a favourite expression of our Prime Minister.

Charles Adams spoke from his experience of the Cabinet Office: "He won't do either. When defeat stares the good bureaucrat in the face he gracefully closes the argument by saying, 'But that's exactly what I've been saying all along, my dear chap.'"

That is precisely what Rawson did. On September 22nd, he wrote to Garwin reversing his position:

"Like other dynamic forces the lift coefficient will be the same for geosims [geometrically similar models] *at the same Froude Number. This, I believe, is what you are saying and, if so, we are in agreement."* The letter ended, *"I regret that I have not the time and resources to continue this correspondence."*

Despite this offensive brush-off, Garwin's reply of October 6th was typically courteous: "I am glad that our correspondence comes to such an agreeable and harmonious end ... We agree that at the same Froude Number (and attitude) the lift is as important in comparison with weight on a large ship as on a small ship ... Therefore the laws of physics (and in particular the 'square-cube relation') in no way prohibit this happy result."

Shortly afterwards, I received a second letter from Sir James Lighthill, also reversing his former position by supporting Garwin and rejecting Rawson's argument. He gave the final verdict:

> "Over the last few months I have much enjoyed the technological argument that has been in progress between Mr Rawson and Dr Garwin. I assure you that I have been reading every letter in the correspondence carefully even though I have not hitherto joined in ... Now, however, I will just make the comment to you that, in my own carefully considered opinion, Dr Garwin's concluding letter of October 6 sums up the position perfectly correctly from the standpoint of hydrodynamic theory."

So, in Lighthill's revised and more authoritive opinion, Garwin was right, and Rawson was wrong. As BHC's Bill Crago, Chairman of the British Towing Tank Council, wrote to TGA: "It undermines the entire technique of model testing." But Rawson had left a door open: "If there were any significant dynamic lift on the OSPREY form, it really would represent an important contradiction of traditional theory." So, based on the judgement of Garwin and his own Provost at UCL, the Osprey really did represent 'an important contradiction of traditional theory'? Where would this lead, I wondered? As Galileo put it after being forced to renounce his claims of the Earth's movement by the Inquisition: '*Epur si muove.*' 'Yet it moves.' ... Or, in our case: '*Epur si solleva.*' 'Yet it lifts.'

Alas, like the Inquisition, the Powers-That-Be were not about to change their official verdict: The Sun still moved around the Earth ... And, within the MOD circle, the Osprey still sank with increasing speed.

CHAPTER 12

The Birth of the S90, the 'Short, Fat Frigate'

Meanwhile there were winds of change softly blowing in the defence Establishment ... the creaking of yardarms signifying a just-discernible change of wind. In June 1981 I had received a letter from Admiral Eberle that contradicted his earlier 'bum's rush' dismissal of the Sirius concept: "It looks quite probable that we shall shortly be trying to design an entirely new sort of Navy – and we shall need all the help and ideas we can get. I look forward to continuing to keep in close touch."

On 17th September 1981, at the Naval and Military Club, John Moore hosted a lunch party for Geoffrey Pattie, MP, a rising young star in the Thatcher constellation, recently appointed Parliamentary Under-Secretary of State for Defence Procurement. Apart from Pattie, John Moore and myself, the party included Peter Thornycroft and Roy Corlett, a retired Royal Navy engineer officer who had worked for Vickers on advanced submarine technology. At the time he was technical adviser to John Moore as well as working with Ingenieur Kontor Lubeck, the leading German submarine builders, and Gould Corporation, the US manufacturers of advanced torpedoes.

It was clear that Pattie was made of sterner stuff than his predecessors. He was appointed to clear a way through the increasingly impenetrable jungle of defence procurement. Following our meeting, his written instructions to me were precise:

> "Prepare an initial proposal for a 'cheap and cheerful' frigate, based on the Sirius Pocket Destroyer, to be presented to the Controller of the Navy. It would be a stalking horse for the new Type 23 frigate programme. You may be invited to join forces with Admiral Sir Ray Lygo and his men from British Aerospace Dynamics, who make weapons systems: they seem to like your ideas as being an attractive exportable platform for their systems."

John Nott, the Defence Secretary, had threatened 'Navy cuts' that were likely to reduce our surface fleet substantially. There was much talk of selling our remaining aircraft carriers – *Hermes* and the recently commissioned *Invincible* – to foreign navies, such as India and Australia, and cutting our frigate fleet by 30 per cent. *Endurance*, the Antarctic research ship and our only naval presence in the South Atlantic, was due to be refitted in the United Kingdom without any replacement. The surface fleet was fading away.

We agreed to work on Pattie's proposal. Peter was thrilled. Here was something he had only dreamed of: to be working on a new sort of ship for his beloved Navy, just as his grandfather had worked on his 'quick launches', which led to the world's first torpedo boats and destroyers. We spent hours at Bembridge in the *Blackwater* and, in London, in the library of the Royal Thames Yacht Club, deliberating on this exciting opportunity.

The Missing Model

Meanwhile the Osprey Case was simmering on. John Moore had spoken to the Managing Director of the Hall Russell shipyard, who told him that 'a final lines plan' was being prepared for the Hong Kong contract, to be submitted in September. During July and August we waited for documents to be released under the latest Court Order.

BS delayed matters by using every trick in the book: it required several hearings in the High Court to force the issue – which meant further erosion of our limited funds. According to Julian Taylor, their continual question was, "Where does Giles get his money?" The answer was the Danish shipyard who were partners in Osprey Ltd and had a major interest in maintaining its Copyright due to their continuing sale of Osprey variants to the Danish and foreign Navies. Fortunately, both John Moore and Roy Corlett were also prepared to work for nothing.

In a last-ditch attempt to halt 'a time-consuming and expensive trial', the BS solicitors wrote suggesting a Payment into Court of £100 to Osprey Ltd, in full and final satisfaction of any claim we might have for damages. We ignored this attempt to trivialise the value of the design and finally, on 23rd September, we received a huge pile of documents.

Here were copies of some of the more dramatic internal memos, including Moor's personal notes. They demonstrated, first, the glee with which Moor told Daniel that he had arranged 'for an Osprey lines plan to fall off the back of a lorry'; second, the commanding position of Daniel in the financing and building of the models; and third, the extraordinary haste with which they proceeded with the Osprey test programme at the beginning of 1981. Finally, the panic within Vickers and BS when the word got around in February that, as Moor wrote, in a two-sentence memo: "Giles knows: If he arrives at SMET, not a word ... a blank wall!"

There it was at last, in the full light of day, dated 3rd April 1981, just two days after we issued our writ: "As a matter of extreme urgency: testing at St Albans and SMET: Weekend, overtime if required". That was just ten days before Rawson's 'Revenge' letter of April 13th. But they claimed they had destroyed all Osprey models after our writ was issued in April. They must have retained STA 2225 – the Missing Model – at Dumbarton. But, that alone was only hearsay, Frank told us. Stronger evidence was required to prove SMET had tested 2225 after our writ was issued.

SMET was the sinister new name for the Vickers Dumbarton tank which stood for 'Ship Model Experimental Tank', sounding more like a branch of SMERSH. Moor had claimed he was testing the Osprey model (number STA 2219) over Christmas in St Albans and, it appeared, over Hogmanay in Scotland. Security Express was driving the model up the Great North Road to Dumbarton on Christmas Eve and back again in the New Year. Why such haste? Why use two tanks for a single model?

He noted that they were testing at model scales equivalent to 150 feet length up to 400 feet, giving the perfect span of measurements from which to interpolate a ship the size of 'Sirius', or larger – up to over 7,000 tons, the size of a small cruiser. This seemed bizarre after Rawson's grounds for rejection of the design. On the one hand Rawson was saying that our hull form 'doesn't work' at the size of an Osprey, yet, on the other, Moor is looking at scaling it for ships of up to two and a half times the length and eighteen times the displacement of Osprey – and at much higher speeds. It helped to know that there were such contradictions in the enemy camp.

Our confidence increased accordingly.

Major anomalies then arose: The Osprey model STA 2219 we had inspected at St Albans on 8th April was a 'self-propulsion' model: it had propeller shaft brackets, shaft tunnels and mountings for rudders, which, together with the shafts and propellers, had been removed. Why?

Usually models are first run 'naked', for 'raw resistance'. Then the 'appendages' (brackets, shafts, propellers and rudders, roll fins, sonar domes, etc.) are added for self-propelled tests with electric motors.

Instructive as these documents were, they were clearly incomplete. BS had disclosed the results of tests with propellers and rudders that pre-dated the tests with the 'naked' model. When we questioned this anomaly, the reply was that the urgency had made it necessary to do the propulsion tests first, following which the model was stripped of appendages and run naked. This was against all accepted practice. It also prompted the question: Why did they go to all the trouble of restoring the appendages to the model merely for our visit to St Albans on 8th April?

Examining the St Albans and Dumbarton test results for the Osprey, I also noticed that the power-for-speed figures did not tally with the BHC data that we had provided: they were much worse. They also differed from the Vickers power-for-speed measurements of the only Osprey model that BS had disclosed in previous evidence. These inconsistencies, together with the fact that the tests had apparently been run the wrong way around, further raised our suspicions. Their story didn't add up. Testing tanks are extremely methodical, and to test the propulsion model first was putting the cart before the horse.

I told our Junior Counsel, John Mummery, of my fears. He was reluctant to accept that people in positions of such high public trust could be guilty of testing the model after the start of High Court proceedings, let alone concealing a second model. Or was there perhaps a second model that they had kept at Dumbarton and had used for Rawson's final 'Revenge' – and then destroyed? We named such a model 'Christine', after Christine Keeler, the famous 'missing model' who disappeared during the Profumo Scandal.

Forward with the S90

To advance the Sirius 90 project suggested by Geoffrey Pattie, we applied to the Department of Industry for another grant: this time to run comparative tests of sea-keeping ability between self-propelled one-tenth scale models of the Sirius 90 and the Leander frigate. The previous Osprey 1/10th self-propelled tests in the Solent were entirely supported by the later *Indaw* sea-keeping trials, so the technique was fully endorsed in our own experience. The Leander was recommended as an outstandingly seaworthy warship. It had taken part in the 'Solent Series' just completed by the MOD – the most elaborate, and expensive, part of any model-test programme. It was entirely realistic, since it included the effect of wind and natural 'wind-generated' irregular waves, as in the open sea. It became the database for the MOD's new 'SCORES-3' computer programme to simulate the sea-worthiness of future warship designs.

The Sirius and Leander models would be 30 feet and 37 feet long respectively, and of the same scaled displacement: the size of family motor cruisers. Driven by diesel engines of equal power, they would be radio controlled and fitted with telemetry to record motions in all axes, including movement of the helicopter deck. A host of other data would also be collected. At our suggestion, the models were fitted with video cameras at the bridge to obtain visual reference of ship motions from high-speed film. This, when slowed down, would give a precise real-time impression of the vessel's behaviour at full size.

During November, Niels Bach, the Managing Director of Frederikshavn Vaerft, came from Denmark to meet Geoffrey Pattie. Niels stated that, if they were satisfied with tank-test results for Sirius, his shipyard would guarantee price, delivery and performance. He would even do so if it were built at a British yard under the supervision of his people: an important statement of confidence in our design.

Far from condemning our ideas, the intellectual elite in the halls of the Royal United Services Institute, the 'RUSI' – Admirals, academics and Whitehall 'hacks' – were keen to know more about the new 'short, fat' frigate. I was even invited to give a brief paper on the subject in the RUSI Lecture Theatre. But, behind the scenes, unknown to us, Marshall Meek, our *Bête Noir*, was also working on the Osprey Case. As he later wrote in his memoirs, "I myself was called upon from time to time to comment on the technicalities and worked closely with Geoff Mills, who led our legal team in BS."[5]

After Christmas 1981, I visited Jim Eberle at his farm on the edge of Dartmoor. He felt 'set up' by Rawson and his crowd who had drafted his 'bum's rush' letter. He now suspected that there was something more than rivalry behind it. Bath's attitude towards him had been almost in the nature of a threat. Following my disclosure of the Vickers Osprey tests, he had been in touch with Geoffrey Pattie and Admiral Lygo, Managing Director of BAE Dynamics, about funding an initial technical programme on the Sirius concept.

Admirals were not expected to take a serious interest in the design of their ships – they have to take what they are given, as in Beattie's famous outburst at the Battle of Jutland: "There's something wrong with our bloody

5 Meek, p. 194

ships today ... and something wrong with the system." Eberle, known as 'the thinking man's admiral', was now awaiting his possible appointment as First Sea Lord. The current 'First', Admiral Sir Henry Leach, was due to retire at the end of 1982 and was fighting a strong rearguard action against John Nott's 'Navy Cuts'. Eberle, however, doubted the need for the newly proposed Type 23 Frigate, which was steadily growing in size, cost and complexity: 'gold-plating' he called it. Hence his interest in anything that might be of greater or equal capability at lower cost. He made the prophetic remark that "one bomb or missile could knock out a Type 42 destroyer for months" – exactly what happened to HMS *Coventry* and HMS *Sheffield* in the Falklands War the following spring.

We sat in his study yarning about the Navy's future needs as the wind roared down from Dartmoor and the rain lashed against the windows. When I asked Eberle about his prospects of becoming First Sea Lord, his mood darkened. He thought that Vice-Admiral Sir John Fieldhouse would relieve him as C-in-C Northeast Atlantic Fleet at Northwood. Fieldhouse was Controller of the Navy and therefore responsible for funding all Naval design and construction at the time of the BS Osprey tests. Eberle feared he would be sent to Portsmouth to the minor command of the Home Fleet: "Shunted into a siding while the express roars through." Fieldhouse had been CO of the Navy's first nuclear submarine, HMS *Dreadnought*, of which our 'Jack Rabbit', or Daniel, was Chief Designer and later, as BS Board Member for Warship Construction, was co-ordinator of the BS covert tank-tests of our designs. He was also popular with the MOD officials, while Eberle had upset Bath – partly on our account, I feared. Perhaps he had been saying too much in the RUSI and elsewhere: "I tend to say what I think and that's the way I'm made. No apologies!"

I drove off into the stormy Dartmoor night, cheered by Eberle's support, by the possibility of BAE becoming a major strategic partner for Sirius, and by our progress with the Osprey Case.

After the dark and turbulent events of the past year, 1982 seemed to herald a brighter dawn.

CHAPTER 13

The S90 Club and the Opening Battle of the 'Short, Fat Frigate'

Members of the S90 Club, comprising BAE Dynamics and some of Britain's leading Defence and Marine Equipment companies, such as Dowty Group, Graseby Electronics, Northern Engineering Industries, Pielstick Diesels, Stone Vickers, BHC and NMI, Corlett Consultants and the Danish Frederikshavn yard, photographed on the terrace of the Bembridge Sailing Club, June 1983. Peter Thornycroft, fifth from the left, standing behind and between Peter Salkeld (BAE) and Per Holst Sorensen (Frederikshavn Vaerft).

Our partnership with British Aerospace Dynamics and a host of other defence companies that later comprised the S90 Club, opened appropriately with a session in the cocktail bar of the Royal Horseguards Hotel, just off Whitehall, a favourite watering-hole of the Ministry of Defence. I was to meet the BAE's Naval Liaison Officer, Peter Salkeld. At first glance he reminded me of Mr Jackson in Beatrix Potter's *The Tale of Mrs Tittlemouse*, 'sitting all over an armchair'. His ample presence rose to greet me with a broad smile, a huge hand and a jolly gravelly laugh.

"Peter Salkeld, British Aerospace," he said. "What's your poison? Pink? G and T?"

Over lunch he claimed there had been much talk about the Sirius frigate in high places. He mentioned all the right names. Admiral Sir Raymond Lygo, his Chief Executive, liked our ideas; as did Geoffrey Pattie, the new Defence Procurement Minister, and Admiral Sir James Eberle: a powerful trio. Quaffing a glass of claret, he suggested that the people at Bath or BS, with their 'gold-plated' warships, should no longer control the destiny of BAE's growing Naval interests. He would arrange for me to meet Sir Raymond Lygo as soon as possible. As I took my leave, he dropped a quiet remark about the South Atlantic and the RN's Research Vessel HMS *Endurance*, based in South America, which was to be brought home without replacement. He thought there was trouble brewing: something to do with the Falkland Islands.

A few days later I met Admiral Lygo at the BAE Headquarters in Pall Mall. He was a typical naval aviator: relaxed, courteous and refreshingly straightforward. He turned out to be far more interested in our ideas than I had dared to expect. He claimed it was about time Britain came up with a better sort of ship. All the export orders were going to the French, Germans, Americans and Italians, who made excellent weapons systems and built fine platforms. Although the people at BS and Bath kept telling him that they designed the best ships in the world, the world didn't seem to want them.

"We haven't sold a new Bath-designed warship for twenty years ... we couldn't do any worse," he said, "and I think you may be onto something better."

He suggested that I should visit the BAE Dynamics Filton Office, outside Bristol, to present my case to the weapons systems experts. He was not concerned about the Osprey court case – on the contrary, he relished it. Based on what little he had read in the newspapers and heard on the MOD grapevine, there was a lot more to come. "It's rousing stuff and makes good headlines," he said with a chuckle.

Peter Thornycroft was dubious about our proposed visit to BAE at Filton. Why were we going? What could it achieve? "Putting our head in the lion's mouth ..." I told him that it did seem we might get the Sirius one-tenth scale model tests financed with the help of BAE, so it was about time for him to apply some of his inimitable Thornycroft charm.

At Filton we were subjected to a barrage of objections which sounded remarkably like those we had heard for so long from BS and the MOD at Bath, just a few miles down the road from Filton. They had been well briefed. But Peter captivated the solemn accountants and technicians with a flow of comic stories about his grandfather and the sceptics who ridiculed his first torpedo-boats.

"A huge success," said Peter Salkeld when he phoned the next morning. "Your colleague had them all in stitches: pricking the balloon of the high and mighty I call it. Jolly well done ... Now it's full speed ahead with Project Sirius!" He did warn us, however, that behind the scenes many were concerned about marching so forthrightly against the forces of reaction in Bath. There was intense rivalry within the British defence industry, and it was made abundantly clear that BAE were taking a gamble in supporting the Sirius programme. There were, it seemed, plenty of the MOD's placemen lurking within BAE, keen to frustrate our efforts.

Back to Ospreygate: 'To Prison Ye Shall Go ...'

In the High Court, we needed something more substantial to prove the link between Vickers and the MOD. All we could do was sift through our piles of documentary evidence – that part of it which had not been 'filleted', as Frank put it.

"It's all there somewhere in the documents," he said – and there were thousands of them: they filled the bookshelves of an entire wall. We wondered how BS had financed the Osprey operation. There must have been a generous source of funding for such an elaborate test programme, to say nothing of the huge legal costs in which they were now involved. Frank claimed that for every pound we were charged they would probably pay ten.

Now, in their enthusiasm to crush the 'little guy', BS made a serious miscalculation. On January 11th 1982, they applied to the High Court, seeking Security for Costs. If they won their claim, we would have to provide guarantees, not only for our own costs in continuing the case, but also for those of BS and perhaps the MOD: a vicious manoeuvre. Based on our hunch about the existence of a second model, we grasped the opportunity to file a counter-claim: BS now had to swear affidavits that they had disclosed all relevant evidence. It proved a brilliant tactical move by Frank Presland.

Our claim was heard before Mr Justice Whitford, a peppery old High Court Judge. In a rushed confession, BS were forced to admit that 'whole files' of further documents had 'gone missing' and other material evidence had been destroyed. Their Application for Security for Costs was thrown out. Instead, Whitford made a further Order to include 'further and better particulars on whatever they provided' concerning the Osprey test programme.

As usual, the documents that Mr Justice Whitford ordered to be released did not appear on time. To obtain them he had to threaten Moor, Daniel and the Chairman of Vickers Shipbuilding with committal proceedings for Contempt of Court. If they did not produce full details of the circumstances surrounding the missing files and any destruction of evidence within 24 hours, he would apply to have them sent to prison.

Whitford's Order was widely reported in the media and the effect was dramatic. The following morning a letter was delivered by hand at the offices of Frere Cholmeley. It was, as Frank stated, another 'bolt from the blue'. Even he was astonished by the deceit of the senior officials involved. It appeared that 'Christine', the 'missing model' had been alive and well but alas, as we learnt, had recently met a painful death.

She had indeed been the *second* Osprey model, STA 2225, which was destroyed four weeks *after* our writ was issued and two weeks after Rawson's 'Revenge'. BS also admitted to having made and tested a further four models of the Azteca, of the same scale and varying length/beam ratios. This was a typical requirement for a 'standard methodical series', the most elaborate type of test procedure for any design. BS claimed the Azteca design copyright lay with the London Agents who sold them to the Mexican Navy. However, as mentioned earlier, although Peter's company, TT Boat Designs, had passed on the sales rights, these had not included the copyright. Therefore we joined TT Boat Designs Ltd with Osprey Ltd as co-plaintiffs to cover the Azteca testing.

'Christine' was never run at St Albans, but at the Vickers tank in Dumbarton, along with the four Azteca models – Dumbarton having kept the Azteca lines plans since 1970. This was well away from any possible 'whistle-blower' at St Albans, after we had filed our Claim. The original St Albans Osprey model was *not* rushed up and down the M6 between St Albans and Dumbarton by Security Express as the previous documents had claimed: that was cock-and-bull. The fleet of Azteca and Osprey model variants they had constructed, if run intelligently, would have allowed BS to predict the performance of any size and any geometric or proportional variation of the Azteca/Osprey design family.

BS also confessed to 'whole files of evidence' having been destroyed. Finally they admitted to breach of copyright in the Osprey design. Allied with John Coates's 'Bonfire at Bath', the connection here between BS and the MOD now seemed more substantial. But we had yet to establish a palpable link between BS and the testing of STA 2225 for Rawson's 'Revenge': a 'smoking gun' to establish a Conspiracy.

It was something of a Pyrrhic victory. A few weeks later John Mummery resigned as our Counsel. He felt that he should dissociate himself from the Case, having earlier advised us that high Government officials could never have perpetrated such misconduct. He could no longer represent our interests without undermining his position as a Treasury Solicitor representing the Crown, including the MOD. He was being appointed to the Chancery Division so there was a conflict of interest. Another £50,000 down the drain ... 'So where does Giles get his money?'

We therefore had to find a further £30,000 to brief a replacement Junior Counsel. This was to become the norm. We were forced to change Junior or Senior Counsel three more times during the course of our six-year litigation, each time at a higher cost for briefing a successor. *Caveat litigator!*

As part of the Discovery process for this latest load of evidence, yet another meeting was hastily arranged at the St Albans tank for us, together with a hydrodynamics expert from Southampton University, to inspect the lines plans from which David Moor claimed he had developed his Hong Kong final contract design. Mr Alexander Silverleaf, an 'independent' hydrodynamics consultant who, for years, had worked with Moor, attended for the other side. In an early-disclosed memo to his staff, Moor had expressed his concern that I might arrive at the door of the St Albans tank and find out 'what we have done'. On this occasion that was exactly what we proposed to do. But now we had the necessary evidence.

The model of the final Hong Kong model design, STA 2254B, again discreetly shrouded, was lying on trestles in the entrance lobby. Peter again respectfully doffed his Squadron cap as we passed by. We would be allowed to inspect it after hearing Moor's explanation of its provenance. Since this was classified MOD property everyone had to sign a 'Statement on Security Aspects' before the meeting convened.

"Aha," said Peter, with his usual graveside humour, "the Book of Condolence ... but when shall we view the corpse?"

"David Moor has made 'a model' and it has been run in the tank as a minor research item ..." (J Daniel, BS, Board Member for Shipbuilding, April 1981). Six of the eight Vickers models finally disclosed: The four Azteca models of differing length/beam ratios (left), the final 'as-built' Hong Kong model STA 2254B (fifth from the left, with stripes). The previous disqualified HKPC model STA 2230 having been a very close copy of the Osprey hull, was destroyed. This model shows the dramatic changes that had to be made to the design to meet the terms of the July 1981 Court Order against use of any Osprey-derived data in the final HKPC design. Finally, the single surviving 16-foot Osprey propulsion model STA 2219 (right) is shown. 'Osprey' was not painted on its stern, as Ken Rawson claims in his memoir. All of the above were tested between November 1980 and October 1981 at St Albans. STA 2225, 'Christine', the 'missing model' tested at Dumbarton for 'Figure 1' in Rawson's 'Revenge' after Court Proceedings opened, was 'cut to pieces'.

The meeting was largely a dialogue between Presland – who skillfully acted the part of the stupid amateur – and Moor, who tied himself up in knots as he tried to explain how the derivation of the Hong Kong boat had nothing to do with the simultaneous testing of either the Osprey or Azteca models.

Moor launched into a long and complicated account of how he had derived the most successful of three different designs – model number 2230 – from two of his earlier designs, without reference to the Osprey. But since it was made *after* both models had been tested before the July Judgement, he must have benefited from his Osprey-derivative, which, according to his tests, would have made the contract speed comfortably. His first design also bore a strong resemblance to the Osprey. The same model, 2230, had been used as the basis for the first design contract with the MOD, announced by John Nott in the House of Commons in July 1981. Following our Injunction in the High Court, he would have had to abandon any idea of using 2230. He had to design another compliant hull from scratch, *after* the contract had been announced, without incorporating any features of the Osprey. Hence the strange 2254B, with its crude 'wedge' incorporated under the stern, and much narrower hull.

Not surprisingly, Moor claimed that, shortly after that contract was announced, a further 'production design refinement process' had to take place. This involved a series of several different designs and the testing of many more models with completely different stern sections, until they fixed on a final design, 2254B which, according to their estimates, was still barely able to make the 26 knots that were comfortably achieved by 2230. We trooped into the entrance lobby to view the corpse of 2254B. The shroud, when withdrawn by Moor, revealed something that looked quite bizarre and bore no resemblance to 2230. It had a fat belly and an exaggerated 'hollow' in the buttocks, with a narrow flat stern and a prominent 'wedge' beneath the transom, apparently to increase lift at about 22 knots. This model became the basis of the final design of the patrol vessel HMS *Peacock* and her four sister ships in the Hong Kong Squadron.

"Sort of pregnant, I'd say," Peter commented. "I wouldn't want to be associated with such a shape. It has no inherent course stability and with so little beam it'll roll like a pig, even if it does make its trials speed. The next thing he'll do is stick on bilge keels!" Which, eventually, he did. It was not until we were back in my car that we could compare notes. "They got it right with the 2230," said Peter. "It was clearly drawn with the Osprey in mind. They went ahead with that design until they got the contract. But a few days later, when they gave their undertaking in the High Court, the poor things had to go back to basics and produce something else."

"... that made the same speed, but looked different!" we said in unison.

But the roll angles of its narrow hull, as described in Peter's later copy of *The Naval Architect*, were unacceptable (see p. 146).

CHAPTER 14

War … and Peace

The Royal Navy was in a panic. The Defence Secretary, Sir John Nott, had proposed his 'Navy Cuts' in the Surface Fleet, so there was some doubt as to whether there would be a new frigate at all. If there were, it might have to be unacceptably 'cheap and nasty'. But all speculation ceased with the invasion of the Falkland Islands by Argentina on 12th April 1982. At a stroke the prospects for the First Sea Lord, Admiral Sir Henry Leach, and the Navy's Surface Fleet brightened, while the political careers of the Foreign Secretary Lord Carrington and Navy Secretary Keith Speed dissolved overnight – along with the proposed Navy cuts. Leach assured Mrs Thatcher that he could provide the necessary forces to repossess the islands.

Old Hat …

Having obtained the required 'Secret: UK Eyes Only' clearance, I was given an outdated pre-Falklands draft from the Operational Requirements Committee (ORC 2064) not the latest Naval Staff Requirement (NSR 7069). It called for little more than a 2,500-tonne, 25-knot, £75 million vessel. In essence it was for a fast tug, towing a two-mile sonar array of hydrophones for anti-submarine detection in the North Atlantic. It was a much inferior version of the sort of ship the Navy could now expect to acquire. It became the basis for our proposed S80 design, but later grew in size and capability. I kept hearing from John Moore and others, about the campaign being mounted within the Naval Staff and MOD against our project. We learnt from BAE that the latest Type 23 had already become much larger than proposed in our preliminary document: probably by about thirty metres, or twenty-five per cent of its length, with twice the power and a substantial increase in cost. 'Gold-plated', as Admiral Eberle had forecast. Therefore we decided to enlarge the Sirius Corvette from 80 to 90 metres, if only to meet what little we had heard of the final Naval Staff Requirement for the Type 23. Our new variant was known as S90. However it did not approach the final Requirement's size that nearly doubled to 4,300 tonnes. Our S90 was, as Peter put it: 'Old Hat'.

Following the Falklands victory, a former Chief of the Defence Staff, Field Marshall Lord Carver, had written to the *Times* suggesting that most of the Navy's ships were ineffective for future conflicts without air cover. In the House of Commons Royal Navy Debate a few weeks later, The Rt. Hon. Edward du Cann, Chairman of the 1922 Committee, stated, "Let us be realistic. In the South Atlantic we won a brilliant victory, although disaster was never far away ... Against a more competent enemy, particularly an air force under better direction, our ship losses would have been far greater. The Government's task is to see that the practical lessons of that experience are learnt and acted upon."[6] It was just as Admiral Eberle had foretold in December. *Sheffield* and *Coventry*, two specialized Type 42 Air Defence Destroyers were sunk, one by an Exocet air-to-surface missile, the other by a World War II bomb – while *Glamorgan*, an older destroyer, was put out of action by a surface-to-surface Exocet missile fired from a truck parked ashore. With the loss of two older Type 21 Frigates, the Landing Ship, *Galahad*, and damage to *Brilliant*, a Type 22 frigate, and other losses, a quarter of the Navy's active Surface Fleet was sunk or damaged by missiles or Second World War bombs dropped from aircraft of Korean War vintage. The *Atlantic Conveyor* was struck by two Exocet missiles, thus denying our land forces almost all the helicopter support she was carrying. After the disastrous attempted landings at Bluff Cove, the Army joked that, to avoid air attack, the Navy had moved so far East of the Falklands it should be awarded the Burma Star. The lack of air-defence was critical to the losses our forces had endured.

Despite the wise words of Lord Carver, Edward du Cann, Admiral Eberle and many others, any changes in ship design, leading to greater stability, defence and manoeuvrability against air-to-surface missiles – and reduction of cost – were ignored. The conclusion by the MOD was that the Falklands experience was 'unique'. Air cover would always be there to provide future protection, so there was no need to change the tried and tested design of traditional hulls. Instead the MOD was given carte blanche for whatever it wanted. 'Gold-plating' and enlargement became the order of the day. S90 had become a pre-Falklands remnant of the 'Navy Cuts'.

6 Hansard Vol. 28 No. 156, Col. 41, p. 62, 19/7/82

Rough Waters ...

At the end of May 1982 we submitted our fourteen-page Proposal for Validation of the S90 alternative Type 23 Frigate to Bryson's office.[7] It merely proposed a short programme, chiefly comprising tank and open sea comparative model tests with the Navy's most proven seaworthy frigate, the Leander class, again to be financed by the Department of Industry. The controlled model tank-tests for powering and sea-keeping would be undertaken by three establishments – the BHC test-tank on the Isle of Wight, the National Maritime Institute in London, and the Danish Technical University, which had conducted the measurements of the *Havørnen* speed trials. These would fix the initial hull design. It was to last for about a year and cost about £250,000.

Together with BAE and Frederikshavn Vaerft, our Consortium, known as the S90 Club, was to be joined by other defence companies. We asked Geoffrey Pattie to arrange a final 'make or break' session with all the key players of the MOD and the Navy. These should include Bryson, Vice-Admiral Sir William Staveley (Vice-Chief of Naval Staff and Second Sea Lord), Sir Raymond Lygo, Niels Bach and myself. There was no point in proceeding unless a consensus viewed the project as having a reasonable chance of success. Peter agreed that we should put our faith in Pattie and Lygo. We knew that the MOD would do its best to discourage us, but if these two 'champions' of the S90 thought it worth proceeding, we would accompany them down the hard road that was bound to follow. Firm guidelines for the conduct of the so-called S90 Validation Programme should be agreed.

Smooth Admirals ...

On 14th July 1982, in Historic Room 13 of the MOD, the decisive deliberation took place.[8] Admiral Bryson chaired the meeting. In his melodic Scottish tones he smilingly dismissed the possibility of delaying the Type 23 programme in order to accommodate a late bid from the S90 Club. He insisted that we should not expect any funding from the MOD, clearly contrary to the instructions of Defence Secretary John Nott and the MOD's Chief Scientist. Geoffrey Pattie was obviously riled, Ray Lygo bristled and

7 'A Private Proposal for the Type 23 Frigate', dated 28/5/82

8 Letter from Controller of the Navy to DLG, attaching the MOD's Notes of a Meeting held on 14/7/82. All of these excerpts are taken *verbatim* from that document.

Niels Bach puffed away at his pipe looking shocked. What was the point of the S90 Validation, or the meeting, if ours was an impossible task? Pattie confirmed his ministerial position: “It is implicit in my decision that, should the validation phase prove satisfactory, the Consortium will be allowed to continue into the next phase.”

Bryson insisted that any discussions before the results of the full model test programme would not be fruitful. Quite apart from the outcome of the validation phase, he had grave doubts about the capability of the S90 Consortium to undertake the detailed design and production phases. It was highly unlikely he would agree to the feasibility/design phase – or Project Definition – that we sought, even if the validation programme was encouraging.

Becoming impatient, Geoffrey Pattie said, “I reserve the right to proceed with the design contract phase *against your advice*, Controller ... If the validation phase, at the Consortium’s expense, provides enough indicators that it is a viable design, then it is implicit in the Secretary of State’s Statement that your proposal would proceed to the next phase. If so, the MOD will pay for the outstanding work up to the time when the detailed design is submitted for evaluation.”

“Would there be arbitration for the S90 validation programme?” I inquired. Bryson replied that “the only people capable of carrying out such an evaluation are contained within the MOD, Ship Department and the Naval Staff ... It would be totally objective.”

I asked if some independent non-MOD entity might be invited to arbitrate. Bryson put on the most affecting of smiles: “Oh, but Mr Giles, are you now questioning the integrity of the Ministry of Defence ...?” Indeed I was. Finally, Pattie confirmed that the initial validation phase would be at the cost of the S90 Consortium; however, if the results were sufficiently encouraging, the MOD would fund the detailed design phase. Lygo still sought some contribution to the validation programme from the MOD, but that seemed a forlorn hope.

“*Post mortem* or *victoriam*?” Peter asked when Niels and I met him afterwards. It was difficult to be sure. There was something about Bryson’s attitude that we all mistrusted: one detected the presence of his officials in all that he said. He smiled too much when one suspected he should not. We had to make a decision: would we risk our own money and that of BAE, the Danes, the Department of Industry and other members of the

S90 consortium by continuing the programme? Little did we know, in those days of the innocence and enthusiasm of the S90 Club, how heavily the odds had already been stacked against us. Yet the MOD agreed that we should proceed with the validation, partly funded by taxpayers' and BAE Dynamics' money, but largely by TGA. Despite Pattie's forceful performance representing HM Government, we still suspected that 'the tail was wagging the dog'. It was time to go sailing again, to get back to the real wind and water: away from the stuffy offices and mahogany tables of Whitehall and St Albans.

Gentle Breezes …

In the midst of these stirring events: Liberty! I had to move *Minion of Oxey*, the yawl I shared with a former submariner friend, from Majorca to Corsica for our planned summer cruise. The prospect of a week's sailing across the open sea would be the perfect preparation for what was likely to be Battle Royal with the powers within the MOD. We flew to Mahon and from Puerto Andraitx set sail for Menorca.

A break from the battle: Sailing across the Mediterranean.

It was long past midnight as we entered the small landlocked bay of Binibeca, in Menorca. Having spent two summers there with my family, sailing a small Topper dinghy, I knew it well. In the moonlight, between the surge gulping across well-known rocks, we silently ghosted our way into the tiny anchorage, empty of boats and people but for a nude couple splashing about sensuously at the water's edge. We dropped our anchor into a cloud of phosphorescence ... and, in the cockpit, relishing the tranquillity of the scene, sipped a few drams of Talisker.

The next morning we set off across a kindly Golfe de Lyon, for Corsica. After two days' sailing, we were greeted by a more sensory landfall. The air was filled with a bouquet of gummy scents of the coastal 'maquis': cistus, rosemary and curry-plant carried across the water on a faint breeze, as we sailed towards Ajaccio. During the whole voyage we saw only one ship: an old Egyptian tramp steamer, with no sign of life aboard, as she passed close-by, wallowing along in a ghostly fashion, like the *Marie Celeste*. Having left the boat in Ajaccio, we flew to Marseille, returning to London to rejoin the battle.

Back to Business

Notwithstanding the MOD's ill-concealed reluctance, the S90 Club had become an impressive collection of companies. Initially it comprised TGA, BAE Dynamics (for the weapon systems), Frederikshavn Vaerft (shipbuilding), Dowty Group (fuel system and anti-submarine), Graseby Electronics (sonar), Northern Engineering Industries (diesels and propulsion system) and Stone Vickers (propellers – not related to the Vickers testing tank). All of them were leaders in their field, willing to commit funds to the pursuit of novel ideas. Like a club, a bond of common interest, rather than any legal instrument, held us together. The challenge was to maintain enthusiasm for our concept, which was forever being undermined by the forces of reaction within the MOD, Whitehall – and those within BAE and other defence companies that owed so much to the MOD.

In September Admiral Lygo invited me to fly with him to a military airport close to Frederikshavn in BAE's corporate plane. He was impressed by the Danish yard and seemed enthusiastic about the entire project. If he could get just *half* the contribution he sought from the MOD, he would recommend that BAE should support the S90 programme. He was convinced by Bach's support for our design and the praise heaped upon it by those we met, in particular their Chief Naval Architect, Per Holst Sorensen, also a member of det Norske Veritas Technical Committee.

We continued with the initial S90 design, ignorant of its growing irrelevance, TGA covering all the costs, gleaned from the proceeds of the growing sales of Ospreys by the Danes. Peter Thornycroft, Arthur Mursell and I were working with the BHC tank to obtain the best hull form for the S90, which had changed significantly from the original Osprey lines. Unfortunately we were delayed beyond the timetable proposed to the MOD. This was due to the reluctance of BAE to proceed without a signed undertaking by the MOD covering financing. We then heard from Lygo that he was leaving BAE Dynamics – 'kicked upstairs', some claimed, to become Chief Executive of BAE Systems, as BAE became and remains to this day. He was genuinely sorry he had been unable to do more. Yet ... it had been the education of BAE Systems in the mysteries of the 'Short, Fat' ship. Lygo had often mentioned the growing dependence of BAE Dynamics on the success of our warship-building industry. Might this mature into a closer relationship between BAE and the UK's privatised, but struggling, warship-building interests?

"It might ..." he replied enigmatically. And so it did, with the merger of BAE and Marconi-Yarrows in 1999 – obtaining our ideas and the S90 with it.

CHAPTER 15

'Deceit Piled Upon Deceit'

Back to 'Ospreygate' ... during October 1982, whilst trawling through the later series of BS documents, I found mention of a video of apparently a *third* Osprey model. Our Counsel applied for the most swingeing of Orders available under the English Law: an 'Anton Pillar Order', allowing the Plaintiffs' Lawyers to enter the Defendants' premises unannounced. We visited Vickers and it transpired that the model was not of the Osprey, but some commercial design, inadvertently filed under 'Osprey'.

In order to clear the record, BS applied to have the Anton Pillar Order rescinded before the Senior Appeal Judge, The Master of the Rolls Lord Donaldson, and Lord Justice May. As shown by the Court Transcript, in italics:

BS Counsel tried to excuse Vickers' mistake: *"As your Lordships have seen ... the items seized by the Plaintiffs have absolutely no relevance to this case,"* said the BS lawyer.

"And whose fault is that?" asked Lord Donaldson, with a twinkle in his eye.

"That is, with respect, an issue for your Lordships to determine."

After outlining some of the confusion and attempted concealment of the first and second models and the apparent existence of a third, Lord Donaldson made a remark that proved central to the Osprey Case:

"The appearance was one of deceit piled upon deceit: and if the same evidence were before us, we would make the Anton Pillar Order again today. We now know that there was in fact no deceit in the last stage; it was negligence and, had we known that, there would have been no need to make the Order ..."

That was enough. The 'deceit piled upon deceit' statement of Lord Donaldson resonated with the media; and it later proved our strongest line of attack: summarizing, in one short phrase, the whole demeanour of the 'Ospreygate' affair.

The Media Are Pressing

In the aftermath, later in 1982 the BBC wanted to make a *Panorama* programme on the lessons of the Falklands War and its relevance to the 'short, fat ship', including the Osprey Case. Thames TV and TV South were also pressing for interviews, and *Newsnight* was calling. The *Times, Sunday Times, Financial Times, Daily Telegraph* and others were all wanting to get in on the 'Little and Large Show', as some Fleet Street 'hack' described our battle.

How could we use our growing notoriety to help with the case? I turned to Charles Adams, always a wise counsellor. His advice was clear-cut: "It goes straight to the top of the heap, otherwise they would never have made that clumsy attempt to clear their name in the High Court. You have a great opportunity," he said ... "Seize it. I hear they're talking about the 'Giles Jihad' in Whitehall. Up the ante!"

When BBC's *Newsnight* phoned, I was hesitant. I had learnt from the 1979 *Risk Business* that it is better to resist the temptation to jump in at the deep end, whatever the attractions. There was a good story to be told, but I didn't want to risk upsetting the S90 project. We still had no commitment from the MOD about financing, and without that, BAE would only provide a fraction of the sum that was needed. Peter Thornycroft was in favour of bringing the argument out into the open, particularly when he heard the views of Charles Adams. Peter Salkeld from BAE agreed: a short piece on *Newsnight*, reflecting public and political interest in the S90, might push the Powers within the MOD to make up their minds. If not, it could hardly make matters any worse. The *Newsnight* report and discussion was shown just before Christmas 1982, when John Nott's proposed cuts of the Surface Fleet had largely been reversed, and the White Paper on the Lessons of

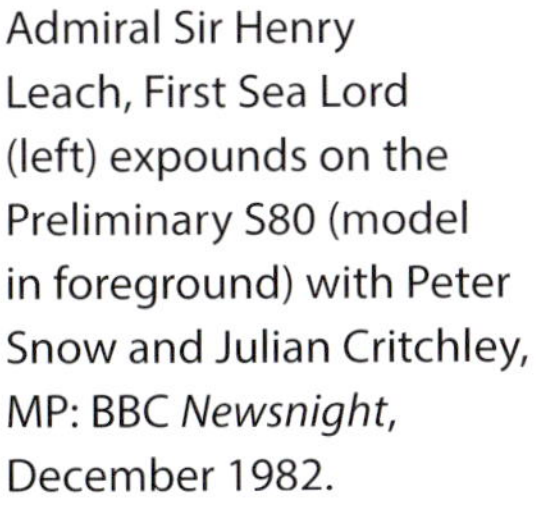

Admiral Sir Henry Leach, First Sea Lord (left) expounds on the Preliminary S80 (model in foreground) with Peter Snow and Julian Critchley, MP: BBC *Newsnight*, December 1982.

the Falklands had just been published. The live participants were Admiral Sir Henry Leach, outgoing First Sea Lord and progenitor of the Falklands Task Force, and Julian Critchley, Conservative MP for Aldershot. Critchley was concerned with the Army's budget, and made a good foil to Leach. Recorded interviews with Ken Rawson and myself were also shown. The presenter was Peter Snow.

Leach singled out the 'stretched' Type 42 destroyer to support his argument that length increases speed: "By increasing the waterline length of the Type 42 by forty-two feet, we have obtained one extra knot of speed," he declared. What he did *not* say was that the penalty for that extra length was a cost increase of about £20 million for added structural strength, or some thirty per cent, and a loss of stability and useful load.[9] An interview with Ken Rawson was then shown, together with some of the computer-aided design work being conducted at Bath. He was asked by Snow if there was any chance that Admiral Eberle, Geoffrey Pattie, and the engineers at the Department of Industry and British Aerospace might be right about the Sirius concept, and Bath wrong. Rawson grudgingly admitted that, although the S90 hull was 'similar to a Victorian river-gunboat', there was such a possibility, but 'remotely so'. In my interview, Snow challenged me with the fact that we were breaking with the tradition of every major navy in the world. I replied that much significant new naval technology – the steam engine, propeller, torpedo boat, submarine and steam turbine – had first been rejected by many of the world's navies, including the Royal Navy.

Back to Admiral Leach, who dismissed the S90. He then made an enigmatic remark: "I'm sorry. Tank-tests have been done and they are not wrong." Then his final remark: "Even if the tank-tests showed the traditional theories were wrong, the Navy would not accept them."[10] The Admiral's selective account of the facts and veiled threats also gave some indication of the battle within the MOD: between those who wished to give the S90 a chance and those determined that it should fail – regardless of the results of the three tanks employed and the Osprey full-scale trials.

9 *Jane's Fighting Ships*, 1985–86

10 This was another reference, I suspected, to Vickers' secret Osprey tests. He continued that, if BAE and others had put any money into this, they should "look to their onions, because they stand to lose it: they're on a hiding to nothing." His own Naval Staff had already insisted that "Giles's ideas should be rejected from the outset," as revealed in an internal MOD memo predating their Agreement for the S90 Validation by a month. Disclosed under the Freedom of Information 30-year Rule in 2013.

CHAPTER 16

1983: Man Proposes; MOD Disposes

The MOD and Navy Top Brass: Keith Speed, Sir John Charnley, Admirals Fieldhouse, Leach, Staveley, Pillar and MOD Officials, watched over by Lord Nelson. The wind-dial on the right wall allowed the Admiralty to direct ships in the Thames and its estuary.

In February 1983, the Secretary of the Defence Scientific Advisory Council – a secret body of 'independent' technical advisers to the MOD – invited me to provide 'a presentation on Sirius 90 and your design philosophy' to its Hull Committee. The meeting would be held in strictest confidence. Its area of concern was 'the construction, hydrodynamics and sea-going performance of ships'. I prepared a short treatise on 'The Action of Water and the S90'. This described the possibly beneficial effects that might be obtained from harnessing the Captive Wave. Due to delays over the BAE/MOD costs issue, we had completed only the preliminary resistance tests on the smaller S80 design and were not expecting to run the comprehensive propulsion and sea-keeping tests for the S90 for another three months.

Until we entered the meeting we were not aware that Marshall Meek was to be Chairman. We knew nothing of his new position as Technical Director of BS, nor of his intimate involvement with David Moor and Vickers in the Osprey Case. To the best of my knowledge he was still the Chief Naval Architect of the eminent Ocean Transport & Trading shipping line.

As I entered the MOD conference room, accompanied by Per Sorensen of Frederikshavn and Peter Knowles of BHC, I encountered a palpable wave of hostility. Meek, in a dark suit, took the chair, with his Committee ranged beside him, most of them directly involved with MOD and BS and selected by Meek as likely to support his views. One was from YARD (Yarrow – Admiralty Research and Development), wholly owned by Yarrow Shipbuilders, who were also proposing an alternative Type 23 design. Two were professors from Newcastle and Glasgow Universities. The only independent member was Bill Crago from the BHC testing tank, but he was in a considerable minority and never spoke a word.[11]

We sat on small chairs in front of this stern Committee, like defendants at a Court Martial.

It was probably a mistake for me to open my presentation with a drawing and a quotation from that great hydraulic engineer Leonardo da Vinci, whose first Commission for the Government of Florence was to divert the course

Leonardo's drawing *Nota il moto del livello dell'acqua,* introducing my talk on the 'Design Philosophy of the S90' that so bewildered the DSAC Hull Committee under Marshall Meek.

11 Today all members of a Defence Scientific Advisory Council committee are bound by a Code of Practice that excludes those who "are involved in work for MOD or another Government Department which is sufficiently related to the matter under discussion as to cause a possible conflict of interest." Likewise, members must "observe the highest standards of impartiality, integrity and objectivity in relation to the advice they provide and the management of this public body." Finally, members "must not misuse information gained ... for personal or political purpose, nor seek to use the opportunity of public service to promote their private interests or those of connected persons, firms, businesses or other organisations ..."

of the River Arno: "When considering the action of water, first consult experience, then reason ..." It seemed that such a union of observation, philosophy, art and hydrodynamics was beyond their comprehension: the huge 'rooster tail' wake of a Type 42 destroyer – or the flat wake of the Azteca at a much higher speed-relative-to-length, or 'Froude Number'. (See illustrations on p. 287.) Nor did the Committee show any interest in my accounts of Sir John Thornycroft testing his extraordinary variety of hull shapes in his private Isle of Wight testing tank. Nor did they want to hear how fascinated he was by the action of water on the smallest of craft operating at higher hull-speeds and how our ideas were a natural progression of his 'philosophy' in the evolution of ship design – from his 'quick launches' up to the largest destroyers and light cruisers. Nor in the opinions of Garwin and Lighthill on scaling, so contrary to those of traditional wisdom. Again: why can one scale-up something that sinks, but not that lifts?

My paper dealt with only the most rudimentary design development of the S90, since we had not at the time started any serious structural design, nor had we decided on the final hull form, having only completed very preliminary tank-tests. My entire paper covered only fifteen pages, mainly illustrations.

Our audience sat mute and expressionless as we presented the arguments for our 'design philosophy' – as specifically requested by the Committee – rather than any detailed design or performance data. There was no more than trivial discussion and we left the meeting in the deepest gloom. Meek was hostile to whatever we might propose and it was apparent that his Committee dared say nothing to contradict him – as typified by the silence of Bill Crago of BHC, the origin of our knowledge of the Vickers' Osprey tests. It had been a sham and we awaited the outcome with profound misgiving.

In his account of this session Meek would later write, "What we found from our eventual meeting with Thornycroft Giles was disappointing, but did not surprise me. *The voluminous documentation* put before us did not explain in recognised technical terms why they claimed such benefits from the short-fat ship ... So Giles had been able to convince the principal Government Minister involved and various other high-ranking people that he had some revolutionary concept long before there were any hard facts and figures available to prove or disprove his case."[12]

12 Meek, p. 198

Meek and his Committee were quite shamelessly jumping the gun – presenting the MOD with a verdict on the S90 well in advance of the results of the test programme agreed with the MOD itself. Of course there would be copious 'hard facts and figures' in the final S90 Validation Report which, seven months later, was placed before some of the most qualified naval architects in the land. Yet, even then, those results were dismissed with contempt by Meek, despite himself coming from NMI, of which he had just been appointed Director.

'To Change the Shape of Ships and Tame the Sea …'

Only a few weeks later, we were obtaining results from tests of the fifteen-foot S90 model in the huge Number Four tank at NMI, formerly part of the National Physical Laboratory and one of Britain's most respected technical institutions. It was chosen by us as a check on BHC's tests, being the only other British tank with solid experience in testing semi-planing hulls. NMI's tank, until its destruction by the Thatcher Government in 1988 to make place for a shopping mall, was the largest in Europe. Its 1,200-foot length was such that measurements had to take account of the curvature of the Earth.

The calm water tests for resistance and powering were consistent with the earlier BHC tests with a smaller model. They correlated to within 3 per cent of the full-scale speed trials of the *Havørnen* by the Danish Technical University that, jointly with NMI, had endorsed them. This greatly increased our confidence in the accuracy of the S90 validation programme.

The sea-keeping tests that followed were extraordinary. In high seas the S90 was slowed down far less than NMI had expected. This created some problems for the tank, which was calibrated on the presumption there would be a speed loss of about ten knots in the worst conditions – but it turned out to be only about two knots, with the model pitching just 2.5 degrees at 30 knots in 22-foot waves. Already operating much further into the area of wave drag, the S90 was not as subject to the effect of increases in resistance that occur as a traditional destroyer hull plunges and pitches through quite moderate seas. A comparison by NMI of its Osprey results, with existing measurements of full-scale frigates of the Leander and Tribal classes in high Atlantic waves, showed impressive reductions in ship motions, shipping of seas and bow slamming.[13]

13 'Full-scale Comparative Measurements of the Behaviour of Two Frigates in Severe Head Seas', by Lloyd and Andrew, RINA Transactions, 16/4/1980

A hydrodynamicist who casually spoke to Peter in the corridor during one of the many tea-breaks, whispered that she had never before seen a hull running through high seas at such a speed and with such apparent ease: "This is something quite exceptional. You've not only changed the shape of ships, you've tamed the sea." Naturally, her words were prefaced by the inevitable statement that she was 'not meant to tell you this', for testing tanks are traditionally as sparing with praise as they are with criticism of different designs. Peter had replied modestly, quoting the words of the great yacht designer, Charles E Nicholson: "Never take water by surprise." These tests demonstrated that we had a success in the controlled conditions now favoured by the MOD. Would this convince them that our ideas were worth considering?

On the morning of Friday 13th May 1983, John Moore phoned. "Have you seen today's *Financial Times*? If not, you should. The MOD's decided to screw you. Put a cold towel round your head, read it carefully and have a long think." The article, entitled 'Last Ditch Attempt to Sink the Frigates', was illustrated by a sketch of the Type 23 that reflected the latest Naval Staff Requirement: a very different vessel to anything we had seen before. The displacement was given as 3,000 tons – 500 tons more than the earlier version with which we thought we were competing. It was also much longer. They'd clearly moved the goal posts again, going for an even larger 'gold-plated' ship once we had fixed our design. In the post-Falklands mood of gratification, the Navy could have pretty well whatever it wanted. The article concluded, "About six weeks ago Mr David Giles, one of the designers, gave a presentation to the Defence Scientific Advisory Council, an influential body for independent technical advice to the MOD which reports direct to the Secretary of State for Defence. The report of this body is highly unflattering to the S90 concept. It finds the claims for lower cost have not been substantiated and says the concept has almost nothing else to offer."[14]

I had been assured that my presentation to DSAC would be in complete confidence. The Committee's very existence was classified. Yet here it was quoted in the press, accompanied by the questionable claim that its members were 'independent'. As expected, the DSAC Report was dismissive of every aspect of the S90 design – on the basis of speculation rather than

14 *Financial Times*, 13/5/83

measured or calculated evidence – ending with outright condemnation: "The short-wide hull form of the S90 has such fundamental drawbacks as a concept for a modern class of frigate that no amount of further testing or detailed designing is likely to affect our conclusions in any important respect. We are, of course, not concerned with the value of the hull form for patrol boats or other small craft."

Geoffrey Pattie had bravely attempted to stand up for the S90 against the Top Brass of the MOD and Navy, but he had been out-manoeuvred – or out-bludgeoned. Like his predecessor, Lord Strathcona, he was 're-assigned' – from the Type 23 and S90 to procurement of the Trident Submarine, long before our validation was complete. His successor, Ian Stewart, was a bland, bureaucratic type who showed no serious interest in the S90.

We decided that the only course of action was to go ahead with the BBC's *Panorama* and TV South's *Bottom Line* programmes whose offers we had previously rejected. The stage was clearly set for our execution, so we decided to make a public show of it.

CHAPTER 17

'They That Go Down to the Sea in Models'

'A Cheap and Nasty Wine for a Cheap and Nasty Ship': Peter Thornycroft launching the 30-foot S90 model with Arthur Mursell (left) and the TVS camera crew.

On a glorious morning in late May 1983, TV South sent a camera crew to film the S90 and Leander models undergoing sea trials. Peter Thornycroft proudly launched the huge yacht-sized model of his latest creation, cracking a bottle of sparkling wine over the bow with the words: "A cheap and nasty wine for a cheap and nasty ship," thus echoing Churchill's 1940 request that convoy escorts should be "cheap for us – and nasty for them".

After it was afloat the TVS camera crew scrambled about the S90 model as if it were a yacht, a cameraman filming from the foredeck while the model went to sea under radio control. Nothing would persuade them, however, to go aboard the narrow-gutted *Leander*. Unfortunately we had to wait until October for sufficiently severe weather in the Channel to test the models in a suitably stormy sea. So these vital tests, based on the technique established by Jack Daniel – and the basis for all future RN sea-keeping assessments – would not be available to confirm the favourable results from the NMI tank before the September 1983 deadline.

The S90 Club Presents its Validation Report to the Navy

On 23rd May 1983, I delivered ten copies of the S90 Validation Report to the Controller of the Navy, Admiral Bryson (see **Appendix 4**, Fig. 3). The Validation Report demonstrated that the S90, while meeting the terms of

The 30-foot 1/10th-scale S90 model on trials with TVS cameraman aboard, April 1983. 'This primitive yellow-painted model aimlessly bobbing about in the waves.' (M Meek)

the outdated document that we had been given as a basis for our design, could still satisfy most of the latest service requirements of the now-enlarged Type 23. A brief summary of the revised Naval Staff Requirements had been released to us only a few weeks before – long after our S90 design had been fixed and tested.

The controlled sea-keeping tests conducted in the huge NMI tank, when compared with the full-scale measurements of Leander and Tribal class frigates, showed the S90 could maintain higher speeds in most open ocean conditions; it was also able to carry a heavier weapon load and sensors much higher in the ship, making for a better defensive platform for a wider horizon.[15] In the written opinion of Sidney Shapcott, the former Chief Scientist of the Admiralty Surface Weapons Establishment, "The increased capacity of this vessel has allowed us to propose a weapon fit on a more extensive scale than has been possible before, in a frigate of this size."[16]

Furthermore, as calculated by Frederikshavn Vaerft, the design had a substantially lower hull-cost than a conventional frigate of similar capability. This was thanks to the wider 'hull modules', or Blocks, used in construction, allowing many more people to work at a time under cover, before the modules were joined together in the building dock. The Danish shipbuilder's production process was much more efficient than anything available in the few British yards that had survived Mrs Thatcher's 'Wind of Change'.

15 'Full-scale Comparative Measurements of the Behaviour of Two Frigates in Severe Head Seas', by Andrew and Lloyd, MOD Admiralty Marine Technical Establishment, Haslar, *Transactions of the Royal Institution of Naval Architects* 16/4/1980.

16 Letter S Shapcott to G Pattie, 16/5/83

After some opposition, the MOD finally agreed to our having a meeting with Yarrow Admiralty Research & Development, or 'YARD', to brief them for an 'independent assessment' of the Validation Report. We agreed that the report could be passed to the DSAC Hull Committee on terms of confidentiality – following the previous debacle with the FT.[17] Owned by BS, YARD had maintained a close relationship with Bath and produced their own design for the Type 23. Indeed Yarrow had already received initial contracts to build the ship. Thus I had given a copy of the S90 Validation Report to a company that was in competition with us and closely connected to all our chief antagonists. When YARD's report was complete we were refused sight of it on grounds of secrecy. However we were able to persuade Geoffrey Pattie, as a final act, to force Admiral Bryson to rescind the secrecy order so that we might have a chance to answer its major criticisms.

The YARD report was simply an endorsement of the MOD's views, with an even more unfavourable power-for-speed than estimated by the MOD (see **Appendix 3**, Figs. 1–3). Jan Neumann, the Managing Director of YARD, visited my home to discuss their conclusions, with which I could not agree. Since they had refused to acknowledge the existence of dynamic lift, and treated the S90 as a conventional 'short, fat' hull, YARD's computer-based power estimates exceeded ours by some thirty-seven per cent. Mr Neumann was clearly embarrassed by 'what his firm had done', confessing that it was the 'worst job he'd ever been asked to undertake'.

The Sixth Floor of the MOD

Through the intercession of Peter, the famous submariner, Johnny Coote, arranged that I should meet his friend, the new First Sea Lord, Sir John Fieldhouse. After walking up twelve flights of wide stone stairs, I was ushered into the anteroom of the First Sea Lord's office. Already waiting there were Vice-Admiral Sir William Staveley, Vice-Chief of Naval Staff, and others. One could not help being impressed by the grandeur of the surroundings. On the walls hung awe-inspiring battlescapes: men o' war sundered by shot, with desperate mortals clinging to shattered spars in boisterous seas. Below these violent scenes stood models of notable ships and submarines. The chintz-covered sofas and homely furnishings, more

17 According to Chairman Meek's memoir, his Committee never saw the final S90 Validation Report.

suited to a country house drawing room, seemed curiously out of place at the centre of Britain's war machine.

When we were called in, I showed the video of the S90 sea-keeping runs in the NMI tank at high speed in heavy head seas, hoping that Fieldhouse, a man of the sea rather than of Bath, might recognise something unusual about the behaviour of our design. As customary on such occasions, the audience sat impassively throughout, making no comment.

It all seemed a waste of time until, just before I rose to go, Fieldhouse muttered to me that there was "something else he wanted to talk about" and beckoned me into his inner sanctum. He shut the door and we were alone, as if in the cramped Captain's cabin of my old submarine. His tone changed completely: "What about this stupid litigation thing over the Osprey?" he asked in a smiling, passing sort of way.

"Are we speaking as man-to-man?" I asked.

"Oh yes. Say whatever you like; take your time. It won't go beyond these four walls."

I replied that we were considering adding a charge of Conspiracy to those already filed. We believed it was in the National and the Navy's interest for us to do so. Based on the conduct of the MOD in connection with the Osprey and the S90 Validation, it seemed the Navy had been misled by the RCNC at Bath for the past five years. I also suggested that he should not believe a word that he had been told unless the MOD were prepared to make a *public* statement, thus giving us an opportunity to respond.

He suddenly became a fellow-submariner again. We chatted happily about the fortunes of my former *Trenchant* Commanding Officer, Tom Clack – 'the Smiling Prince' as he was known – and his son David, Fieldhouse's godson – whom I had known in Malta as an enchanting small boy. We parted on friendly terms. As I left he said, in a solicitous way, "Just let me know if you do have any hard evidence that the Navy was involved in the Osprey tests."

I returned to my car and noted down my thoughts after this extraordinary conversation. I had a feeling Fieldhouse might just be an ally. Of course as Controller of the Navy he must have known everything on the MOD side. I understood what he wanted from me: to find out if we had 'joined the dots' – linked BS with the MOD or Bath in the Osprey testing. We hadn't ... not quite. But what was I to do? Say we had ... and force him into defensive action?

Had Fieldhouse funded the Vickers Osprey test programme? It seemed very likely, but it was possible he had done it in the belief that we had agreed to it. The officials (Atkinson, Moor, Daniel, Rawson and others) would have been as happy to mislead Fieldhouse over the Osprey as they had been to mislead Bryson over the S90. Afterwards, Peter Thornycroft wrote to tell me that, according to Johnny Coote, Fieldhouse had been enthusiastic about his meeting with me; he was impressed with what I had shown him and 'had recommended that the S90 design should be thoroughly investigated without delay'.[18]

The *Panorama* programme, shown in July 1983, was devastating for everyone concerned: the only winners being the BBC and its viewing figures. Fred Emery, the interviewer, had a field day. For the best part of an hour there were interviews with Peter Thornycroft, Rawson, Bryson, Pattie, David Moor and his boss, the Chairman of Vickers. Also interviewed were Admiral Sir Anthony Griffin, the former Chairman of BS, Lord Strathcona and Admiral Fieldhouse. I was filmed saying, "Someone's out to screw us at every turn." Marshall Meek and Jack Daniel were notably absent.

The issues were presented in stark relief, the major one being the lessons of the Falklands War on future warship design and whether new ideas should be considered: first for improving self- (and fleet-) defensive capabilities and, second, for reducing the cost of ships and thus increasing the size of the surface fleet. Inevitably this led to the question of hostility without – and low morale within – the Royal Corps of Constructors at Bath. Rawson regretted his forthcoming 'early retirement' in a resigned way. Although he did not employ the 'rape' imagery he was careful to follow his line that the S90 was "no breakthrough ... no breakthrough at all".

Many of those interviewed described the S90 Club as a group of amateurs working on a project that was far above their heads, regardless of the technical calibre of their partners. It was constantly repeated that there was 'nothing new' in the S90 design.

"For a hundred years," Rawson claimed, "... we have known the characteristics of the short fat ship. Such hulls were used for river gunboats in the last century; they require higher power for speed and have severe shortcomings in roll behaviour." He was again publicly contradicting his final agreement in his correspondence with Dr Garwin, Sir James Lighthill and

18 Letter, PT to DLG, 23/6/83

other academics, that the Osprey, thanks to Garwin's argument confirming the presence of 'dynamic lift', thereby represented: 'an important contradiction of traditional theory'.

Geoffrey Pattie, although recently 'reshuffled' within the MOD, stood up bravely. He was in favour of the need for more detailed views of innovative ideas than merely listening to the internal opinions of the MOD's advisers. He cited examples of previous technology such as the Navy's initial refusal to accept the steam turbine: "And how do I know that the man sitting in reception with a brown paper parcel under his arm is not another Frank Whittle?"

Strathcona suggested that the Hong Kong contract was a put-up job between the MOD and BS yards – probably intended to exclude TGA from the start. Fieldhouse broke out his true colours publicly, contradicting precisely what he had privately told Peter's friend, Johnny Coote, a few days before: "It is important that the Navy should not be subject to every idea or whim put before the politicians: it is essential for the Navy to have access to its own independent experts."

Admiral Griffin, the former Chairman of BS, described me as 'a very irritating young man' who wasted too much of the taxpayers' money by refusing to listen to those who knew better. Charles Adams phoned immediately after the programme was shown: "You're on your own now – and about time too," he said.

To him it was crystal clear that the purpose of the MOD's destruction of the S90 was threefold: first to put Pattie and like-minded political upstarts in their place, whilst hastening the political priority of privatising BS through letting the Type 23 contract continue with Yarrow's ... Second to reassert the dominance of Bath in ship design after the embarrassment of the Falklands ... And third, to entice us into a trap that would force TGA to waste its limited funds and drop the Osprey Case. The game was over. The S90 would be thrown out.

Tony Morrall called from the National Maritime Institute a few days later, advising me to get around there at once and collect all our documents and S90 tank-test data before they 'went missing'. It was particularly important, he said, for us to keep the originals of the Danish Technical University reports on the *Havørnen*. These had compared the results of tank tests with speed trials at sea. The exceptional performance of the *Havørnen* at sea had led the Danish University to correct its figures for calculating the speed-power propulsion factor, which NMI had thereby fully accepted. This was an essential part of our evidence for the S90 Validation Report. The incoming Director of NMI, Marshall Meek, might well attempt to withhold it.

Morrall was upset by the conflicting views over his NMI tank tests. YARD had refused to accept them, preferring their own computer-powered predictions for a conventional hull of the same dimensions and displacement. He met me at the NMI gate and accompanied me into the building. The original reports and letters were on his desk in a brown envelope, together with the final NMI invoice for £51,315.50. As I crammed the documents into my briefcase, he muttered, "There are big changes afoot at NMI: Marshall Meek is taking over as Director next year – and we shall have a new name: British Maritime Technology, or BMT."

"But if YARD has rejected your tank tests, Meek and the DSAC won't accept them either. Meek will probably say they were from an Institute which no longer exists."

"So he won't get a penny of the £51,315 owing by TGA," I said, as I drove away.

*****

I told Geoffrey Pattie at the MOD about the refusal of YARD, MOD and DSAC to accept the NMI tank tests (see **Appendix 3**, Figs. 1–3). Although he had relinquished his interest in the S90, he called a meeting with the DSAC, including Meek, to discuss the question of the discrepancies between the tank tests and the computer predictions of the 'independent assessors'. Since I was not allowed to attend, Pattie was joined by Nigel Ling, a naval architect and marine surveyor advising John Moore at *Jane's Fighting Ships*.

Ling pointed out that, on speed trials, the first Hong Kong Patrol Craft, HMS *Peacock*, had achieved a speed 3 knots faster than Moor's estimates, based on his tank propulsion tests using the normal correlation factors for conventional ships. This confirmed the inaccuracy of Vickers' tank tests using their traditional (1+x) propulsion factor for predicting the performance of semi-planing hulls of which, in Moor's words, 'they knew very little.' He suggested that the YARD computer predictions for Correlation Factor were therefore wrong and, had they been applied to HMS *Peacock*, they would have predicted failure to achieve her trials speed. This created some uproar. Ling called me the next day: "The NMI tank tests were rejected by the DSAC, and the YARD computer estimates were formally accepted."

"Who made that recommendation on behalf of the DSAC?" I asked. "Marshall Meek," said Ling. "And he seemed very aggressive."[19] Meek refused to admit the difference between the S90 and a conventional hull, putting it down to 'Giles misinterpreting the correlation factor'[20] ... But NMI, BHC and the Danish Technical University – and most significant, Moor at Vickers, had all calculated a *negative* factor: (1–x). So the ship went faster on trials than calculated by tank-testing for conventional hulls. There had to be a difference. (See **Appendix 3**, Fig. 3.)

On the DSAC's behalf, Meek had formally recommended outright rejection of the tank tests performed by NMI – the very institution of which he was to be the next Director. How he squared such actions with his conscience is a mystery to me. Was he too lazy or arrogant to consider new ideas? Too fearful, or stuck in his ways? Or perhaps just an obedient servant of the System? Or was it a matter of religious conviction ... or Public Honour, which he did achieve? Whatever his reasons, he took care to disguise the truth in his published version of events. In his book *There Go The Ships*, he claims that TGA, not he, rejected NMI's findings: "After making part payment, I found that he (Giles) had refused to pay the balance because he did not agree either with the actual results, or the way the NMI staff analysed them. In other words they did not support his claim of having found some unique hull form."[21] But our whole point throughout this saga was to accept those results that he, in a recorded MOD meeting, had so strongly rejected. So why did he and NMI's successor, BMT, make no serious attempt to recover the £51,315.50 owing by TGA?

But now it was time to escape from the battles within the Corridors of Power that were so taxing to my life and that of my family. During August we planned to drive to Nice and take a flight to our yawl, *Minion of Oxey*, now moored in Ajaccio, Corsica. But we had a prior task: to purchase a case of that noble claret, Château Beychevelle, in the St Julien region. The

19 In a 30/3/84 letter to Geoffrey Pattie, Nigel Ling states: "You may remember that, at the DSAC presentation on the S90 that we attended last year, I asked whether the NMI predictions for propulsion requirements were accepted or rejected, I obtained the statement in reply that 'they are rejected.' The YARD/MOD computer predictions were accepted."

20 *There Go The Ships*, p. 195

21 *There Go The Ships*, p. 195

Château and its estate and vineyard once belonged to the Duc d'Epernon, traditionally the Lord High Admiral of France. As a mark of respect, all ships sailing on the Gironde, which flows below the estate, had to lower their sails: hence *'baisse les voiles'*, over the years becoming 'Beychevelle'. Apart from its naval significance, the wine is slow to mature: described by André Simon, in his Gastronomic Encyclopedia, as a 'marathon among wines', taking up to twenty years to reach maturity. I had agreed with Peter that we would keep this case in reserve until the first S90 or other 'big ship' made its trial speed – knowing it might take enough years for the wine to mature nicely. In fact it took twenty-eight years ... with the successful 47-knot trials speed of USS *Freedom*. At a celebratory blind-tasting, a leading wine-merchant pronounced it "superb with the typical barnyard note of a fine St Julien vintage." Sadly, by that time Peter was no longer amongst us and able to enjoy it – as he surely would have done.

Our tour of the vineyards and cellars of the wine country, of St Emilion and the Médoc, was an education as much for me as for my three teenage sons. We duly bought our case of vintage at the Château de Beychevelle for £10 a bottle and packed it carefully in the back of the Peugeot, for the drive to Nice. I made a point of parking on the lowest and coolest level of the Nice airport underground car park: outside the heat was stifling.

After the short flight to Ajaccio, where we had sailed from *Minion*'s winter quarters in Porto Santo Stefano, we steered our good ship north to the remote anchorage of Girolata, dominated by the mysterious deserted castle of the notorious Corsair Duarte. A wild place, surrounded by high mountains, still accessible only by boat, mule or helicopter, the whirlwinds eddying down between the precipices, swivelling the boats about their anchors and seething the waters of the small cove beneath the castle ruins.

So we explored the small ports and anchorages of Corsica, the Straits of Bonifacio and the Isles of Lavezzi, haunted by the incessant bickering of flocks of shearwaters. Then, afterwards, for light family relief, the Costa Smeralda. This was something of a pilgrimage for me: revisiting the waters in which I had lurked in my old submarine, viewing, often by periscope, these tempting cruising grounds – and vowing to explore them at a later date. Then, on across the Tyrrhenian Sea by way of the islands of Monte Cristo, Giglio and the perfectly-preserved Roman port on the island of Giannutri – and to Porto Santo Stefano, to winter in 'Pit' Gaspari's boatyard.

Such joys and some challenges of family cruising were the perfect foil to the mind-battering tribulations of struggling with the MOD and the Osprey Case. In those days, before the mobile phone and e-mail, even with three or four teenagers aboard, one was well insulated from most of the cares of the great wide world, without all of today's associated chatter. The odd glance at a three-day-old British newspaper or *Corriere della Sera* was our only contact. Our only radio was an old Brooks & Gatehouse Homer/ Heron, chiefly used for navigation but we could sometimes pick up the BBC World Service.

There were the inevitable minor catastrophes of a broken-down oil pump, non-functioning fridge in a heat wave, leaking hatches, torn sails, chronic shortage of water – or the lost passports and ship's papers for the minor customs official who strikes unexpectedly at dawn. Such trials are an inevitable, if transitory, part of cruising – but they do provide a comfortable sense of achievement when resolved. The same could hardly be said of the seemingly endless succession of disasters that greeted me upon my return to London in late August. But, first ...

... Cabaret!

During those final days of the S90 battle, in early September I was phoned by a young Navy Captain organising the Royal Naval Equipment Exhibition at Whale Island in Portsmouth Harbour, one of the Navy's 'Holy of Holies': its Gunnery School, the 'Stone Frigate' HMS *Excellent*. It was no surprise when he said: "you'd never believe the number of Navy people who support your ideas – we've really had it with the long thin ships: they're just too bloody delicate." So Peter suggested that BHC might provide a short cabaret – and the Captain thought it a great idea.

On the main day of the Exhibition, BHC radio-controlled the S90 one-tenth model across Spithead from Bembridge to Portsmouth Harbour. Just as I arrived at the Whale Island marina, where various small craft were waiting to show off their paces, she hove into view. Under her own power, she was skilfully radio-controlled around the west corner of Portsmouth Dockyard into her berth opposite HMY *Britannia*. This caused quite a stir – for it was our intention to show the Navy that this was a sizeable model, larger than some of the boats in the exhibition ... not a small 'bobbing model' as Meek described it. To drive the point home, we invited Admiral Bryson, Admiral TK Khan of the Pakistan Navy and Bryson's Naval Secretary to stand on the model's helicopter deck: two Admirals and one Captain. After some hesitation, they stood there chatting away without the S90 showing any tendency to heel over. Peter bounded up, saying: "Well gentlemen, you certainly wouldn't want to try this on the Leander model – you might get rather wet!" Then, looking up at the Royal Boats carried in davits along the upper deck of Royal Yacht *Britannia*, he pointed out two of his designs – Nelson 34s – of about the same size. "Big enough for the Queen and Prince Philip ... so big enough for a couple of Admirals," he joked.

Khan roared with laughter, while Bryson looked perplexed. He was clearly impressed, but couldn't say so, as that would be against the 'line to take'. Plenty of young Royal Navy officers and others witnessed the occasion: they surrounded the model, asking for a ride. It was a small gesture of peace on our part – but we doubted it would change the outcome for the S90 or the attitude of the MOD.

CHAPTER 18

Sentenced to Death

The decisive meeting of the S90 Club with the MOD, on 15th September 1983, was chaired by Bill Sanders, the Type 23 Project Manager. Sanders was a dominating personality, determined to drive the meeting to a pre-arranged conclusion. Selected evidence was put before the delegates, created by YARD and the MOD through computer estimates for existing designs of similar size and proportions. Our S90 estimates, as agreed a year before by the MOD, were based on thorough analysis by BHC, NMI and the Danish Technical University of Osprey model tests and full-scale sea trials. Then these, further validated by rigorous resistance and propulsion tests by BHC and NMI, were totally ignored. The YARD and MOD estimates of power came out some thirty to forty per cent higher. Of course, they were based on S90 as a 'conventional' hull that continued to sink with speed.

Worried by the way things were going, I quoted a remark by the Chief Scientific Advisor, Sir Ronald Mason, about the importance of "the (S90) study being perceived, from the outset, as unbiased and independent".[22] In response, Sanders lost control and started shouting about our reluctance to accept the decision of 'acknowledged experts' and our misinterpretation of the Correlation Factor. It was obvious that no test results we could provide, nothing we could say, would change their minds from the original script that Marshall Meek and his Defence Security Advisory Council had written for them. Further comment by us was pointless. Meek summed up the day's proceedings: "The S90 hull form has no development potential whatever. TGA cannot substantiate their claims. The so-called radical new shape behaves like all other RN hull forms in generating negative lift as a function of speed." (See **Appendix 5**, Fig. 2 for NMI's measurements of positive dynamic lift.[23] Also see p. 285 for further evidence from the 'rooster-tail' wake of a sinking stern and the flat wake of a lifting stern.)

---

22 Memorandum from Sir Ronald Mason, MOD Chief Scientific Advisor, to Sir John Charnley, copies to Admiral Bryson and G Pattie, 17/6/82.

23 For discussion of MOD/YARD technical evidence, see **Appendix 3**: The 'X Factor', Figs. 1–3.

The MOD's Verdict

The Controller of the Navy, Admiral Bryson, closed the MOD's pronouncements by stating that the advice of his highly qualified staff was unanimous: the S90 was "unable to meet the Naval Staff Requirement for the Type 23". Of course we knew nothing of the increasing size of the Requirement since we were given the original a year earlier: chiefly the 'gold-plating' as lamented by Admiral Eberle. We were told that we should accept their decision like good boys and pin our hopes on a new project that was being mooted by the MOD: the Enhanced Offshore Protection Vessel, or EOPV, later the OPV-3.

The S90 Club had been told from within that a slightly smaller S90 might be suitable for that application. Once again it was 'jam tomorrow': a repeat of the Hong Kong Patrol Craft ... Were they leading us up another blind alley: blind for us, but not for them? Had we become a secret source of further Research and Development for the MOD?

Hugh Metcalfe, the new Chief Executive of BAE Dynamics, wrote to Ian Stewart, who had replaced Geoffrey Pattie: "We have always been concerned that the controversy surrounding this venture might affect the progress of the Type 23 and our part in that programme."[24] How unlike his boss, the redoubtable Ray Lygo, who had been 'kicked upstairs' as

S90: Cancelled because of TGA's refusal to close the Osprey Case.

24 Letter from H Metcalfe to Ian Stewart, dated 9/9/83

Chairman. Other members of the S90 Club advised us to accept defeat and wait for the possibility of the EOPV.[25] A few days later, I met the respected political journalist, historian and Whitehall-watcher, Peter Hennessy. He suggested I should give up the unequal struggle. I told him that was not an option. The MOD had shown that the only way to argue against us was to counter fact with fiction, science with 'spin'. At least on those grounds, the S90 had won a Pyrrhic victory.[26]

"I feel sorry for you, David," he said. "If they can't destroy your ideas by technical argument, they'll destroy your reputation." Which is exactly what they did.

Disbandonment of the S90 Club

On October 18th 1983, I was at a meeting of the entire S90 Club at BAE's Bracknell headquarters, called to discuss initial plans for our EOPV proposal, based on a slightly smaller version of the S90. I was called away to take an important message: would I phone Ian Stewart's office immediately? I did so and was put through to his Private Secretary: "The MOD has decided against the S90 and in favour of the Yarrow design for the Type 23. The Minister will be making an announcement in the House of Commons this afternoon." I told the S90 Club of the decision. On the whole, being professional defence contractors used to such contrary decisions, they took it with indifference. We tried to agree a Press Statement, but it was decided to wait until it could be vetted by the chief executives of the companies concerned. The meeting broke up with the decision that we should meet early in November to plan how to proceed with the EOPV.

A short letter from Ian Stewart was waiting for me when I got home, delivered by messenger. It was brief: "My conclusion is that the shortcomings

25 Letter, D Lait, Managing Director of Graseby Electronics, to P Thornycroft, dated 21/9/83: "We should not continue to offer S90 as an alternative to the Type 23, to do so merely plays to the prejudices of our customer ... We have been given a clear steer by the Controller and others towards a smaller vessel of corvette proportions and such a vessel would have export potential."

26 Some years later, as Lord Hennessy, he dedicated his book, *The Secret State*, to Sir Frank Cooper, a wartime Spitfire pilot, who was Permanent Under-Secretary of the MOD from 1976 to the beginning of the S90 validation programme, in December 1982, which he must have agreed the previous June on the advice of Sir John Charnley, the Controller of Research Establishments and Sir Ronald Mason, the Chief Scientific Adviser to the MOD. As we now know, this was in extreme opposition to the recommendation of the Naval Staff. He retired, somewhat unexpectedly, at the end of 1982, to become a Director of Westland Aircraft, with whom we had a long relationship. Its subsidiary, BHC, was at that time responsible for the analysis of helicopter operations during the S90/Leander one-tenth scale open-sea comparative sea-keeping tests. Had he suffered the wrath of the Naval Staff?

of the S90 against the requirements for the new Anti-Submarine Warfare frigate are so fundamental that the Ministry of Defence would not be justified in supporting the development of the proposal ... We have gone to exceptional lengths to ensure that the proposal was investigated thoroughly and objectively. I was intending to announce the decision tomorrow but in view of my new appointment at the Treasury you will understand why I felt I should make the statement this afternoon."[27] Of course we had no knowledge of the latest requirements: the S90 was overtaken by events. In such circumstances their decision was probably justifiable.

With Stewart out of the running, Ray Lygo promoted to Chairman of BAE Systems, and Geoffrey Pattie in a new post, the industrial-political rug had been pulled from under us. The next day, the newspapers carried the announcement – just the bald statement by the MOD and a few comments. But that expression: "The proposal was investigated thoroughly and objectively" ... For whose benefit? Over the past three years, they had acquired a substantial data-base, covering the hull-design, construction, powering and sea-keeping of the 'Short, Fat Ship' – much of it at TGA's cost and that of the Frederikshavn Yard, DOI and BAE Dynamics. Apart from their involvement in the Vickers' model tests, the MOD itself had not contributed a penny – apart from their probable financing of the Vickers Osprey tests.

Act of Grace by the BBC?

That morning the BBC phoned. Could I be at Broadcasting House by 12.30 to be interviewed about the S90?

I arrived at Portland Place a few minutes late and ran through the big bronze door to be met by a frenzied lady: "Come on, Mr Giles, hurry up; Sir Robin is waiting." Ushered into the studio, I was faced by Sir Robin Day, the most feared interviewer at the BBC, wearing his customary blue and white spotted bow tie. We were to appear live on *The World At One*. I had thought it was only to be a short tape-recording for the evening radio news. I steeled myself for a verbal assault. Day had caused Sir John Nott to walk out of the studio, and held stirring public battles with other dignitaries, including Mrs Thatcher. He was mild enough before we went on air, talking about my father's work and Lymington, which he knew well, having started out as a reporter on the local newspaper. Then the red light went on.

27 Letter from Ian Stewart to DLG dated 18/10/83

Day was well briefed. "Why have you been so certain that you were right," he asked, "when the experts opposed your ideas so energetically? You aren't even a naval architect!" He was going for the jugular.

"No," I replied, "but Peter Thornycroft is one; and we have others working with us. Anyway you don't have to be a naval architect to understand the basic principles of fluid motion – any more than did Archimedes, or Leonardo da Vinci, originally employed as a Hydraulic Engineer."

"But trying to scale up a small hull to such a size as the S90 was your idea, wasn't it?"

"Partly ... based on my experience with aircraft. And I suppose you could say that I might not have suggested going that far, if I *had* been a qualified naval architect. Judging by the MOD's reaction, I'd hardly have dared advance such a heresy. Look at the 747: to a degree that depends on the same pressure-based rules of fluid dynamics as the S90. If these rules didn't scale to larger sizes, a big plane wouldn't fly as economically as a small one – the opposite of what actually applies."

He seemed intrigued: "So you mean to say that naval architects don't understand the laws of scaling?"

I suggested that many naval architects did not seem to appreciate all the consequences of the pressures exerted by an object moving through water; although the 'miracle' of flight through another fluid – air – might have been identified by Scott-Russell or Froude, long before the Wright brothers. But many, like Peter Thornycroft, Alan Bond's designer Ben Lexcen (who had just won the America's Cup) or my father, have spent their lives playing around in testing tanks with models, trying to find better ways of increasing stability and speed – particularly in waves – without excessive power. That meant overlooking some of the so-called 'Laws', although there was always uncertainty expressed by certain leading naval architects. Forced to improve their designs for smaller craft in a highly competitive commercial or racing environment, they *have* to look further into the realm of wave-drag, or 'residual resistance'. They must consider its advantages and disadvantages – far more than the designers of big ships, for which propellers present a barrier to increasing speed above about 35–40 knots. I believed, as did some others, that such designs could be scaled up for ships, just as for aircraft – but only given adequate power in proportion to the increase in drag, using a new form of propulsion like the jet engine.

"Do you think your ideas had a fair hearing?" He glared at me hard through his black-rimmed glasses.

I paused, thinking of our agreed policy of not rocking the boat. What could I say?

"Not entirely," I replied.

He pounced like a prosecuting lawyer: "Do you mean, Mr Giles, after all the time and public money that has been spent, you still reject the findings of the independent bodies of experts that have done their evaluation for the Ministry of Defence?"

"No, I acknowledge their decision: I've done all I can to put forward my point of view and have obviously failed. But I still believe that someone, somewhere will get the message." I repeated a prophecy I had first made on a programme for TV South: "It might be three years; it might be thirty – but it will happen."

He was becoming more aggressive: "Mr Giles, do you or do you not believe that your ideas have been given a fair trial?"

I decided in favour of winning other battles: "I have to accept their decision," I replied, "even if I don't agree with it."

"Thank you Mr Giles." The red light went out.[28]

28 This is my memory of the interview. Later my son Tom, as Editor of BBC's *Panorama* in 2014, confirmed that the BBC no longer retained a recording of the conversation. Of course 'it' did happen within 30 years – and in two of the world's leading Navies.

CHAPTER 19

Reprieve – An Admiral of the Fleet Intervenes

October 20th 1983: two days after our Execution. A friend had invited me to a celebratory dinner on the eve of the 178th anniversary of the Battle of Trafalgar in the Painted Hall of the former Royal Naval College at Greenwich. Coincidentally, it was also the 100th anniversary of the founding of the Royal Corps of Naval Constructors.

I dressed up in white tie and tails and headed to what is now the Joint Services Staff College, at which my host, a Royal Air Force officer, was attending a course. We met in the anteroom where members of the Services and their guests were milling around, many of them sporting the blue and green Falklands campaign medal. After the recent *Panorama* programme, the rejection of the S90 and the many press comments, I was conscious of glances and whispered remarks, particularly among the naval officers present. Putting on a bold front, I accompanied my friend up to the Painted Hall, where I was ushered to a seat at the High Table. The scene was magnificent. The band of the Royal Marines played sea shanties, the British and foreign regimental and naval uniforms created a striking spectacle, medals sparkled. Mess silver, trophies and glasses glinted on the long tables, while huge glass chandeliers glowed overhead. Crowning the scene was Sir James Thornhill's vaulted ceiling, a cloud-capped fantasy of ships, swains and sea-nymphs tumbling around the sky in glorious confusion.

The Hall fell silent as the Admiral Commandant and accompanying dignitaries strode up the centre aisle to the strains of 'Old Comrades' by the Royal Marine Band. In their midst I recognised our Guest of Honour, Admiral of the Fleet Lord Hill-Norton: an impressive figure with shining black hair, beetle-brows and piercing dark eyes, shooting laser glances around the assembly as he strode towards the High Table.

Hill-Norton was something of a legend. As Second Sea Lord, in 1967, he abolished the traditional Rum Issue, on the grounds that large numbers of serving seamen who could be officially categorized as alcoholics, were operating highly sensitive electronic and technical equipment within minutes of receiving their tot of high-proof 'grog'. As the Vice-Chief of Naval Staff, in 1968, he had been chiefly responsible for ordering a new class of warships, known as 'through-deck cruisers', to confound the RAF who opposed any future need for aircraft carriers. All present knew that, without the new 'through-deck' carrier *Invincible* and the old aircraft carrier *Hermes*, Britain could never have recovered the Falklands. With typical foresight, only a month before the Argentine invasion of the Falklands, Hill-Norton had publicly castigated the Defence Secretary Mr Nott for his proposed sale of *Invincible* to Australia. Without her we might have lost the campaign. So the occasion was something of a celebration of the battles he had fought with the politicians and officials. He had also recently presented a series of BBC programmes on 'Sea Power'. The Navy owed him a debt of gratitude and he knew it.

After Grace, we sat down to the banquet. A baron of beef on a huge trencher was borne through the Hall on the shoulders of four sailors, accompanied by a long drum-roll. I had never attended a naval occasion of such high ceremony before: it was just as it might have been in Nelson's day. Dinner over, Lord Hill-Norton was introduced as the main speaker. He was known to be a man of forthright views. In appearance, speech and manner he bore an extraordinary resemblance to Jeremy Brett playing the part of Sherlock Holmes in the concurrent television series.

He went straight onto the attack over the diminishing number of ships in the Fleet, praising Sir Henry Leach for having forced the issue with Mr Nott during the Falklands campaign. To my relief he made no mention of the 'short, fat ship', although, to judge by glances in my direction, there were some who were expecting it. On the contrary, his message seemed to suggest that something had to be done to reduce the cost, and increase the numbers and capability, of our ships. This was exactly what many from the Navy were waiting for, and there was much applause from the Falklands veterans. The speech was typical of the man: short, sharp and to the point.

We sat for the loyal toast – in the best naval seagoing tradition, still observed on dry land – and were entertained by jolly Jack Tars dancing the hornpipe in Nelson-era garb. Later we repaired to the wardroom for port, brandy and cigars. My friend suggested that I should meet the Admiral Commandant, Admiral Conrad Jenkin. As I was introduced, he looked me sternly in the eye and blurted out: "Oh, you're that 'short fat' bastard – what are you doing here?"

My host – an RAF Squadron Leader – corrected the Admiral, saying, "He is my guest and you, Sir, should apologise." He did so, his eyes sparkling: he sensed the chance of an engagement. He asked if I would like to be introduced to the Guest of Honour. "Or perhaps you don't feel sufficiently confident of your ground to stand up to a Hill-Norton broadside? He's a Gunnery Officer, you know."

"So was I," I said. "Of an old 'T' class submarine," taking my unqualified advantage of the occasion.

He led me to the far corner of the room where Hill-Norton was standing, surrounded by a retinue of young naval officers and civil servants. Interrupting him, Jenkin said: "I'd like to introduce you to the 'short fat ship' merchant." This riled Hill-Norton, who was clearly absorbed in another conversation.

"I had no intention of interrupting you, Sir," I said. "I must just congratulate you on an excellent speech."

There was a pause. It was as if a pair of sixteen-inch guns was being trained round, aiming straight between my eyes.

"Mr Giles, what are you trying to do?"

"Provide the Navy with more ships – and perhaps better ones," I said without hesitation.

The surrounding roar of conversation ceased. All eyes were fixed on the two of us, waiting for a broadside that would blow Giles away from Greenwich and into oblivion. He turned to me with a cold stare from his beetle-browed eyes. In measured tones, Lord Hill-Norton spoke out: "I have been meaning to say this for some time ..." His tone was ominous and I prepared myself for the worst.

"... and it is as well that I say it here and in this company ..." He paused for effect before firing his broadside:

"I'm beginning to think – and I've never said this before – *that you may be right.*" There was a gasp from the assembled company.

He continued: "Do not give up. Your problem is money and, if I had any, I might help out. You need a million pounds. Getting one ship into the Navy is *not* the problem: your problem is Bath. Do not be kind to them. *Beat the buggers!*"

This must have upset some of those officials in the vicinity who were sure to have come from Bath and the MOD. The Admiral Commandant looked uncomfortable too. Perhaps he was a political animal and regretted making an introduction that inspired such an indiscretion from his Guest of Honour. Hill-Norton continued: “If you have any problem with those abominable Bath-men, do consult me.”

“May I come and visit you if I wish to take you up on that?”

“By all means.”

Turning to me, my RAF friend inquired: “Satisfied?”

It appeared that we had found ourselves a friend. As an Admiral of the Fleet and former Chief of the Defence Staff, Hill-Norton had the seniority to obtain an audience with the Prime Minister at any time. He would make a formidable ally. I went straight to my car and wrote down every word I could remember in my notebook.

CHAPTER 20

Enter Bond – Alan Bond

In September 1983, Alan Bond had won the America's Cup in *Australia II*, with the help of his designer Ben Lexcen and his revolutionary winged keel. For the first time in 132 years the 'Auld Mug' was unbolted from the floor of the New York Yacht Club and carted off in triumph to Perth, Australia.

I had a call from Jon Bannenberg. He was now a highly successful designer of mega-yachts for the super-rich. He had excited Bond with the idea of a flagship for the next America's Cup series to be held in his hometown of Perth, where a huge revenue was expected from his Swan Brewery, much of it flowing, via the local community and publicity, into Bond's pocket.

Bannenberg had followed the S90 battle with interest and was convinced that we were right: he had been impressed with the images of the one-tenth sea-keeping model tests shown on a TV programme. But his chief interest in our designs for his new Bond flagship was to carry the greatest load on the shortest length with a top speed of over 20 knots – very similar to the Osprey. Bond liked to entertain lavishly and the best place to do so was aboard his yacht with his Swan Lager – large quantities of which would be carried aboard, plus many other heavy luxuries such as a meat trolley in the style of Claridge's and a grand piano. So his dream ship had to be wide and stable. And, speeding around the world's oceans, she would also have to carry plenty of fuel and survive the worst weather. Jon believed that we had the best hull design available. She would be about the size of an Osprey, but more of an S90 in hull form, and built in Japan entirely of carbon-composite reinforced glass fibre, moulded into his characteristic sculptured shapes. It was an exciting prospect.

Coincidentally, Frederikshavn Vaerft's sister-yard in Aalborg was working on a 180-foot, 450-ton fibreglass (GRP) version of the Osprey which became the StanFlex 300, or 'Flying Fish' class of the Danish Navy. Fitted with gas turbine/diesel propulsion for a speed of 30 knots, it was to be loaded with different weapon-modules, according to the operations anticipated.

HDMS *Viben*, one of 17 StanFlex 300 'Flyvefisken' 30-knot corvettes of the Danish Navy, based on the Osprey hull design.

Experience with the StanFlex gave us much-needed confidence in the use of similar material for the hull of Jon's new Bond-yacht, of similar size and proportions. Meeting at the Chelsea Potter or Club dell'Arethusa in London's King's Road, we discussed our ideas, with Jon sketching away as we talked. On the back of an envelope, the bill, the tablecloth or a napkin, he took our hull and added flowing superstructures, a long curving bow, huge porthole windows and a multi-tubular mast looking like an advanced communications tower. It was an unexpected flash of sunshine in those dark days of rejection, rekindling hope in our lives. If we couldn't build the real S90, we could at last derive some benefit from all the years of testing. Jon had no hesitation in accepting our BHC tank tests – nor did Alan Bond: without accepting tank tests for *Australia II* he might never have won the America's Cup. The design for Alan Bond also provided income for the Osprey Case.

'Bobbing Models'

We still had a degree of support from the S90 Club members in the EOPV, or OPV-3 Project. We spent much time and money on this, employing the services of Conran Design Group to provide a new way of life for the RN crews, accommodated in personal 'pods', with built-in entertainment systems (see **Appendix 4**, Figs. 1 and 2): a contrast from the open mess-decks and hammocks I had experienced as an Ordinary Seaman in basic training aboard the old carrier HMS *Indefatigable*. Meanwhile HMS *Peacock*'s trials results had opened a new chapter in the Osprey Case, involving the Navy; and we were considering taking new action on that account.

The 'SCORES-3' sea-keeping tests comparing the S90 and the Leander class frigate had not been completed by the end of the S90 validation programme, but the October gales provided the heavy weather we needed. The scaled conditions were far worse than the validation programme required, with waves equivalent to 27–36 feet in height, and average wind speeds of 60–63 knots, or Force 12. These conditions were a revelation, demonstrating the behaviour of the two designs in some of the worst seas one might expect, even in the North Atlantic. The S90 exulted in this weather, forging across the high seas at a scaled speed of about 23 knots, while the Leander was unable to keep pace. The support boats accompanying the models had difficulty keeping up: one of the cameramen was nearly lost overboard.

The Leander model wallowed along at under 20 knots, either with her bow buried and seas breaking over the bridge, or showing half her length of keel above the water, as she rolled 35 degrees to either side. She went crashing down into the next wave with her stern going underwater one moment, her propellers racing in the air the next. In beam seas she was just as bad, lying over at an average angle of heel of about 50 degrees – known to sailors as 'lolling' – while in following seas she became almost uncontrollable: yawing wildly and heeling over so far that her mast seemed almost level with the sea. In one shot we could see the entire length of one bilge keel and some of her main keel. In its clinical way, the BHC Report described this as 'the possibility of an onset of broaching'.[29]

The S90 took the seas confidently, without serious rolling and no trace of 'lolling', or lying over on one side for extended periods, as typical of traditional 'long, thin' destroyer hulls. Both models had engines of the same horsepower. In calm water, the Leander showed a full-scale speed advantage of 2 knots, as expected. In normal wave conditions, their performance was similar; but, in higher seas, the Leander was left far behind – exactly as was to be expected from NMI's controlled sea-keeping tank tests. No later consideration was given to this important aspect during the S90 assessment by the DSAC, YARD or the MOD.

BHC also made a comparison of the motion of the helicopter deck of each model. The technique employed was devised by Westland Helicopters

29 BHC Report No. X/O/3204 'Comparative Seakeeping Tests on Two Frigate Designs' with accompanying videos of the models themselves and also taken from the bridge of each, slowed to full-scale. Dated March 1984.

Ltd, of which BHC was a subsidiary. The shorter the period the helicopter had to wait between sufficiently 'quiescent' periods for landing, the better. In general the S90 'Time Between Quiescent Periods' was shorter than that of the Leander, thus affording a greater window of opportunity for helicopter take-off and landing.[30] In general the S90 was a superior platform for helicopter operation – an essential requirement of the Type 23. This was again contrary to predictions and pronouncements by the MOD.

It was important to have the hours of video from the one-tenth model tests made into a coherent short film with the shots from the bridge of each

1983: 'Bobbing models', as Meek contemptuously described them – that impressed Bannenberg and the Captains of the 5th Leander Frigate Squadron in Devonport. The 30-foot S90 (far) and 36-foot Leander (near) models at full power in scaled heavy weather. The Leander model pitches heavily in the scaled 26-foot waves and slows down, while the S90 maintains a more level trim, and steady speed, her bow never shipping green water, thanks to the high pressure under her stern. As confirmed by her bridge video, the Leander bridge was often obscured by seas breaking over the fo'c'sle as far as the bridge and spray covering the superstructure.

30 Addendum to BHC Report X/O/3204 'Quiescent Period Analysis' 12/84, pp. 1–3

ship slowed down to full-scale speed and shown within the same frame for a true comparison. Thus an audience could obtain a realistic impression of the often quite dramatic difference in the behaviour of the two frigates. The Controller of Programmes at TV South offered to make a professional video using the latest techniques available, which I retain.

The majority of his viewers lived in the Solent area, where the S90 had become something of a *cause célèbre*, with Peter a local hero. This video gave the message in a more dramatic form than any number of words or graphs and would later prove helpful in other ways.[31]

At about this time, I was introduced to Charles Hoste, who had worked in military intelligence during the Falklands campaign. Like many others, he was fascinated by the battle over the S90 and was prepared to spend time, money and effort supporting the project with his boundless enthusiasm. A friend of his, the Commanding Officer of the Leander class frigate HMS *Jupiter*, had seen some early sea-keeping videos that I had shown to a group of Leander COs in Devonport Dockyard. Like them, he had endorsed the behaviour of the Leander model as being realistic. He later sent a postcard to Hoste from Stavanger, in Norway, about his voyage out from Devonport: "Two dreadful days getting here in gales and a large beam sea: rolled 35 degrees either way all the time. D Giles has got to be right!"[32]

The End of the S90 Club – and of the OPV-3

The fate of the S90 Club was finally sealed by a letter from BS. Any possibility of building our proposal for the OPV-3 in a BS yard depended upon our withdrawing from the Osprey Case: "It is essential that this litigation is withdrawn without reservations before we can proceed." The minutes of the next meeting of the S90 Club (to which I was not invited) were ominous: "Osprey Ltd will proceed with their court action over design rights, but BS have stated that they are unable to consider any work arising from the S90 or OPV-3 unless the court action is settled." The Managing Director of Dowty Fuel Systems, Colin Cox, wrote, "There is little doubt that the Navy is still very upset about the affair and if the matter is to be

31 Computer analysis by MIT and the Danish Technical University, as well as testing tank measurements at Southampton University, SSPA Gothenburg and Trondheim University, confirmed these results, which were accepted by det Norske Veritas and the American Bureau of Shipping in their 2001 Classification of the 32,500-tonne FastShip design.

32 Comment on postcard from the CO of the Leander frigate, HMS *Jupiter*, to C Hoste, dated 16/1/84

brought to the public gaze again via court proceedings the chances of the consortium being funded are slim. Without this none of us will be able to recommend the acquisition of a shipyard or other similar arrangement to our Group Boards. My personal advice therefore is that you should settle out of court on the best terms you can."[33]

The final meeting of the S90 Club took place at BAE Dynamics, Bracknell, on 27th February 1984. The result was a foregone conclusion. As the meeting disbanded, Julian Taylor asked if I could have a quick lunch with him. For several years he had been a part of my life – and my family's. His conversation was typically light-hearted until, as we were walking back from the pub, he asked, "Just how much money would it take for you to drop the case, David? There must be *some* figure you have in mind."

"They've already offered a hundred pounds," I joked. "But they'd offer a lot more than that today ... But the Danes and Osprey Ltd want to retain their copyright in the Osprey design. That means bashing on with the case for as long as it takes to find out what they were really up to. Today we still lack the key evidence of Conspiracy between BS and the MOD; and we should seek Hill-Norton's advice – it could be crucial." He was disappointed with my answer: "Is this your idea," I asked him, "or did it come from BS?"

"Oh, neither: I was just testing the water." So saying, he walked away to his car – and out of my life with as little ceremony as he had entered it seven years before. I have neither seen, nor heard from him since. The matter finally ended in March 1984, with a formal letter from the Chief Executive of BAE Systems confirming they would make no use of the latest design developments without TGA's written agreement.

Apart from Ray Lygo, the only person from BAE who seemed upset over the demise of the S90 was Peter Salkeld. He was close to tears as we had a valedictory session at the Horseguards Hotel, telling me it had nearly broken his heart. It had also affected his position within BAE and the MOD. He was ailing with cancer and, shortly afterwards, his wife called me to his deathbed to hear some last revelation he wished to make. Alas, I arrived too late and he took his last breath in my presence. At his funeral, his coffin was borne by six Royal Navy Captains – an indication of the respect and friendship that he enjoyed in the seagoing branch of the Royal Navy.

33 Letter, Colin Cox to DLG, dated 10/11/83

The S90 battle was a frustrating and harrowing experience; not just for us but the other side as well: David Moor, Ken Rawson and Jack Daniel all took 'early retirement'. Others were sacked or reshuffled: Strathcona, Eberle, Lygo, Pattie, Stewart ... Long afterwards, in 2003, Lord Hill-Norton wrote to me:

> "You were faced with a carefully thought out and well-executed plan to do you down. The very care with which their disgraceful plan was conceived and executed does, I believe, make it clear how really frightened and worried they were."[34]

Others fared better. Having awarded the Type 23 to Yarrow Shipbuilders, Admiral Bryson, Controller of the Navy and Third Sea Lord, went on to become a Director of GEC-Marconi – which took over Yarrow's soon after they received the Type 23 contract. Later they were absorbed into BAE Systems.

Did we play the card of the Osprey Case too strongly? Should we have dropped it? If we had, would the MOD and its supporters ever have given our ideas a fair trial in a real ship?

How could we trust them? After all that they had claimed publicly – in Parliament, on TV and in the Press – it was too late to unsay what was already said, without serious political consequences. Our paymasters, the Danish yard, did not trust the MOD and, with continuing Osprey sales, were determined to maintain their interest in the Osprey copyright.

The answer to every solution we considered was 'No'. We sought neither weasel words nor hush-money. Now, apart from the Osprey Case, we had but one course to steer: Two seven zero ... Due West.

> Sail forth, steer for the deep waters only,
> For we are bound where mariner has not yet dared to go,
> And we will risk the ship, ourselves and all.
>
> Walt Whitman, 'Passage to India'

34 Letter, Lord Hill-Norton to DLG, dated 3/11/2003

CHAPTER 21

1984: Attack and Counter-Attack

An unexpected possibility suddenly brightened our dark horizon: John Moore phoned to say that he would like to visit me with Lord Hill-Norton. They wanted to watch the video of the comparative sea-keeping tests between the S90 and Leander models and discuss future plans. I drove into London to pick them up at the Army and Navy Club and enjoyed the salty comments of both, expressing their concern at our treatment at the hands of the MOD Officials led by what they called the 'Royal Obstructors' at Bath. Later, sitting in my ex-stable in Wandsworth, they were fascinated as they saw the video adjusted down to full-scale speed. They agreed that what they had seen was consistent with their own experience of the Leander class and of other RN frigates and destroyers. To them, the superiority of the S90 at higher speeds in severe conditions of wind and sea – the most challenging environment for any fast Naval vessel – was obvious.

When I described the process leading to the rejection of the S90, Hill-Norton fired a broadside of questions: When and where had we found the greatest resistance? What were the arguments proposed against us? Who at Bath and British Shipbuilders was involved in the Osprey tests? As he interrogated me, with his laser-like glances, he again reminded me of Jeremy Brett's TV depiction of Sherlock Holmes on television.

It was 'not the Navy's doing', he stated tersely: it was simply that no serving officer dare criticise the Royal Corps of Naval Constructors without serious risk to his career. By his account this could apply to any naval architect or designer, or anyone from the First Sea Lord down – as it later appeared to do in the way he was insulted by the MOD's chosen technical advisor, Marshall Meek.

Driving back into London, we discussed what might be done. More powerful forces were required for an objective assessment of the S90. Hill-Norton said he would first have to consult others on how to obtain a body of opinion that would be sufficiently robust to stand up to the combined hostility of the Ministry of Defence, the Defence Scientific Advisory Council, the Navy's Ship Department in Bath and the Royal Institution of Naval Architects. It would be an almost impossible task: those who had opposed us in the past made up almost half the Council of the RINA. But he would pursue the idea of setting up his own Committee to produce a report for Mrs Thatcher.

*****

Meanwhile the Osprey Case was proceeding at a snail's pace, but more reports were coming in about the unusual performance of HMS *Peacock*, British Shipbuilders' attempt to produce a Patrol Vessel for Hong Kong based on their covert model tests of the Azteca and Osprey. On a fine spring morning I visited Peter in a friend's Thames-side garden. He was brandishing a copy of *The Naval Architect*, the official RINA magazine, with an article by its editor about a trip through the Irish Sea aboard HMS *Peacock*.[35]

"Goes like a train but rolls like a pig: too narrow for a semi-planing hull, thanks to the inbred restraints of their Traditional Wisdom," Peter joked, repeating the prophecy made at his last visit to St Albans to view the *Peacock* model. He opened a page of the article with a photo of the engine-room inclinometer showing roll angles of up to 65 degrees to one side: some wit had written 'Getting Off' on the glass. The engine manufacturers had to strengthen the engine mountings, and bilge keels would be added to reduce the roll, but with a loss of speed. The extreme recordings of the inclinometer were explained as 'inertia swing' – but neither Peter nor I had ever heard of that one before. *Peacock* seemed fine when head-to-sea, maintaining a speed of 20–22 knots in Force 9 conditions without serious slamming: typical behaviour of a semi-planing hull. Despite an estimated speed of 25.6 knots from the Vickers propulsion tests, *Peacock* had exceeded 28 knots on sea trials. This was no surprise to us. It matched the difference in performance between tank and open sea trials on the Osprey *Havørnen*. The article further increased our resolve to obtain release of the official *Peacock* speed-trials results. This seemed unlikely, however, since it would contradict the YARD-MOD claim of 'Giles's misinterpretation' of the S90 propulsion tests.

Driving home a few days later, I turned off the Meon Valley road at the Pig and Whistle, towards the village of Privett, where Charles Adams had

35 *The Naval Architect*, February 1984, pp. E67–71. Probably due to being restrained by the old 'Laws' of 'traditional wisdom'. *Peacock* could have been 20–30% wider for its length to obtain the full benefit of a semi-planing hull – with about the same power-for-speed. The later US Navy LCS suffered from the same traditional instinct for a narrower beam than was necessary.

converted an old wood mill into an idyllic place of retirement for himself and his wife Oriana. Shortly aftterwards she was diagnosed with cancer and died soon after. Charles having just begun to cope with this unexpected tragedy, I dropped by to cheer him up with news of our latest Champion.

Sitting before his flickering log fire, gazing into a glass of whisky, he was understandably cynical about the world and its foibles. He seemed reconciled to his lot: the mill was a hole into which he could 'crawl away and die'. In the company of sepia-tinted photos of the China Station, models of the wartime corvettes and destroyers he had commanded, and a picture of his last ship, the Coastguard Vessel *Miranda* alongside an iceberg, he could sink into oblivion.

He yarned about the War: Narvik, Dunkirk, the Murmansk convoys, hunting U-boats south of Iceland; and his DSC or 'Distinguished Swimming Cross' as he called it, having been sunk three times. Then we got down to business.

I told him of a session at the RINA due to take place in June, where Admiral Bryson was to present his Learned Paper, 'The Procurement of a Warship', about the Type 23 selection process. He recovered some of his former spirit when I told him of Hill-Norton's entry into the fray.

"You'll need all the support you can arouse from him – and he's probably the best ally you could get ... But you'll need his unqualified support. D'you realise what you've done?" he said, gazing at me sternly. "You've brought dishonour on the Royal Corps of Naval Constructors in their centenary year: challenged the integrity of the Great and Good, which is getting pretty close to Royalty. That's intellectual High Treason! Once they've sunk your dream ship they'll want to destroy your reputation with a public spectacle: Christians to the lions! That's the point of Bryson's Paper. But you must take it like a man: say as little as possible and stick to your case. Be charming. Turn the other cheek. You're right and many of them know it! You have time on your side."

The Diet of Worms

Initially, Peter and I were reluctant to attend what was clearly to be the public burial of all we had worked for over the past eight years. But to be absent might suggest a guilty conscience, and we both rather relished the idea of being branded as Unbelievers; better to take Luther's approach at the Diet of Worms: "Every man must do his own believing and his own dying." The hull lifts with speed! That was our believing and, today, it was to be our dying.

So, on June 7th 1984, we entered the Weir Lecture Theatre of the Royal Institution of Naval Architects in London, to take our punishment. The President of the proceedings was the former Third Sea Lord, Controller of the Navy and Chairman of British Shipbuilders, Admiral Sir Anthony Griffin. He started with a dig at us, claiming that "Throughout its century, the credibility of the Royal Corps has been challenged by the amateur inventor claiming to be able to design ships which are smaller, cheaper, more heavily armed and faster than those designed by the Admiralty." He mentioned the fable of the Kaiser who had designed a warship of such amazing speed, armament, interior appointments, etc., that she would have done everything claimed by her designer except float.

This raised a great roar of laughter from the assembly, with all eyes turning on Peter and me. We kept smiling and Peter whispered something about the inferior performance of some of our ships against those of the Kaiser at the Battle of Jutland. He said nothing of the loss of HMS *Hood*, possibly in part due to basic deficiencies remaining uncorrected for a quarter of a century, until her sinking with only three survivors in 1941.

Bryson spoke next. He was dismissive of the S90, claiming that "TGA believed a full-scale hull would give results better than tank propulsion tests, thereby suggesting that established laws were incorrect." This was a cunning reversal of the truth: it was the MOD and their advisors who decided that full-scale hulls would inevitably perform worse than models in testing tanks, since this was what 'established laws' predicted – never mind that the agreed S90 Validation had shown the contrary. And so it continued through a catalogue of the usual criticisms, ending with: "There are many Naval vessels afloat of low length/breadth ratio." In other words our hull was nothing new; it might as well have been the same hull form as a tugboat. The assembly was cautioned on the perils of adopting the ideas of the enthusiastic amateur. We had heard it all before. "A diet of worms indeed," said Peter, "... feeding off the carcass of the S90."

We had one defender in the face of this onslaught – Commander Michael Ranken, a respected naval engineer and President of the British Maritime League. "Professional ideas," he said, "have had a bad time down the years at the hands of the arrogant, the complacent and the 'not invented here' brigade. The 'short-fat ship' is only one of them: the latest of those ideas where not a shred of proof has actually been forthcoming to show that it cannot be made to work – perhaps even better than its own protagonists presently claim."[36] There was an audible gasp of surprise.

Frank Presland had warned us against presenting our own Learned Paper, as we were frequently urged to do by our detractors: it might erode our Copyright, and thus reduce our strength in the Osprey Case and any future Patent rights. In the end we were quite satisfied with the outcome: once more it had provided us with good evidence that the public arguments deployed against the S90 were largely ill-founded.

Later, in a written response to the Paper, Marshall Meek wrote what he must have hoped would be my epitaph, in his typical evangelical style:

> "We need to differentiate between inventiveness and creative design. No new invention is involved in the S90 that could drastically alter either basic design thinking or the basic design principles of naval architecture. It is quite different from, for example, the effect that Parsons' Steam Turbine had on ships; or Whittle's invention of the jet engine had on aircraft design. Therefore, since known theory and design practices apply, it should have been reasonably easy for a competent designer to present a preliminary concept that could stand up to professional scrutiny. The history of this case proved otherwise ...
>
> "It is difficult to conceive a similar state of affairs applying in other branches of engineering where, as in this case, even Government Ministers seem to be prepared to be misled by what was a concept only, against the advice of recognised experts in the field. In the medical field, someone who presses a novel idea in public and in influential circles without having the necessary evidence to support his claims and who does not accept criticism of the recognised medical authorities tends to be labelled a quack."[37]

It is clear that Meek knew nothing of the rejection in 1929 of Whittle's idea for the jet engine by the Air Ministry's Chief Advisor, Professor AA Griffith – a man in a similar position to Meek – as being 'impracticable'. Nor, in 1935, of the Air Ministry's refusal to supply £5 for the renewal of his 1930 Patent, so that it lapsed, leaving the way open for his German

36 Transactions of the RINA, Transcript of the proceedings on 7/6/84 at the CON's Presentation

37 Written discussion to 'The Procurement of a Warship', Transactions of the RINA, July 1984, two years before Meek and his DSAC, with the Inquiry's Secretary, crafted the conclusions of the Lloyd's Warship Hull Design Inquiry before it ever convened.

rival Dr Hans von Ohain to develop his own jet engine without restraint, allowing Germany to operate jet fighters before the end of the War. Nor of the personal, technical, mental and financial privations Whittle endured over some twenty years to establish his small company, Power Jets Ltd, and see it through to the first flight of his first jet-powered aircraft,[38] two years after von Ohain's. I had met Sir Frank Whittle in 1972 at Pan Am's Sky Club in New York, after he moved to the USA with a handsome stipend from the ever-grateful Pan Am. He became friendly with his former German rival, von Ohain, who had also moved to the USA. He warned me of the cupidity of large companies, the difficulties in getting new ideas accepted, and the manoeuvring of bureaucracies to benefit from the ideas of the lone inventor. This was the man who made perhaps the greatest contribution to human mobility and communication since the invention of the wheel, yet he still retained some trace of the bitterness carried over from his former battles to have his invention accepted.

In his written summary of the Proceedings, Bryson made a highly misleading statement: "We fully accept the test data produced by BHC and NMI, but we have not accepted all the interpretations Mr Giles has placed upon them." This was worthy of the final *volte face* in Rawson's correspondence with Garwin. The written and recorded evidence confirms that Meek, the MOD and YARD only considered the initial S90 tests, which took place months before the Validation Programme, including NMI's Report comparing the tank-tests and sea trials of the Danish Osprey *Havørnen*, were complete. 'Interpretations by Mr Giles' were anyway irrelevant. The full measured evidence was in the NMI Report, but Meek and his colleagues chose to ignore it. The evidence did not suit their cause.

The Death of British Shipbuilding

In association with the Frederikshavn yard, we completed our OPV-3 Proposal for the MOD at the end of 1984. The plan was to build it in the

38 See Sir Frank Whittle, *Jet*, published 1953 by Frederick Muller Ltd. It tells of his many challenges in dealing with Official intransigence in the tortuous years leading to his invention being accepted too late to seriously influence the outcome of World War II. It might have done so earlier, had the Air Ministry had greater foresight. Otherwise the Luftwaffe could have been faced by an RAF with 600 mph fighters well before von Ohain's superior axial-flow jet engine was used in the Messerschmitt 262. In 1944/45 Me 262s shot down 542 Allied planes. The RAF's first Meteor jets entered service later, destroying 46 German aircraft, chiefly in ground attack.

abandoned Chatham Dockyard under the management of the Frederikshavn yard, with Per Holst Sorensen as Managing Director.

OPV-3 was a version of the S90 but a far more developed design in both weapons specification and internal layout, with revolutionary accommodation arrangements by Conran Design. The weapons systems data by Consep of Portsmouth was provided in great detail with the co-operation of Sidney Shapcott, formerly Head of the Admiralty Surface Weapons Establishment. This exercise had sorely depleted our diminishing funds – by about £250,000 – but provided us with further evidence of the potential of the Sirius design family.

Conran demonstrated the advantage of the wide beam in locating most of the crew accommodation and working spaces further aft, where motion would be reduced. The crew would be divided between several smaller mess-decks, and each member would have an individual personal 'cuddy', with sliding doors, its own TV and entertainment system: far greater privacy than usual in RN crew quarters. Consep Ltd endorsed the advantage of our 2,500-tonne OPV-3 by placing the air-warning radar at the same height as that of the 26,000-tonne carrier HMS *Invincible*. (See **Appendix 4**, Figs. 1 and 2, for General Arrangement and Accommodation layout.)

Frederikshavn Vaerft, now 'Danyard', led by Per Sorensen as Managing Director, considered investing in an alternative to Chatham Dockyard, such as a privatised British shipyard, for construction of the OPV-3 and military versions of the Osprey. Labour rates were about half those in Denmark and many incentives were available under privatisation. They tried to make a bid for the Hall Russell Yard but, lacking the necessary security clearance, the Danes were not allowed to inspect Hall Russell's alternative OPV-3 proposal and were then discouraged by British Shipbuilders from further interest.

Sorensen believed the MOD would never order the vessels because of their antipathy to anything associated with TGA. He proved right. The entire OPV-3 project was scrapped only a few days after our three-volume Proposal was submitted to MOD Ship Department, in October 1984. Although we had shown how much could be done at a relatively low cost, it seemed that the OPV-3 was no more than a charade to exhaust TGA's share of the funding of the Osprey Case and interfere with the privatisation of British Shipbuilders.

This was the high day of Thatcherism, when the whole of British industry was up for grabs: the era of privatisation, downsizing and dismantling of the great industries – coalmining, automotive and shipbuilding. Inspired by a belief in free markets, competition and efficiency, the major public utilities and infrastructure (electricity, water, gas, telephones, railways, postal services) were re-cast as corporations and their shares were sold on the stock market. The London Stock Market itself was privatised. The free-market zeal stopped only at the BBC and the National Health Service, as being too popular to sell off without public outcry ... Behind the claims of greater efficiency, however, there were some dubious deals going on. In order to sell a publicly-owned company to private investors, the company had to appear profitable. A new phrase entered the language: 'creative accounting'. This was the art of manipulating the figures so that they looked more attractive.

Structural changes were also made. Loss-making activities would be 'hived off' (i.e. shut down), and only profitable areas of the business would be allowed to survive. Government subsidies, though officially taboo, took many subtle and not so subtle forms. British Airways was privatised on the basis of profits inflated by contributions from North Sea Oil revenues. In the case of British Shipbuilders, a sprawling conglomerate of 19 shipyards, government contracts were put in the way of parts of the company that were designated for sale. So Yarrow Shipbuilders, for example, were hastily given the contract to build the Type 23 frigates, and the inflated cost of these ships made the company look highly profitable. Of course there is nothing 'efficient' about such a system, nor is it genuinely competitive. Nor indeed is the market 'free'. But it did prove fortunate for those lucky individuals involved in 'management buyouts' of BS yards that were favoured by the MOD with promises of contracts that *did* come to pass – thus providing them with substantial profits. But there were several that *did not*. All three Clyde yards that built the Aztecas had been closed. Great names like Fairfield, Swan Hunter, Lithgow and Robb Caledon were either on death row or already extinct. British shipyards were using twice the number of man-hours per ton of built steel as the Japanese and Danish yards. Official policy was to sell them off, perhaps with the enticement of a small order. If not, the gates were closed and the workforce paid off. The haste with which this happened was brought home to me during a visit with Per Sorensen to the Robb Caledon yard in Leith – the very same yard where Marshall Meek started his career. There were the usual dilapidated buildings, dark erection halls with broken windows and earth floors. But in this case, the 'bothies', as they were known in Scotland, where the boilermakers, steelworkers and pipe fitters made their tea and 'jammy-pieces', were still filled with packets of tea, old bottles of milk, biscuits and the newspapers of the day when the gates were locked. There, too, were the work coats neatly hanging on their pegs, some with wooden clogs ranged in pairs beneath, as if awaiting the return of their owners. The yard seemed manned by ghosts. All true to life – or death, perhaps – as in Sting's touching musical, *The Last Ship*. (Sting's father had been a shipyard worker and milkman on Tyneside.)

Britain was to become a service-based society, living off banking, hedge funds, property speculation, fashion and celebrity. Thus, in only thirty years, did the twin magic wands of North Sea Oil and privatisation reduce our UK-owned manufacturing sector by fifty per cent. The much-vaunted 'New Elizabethan Age' of the 1950s, characterised for me by the Comet 1 and other examples of Britain's great contribution to aviation, had declined into an age of short-term speculation and greed.

The once-proud de Havilland factory and airfield at Hatfield, the birthplace of the jet airliner, was just another housing estate. The great Number Four Tank at the National Maritime Institute, or NMI, opened in 1967 during Harold Wilson's 'White Heat of British Technology', was pulled down and is today a shopping mall.

*****

Meanwhile we had captured Lord Hill-Norton's growing interest and were making progress with the Osprey Case. The continued sales of Ospreys by Danyard, Alan Bond's yacht and other projects were developing with greater certainty of being completed. So there was plenty of work and income, if little profit, thanks to the cost of the S90 validation project, the OPV-3 exercise – and, of course, 'Ospreygate'.

In the High Court our attempts to obtain the speed-trial results for HMS *Peacock* were meeting with continued obstruction by both the MOD, on grounds of 'public interest immunity'; or of 'irrelevance' since they supported Mr Giles's 'misinterpretation' of the S90 (1-x), or propulsion tests. Defence Secretary Michael Heseltine quoted a claim by MOD solicitors that the *Peacock* trials results could only be disclosed *in camera* since they contained 'classified information relating to the operational performance of ships currently in service'. Lord Hill-Norton, interviewed by the *Times* Diary, voiced his suspicion that this was a ruse to hold the full Osprey trial *in camera*, as "a deliberate attempt to save the reputation of top MOD ministers and officials".

Frank Presland likewise warned us that, on security grounds, the MOD might apply for the whole Osprey Case to be heard *in camera*. The *Times* wrote a Leader on the subject, under the title 'The Osprey and Mr Speaker'.[39] It reminded readers that, since the Osprey Case had not been set down for trial, under Contempt of Court rules the press were at liberty to write about it:

> "Therefore it is open to reasonable discussion. If the press is in a position to write about it, it is wrong that the House of Commons cannot discuss it."

The Leader continued:

> "Although sensitive individual documents clearly cannot be discussed in the House, Mr Heseltine should be prepared to answer some of the questions raised in the case so far. Why did BS, who had agreed to examine Osprey's patrol craft designs under guarantee of confidentiality and copyright, carry out secret tests of those designs without Osprey's consent or knowledge? Was there an inquiry into BS's admitted breach of copyright in these tests and the subsequent destruction of key evidence? If so, what were the results?
>
> "Mr Heseltine does not need to break the sub-judice rules in order to guarantee to Parliament that State industries are as liable to the law of copyright as any other."

After a series of expensive and protracted hearings in the High Court, on 10th May 1985 British Shipbuilders were ordered to release the *Peacock* speed trials results.

This meant there could be no attempt by the MOD to apply for the full trial to be heard *in camera* – an important victory for Osprey Ltd. The Osprey Case now moved onto a new level, in the Press and in Parliament. We were playing for high stakes.

The Man From Number 10

"Monckton, from Number 10," he said cheerfully, tipping his bowler hat as he stood on my front doorstep.

---

39 The *Times*: Editorial Article and Parliamentary Reports: 5/2/85.

It was Christopher Monckton, a journalist who worked with the Prime Minister's Policy Unit in Downing Street. He was clad incongruously with the bowler hat atop his garish Ducati motor-biking garb, which he stripped off with a flourish after striding through the front door. "I've parked the bike in your drive – hope it'll be OK," he said, pointing to his monstrous machine that threatened to topple onto Vanessa's precious *Acer* to render it even more *prostrata*.

Once in his sober city suit, he started to explain the urgency of his unexpected visit. From his enquiries at Number 10 it was apparent that some of the PM's advisors believed the Osprey Case could become more of a time-bomb than yet another banana-skin. The 'short-fat frigate' battle had created ill-will between politicians and MOD officials which might, at the very least, delay the BS privatisation programme.

"So you've been told to sort it out, I presume," I remarked.

"Quite – that's what the Policy Unit is for!"

His first priority was to establish if there were merit in our ideas. Whatever had been claimed by the MOD, it was necessary to get a well-reasoned independent view: to have a qualified professional opinion that could withstand the scrutiny of the cleverest MOD officials. In my view he was unlikely to get any respected naval architect to take on the combined might of the MOD, BS and RINA on our account. But I suggested that there was a retired Admiral of the Fleet who seemed interested in establishing the truth about our ideas. He had no axe to grind and had only the interests of the Navy at heart. He might consider collecting a few independent thinkers, probably from other scientific disciplines. I mentioned Professors Garwin, Jones and Lighthill.

He seemed familiar with all of this. "Old Hill-Norton is known to be a bit of a firebrand," he said, "although he does have a reputation for getting to the heart of the matter."

He wondered if the Admiral might be prepared to set up an independent inquiry to look into it? Could this be the answer to our dilemma?

I suggested that he speak to Hill-Norton himself, but the cost of getting all the participants together and producing a report could be prohibitive. RV Jones was at Aberdeen University and Dr Garwin was at the IBM Research Centre in upstate New York. Per Sorensen, the only credible non-UK naval architect – and a member of the Norske Veritas Technical Committee (the Scandinavian equivalent of Lloyd's Register) – was in

Denmark. There was no chance of any funding from the Government, while we needed all our limited reserves to fight the Osprey Case.

"Ah well, now; you may have to consider dropping *that* for the time being."

So that's his game, I thought: the S90 Club all over again. Kill them with kindness – and take their Copyright.

"Not a hope, I'm afraid," I replied. "We've already turned down that idea. Every time it crops up it increases our resolve to press on with the Case – if only to get to the bottom of this unholy alliance of MOD and BS."

At his departure he seemed less genial. There was something slightly sinister about his sudden appearance on the scene. I wondered how his entry might affect our battle, as he roared away towards the Wandsworth one-way system and on to Downing Street.

Later I phoned Charles Adams: "I've been visited by a man from Number 10 who's also a journalist." I told him of our conversation.

"Keep him at arm's length. As a journalist he may be after a story. See what Hill-Norton thinks. He could be doing it for all the right reasons. Or perhaps he's trying to apply another lever to break your Case. You never can tell with these political gadflies ..."

*****

From his statements to the *Times* Diary in January, it was apparent that Hill-Norton was interested in setting up the sort of advisory group that I had discussed with Monckton. He mentioned the possibility of arranging his own Committee to consider the grounds on which the design had been rejected by the MOD. He was emphatic that any such Committee should be independent of the politicians, the Government or the MOD – and thus impervious to any pressure that might be applied. As a former Chief of the Defence Staff, Hill-Norton's concern for independence and correct procedure was understandable. He accepted Monckton as Secretary of his Warship Design Committee, 'to arbitrate', thanks to his strong links with both the politicians and officials – and therefore 'pig in the middle', as he so characteristically put it.

The Hill-Norton Committee Convenes

The first meeting of the putative Hill-Norton Committee was a lunch at Brooks's Club on March 27th 1985. Naturally I was not invited but, based on Christopher Monckton's account and a report in the *Times*, it included Dick Garwin, RV Jones, Lord Strathcona, Sir Terence Conran and Charles Hoste. Per Sorensen was there to give the view of a professional naval architect. Charles Hoste had experience of MOD intelligence practices, and took the precaution of having the private dining room 'swept' to see if it were bugged. Apparently it was: with the most advanced type of device, which changed its frequency as soon as it was detected. This had the effect of further convincing the members of the importance of the Committee – and reminding them of the apparently unbridled power of the MOD to thwart it.

All were shocked by Garwin's straightforward account of his debunking of Rawson's flawed theory on the question of dynamic lift and the Osprey. Monckton asked Garwin for a note on the matter, which was delivered to him at Number 10. Monckton wrote a short covering note and arranged for it to be forwarded to the First Sea Lord, Admiral Sir John Fieldhouse.

Hill-Norton was suitably impressed. I was asked to draft a summary of the differences between the Leander type of frigate and our alternative proposal, based on the full results of the S90 Validation Programme and our work on OPV-3. For comparison we had much data on the Leander taken from published sources such as the RINA report comparing Leander and Blackwood traditional frigates in head seas. Also, records of personal experience told me by Leander COs of the 5th Frigate Squadron in Devonport and their endorsement of the BHC trials video. Finally, we had the NMI comparative controlled sea-keeping tank-tests and the BHC report on the one-tenth scale 'SCORES-3' open sea model tests. Per Sorensen and George Ley of Danyard could make a comparison of structural issues and construction costs. Our summary would then be checked and used as a working document by Hill-Norton and his advisers, chiefly Garwin, Jones and Strathcona. If the results were still interesting, a final version would be passed, via Christopher Monckton, to the Prime Minister. Charles Hoste was appointed acting unpaid Secretary to supervise and co-ordinate these activities.

It was crucial to all our plans that the report prepared for Hill-Norton should be an objective document. For the week required to assemble all this data into a Report, Hoste and I rented a remote boathouse on the River Erme in Devon. The result was a sixteen-page *précis* called *The New*

Dreadnought: An Examination of The Relative Merits of 'Long Thin' and 'Short Fat' Warships.

After a long night spent working on the final draft, we arrived in Fordingbridge with ten minutes in hand. Hoste dropped me at the Bridge Hotel, then drove flat-out to reach Hill-Norton's thatched cottage at The Hyde, about three miles away on the edge of the New Forest. At the appointed hour Lord Hill-Norton was standing at the gate, glaring at his watch.

"You're three minutes late, Hoste ... but I'll forgive you for that."

He liked our draft and decided to go ahead with his Committee. Monckton had confirmed he would see that their Report went right to the top. There would have to be a series of sessions between Garwin, Jones and Strathcona and some of our opponents, and that depended upon the availability of the necessary people in London. Frederikshavn, Conran Design, Consep, our partners on the OPV-3 project, would be consulted on the structure, propulsion, overall costing, crew accommodation and layout and weapons systems, for which they would not be paid. Hill-Norton hoped to have it all wrapped up within a year.

At last, it seemed we might get a fair trial. I had no reason to doubt the integrity of Lord Hill-Norton. If we had received the same measured, logical reaction from the MOD at the outset, how much time, money and trouble might have been saved.

*****

At our meeting in 1983, Admiral Fieldhouse, as First Sea Lord, had asked me to let him know if there was any hard evidence of the Navy being involved in the Osprey Case. Based on recently disclosed evidence of the goings-on between BS and Bath over the Osprey, as well as the HMS *Peacock* speed trials, it was increasingly clear that such a link might exist. This was a matter of great interest to Christopher Monckton and the Policy Unit. The involvement of the Navy could pose an embarrassment to the Government and, in particular, Mrs Thatcher's hero, Admiral Fieldhouse, Victor of the Falklands. Fieldhouse, as Controller of the Navy at the time of their commencement, could have financed the Vickers tests – albeit unwittingly.

Both Christopher Monckton and Frank Presland encouraged me to discuss these matters with Fieldhouse personally. It might spur some new initiative to settle the Osprey Case on reasonable terms, disclosing the MOD/BS 'Axis' before it became more public and, thus, more embittered.

My first letter to Fieldhouse was ignored but my second, copied to the Policy Unit, had immediate effect. He replied saying, "I firmly recall that you raised this possibility, which I challenged, and specifically denied any knowledge of RN involvement. This remains the position and that is the limit of my concern."[40]

According to my records, it was *he* who had taken me aside at the beginning of the S90 Validation Programme and asked me to let him know if there was any hard evidence of the Navy's involvement in the Osprey tests. Nevertheless he now agreed that I go and see him. For me it was an extraordinary conversation and perhaps equally difficult for Fieldhouse himself. Most unusually, he met me unaccompanied. As at our previous session, there were just the two of us sitting in his inner sanctum.

It was easy to see why he had achieved his position. He was, essentially, a submariner – the type I had come to respect during my short time in that branch of the Service: humorous, intelligent, reflective – and decisive. He knew how tricky his strategic position as Controller had been: with responsibility to finance the recommendations of the Navy's Ship Department, the final authority on all matters of ship design.

It was barely within his power to question the decisions of the professional naval architects of Bath. That was his dilemma: if there was such a link he had to take the responsibility for conduct that he might not have condoned, had he been given the true story at the outset. Yet the officials, being non-accountable, would pass the buck up to his desk at the end of the day.

Now it was a matter of tactics: how to extract the Navy, silently and without any trace on the surface, from the embarrassing waters into which the MOD officials had steered it. Whatever Fieldhouse's own interpretation of the matter, it was clear, from certain things he let slip, that he had been meticulously briefed by the Treasury Solicitor's office beforehand. He carefully followed the 'Line to Take' and I learned nothing. Like a good submariner in a tight spot, evading an attack from depth charges., he knew exactly what to do: 'Flood 'Q' tank; go deep ... Silent routine.'

40 Letter, June 17 1984, Sir John Fieldhouse to DLG

CHAPTER 22

1985: Yachts and Challenges

Alan Bond's *Southern Cross III*, at 20 knots. Note the characteristic high bow-wave and flat wake of a semi-planing hull.

It was a welcome relief, in July 1985, to be invited to Japan to visit the Nishii Shipyard in Ise City, outside Osaka, on the shore of the Ise Wan, a great inland sea, where Alan Bond's new mega-yacht, *Southern Cross III*, was taking shape. She was a smaller 55-metre version of the Osprey, of almost identical scaled speed, hull proportions and displacement. The many resistance tests for both hulls showed we could achieve the required speed of 25 knots at a contract displacement of 400 tons, subject to final propulsion tests due to start at BHC shortly.

We insisted on supervising the construction of the yacht, without which we would not release the plans to the yard. The shipyard's American owner had to take our supervision and travelling costs out of his profit – he had already signed a contract with Bond, ignoring our earlier requirement for these terms. This reduced him to near telephonic tears. Inevitably, perhaps, he had tested the design in the Mitsubishi tank without informing us. It had predicted a lower speed than BHC after towing the model as a normal displacement hull. He tried to retain our final payment for the design on that account, but fortunately we had the *Havørnen* speed trials to back up the BHC test results: good enough for Alan Bond, if not for the MOD.

The Nishii Yard agreed to all our terms, and the tank tests, and construction went ahead. David Jenkins, as consultant naval architect, and David Bridges, our newly-appointed contract engineer – former Chief Engineer of Vosper Thornycroft – spent much time over the next year in Ise City. They quickly became known as 'the two Davids'.

Southern Cross III was at that time the largest non-naval composite fibreglass and carbon-fibre hull ever built. Her weight grew alarmingly and the 'Davids' had to re-design our area of responsibility, the hull structure, in order to keep it within contract estimates. However her superstructure (the yard's responsibility) finished up about eighty tons heavier than estimated. The addition of heavy necessities like a grand piano, a gym, a speedboat, an enormous meat trolley, plus several tons of Bond's own brew, Swan Lager, and other 'toys' and an ocean-going fuel capacity, meant that she was about 100 tons, or twenty-five per cent, overweight. Our warnings of the consequences of all this upon her speed were ignored by Nishii. But, as Jon Bannenberg had pointed out to me many years before: "The decisive factor in any yacht is her owner and what he wants; not the sea." To Alan Bond, the beer, the piano and all the other accoutrements meant more than a couple of knots of speed. After her delivery voyage from Japan to Australia, Jon Bannenberg wrote that she was the steadiest vessel at 20 knots in rough seas that anyone aboard had ever experienced. Bond was delighted.

Upon arrival at the shipyard, I was given a blackboard and chalk in order to explain the design of *Southern Cross III* and my confidence in our tank tests. The yard's first concern was the poor results obtained from the highly-respected Mitsubishi tank. They listened inscrutably as I analysed the towing process and explained how the model's towing method and accurate measurement of dynamic lift was critical to assessing the full-scale performance of the tank model.

"And what other boats has your firm been testing and building, Mr Gires?" asked the interpreter, after a remark by one of the Japanese technicians.

"Pilot boats," I replied through the interpreter.

He quizzed me with a strange look. "Ah ... *pirate* boats, you say?"

"No: *pilot* boats," I said emphatically. But he seemed determined to sustain the illusion.

Solemnly, old father Nishii repeated: "Pirate boats, Mr Gires ...!" He laughed, a real 'belly-laugh', and made a long, rising, guttural sound that clearly implied approval. I had no further problems with the technical presentation of my case.

The next morning, David Jenkins and I were invited to witness their extraordinary production method. A large section of the hull mould was being removed. It was like an enlarged version of the sections used in the construction of the tunnels of the London Underground. This great object, like a piece of modern sculpture weighing several tons, was being manhandled by dozens of women who supported it from the side with long poles and gently guided it into place. So skilfully had the moulding of the section been carried out that, even from close to, it was difficult to see where it joined the adjacent section of the ship's side. Fascinated by this, David and I were openly taking photos of the procedure when the yard manager arrived. Old Mr Nishii was offended. It was a secret technique. I agreed to let his people remove the film and destroy it. He walked away muttering "Pirates", and emitted a fierce guttural expletive.

It took a lot of whisky and *sake*, to say nothing of chicken knuckles and quivering slivers of live fish, to recover my reputation with Mr Nishii and his Board. A night at the Toyota corporate hotel in Toba and a morning watching girls diving for Mikimoto pearls finally restored our mutual trust.

*****

At about this time, we were introduced to Bob Bell, an insurance tycoon who had set the Atlantic record for a crossing under sail with his 'maxi' ocean racing yacht *Condor of Bermuda*. His next goal was to win the Hales Trophy for the Blue Riband of the Atlantic. Established in 1933 by Harold K Hales, the trophy was to reward the fastest crossing of the Atlantic both east and westbound. In Hales's words, it was intended "to improve the craft of speed in nautical engineering for commercial shipping". Bell challenged us to design a small powered vessel for commercial use that could break the speed record set by the SS *United States* in 1952 without refuelling. The record stood at an average 35.59 knots over the 2,870 nautical miles between the Ambrose Light Vessel, off New York, and the Wolf Rock lighthouse, off Land's End.

Richard Branson had just failed in his record attempt with a 36-ton water-jet-powered catamaran. He and his crew had to be rescued when the boat foundered in the Western Approaches. Bob Bell wished to investigate the possibility of using our semi-planing monohull as an alternative. For the required speed it would have to be driven by water jets, and only a 'short fat' hull could provide enough beam at the stern to accommodate the number and power of jets needed. We called the programme 'Indian Summer' in recognition of the beautiful weather that prevailed, and our hopes that it might provide a little extra income for Peter Thornycroft in his retirement.

Having *Southern Cross III* under construction was a useful spur to the interest of Bob Bell and his ideas for winning the trophy. He agreed to finance an initial series of model tests to establish the possibility of achieving higher speeds over greater open ocean distances than had ever been considered possible for small ships. It was our first experience of water jets.

'Some New Type of Propulsion ...'

Thirty-eight years before, I had asked my father about the possibility of semi-planing in a big ship like the old *Aquitania*.

"The increased horsepower," he replied, "would require some new type of propulsion: much lighter, more powerful and efficient than anything available from steam turbines and propellers – something like the new jet engine just being introduced into aircraft." Perhaps he was foretelling the arrival of the Water Jet.

Water jets are based on the turbines in hydroelectric dams but, instead of the water passing through the turbine to generate power, power turns the turbine causing it to suck in water and expel it at high pressure, thus propelling the ship: the same difference as between a windmill and an aircraft propeller. There is no heat involved, as in an aircraft jet engine. Its power comes from a turbine increasing water pressure, from its outlet nozzle, which propels the ship. These outlet nozzles swivel to provide steering, thus removing the need for rudders; while curved 'buckets' are lowered over the outlets to provide reverse pressure.

Given high pressure at the jet inlets beneath the stern, above a certain speed the ship provides the pressure equivalent to the deep water at the

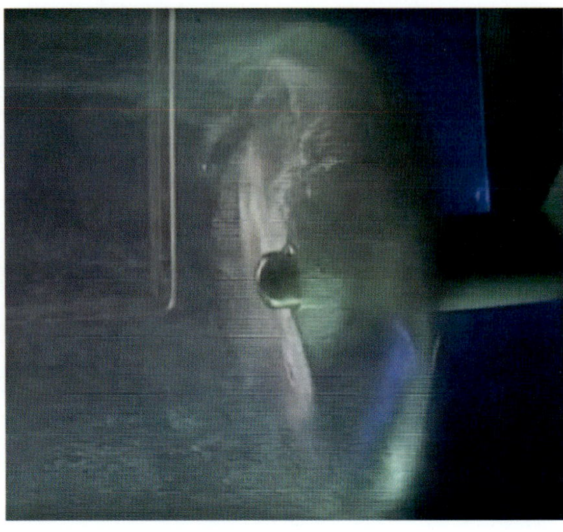

A ship's propeller 'cavitating', or making bubbles, above about 35 knots. This is similar to an aircraft propeller that loses efficiency due to blade-tip stalling in the increasing air-density as a plane's speed approaches the Sound Barrier.

base of a dam. Thus the high pressure from the captive wave beneath the stern of our designs creates exactly the environment needed at the intakes to make the water jet perform efficiently. Water jets do not cavitate, or create bubbles, like propellers at their limit of efficiency. This is because they are enclosed in a pipe as in the use of jets rather than propellers in high-speed aircraft. Beyond the speed of propeller cavitation, they are increasingly efficient: another indication of the similarities between aero- and hydrodynamics, controlled by the relative densities of air and water.

Based on our conversations with Rolf Svensson, Head of Hydrodynamics at KaMeWa, the Swedish makers of turbines for hydro-electric dams and water jets, our semi-planing hull appeared an ideal application. Unlike the MOD, KaMeWa did not regard ours as a 'conventional' hull form, which would have been unsuitable for such an installation because it would generate low pressure at the intakes. The Italian Navy had tried it out with one of their traditional 'Lupo' class frigates and it had not worked. However, even if high speeds could be attained in calm water, we had another important task: to see if we could maintain sufficiently high speed in average Atlantic conditions.

The new model tests at BHC were crucial to Bell's objective: and they were challenging. However, we reduced the resistance of the model significantly, experimenting with a variety of stern flaps, differing stern shapes and transom cut-offs. Crouching on the tank carriage as it whizzed along the rails, the wind tearing through our hair, we experienced terrific acceleration and braking forces. Within its 500-foot length, the carriage achieved speeds it had seldom reached since tests of take-off speeds for the huge 350-ton 'Princess' flying boat in the 1950s. The model tore down the tank, the water licking up its bow and rushing off the stern in a wide, flat wake. The hull seemed a living thing, with its own personality, exultant at the highest speeds. We concluded that the gentle concave stern profile was the optimum for both high and low speeds. This was possible without stern flaps, as employed by the Vosper-Thornycroft Type 21.

At 40 knots, scaled to the full-sized hull, the Prelude model's centre of mass, or Longitudinal Centre of Gravity, was between 30 and 45 centimetres above the static waterline, depending on the displacement. This was the measurement of dynamic lift that, many claimed, could not happen. But it was happening, for all to see. Following the final tank tests, BHC concluded that it should be possible, with our 53-metre hull, to average 40 knots across the Atlantic.

Now we had to test our model at BHC in Atlantic sea conditions; this meant more funding from Bob Bell. Unfortunately, by the year's end the foundation of his wealth, the C.E. Heath insurance business, was crumbling. He was no longer able to finance the work we had commissioned at BHC. We were out on a financial limb with the 'Indian Summer' programme, and there was more heavy expenditure required for the Osprey Case.

Needing to raise money fast, I increased the mortgage on our Wandsworth house and, with great sadness, sold our 36-foot yawl, *Minion of Oxey*, in which, with Vanessa and my three boys, we had enjoyed such

1987: The tank test of the 258-foot *Prelude* at a scaled 48 knots and 1900 tonnes displacement. The beginning of the 'Short, Fast Ship'.

wonderful family holidays. The proceeds also went towards the briefing of our fourth and final Senior Counsel for the Osprey Case, Tom Morison QC.

At his second attempt, Richard Branson had made the fastest-ever Atlantic crossing with his aluminium monohull *Virgin Atlantic Challenger II*, although he had needed to refuel *en route*. Meanwhile, the Hall Russell yard, builders of the five Hong Kong Patrol craft, had been privatised as 'Aberdeen Shipbuilders'. Its new owners, introduced by Charles Hoste, would finance research into an enlarged version of 'Indian Summer' to beat his record. It was to be called *Prelude* and the intention was to raise the record from the 36.8 knots achieved by the Branson's speedboat to well over 40 knots, travelling non-stop.

Being almost the size of the S90 and convertible to a mini-cruise ship, *Prelude* might qualify as a commercial vessel for the Hales Trophy, but the most likely source of interest was Admiral Metcalf of the US Navy, who was seeking a new sort of fast frigate, the 'FF-X'. *Prelude* could herald a completely new type of warship. And it did, in principle, as the US Navy's Littoral Combat Ships, or 'LCS', in 2010, although a disappointing and unsatisfactory interpretation of my Patent.

CHAPTER 23

1986: Hill-Norton Reports and Mrs Thatcher Intervenes

The Hill-Norton Committee's intention was to judge whether the conclusions of the DSAC Hull Committee, which condemned our ideas, or the S90 Validation Report, which supported them, were correct. In the latter case, had the results of the model tests been 'misinterpreted by Giles', as claimed by Meek and Bryson? Or did the Sirius design family have sufficient merit to justify further consideration – not only for the Type 23? Anyone with a grounding in physics or fluid dynamics, Naval architecture and a fair knowledge of the needs of the Navy could understand the simple issues involved.

As Honorary Secretary, Charles Hoste was interviewing the people at Frederikshavn on questions of building and operating costs. He was also setting up meetings between Garwin and RV Jones and some of those involved on the MOD side, such as Admiral Bryson and Professor Bishop of Brunel University, also a member of the DSAC. Surprisingly enough, the latter did not provide any fundamental opposition to our ideas: in fact he seemed to have some understanding of the advantages of the concept of the 'short, fat ship'.

While Dr Garwin was in London, I attended a dinner for him and Professor RV Jones, arranged by a friend in a private room at Buck's Club. We listened as they discussed, in mathematical terms, the possibility of obtaining a hit with Reagan's 'Star Wars' system on an incoming missile. The mathematics was far beyond our host and myself, but it was enthralling to witness. They agreed that the project was impossible – and so it proved.

In November 1985, Christopher Monckton and Richard Owen, both of the Downing Street Policy Unit, had visited Bath and questioned a deputation of leading members of the now grandly re-titled 'Sea Systems Controllerate', comprising a completely fresh assortment of officials led by Rear Admiral Thompson as Director General of Surface Ships (another new title). It seemed that we had changed the shape of Ship Department, even if we had failed to change the shape of its ships.

For the Policy Unit, Monckton wrote a lucid explanation of the errors perpetuated by Bath, but the MOD would not countenance any discussion on points of physics. However hard he pressed for satisfactory answers, using the full authority of the Policy Unit, their replies were evasive and often misleading. Finally, on 13th November, after meetings at Bath, he was forced to send recommendations to Professor Brian Griffiths,[41] Head of the Policy Unit.

Professor Griffiths submitted their joint Minute to the PM, but it was intercepted by her Principal Private Secretary, Nigel Wicks. He in turn contacted Sir Clive Whitmore, Permanent Secretary of the MOD. Wicks, with the agreement of Whitmore, suppressed the Minute and refused to pass it to Mrs Thatcher – the whole story a fine example of the Bureaucratic process in action.

Wicks reprimanded Monckton for not having sought permission from a higher authority before conducting such investigations. Monckton claimed that he not only possessed the necessary authority to bring his researches to a conclusion, but also that, in doing so, the Policy Unit was complying with instructions to investigate anything that might bring needless and unexpected trouble to the government. He finally insisted that, notwithstanding Whitmore's 'understandable' objections, the Report should be put before the PM without further delay, on pain of his resignation. Wicks apparently capitulated and said he hoped that "you don't think we're trying to cover this up". Monckton replied that it was no longer possible to cover the matter up: "unless action was taken, the whole story would come out in court, since the MOD's application to have the whole Osprey Case heard *in camera* has been thrown out".

In an attempt to find some mutually acceptable solution to the deepening row over the Osprey Case, Monckton invited me to join him for dinner with the Director of Naval Public Relations and the MOD PR chief. However they insisted that the Navy was not involved and seemed contemptuous of Monckton's concern that it presented a possible embarrassment to the Government.

Monckton's Report was sent to the PM. She at once ordered Michael Heseltine, the Defence Secretary, to inquire into the matter. This was thwarted by Heseltine's resignation over the Westland affair. His

41 Later Lord Griffiths of Fforestfach, a Director of the Bank of England, Trustee of the Archbishop of Canterbury's Lambeth Trust and International Adviser to Goldman, Sachs.

replacement, George Younger, appointed a senior retired official from the MOD Procurement Executive to investigate the provenance of the Hong Kong design and the rejection of the S90. The investigator did not even take the trouble to question Monckton over his discussions with the people at Bath. Not surprisingly, the MOD's actions were exonerated by the report of their self-appointed inquiry. However Monckton's detailed account of these events gives a very different picture.[42]

In January 1986, the *Times* published a piece by its Defence Correspondent, Rodney Cowton, disclosing the existence of the Hill-Norton Committee and the likelihood that it would be critical of the MOD's decision to go ahead with the Type 23 without due consideration of the S90.[43]

A further unexpected development came the following week. The *Times* again let fly with an 'Op Ed' feature – most of a page – on 'Election pluses and pitfalls for the Tories'. It described the problems that might afflict Mrs Thatcher's Government in the run-up to the forthcoming election, expected in the early summer or autumn of 1987. One of those was the Osprey Case:

> "Major embarrassment for the government could come, for instance, in the High Court action being brought against British Shipbuilders, that could begin in January. Osprey claims BS may have used plans for their Osprey patrol boat to assist in the design of a new Royal Navy patrol boat, HMS *Peacock*. The government applied – unsuccessfully – to have part of the trial to be held in camera.
>
> "Admiral of the Fleet Lord Hill-Norton, an Osprey supporter, believes 'it wishes to suppress something which may be dangerous to the reputation of government ministers and officials'."[44]

Could this be the Policy Unit's response to the earlier leaks about the Hill-Norton Committee? Was the Policy Unit applying pressure to Mrs Thatcher to take action? If so, we had to ensure that the case came to Court as close as possible to the mooted election date – so we needed to move quickly. Frank Presland and Tom Morison QC, our latest leading Counsel, had both suggested that late 1986 was the earliest possible date for the full trial, but everything depended upon our obtaining the services of a qualified Expert

42 Christopher Monckton, Draft for Report to Serious Fraud Office, 24/8/88
43 'Choice of frigate design is heavily criticised', Rodney Cowton, the *Times*, 13/1/86
44 Martin Fletcher, the *Times*, 20/1/86

Witness to support our case. The papers were talking about an April '87 election and it was likely to be hard-fought. Time was of the essence.

*****

Lord Hill-Norton wrote to me on 19th May 1986, confirming that he had sent his Report to the Prime Minister recommending that an Official Committee of Inquiry should be set up "to examine the causes of the events of 1983" – the rejection of the S90. He also wrote: "This achieves about 70% of what I intended. I hope the Committee will achieve the other 30%." He added: "The PM tells me that she intends the Chairman to be 'a professional expert of recognised impartiality', and I have suggested (and very much hope) that it should be Lighthill."[45] However he still favoured the appointment of a Judge or QC.

I received a copy of the Hill-Norton Committee Report, *Hull Forms for Warships*, on 28th May, the day before it was publicly released. It was a 36-page document of admirable precision and simplicity that had been sent to Mrs Thatcher a month earlier. The Summary and Recommendation were clear, concise and to the point. I shall quote them in full:

> **Summary**
>
> The conventional design of destroyers and frigates is based upon a long/thin hull, developed since the beginning of this century to optimise speed and fuel economy.
>
> A radical alternative design concept has been proposed based upon a short/fat hull form which, it is claimed, would be much cheaper to build and would present several other substantial advantages, with no operational penalties.
>
> It is high time that this controversy was resolved, in the interests of the operational capability of the Royal Navy, with less important but substantial effects on the warship building industries and, probably, exports.
>
> An unofficial committee, whose report this is, was established in April 1985 to look at the whole problem. The most important findings are:
>
> The claims of the proposers of the short/fat concept (Thornycroft, Giles & Associates Ltd) were examined by the Defence Scientific Advisory

45 Letter PH-N to DLG, 19/5/82

> Council (DSAC), in 1983. We find that some of the statements in the paper prepared by its Hull Committee, and on which the deliberations were based, were wrong in fact, and that in others the opinions expressed were not well founded.
>
> We therefore conclude that the rejection by the DSAC of the claims by TGA was unsoundly based.
>
> We believe that, certainly for ships up to destroyer size, the short/fat hull form offers enough advantages in the important elements of construction time, habitability, between-deck and weather-deck layout, stability and seakeeping, and weapon-siting, to merit much more serious consideration than it has so far been accorded.
>
> We find that the short/fat hull form may offer a significant increase in top speed over the maximum which can be realised in a long/thin hull of similar size. If this is confirmed it is a most important military advantage.
>
> If the 25% saving in the unit cost suggested by Frederikshavn Vaerft can be confirmed, a wide degree of flexibility is at once offered to the Royal Navy with the choice of a major increase in military capability or a corresponding reduction in the procurement budget.
>
> **Recommendation**
>
> We recommend that an Official Committee of Inquiry be established under the Chairmanship of a learned Judge or Queen's Council, with independent members expert in the relevant fields, to validate or reject our detailed conclusions and report the results urgently to the Prime Minister.

In the words of David Fairhall, the defence correspondent of *The Guardian* newspaper, "Mrs Thatcher had forced a reluctant Royal Navy design establishment to take a fresh look at the basic shape of its warships." The news was carried on the TV channels that evening and, prominently, in the major newspapers the following morning. Fairhall mentioned that "Mrs Thatcher has now ordered the Defence Ministry to set up the official Inquiry, chaired not by a lawyer – as the report recommended – but by a 'professional expert of recognised impartiality'."

Hill-Norton saw the significance of this sudden change in the terms of the Inquiry.

The MOD had been instrumental in the Vickers Osprey tests and the design of the *Peacock* class of patrol vessels; they had thwarted us throughout the saga of the S90, putting it on trial and condemning it on partial evidence, then organised its execution and written its obituary. They were now to re-try us. Once again they would appoint an 'independent' expert to adjudicate – more independent, one had to hope, than the members of the Defence Security Advisory Council and the team from YARD, a subsidiary of our rival Type 23 contractors at Yarrow Shipbuilders, that had advised the MOD over the S90. The MOD issued a Press Release stating its own Terms of Reference for the Independent Inquiry. They were:

> "To review the reasons for the rejection by the MOD in October 1983 of the S90 hull form as proposed to meet the naval staff requirement (NSR 7069) for an anti-submarine warfare frigate, taking account of the independent assessments made at the time by YARD and the Marine Technology Board of the Defence Scientific Advisory Council, and of the Hill-Norton Committee Report Hull Forms for Warships published in May 1986."

The Press Release concluded, "The MOD would not wish to miss any opportunity for improving capability, and welcomes the opportunity of resolving this controversy."[46]

Reports in the Press, notably *The Guardian* and the *Times*, were remarkably detailed, accurate and well-balanced. David Fairhall in *The Guardian* noted:

> "Apart from its political background, setting Number Ten against the naval design establishment at Bath, the most striking feature of the report is its conclusion that the Royal Navy has been led astray by a simple mathematical error. According to Dr Garwin, better known for his opposition to President Reagan's Star Wars programme, the traditional naval architects have always assumed that scale model tests which made the short, fat hull look more efficient could not be translated to the design of a full-sized frigate. This was said to be because the hydrodynamic lift on which the short, fat design's efficiency depends would increase only as the square of the length whereas its displacement increased as the length cubed.
>
> "The flaw in this thinking, the American scientist said yesterday, was that lift would also increase as the square of the larger hull's

46 MOD Royal Navy News Release No. 39/86, dated 29/5/86

> higher speed. That such a basic point should be in dispute may seem inconceivable ..."[47]

On May 31st 1986, the *Times* fired its final broadside: a First Leader in its editorial section. I could not have asked for a more even-handed statement on the situation:

> **Naval Manoeuvres**
> "Seldom do issues of naval procurement so grip the imagination of the press – or provoke so much bitterness in Whitehall. It has rumbled on and off for most of this decade and has antecedents stretching back ten years and more. Against this background, the Prime Minister's decision to set up an independent inquiry, following the latest salvo from the short fat warship's guns, is unusual but unquestionably right."

After summarising certain background details, the *Times* continued:

> "This newspaper has previously reported an allegation that the results of a successful series of tests carried out for Thornycroft, Giles by, among others, the National Maritime Institute were rejected out of hand by the ministry ... Now, it is extremely difficult for people outside the specialised area of marine engineering to make a judgement upon such an issue. Comparisons of seakeeping performance and cruising speed demand special expertise. The Hill-Norton committee contains a number of highly successful men, including Lord Strathcona, a former minister of state at the Ministry of Defence, Professor RV Jones the Second World War scientist, and Dr Richard Garwin, the eminent American defence physicist and strategic analyst. But even they would probably lack specialised knowledge on the central issues. They are distinguished enough, however, and their report is detailed enough, to underline doubts which have continued to surround the Type-23 controversy. The cost factor alone, if correct, is enough to give pause for thought at a time when the defence budget is coming under particularly heavy pressure. The Type-23 was, after all, conceived as a relatively inexpensive frigate ... But who is to conduct the inquiry? As all the country's naval architects seem to be divided on the issue, and as few people outside their chosen speciality can understand the technicalities, the pool of available names must be considered very small. A marine engineer is understood to have been approached, with a view to his

47 David Fairhall, *The Guardian*, 30/5/86

> producing a report before the end of this year. It is arguable, however, that Whitehall should have endorsed the Hill-Norton committee's recommendation of a high court judge to conduct the inquiry, with one or two technical experts to advise him.
>
> "This country has a sorry reputation for wasting its inventive genius by bureaucratic ineptitude and lack of official imagination. This is not necessarily the case this time. But the Government has a duty to make sure."[48]

Was it significant that the doyen of naval correspondents, Desmond Wettern of the *Daily Telegraph*, alone did not put his name to any report in his newspaper? Perhaps, considering his respected position as a reporter on Naval affairs, he did not wish to be associated with such an extraordinary development.

48 The *Times*, 31/5/86

CHAPTER 24

1987: The Osprey Case – The Quest for an Expert Witness

The Inquiry ordered by Mrs Thatcher could not convene until the Osprey Case was concluded, therefore the question of deciding on a Chairman of sufficient impartiality was put on hold. A similar question of impartiality applied to the Case itself, which had reached the point where an exchange of the Reports of Expert Witnesses was required.

Our efforts to find one British naval architect, hydrodynamicist or marine engineer, with suitable qualifications to represent us as an independent 'expert witness', were unsuccessful. To stand on our behalf against the combined might of the Ministry of Defence, British Shipbuilders and the Royal Institution of Naval Architects would be tantamount to professional suicide. Tempers had become so frayed – and arguments so tortured – that, as the *Times* had said, "all the country's naval architects seem to be divided on the issue and ... the pool of available names must be considered very small".

It seemed better to seek such assistance outside the British Isles. I decided to contact a great yacht designer with a reputation for integrity: the American, Olin Stephens. I had never met him, but knew he was an old friend and competitor of my father in the world of ocean racing. Each had the honour of designing the only yachts that have ever won the biennial Fastnet Race on two consecutive occasions: *Dorade*, in 1931/33 and *Myth of Malham*, in 1947/49. He had written a very kind letter to my mother after my father's death reminding her of their mutual respect and friendship.

Stephens replied to me promptly, with sympathy and the open-mindedness that characterises most designers of yachts. He demurred at becoming personally involved in such a technical argument, being, he wrote, "not much of a technician and, at my age, seeing things in shades of grey rather than black and white". But he thought he could find someone who might help. It was a stupendous offer: his letter ended with an invitation to visit him in Vermont during September. His modesty belied the fact that he was on the Technical Committees of the Cruising Club of America, the America's Cup, the New York Yacht Club and the International Yacht Racing Union.

I flew to New York, and on to Keene in New Hampshire. From there I drove to his stone and timber hideaway set on a high Vermont ridge with spectacular views across the undulating hills of the high Alleghenies: the Earth's once-molten crust fixed in time, like a giant seascape. Stephens emerged to welcome me, blinking in the sunlight. Slight and slender with thinning grey hair, dark-rimmed glasses and a bow tie, he reminded me of my father, who described him as the greatest of all yacht designers. And here he was taking an interest in my crazy ideas.

Stephens was suspicious of laws or hard and fast rules. Yacht design was not a precise science; it had not reached its limit, the vanishing point in the struggle for improvement. Had ship design done so? He thought not, and his view was straightforward: two opposing forces limit the speed of a hull through the water – power and resistance, or drag. In yacht design, as long as we can increase the former by improving the sail-plan, or reduce the latter with subtle hull design, we should be able to make vessels go faster – and with improved seagoing qualities. His whole life had been spent working on both sides of the equation, with some success. He saw no reason why this process should not continue beyond the end of his life for both sailing yachts and bigger ships, given the required power. He had no views on what power was 'reasonable'; it was merely a question of the circumstances of its application, what it cost and what people were prepared to pay. By the same token reductions in resistance might be achieved – also at increased cost and complication. In this case the sea would be the final arbiter as to what was 'reasonable'.

After consideration of the question of an expert witness, he recommended Karl Kirkman, a naval architect and expert in hydrodynamics who worked for various technical committees with which Stephens was involved for the America's Cup and ocean racing. Kirkman was also concerned with submarine design, had been the Director of the Webb Institute test tank in New York and now worked for the leading American naval architects, M Rosenblatt and Sons, of Arlington, Virginia, on the design of US Navy ships.

Refreshed and reassured, I flew to Washington and caught a cab to Crystal City, across the Potomac in Virginia, the Mecca of the US Defence

Industry. Dick Garwin was in Washington and agreed to join me the next day for the meeting with Kirkman. It would help to have his intellectual horsepower at my side on this crucial occasion.

We met in the subterranean Mall beneath the office blocks of Crystal City, hard by the Pentagon and the Mecca of Washington's Defence Community. We took our breakfast in a hallway, surrounded by scores of uniformed servicemen striding between the shops, carrying plastic coffee cups and paper bags of bakery goods.

Naval officers wore their summer whites, with gleaming gold rings around blue epaulets – a contrast to the drab civvies worn by their counterparts in the MOD where, thanks to the IRA, the sturdy chap in a suit standing next to you in the lift could be anyone from a Chief Stoker to the First Sea Lord.

Civilians also scurried to and fro in sleeveless white shirts, with bundles of security cards dangling from slender chains around their necks. I felt myself in a new world, reminiscent of Kafka's *America*: a scene of bustling urgency and movement. It seemed to reflect the contrast between the conduct of American and British defence policy: the one intense, focussed, active; the other sometimes deceptively relaxed – superior ... even indolent.

Karl Kirkman, a jolly, rotund fellow, met us in the mall lobby and escorted us to the lift.

His office was a glorious mess dominated by pictures of sailing boats. It was clear from our first exchange of views that he was another kindred spirit. Garwin explained the basic physics of dynamic lift above a certain speed and described his argument with Rawson. Karl found it incredible. After running Hydronautics Inc., one of the most respected testing tanks in the USA, he could not understand how anyone of Rawson's professional calibre could suggest that the whole technique of tank testing fell apart for a certain type of hull above a certain size and speed or Froude Number. He was even more appalled by the behaviour of BS and the Vickers crowd.

Karl agreed to apply himself to answering the key question: 'What were they trying to do?' And, more important, was there any link between the Osprey and Azteca tests and the refinement of the Hong Kong Patrol Craft design? It would be an expensive exercise, involving scrutiny of many of the thousands of documents disclosed by BS. I should prepare myself for a bill of more than $20,000, plus the expenses of a trip to London, to discuss his report with our lawyers – plus their time, of course. Somewhat daunted by the cost, but encouraged by Karl's interest in our Case and the synergy that apparently existed between physics and naval architecture in the USA, I left Karl and Dick to talk on while I went off to another appointment.

I was due to meet Dr Anthony Wells, a former Royal Navy officer, now a US citizen, working on sonar technology for an electronics firm with an office in Arlington. He had been involved with John Moore of Jane's and with our anti-submarine consultant, Roy Corlett, in submarine development. Wells told me about a new programme in US naval circles, known as the 'Fast Sealift Initiative'.

Gorbachev and Reagan were due to meet in Reykyavik the following month. With some prospect of an end to the Cold War, there was talk of reducing the huge US military presence in Germany through greater dependence upon improved logistics. This meant an interest in faster ships with roll on roll off ('Ro-Ro') capability, suitable for carrying large quantities of heavy armour and other self-propelled equipment back to Europe. A wide, stable vessel was called for, capable of much higher speed than hitherto considered practical: perhaps a huge enlargement of our sort of ship.

Wells had two questions. First, what was the largest size to which our ships could be scaled? And, second, would they have sufficient stability to carry the quantity of heavy vehicles required? Having discussed the first question with Garwin and Kirkman that morning, I gave him our estimate of 25,000 tonnes displacement with a speed of 40 knots. This was bound to involve some quite sweeping assumptions of how far our hull design, marine gas turbines under development, and, for that size and speed, whether the emerging technology of water jets could be scaled up. But, to Kirkman and Garwin, it did not seem beyond the realms of possibility.

In answer to Wells's second question, I said that the wide beam would be a real advantage: there would be no need for multiple decks linked by short or spiral ramps or bulkheads and watertight doors. Instead there should be adequate stability for two large unobstructed internal cargo decks: like a huge car ferry. It was worth a try.

In Washington at this time there was also talk of developing a high speed frigate, a concept promoted by Vice-Admiral Metcalf, the US Navy's Deputy Chief of Naval Operations. He had cancelled the Navy's Future Frigate, saying that "The Navy must stop designing ships which are simply updates of their World War II forerunners." The problem was the greatly increased submerged speed and capability of Soviet nuclear submarines and torpedoes; and intercepting them in higher ocean seas on the surface of the stormy North Atlantic.

The USA seemed a possible market for our ideas, but the first priority was to win the Osprey Case, for which Karl Kirkman seemed the ideal expert witness. He understood the problem and was sympathetic. It was unlikely that anyone so forthright could be nobbled by the powers of BS, the MOD or RINA: the UK's controlling force in all matters of hull design.

*****

Back in London the pace quickened. The full trial of the Osprey Case was set down for January 12th 1987, so the last months of 1986 were taken up with Court matters.

First there had to be a more extensive briefing of our leading Counsel, Tom Morison QC. Sunk in battered antique leather furniture in his Fountain Court chambers at the Inner Temple, we heard him express horror at the misdemeanours of leading Civil Servants and professionals. We had a strong case, he believed, reinforced by the admitted concealment and destruction of evidence by the Defendants, and their apparent deceit. Such behaviour by people whose integrity should be above suspicion shocked him deeply.

We also had to provide a response to the technical issues outlined in the Report of Alexander Silverleaf, the Expert Witness for British Shipbuilders, whom we had met at the last session with David Moor and his colleagues at the St Albans tank. He was a former Superintendent of the National Maritime Institute and a long-term associate of Meek, who wrote fulsomely about him in his memoirs. The list of his technical qualifications was impressive, although all his experience had been in merchant ships. Karl Kirkman, acting on our behalf, carried fewer letters after his name but his professional achievements were more relevant. He had worked on all types of US Navy projects, from fast patrol craft to nuclear submarines, aircraft carriers and multi-hulls. He had also been a member of various Technical Committees on yacht design. He had a life-long devotion to sailing and racing in yachts, from ocean racers and Twelve Metres to racing dinghies, including designing and building his own fast semi-planing motorboat. Having a background devoted to the design and tank testing of novel hull forms, as well as a lifetime of practical experience at sea, he knew of the significance of a hull lifting to its captive wave – and was prepared to confirm its potential benefit.

Each witness would present the Court with an Affidavit: Silverleaf, his justification of the three simultaneous test programmes carried out by British Shipbuilders on TGA's Azteca and Osprey – and the RN's HMS *Peacock*. Then Kirkman, would deliver his explanation of the significance of the Azteca/Osprey design family.

Silverleaf gave the MOD's exacting Contract Specification for *Peacock*, or the Hong Kong Patrol Craft, as the reason for Moor altering the hull shapes of the various 'parent designs': Moor was merely doing a separate assessment of the Osprey design, as instructed by Daniel, and was interested in the performance of 'slope-sided' hulls like the Osprey, as opposed to 'wall-sided' vessels, which he usually tested. Silverleaf dismissed any relevance of the Azteca or Osprey in the critical matter of establishing a correlation factor for propulsive efficiency of the propellers: the (1+x) – or in our case a (1–x) factor. As Rawson claimed, the 'single Osprey model' was intended to establish '*How bad it was*'.

Such a simplistic explanation could hardly justify nearly two thousand runs in two tanks, on six models, with 'utmost priority' as required by Moor, and work continuing in Vicker's two test tanks in St Albans and Dumbarton: over Christmas in England and Hogmanay in Scotland. In Kirkman's view, as a tank-testing professional, the costs of the whole programme, together with the value of being able to meet an unprecedented performance target, would have amounted to hundreds of thousands, if not millions, of pounds.

After a quick visit to London to discuss his findings with our legal team, Karl Kirkman presented his Affidavit. It opened with these words: "Before undertaking a study of this matter, I was not acquainted with any of the Plaintiffs or Defendants beyond my awareness of their activities gleaned from a normal review of published literature." The British Shipbuilders team could hardly claim such professional disinterest for their expert witness.

Back to the 'X-Factor'

Kirkman's dissertation went into much greater detail than Silverleaf's, exploring the reasons why the combination of the Hong Kong boats' requirements for speed and hull specifications were contrary to both traditional wisdom and Moor's experience. The Azteca and Osprey tests were

Hong Kong Patrol Craft HMS *Peacock*. A semi-planing hull 'trying to do the same while looking different', as ordered by the High Court. Her speed was 2.4 knots faster than Vickers' tests predicted: a 21% reduction in power for speed, compared with 'conventional theory'. This was proof enough of the improved propulsive efficiency of her semi-planing design to cause the MOD to attempt to put the entire Osprey Case in camera. Fortunately the idea was quashed in Parliament. (See Appendix 3, Figs. 1–3.)

therefore highly relevant to a programme with such swingeing penalties for non-compliance with the specified speed. In particular why an unusual *negative* correlation factor, or (1-x), between model and full-scale trials for the type of hull under consideration, was vital for the accurate prediction of speed.

Silverleaf and his former colleague, Marshall Meek, belittled its importance: "Giles ... seemed to argue that it was itself a measure of success of performance of a hull design. In actual fact it is only an empirical correction factor, obtained from years of experience ... It is always very near to 1.0 and only varies between extremes of 0.9 and 1.1."[49] Yet Moor's man had spent a month examining his calculations of the Osprey, which showed a variation of between 0.8 and 1.11: a variation of over 30 per cent – not quite as insignificant as Meek implied. Kirkman, on the other hand, was emphatic on the importance of correlation factors being linked to the type and proportion of the particular hull under consideration, rather than 'out of the book'. (See **Appendix 3**, Fig. 3.)

Having read Kirkman's report, we felt more confident. Mr Justice Whitford, who was to be the final trial judge, was of similar temperament to Lord Hill-Norton and he would have no difficulty in comparing the

49 *There Go The Ships*, p. 195

simple technical issues of the Silverleaf and Kirkman reports. Whitford had issued the Court Order against BS five years previously, in January 1982, demanding further disclosures, upon pain of imprisonment, which included the Osprey 'missing model', the four Azteca models and a host of documents. Tom Morison told us with some glee that it was Whitford's last case and he was much looking forward to it.

Morison was pleased – almost delighted – with Karl's work. But there was a hint of doubt in his voice as he advised us on tactics: a cloud across his otherwise sunny demeanour. Frank Presland was also becoming tense. I presumed it was because of the gnawing fear that they might never get fully paid for their labours, if we went through the full trial. Frank was still disappointed that we had no 'smoking gun'. Was it just about money, or did something else loom in the background?

CHAPTER 25

1986: The Search for a Smoking Gun

Since our invitation to the Vickers tank almost seven years ago we had spent over a quarter of a million pounds on legal costs, to say nothing of the personal time, effort and spirit expended by Peter and myself. Karl Kirkman had provided a convincing account of the reasons for Moor's testing of the Azteca and Osprey at Vickers: semi-planing hulls were '*an area of which we have very little knowledge*', as Moor had put it to Bill Crago at BHC in 1981. Now I had to connect the work at Vickers with the Ministry of Defence, despite the attempted destruction of compromising evidence by the Defendants. Or, as described in more graphic terms by Tom Morison QC in his proposed 'short opening' for the Trial:

> "It is our submission that BS have ruthlessly exploited the information which they either obtained unlawfully or in confidence and thereafter misused that information for their own benefit ... That they have sought to cover up what they had done by deliberately misleading my clients and their solicitors; by misleading this Court; by lies on oath; by putting in a knowingly false and carefully filleted list of documents; by destruction of material evidence after they knew the writ had been issued and after my solicitors had expressed great concern that documentary evidence should be preserved."

Somewhere in the thousands of documents disclosed by British Shipbuilders after 'filleting', there must lurk the vital link between Vickers and the MOD. We already knew of circumstantial links, but what we sought had to be undeniable, either mathematical or in writing, directly connecting Bath with the St Albans tests. Frere Cholmeley had set up an Osprey Case 'war room' in which all documents, both theirs and ours, were neatly lined along three walls in large box-files in chronological order. I spent many days going through them looking for any scintilla of evidence that might convince the Court of the link between Moor and Rawson: as between Jack Daniel, ex-MOD and the responsible Director of British Shipbuilders – and the Ministry of Defence.

We had to agree tactics with Morison and Presland. There were meetings almost every day leading up to the trial, and we noticed that our legal team was becoming less sanguine about the outcome. There was a risk that, after the bloodshed of the first day or two, the Court (i.e. Mr Justice Whitford – there would be no Jury) would revert to the grind of cross-examining witnesses. BS would put up honest down-to-earth professionals: draftsmen, mathematicians, model makers and test-tank technicians. Reputable naval architects from St Albans and Dumbarton would be called, and they were bound to swear they had developed the Hong Kong design precisely according to Silverleaf's Affidavit. They would surely deny any connection with the many Azteca and Osprey models, or '*an Osprey model*' used as '*a minor research item*', to use Jack Daniel's deceitful words – of no relevance to their MOD work. The final advice of our Solicitor and Counsel was that the Court would tend to believe them.[50]

With the constant repetition of abstruse hydrodynamics, the Judge, the Press and the public would suffer from 'detail fatigue'. Giles would become the conspiracy theorist, the amateur enthusiast and 'quack', the 'vexatious litigant'. What hope would he have, supported by no more than a little-known American technician and a British naval architect who, on his own admission, designed ships 'by the seat of his pants'? Why was no professional from Britain, the world's premier inventive maritime nation, prepared to defend these wild theories?

The question of money kept cropping up. Frank insisted on further financial guarantees: the trial could drag on for nine weeks, with soaring costs. He also wanted Peter's written agreement to co-operate with TT Boat Designs standing as Co-Plaintiffs. Finally he was 'obliged' to tell me that Morison had had a recent meeting with the Defence Secretary, George Younger. That sounded sinister, although he tried to reassure me that it was only 'Cabinet business' which, according to Christopher Monckton and the PM's Policy Unit, it most certainly was. Morison and Younger, former members of Winchester College, had also been grouse shooting together in Scotland during the autumn. Was this just a couple of Old Wykehamists enjoying each other's company or was a secret deal being forged out on the moors?

On January 11th, the day before the trial was due, providence at last intervened in our favour. It snowed, then froze: the coldest January for

50 Mr SJ Phillips, who led the St Albans testing of the Osprey, later wrote to the Author (20/10/94): "I worked in the St Albans tank over the period 1979 to 1981. I do not think that anyone there really understood your technical message, which I believe was an interesting one."

years. Mr Justice Whitford was snowed up in his Kent home, the Royal Courts of Justice were closed – the heating had broken down – and half of the court staff were unable to get to work. The trial was postponed until the following Monday, January 19th. We still hadn't found the smoking gun, but this gave us our last chance.

During that providential week, I had several conversations with Charles Hoste. He had joined a financial team to take over the privatisation of Hall Russell, builders of the five Hong Kong Patrol Craft, as Aberdeen Shipbuilders. He found that the remaining members of the former management refused to admit that Moor's final design by Vickers had anything to do with the Osprey tests. Frank Presland also travelled up to Aberdeen for a day and was unable to obtain any such admission. In fact some of the former technical team claimed to have designed HMS *Peacock* entirely by themselves, knowing nothing about the Osprey tests other than what they saw on the 1983 BBC *Panorama* programme or read in the papers. They were well tutored: I was reminded of Moor's famous telex: "Giles knows ... If he appears, tell him nothing: a Brick Wall!" So we would get nothing out of Hall Russell. Where could we turn?

I had a meeting with Presland and Morison just four days before the new trial date.

Years before, John Mummery had recommended we should apply for a million pounds, in order to include a substantial payment in respect of exemplary, or punitive, damages. Today Morison's message was very different: "The last thing you want is for the trial to turn into a slogging match. If you are offered a settlement of £100,000 over your legal costs, take it! You'll be laughing all the way to the bank." It seemed that he had changed his tune from his robust confidence of the autumn to midwinter pessimism. His meeting with Younger was surely about Cabinet business – to agree the veil used to conceal the influence that our Case might have over the forthcoming general election, expected in April or May? This had been suggested in the *Times* 'Op Ed' published almost exactly a year earlier. It was becoming ever more important that I should 'connect the dots'.

The Glorious Resurrection of Peter Thornycroft

Were the forces of the Crown also soliciting Peter behind my back? Frank Presland wanted a private meeting with him on 17th January. Were they trying to divide and rule just to resolve the case and assure their fees?

To restore Peter's resolve I had to persuade him to sign a piece of paper instructing Frere Cholmeley to continue the Case with TT Boat Designs as Co-Plaintiff. If he caved in we were finished.

Vanessa and I met Peter off the train at Waterloo, catching him before he went to see Frank. Due to domestic upheavals at his home, his long-suffering secretary Jill Swallow, with other locals, had persuaded him to move from the arctic conditions of the *Blackwater* hulk to a small flat above The Pilot Boat pub beside Bembridge harbour. It was very cold, and he looked awful.

Peter brought with him four small models, three of them fashioned from the disclosed lines plans of the St Albans models (numbers STA 2224, 2225 and the best of them, STA 2230, the Osprey look-alike that was discontinued after the July 1981 High Court Injunction blocked its use for the Hong Kong contract). They were carved by Arthur Compton, the retired foreman of Peter's old Keith Nelson boatyard. The fourth model was a 'half-and-half', the Osprey hull on one side and the 2230 hull on the other: the two halves were almost identical. Beautifully made, as with everything Arthur did, they might show the Judge how 2230 was much the closest to the Osprey of all the Vickers models: great evidence against Silverleaf's written testimony denying it incorporated any Osprey features. Unlike a lines plan, which could only be interpreted by an expert, most people could see, and more importantly *feel*, the startling similarity between STA 2230 and the Osprey, in three dimensions.

We drove to Frere Cholmeley, where Frank was grim-faced. He and Peter went into their private session. Later we all met and Frank asked us to consider a settlement of £100,000 over Frere Cholmeley's costs. I objected, saying this would not even cover our own expenses over the past six years. Peter looked bothered: he knew how little would be left over for him. We asked to be left alone to decide our requirements.

We drafted two letters. In one we set out our acceptance of the Frere Cholmeley recommended settlement figure of £100,000 after costs. The other stated that we did not accept their recommendation and specified £650,000 over costs as the minimum amount we would require. We agreed that, if offered £400,000 over costs, we would only consider it, but made no commitment to any final figure. That was the one we chose.

We also agreed that I, personally, would be responsible for any costs arising from my refusal to accept the figure recommended by Frank. I was prepared to put everything on the line.

This was not popular: Frank strongly advised me to accept the Frere Cholmeley figure. Worse, Peter's fortitude appeared to be crumbling. After his meeting with Frank, he seemed dispirited. We took him to an old pub just off Lincoln's Inn Fields and, once we got him seated in the bar by a warm fire, he cheered up. After a merry session, his confidence recovered. I produced the two letters we had drafted at Frere Cholmeley and wrote them out longhand on an Oxford pad. Peter signed the letter demanding £30,000 over costs for TT Boat Designs. We would co-operate fully and Peter would stand by the participation of TT Boat Designs as a Plaintiff with its copyright in the Azteca. He would get a good cut of any winnings. Warm, wined and dined, he was more like the Peter of old.

I phoned Frank at home that evening and told him that Peter had agreed to go ahead with the case, and would continue to allow his own firm, TT Boat Designs, to be a Co-Plaintiff, thus allowing us to benefit from the testing of the Azteca models. I would give Frank the agreement Peter had signed to that effect on Monday morning. Frank's reaction was impassive: he was pessimistic about our chances in Court if we rejected the Defendants' offer. Like Tom Morison, his view was that, once the first blood had been shed, it would be a long slog with our position being steadily eroded by the Judge's loss of patience in a maze of technicalities.

The Eve of Trial – Our Last Chance

I allocated the whole of the final Sunday and, if necessary, the entire night, to a last attempt at finding the palpable link between Vickers and the MOD. During the previous days it had occurred to me that activities linking Moor and Rawson were going on over the weeks leading up to 13th April 1981 – the day on which Rawson sent me his *coup de grâce* letter. Bill Crago, Principal of BHC, had visited St Albans on 19th February, when Moor asked him for assistance in the interpretation of the Osprey tests: "this is what I have been told to do – but I don't know much about it", as he says on his tape recording which I still possess. Did his now-Technical Director, Marshall Meek, or the BS Board Member for Warship Building, Jack Daniel, tell him to do it? It is also as summarized in Chapters 9–10, pages 70–82 above.

The Vickers test programmes on our Azteca and Osprey and their Hong Kong designs were run concurrently in the St Albans and Dumbarton tanks over a period of five months. A total of 1,883 runs were made on the four

Azteca and two Osprey models, while only 296 runs were made on the three Hong Kong models. This was not 'a minor research item' as BS had claimed.[51]

I re-examined the letters between Rawson and myself. From the end of 1980, we had been corresponding quite amicably comparing the performance of the Osprey with traditional hulls. On 17th January 1981, I had sent Rawson basic Osprey resistance measurements at 380 metric tonnes displacement. In his reply dated 25th March, he had written that he "would try to address some of the points made in your letter, but we are rather heavily pressed at the moment ... You do need to produce evidence which irrefutably supports your claims ... Theoretical extrapolations do need hard evidence from tests to support them if they are to be convincing."

On 27th March, Jack Daniel had written from BS in his sudden confession of the Vickers tests of 'a model', telling me that he needed 'more and more methodical data', including powering data on Osprey as part of a 're-assessment' he was doing.

Rawson's next letter, dated Friday 10th April, was couched in the most offensive terms – untypical of the man I had met in Bath. It hardly seemed possible that it came from the pen of that affable and reasonable gentleman who had kindly donned his deer-stalker hat and driven me down to Bath station in his ancient Morris Minor: *"I have yet to complete the study promised to you in my letter of 25th March ... While understanding that your objective is to keep your proposals in the public eye at any cost, I believe that rape of this department's views, which have been consistently fair and objective, will not help to dispel the widespread public distrust of your claims among professional engineers and scientists."*

Three days later, on Monday April 13th, Rawson wrote: "We have now had a look at some of the data contained in your letter ... It seems increasingly unlikely that evidence of Osprey will now be revealed which represents important advances in naval architecture as you have been claiming for some years ... I have to say that the demands on my depleting staff are such that, having an adequate knowledge of the advantages and disadvantages of the Osprey type form, I am unable to devote any more time and effort to the subject."

Something must have happened over that weekend of April 11th–12th to allow him to complete his study. I had another look at Figure 1 of his *coup de grâce* – or 'Revenge' letter.

51 See Chapter 9, p. 71 (RJD 'Jack Rabbit' letter, 26/3/1981: BS formal admission of St Albans tests.

I suddenly remembered that, among the thousands of disclosed BS documents, there was a log kept by the technician who ran the Dumbarton carriage used to tow the model through the tank. It was one of the few STA 2225 documents that were not destroyed – and it would record the precise time and date on which the tests at 373 TSW (Tons Salt Water) had taken place. Test-tank technicians tend to work on their own, particularly at weekends, and they do not falsify their logs. That must have been the significance of Moor's telex to Dumbarton on 3rd April 1981: 'Extreme urgency: Weekend, overtime if required'.

After much searching, I found exactly what I sought. The Dumbarton carriage log confirmed that Osprey model number 2225 was run at 373 Tons equivalent to 380 metric tonnes, on Saturday 11th April – not on Tuesday 14th April. This was out of sequence with both the original schedule requested in Moor's disclosed telex to Dumbarton, and with the dates given in his diary, in which the run at '373' (the Imperial tonnage used for all internal memos at both Vickers tanks) was shown to have taken place on Monday 13th April. This would have been too late for Rawson's graph, which was drawn and annotated in his own fair hand on Sunday 12th, to be ready for typing-up and mailing on Monday 13th. Moor's telexes and diary notes conflicted with the only entirely reliable source of the timing of the tests: the Dumbarton carriage log (See **Appendix 13** for Rawson's Figure 1 and explanation). That was his fatal error.

To the many officials and experts to whom it was circulated, Rawson's graph would no doubt appear 'methodical', but close examination showed that it had been cobbled up from data exchanged between Osprey Ltd, the British Shipbuilders' tanks at St Albans and Dumbarton, to much of which Rawson and the MOD denied having had access.

Now we had not one, but four 'smoking guns', the second inadvertently provided by Admiral Bryson:

1. The falsification of Moor's telexes over the true date for the tank test of 2225 at '373' and its obvious link with Rawson's *coup de grâce*, or 'Revenge', based on the Dumbarton carriage log for 11th April – and the subsequent destruction of 2225.

2. The 1982 statement by Admiral Bryson in the memorandum from Sir Ronald Mason to Sir John Charnley concerning the inferiority of the Osprey: *"These contentions have been based on a standard methodical series for resistance and sea keeping trials on the Osprey in model and full scale."* This

could only have referred to Moor's 1980–81 standard methodical series on the six models of the Osprey and Azteca, together with the full-scale measurements by MOD and DOI of the *Indaw*, the subject of a detailed analysis by Rawson.

3. Likewise, Admiral Leach's 1982 *Newsnight* gaff that *"tests have been done and they are not wrong"*, could only have applied to the Vickers tests. Likewise his remark: *"Even if the tank tests showed the traditional theories were wrong, the Navy would not accept them."*

4. At one o'clock on that Monday morning of the trial, while I was still at work, Karl Kirkman phoned from Washington in a state of some excitement. Going through all the tank tests, he had determined that, while model number 2230 was being tested, the Osprey was being run at precisely the same beam-to-draft ratio as all the Hong Kong models: it could only mean a conscious effort to obtain good comparative data. Karl said he would now be prepared to go on oath that Osprey and 2230 were directly connected: a final smoking gun.

We had established the link between Vickers and the MOD, as between the Osprey and the Hong Kong Patrol Craft: four smoking guns, a good night's work. After putting my conclusions into a final Proof of Evidence,[52] I slept soundly for a couple of hours before waking to face the full trial, due to start at 10.30 that same morning.

52 'The Significance of STA 2225', dated 19/1/87

CHAPTER 26

1987: Mr Justice Whitford's Last Case

At 8.30 on the morning of Monday 19th January 1987 my wife Vanessa and I picked up Peter Thornycroft at the Royal Thames Yacht Club, and drove to the Frere Cholmeley office in Lincoln's Inn Fields. Frank was there, with his assistant Jonathan Cornthwaite and several clerks trundling porters' trucks laden with documents. When I told Frank that I had found those 'smoking guns' he seemed unimpressed. It was too late to be bringing in new evidence, he said; we had provided all we needed. We walked, a silent little procession, around Lincoln's Inn Fields, down the alley and into the back door of the High Court. Frank was chain smoking, and Jonathan Cornthwaite was not his usual cheerful self.

We were in good time, so Frank suggested we go to the basement cafeteria for a chat and a cup of coffee. Glancing sideways at the correspondents collecting outside the door of court number 25, he beckoned us to follow him downstairs. There was clearly something important he had to tell us, even at this late stage. As he puffed away at another cigarette he was shaking like a leaf, his coffee slopping onto the table. Vanessa asked him if he was feeling all right.

He insisted that we should reconsider our position. We could not afford to let this go ahead – even now. It could degenerate into a slogging match, as Tom Morison had said. We must accept the BS offer. We should consider how much we would be in debt to Frere Cholmeley if we did not settle now. It could be hundreds of thousands of pounds; even our house wouldn't cover it.

Peter at once joined in: he was prepared to throw in his own beloved Nelson 40, *Grand Espoir*, if it would help. Anything to get level with those robbers! "Robbers they may be," said Frank, "but after a few days they'll make you look like the villains." The High Court hearing would be portrayed as the Nation wasting public funds at the behest of Giles's conspiratorial fantasies. We were facing some of the most highly respected institutions in the country. This was not about Giles and Thornycroft, nor

Ken Rawson, nor Jack Daniel, nor David Moor, nor even British Shipbuilders. It was about two small-time amateurs and the Ministry of Defence, and, above all, the integrity of The Crown. The Defendants would wriggle out of it and leave us on the street with our reputation in tatters. They had asked for *nine weeks* to be set aside for the trial.

Despite Frank's entreaties we remained determined: with Kirkman's latest revelation and the carriage log we had established clear links between BS and the MOD and, with his Expert Witness Report, we were ready to stand against all comers. If the full trial went ahead we knew the Government would have every reason to flinch. This was about three of the most powerful forces in the land: the Navy, the Shipbuilding Industry and the MOD covertly conspiring to use our designs, whilst discrediting us unjustly. It was indeed a 'Conspiracy' between BS and the MOD, to use Rawson's own expression.[53] As Martin Fletcher of the *Times* had foreseen a year ago, the Osprey scandal could infect the public perception of the MOD and the naval community in the run-up to the general election. It might even cost Margaret Thatcher her job!

"Sorry Frank," I said. "We're not giving an inch."

I only had to look at Peter and Vanessa to know that they, too, were inflexible. "Very well. On your heads be it," he muttered as he stumped back upstairs to the throng of clerks and reporters milling around outside the Court.

*****

The BS crowd were there: dozens of them. Jack Daniel was looking greyer than his years; Moor a whiter shade of pale. Surrounded by newly acquired, independent lawyers – due to a split between the camps of different involvements and priorities – they were carefully avoiding each other. Each group of defendants had its own legal team with the BS Lead Counsel, Christopher Clarke QC, striding about importantly. Trailed by a posse of underlings, he doled out words of wisdom to his followers.

Frank was chatting with one of the BS solicitors: George Martin, an old friend of his. I suspected he was trying to stitch up a last minute deal.

53 KJ Rawson, *Ever the Apprentice*, p. 115: "What his lawyers sought to do was show a conspiracy between the MoD and BS which would much enhance the damages payable and, although I was to leave the MoD in 1983, I was pursued by them and the MoD until the matter was resolved in 1987. There had been no such collaboration between BS and MoD but it is not easy to prove a negative and I was subject to severe cross-examination by the barristers in 1981 and in 1986."

Marshall Meek was notably absent. Despite having advised Moor and the BS legal team over the Osprey tests, he was not part of the Defence, although he might have been called as a witness.

Peter was relishing his notoriety, seeing the *Times* legal correspondent and other journalists talking in hushed tones to Frank and Tom Morison. He beamed at all and sundry in his usual cheerful way, clad in his threadbare Yacht Squadron suit with the usual spotted hanky in his top pocket.

"And a very good day to you too, my friend!" was Peter's reply to greetings from reporters and others. His demeanour was in strong contrast to the murmurings and furtive looks of those representing the Defendants.

*****

A gowned flunkey threw open the ornate mahogany and glass doors of the Court. "Court's in session," he announced, and we began to file into the arc of benches and polished woodwork.

Our own numbers were sparse: Tom Morison, Kevin Garnett (our Junior Counsel), Frank Presland, Jonathan Cornthwaite, a couple of junior and articled clerks, Peter, Vanessa and myself. That was all. The other side must have numbered thirty or forty people. Grimly trying to seat themselves, their mutterings soon confirmed not only dissension in their ranks, but also that they were too numerous for the Defendants' benches. By contrast we were surrounded by an embarrassment of empty seats. Frank asked if we could make room for some of their people.

"Move over in the boat," cried Peter. "Prepare to receive boarders."

His bonhomie was in stark contrast to Frank's visible tension, as he rushed out for a quick chat with Morison, and a final cigarette, before the hubbub died down. Frank and Morison reappeared.

"Silence in the Court ... All rise for Her Majesty's Counsel!"

Behind the Judge's throne a heavy mahogany door swung open. A bewigged head could be seen preparing to enter the Court. The stern face of Mr Justice Whitford appeared around the door. The determination of our fate was about to begin.

My heart was pounding. Was Frank right? Should we capitulate?

I looked at Peter. He sat there with an expression of supreme confidence on his fine old craggy face. Vanessa was cheerful and suitably glamorous, in a smart dark dress. Frank looked extremely unhappy, Jonathan Cornthwaite pale, Tom Morison unusually glum.

At that moment Christopher Clarke QC, the Lead Counsel for British Shipbuilders, jumped up and asked for leave to cross the court. He wished to seek an adjournment from Morison. Frank went into a huddle with Morison and Clarke. Mr Justice Whitford retreated behind his door.

After a moment Frank turned to me. "They're asking for a fifteen-minute adjournment. Do you consent?"

The whole act seemed over-rehearsed. "Only on the terms stated in the letter I gave you," I replied. "£650,000. Not a penny less than we determined on Saturday."

He returned to his huddle and there was further muttering. He came back to me: "All right ... They've agreed to discuss your terms."

"There's no discussion," I stated emphatically. They must understand that if they go outside the terms of our letter, we'll go straight back into Court. Those are your instructions."

"And remember, we don't trust them an inch," Peter added for good measure. Frank reluctantly went back to his huddle. After a moment he returned.

"OK, they've agreed."

"Well, that's it," was all I could say. The conclusion of six years of struggle, expense, frustration and nervous exhaustion. We tried to keep straight faces: we were not sure whether to laugh or cry, but there should be no display of emotion. Above all, show no signs of relief: there was still some negotiating to be done over the final settlement. But we had won; and on our terms. But how much compared with the award, had we persevered against the advice of our Counsel? Surely only a fraction.

Jonathan Cornthwaite leaned across from the high-backed mahogany bench below me and shook my hand warmly.

"Nerves of steel," he whispered with an understanding smile.

Tom Morison handed me a copy of his 'Short Opening', a withering 14-page indictment of the Hong Kong tender process that he was about to read out in Court. To judge by his expression, he felt we had done the right thing, though I still felt cheated. But that's the Law and Lawyers, and what passes secretly between them.

Most of those concerned – the Defendants and Plaintiffs, lawyers and executives of BS, as well as our own people – were up much of the following night, closeted at Frere Cholmeley, forging the Terms of Settlement. Their cars and drivers remained parked in Lincoln's Inn Fields, with exhaust fumes rising in the freezing night air. The Terms remained confidential, apart from that strange phrase "each side recognises the professional integrity of the other".

For many years afterwards, this mutual recognition held good. Then the Defendants and their supporters started publishing their memoirs. Disparaging remarks about Thornycroft and, particularly Giles – 'the Quack' – and the 'short, fat ship' began appearing in print. This hardly seemed fair play. Having taken early retirement, Rawson and Daniel sought to justify themselves by distorting the historical record in their memoirs – as did Meek. They showed no sign of having learned any lessons, no regret for wasting public money or bringing other distinguished careers to a hurried end; certainly no recognition of our Plaintiffs' 'professional integrity', as claimed in the Settlement. If they insisted on telling their side of the story, I feel justified in telling mine.[54]

Newspaper reports claimed we had won up to £1 million. We had not, but we had to be satisfied with what we won – although if, as implied by Rawson in his memoirs, as Conspiracy, we could have been awarded huge sums on moral and financial grounds. We should perhaps have gone for much more, but the lawyers weren't interested. The Danes supported a quick settlement. In retrospect I can understand why. After all, Queen's (or King's) Counsellors are all appointed by the Crown. They have their own personal priorities. The consequences of what some had even called 'Grand Conspiracy', for the Government, Royal Navy and the MOD, could have been disastrous for all concerned.

At last, with the prospect of real income, we could start work on bigger and faster ships for America, Mrs Thatcher could get on with her privatisation and election, and the Lloyd's Inquiry could begin.

---

54 (1) "After all this time, matters were settled out of court in 1987 on terms that were not to be disclosed but which rumour had it were close to three quarters of a million pounds, plus lawyers' fees that must have been considerable". KJR, *Ever the Apprentice*, p. 114
(2) "We were told that an embarrassing forty-five day trial scheduled for January 1987 (over a year away) was to open and this would affect Mrs Thatcher's expected general election date because senior Government ministers would be involved. There never was going to be such a trial, but the arguments – and costs – dragged on until January 1987 when British Shipbuilders settled. I feel it was probably sheer weariness that prompted the settlement which was on the basis of each side recognizing the professional integrity of the other, and there was no order sought as to damages or costs. There was, of course, reputed to be some undisclosed payment by British Shipbuilders, which I think they were wrong to make. By that time the final conclusion that there was no value in the Giles design proposal was about to be announced." Meek, *There Go The Ships*, p. 194
(3) *The End of an Era* by RJ Daniel RCNC, Periscope Press, 2004

CHAPTER 27

1987: The Anatomy of a Public Inquiry

> "All official committees of inquiry know what they're expected to find. Some of the less intelligent do it a little over-enthusiastically."
>
> Detective Inspector Piers Tarrant in *The Lighthouse* by PD James, formerly a senior Home Office Official

Had we achieved our objective? If mention of Thornycroft or Giles had set the MOD's teeth on edge in the past, our January 1987 High Court quasi-victory must have ground them to dust: '*Grist-bitung*', as an Anglo-Saxon might have said. The Settlement had shown us to be guilty of the unforgivable sin of 'being right' in our suspicions of grave misconduct by the Defendants. It had done nothing, however, to advance our cause on the technical front.

The opposing forces were now even more determined to crush us, and what better instrument than the forthcoming 'independent' Public Inquiry, fought for so hard by Lord Hill-Norton and handed to the MOD on a plate by Mrs Thatcher the previous summer?

However sunny our prospects appeared to be as we trooped out of the High Court into Fleet Street on that cold January day in 1987, we were filled with misgivings.

The warning signs were there the previous June when it was announced that Professor Caldwell had been chosen as Chairman of the Public Inquiry. He was hardly 'independent', being President of the RINA, a Director of BS Engineering and Technical Services Ltd, and, like Marshall Meek, a Director of NMI Ltd, now BMT, with whom we were in dispute over their refusal to endorse their own tank tests.

When asked by the *Times* for an opinion, Lord Hill-Norton had responded: "The choice of Professor Caldwell was most unfortunate. I'm sure he is an honourable man. He is very eminent, well qualified and no doubt will be impartial. But other people will say: 'it is funny they should choose someone who has to do with British Shipbuilders – he is the wrong person to be leading the inquiry'." He repeated his view that a Judge or QC would be preferable, as recommended in his committee's report.[55]

Sir Edward du Cann MP, who as Chairman of the 1922 Committee was instrumental in electing Mrs Thatcher as Party Leader, wrote to her with similar objections. He supported Lord Hill-Norton's suggestion of an impartial person such as an eminent Queen's Counsel, or Judge.[56] The conflict had become as much a human as a technical matter ... "A lawyer is qualified to pass judgement on human conflict: a naval architect is not," he wrote.

In November 1986, the MOD had finally appointed Lloyd's Register of Shipping to undertake the Inquiry. Hill-Norton called me about a preliminary meeting he had with Roderick MacLeod, the Lloyd's Chairman; Garry Beaumont, the Chief Engineer who would serve as Secretary; and Dr Ross Goodman as the lead Naval Architect. It had been agreed that the objective of the Inquiry would also be 'to establish whether an S90 type solution for the Royal Navy's Type 23 frigate programme has advantages and, if so, what they are.'[57] "It's not exactly what I wanted," he added, referring to his view that the significance for future designs should also be considered, "but it's better than a kick in the arse."

I told him of my concern that the MOD was Lloyd's biggest customer. However, Hill-Norton had done his best and there was little more one could say. Lloyd's Register would decide the matter.

*****

Hill-Norton wrote to the *Times* endorsing Lloyd's as "a body whose authority and impartiality are widely respected, and I am in no doubt that they will produce a report which should be accepted by all those who have been involved in this controversy".[58] He called for those concerned to forebear making any public statements until the Inquiry presented its findings.

At the particular request of British Shipbuilders we had agreed that the first formal session of the Lloyd's Hull Design Inquiry should be delayed

55 The *Times*, 28/6/86
56 Letter from the Rt. Hon. Sir Edward du Cann MP to the Prime Minister, 1/7/86
57 DLG diary note of conversation dated 27/10/86
58 Letter, Lord Hill-Norton to the *Times*, 18/11/86

until after the conclusion of the Osprey Case. The BS solicitors wrote: "In view of the likely, perhaps inevitable, duplication of evidence in the Osprey case, it would be inappropriate for anyone concerned with the Defence of the Osprey case to give evidence to your Inquiry for so long as the case is *sub-judice*."[59]

Dinner Party Doubts

The likely attitude of the Inquiry was first suggested at a dinner party given by an ex-submariner friend on February 9th 1987, six weeks before the Inquiry opened. A fellow guest had been at a recent dinner where Roderick MacLeod, the Lloyd's Chairman, had claimed that 'he had no time for the short, fat ship', or words to that effect. Perhaps realising his indiscretion, he tried to retract it, although my host and hostess, Vanessa and others, were also witnesses to the remark. Hill-Norton was appalled when I told him. It was precisely what he had most feared from an MOD-appointed Inquiry: prejudice from the start. Following two weeks' exchange of letters MacLeod wrote: "I can understand that Mr Giles would be upset if he believed that I had formed an opinion on the short/fat ship before the hearings had started. He can be assured in the strongest and most categorical terms that I have done no such thing. Lloyd's is wholly impartial and I have a completely open mind."[60]

The Permanent Under-Secretary of the MOD, Sir Clive Whitmore, wrote to Lord Hill-Norton of "the over-riding importance of the Lloyd's Inquiry being, and being seen to be, absolutely impartial".[61] So the perception of impartiality appeared more important than the reality. Should we have objected more strongly and fought harder for a truly objective arbitrator? I suspect it would have made no difference, incurred further wrath within officialdom and upset Hill-Norton who had done all in his power to arrange the Inquiry. The stage was set for a classic British tragi-comedy of near-Shakespearean proportions – A Tragedy of Errors.

59 Letter, Ince & Co. to Lloyd's Register, cc: Frere Cholmeley, dated 28/11/86 and JGB/HDI to DLG, dated 10/12/86

60 Letter, HR MacLeod to Sir Clive Whitmore, Permanent Under Secretary of State, MOD, dated 24/2/87

61 Letter, PH-N to DLG, dated 2/3/87

The Lloyd's Inquiry Battle-scape

Lloyd's Register of Shipping was founded in 1760, to 'examine merchant ships and "classify" them according to condition'.[62] The purpose of such classification was to ensure reasonable standards of seaworthiness and structural integrity on behalf of the members of the 'society' of insurance speculators that had originally met in Edward Lloyd's coffee house almost a hundred years before. As a consequence of the growth in empire, trade and shipping, Lloyd's Register had moved into an impressive Victorian building in Fenchurch Street in the City of London.

On March 12th 1987, David Jenkins[63] and myself attended the first formal session of the Inquiry at Lloyd's Register. The lofty double doors of the main entrance, of cut glass and mahogany, were surely calculated to impress. We presented ourselves to the uniformed commissionaire who guarded those portals; we felt like interlopers in this inner sanctum of traditional wisdom. We were shown into the famous Board Room, a huge, dark, oak-panelled space without windows, haunted by the ghosts of past shipwreck, disaster, incompetence and blame: it boded ill for the chances of salvaging the S90. Yet we were greeted with good cheer, there being ten of them and but two of us.

Rod MacLeod, the Chairman, exuded goodwill, perhaps conscious of the need to mend fences after the dinner party incident. It was clear to us from the start that the Secretary, Garry Beaumont, would act as Devil's Advocate: no surprise considering he was a member of the RINA Council and Secretary of its London Branch. On that same Council sat many of our most implacable antagonists, including Jack Daniel, David Moor, and Marshall Meek.

We were given the revised Terms of Reference for the Inquiry. These had been finally agreed after Hill-Norton had negotiated revisions directing the Inquiry to examine the possible future significance of the *Sirius* design family, thus allowing it to be adapted to the final Naval Staff Requirement for the Type 23 that had nearly doubled in size and cost from the 25-knot 'armed tug' against which we were competing with the original S90 Validation five years ago. The new Terms of Reference were:

---

62 LR website

63 The naval architect who first worked for us on the design of *Southern Cross III* in 1984 and has led our design teams from then to the present-day. With a profound understanding of the Sirius design family, he is also a most tenacious and scrupulous technician with a delicate sense of humour.

> "To consider the advantages and disadvantages of the S90 hull form for the purposes of meeting the Naval Staff Requirement (NSR 7069) for an anti-submarine frigate (insofar as the current state of development of the S90 permits), taking account of independent assessments by YARD and the Marine Technology Board of the Defence Scientific Advisory Council, and of the Hill-Norton Committee Report 'Hull Forms for Warships' published in May 1986, and to identify any implications for the design of future destroyers and frigates for the RN."

This first full-day Session was used to go through the complex background to the Inquiry: the maritime achievements of the Thornycroft family, the Nelsons, Azteca and Osprey, the DSAC, the OPV-3 ... in fact, everything *but* the S90 Validation and its later rejection by its own source, NMI, on Meek's orders. Mention of the Osprey Case was expressly forbidden from the start. It was clearly reminiscent of our appearance before the DSAC in 1983.

Once the initial pleasantries were over, I was faced with a silent, expressionless audience who made little or no comment on my account of Sir John Thornycroft and the origins of the semi-planing monohull. It was like talking to a blank wall. With the exception of a clearly embarrassed Bill Crago of BHC, their attitude became negative – even hostile.

As Chairman, MacLeod was only concerned with directing the meeting, while Secretary Beaumont was noticeably prickly. Dr Goodman remained silent and expressionless. As the presiding naval architect we expected him to show more interest – as he was bound not to do, considering, as we later learned – that he had already assisted the DSAC in deciding what the Inquiry's conclusions should be. So he bided his time.

MacLeod was wrapping up the afternoon's proceedings when Goodman, surprisingly, interrupted at the last minute: "If I could just double back on one point on the resistance and propulsion calculations and model experiments. Is there anything at all in the S90 hull form, or any reasonable variation of it, which is sufficiently innovative that could not be represented in your view in a model experiment?"[64]

David Jenkins replied, "The only thing that cannot be represented in a model experiment is a comparison of the model with a full-scale ship."

64 The transcripts of the Inquiry Sessions – all of which I have on file – are not a complete record. The 'Formal' morning sessions, which were transcribed, were often followed by an 'Informal' afternoon session to implement whatever had been agreed that morning. At the informal sessions, no official transcript was taken, so there could be no jointly agreed record, only the Inquiry's version.

Goodman interrupted: "Excluding the correlation, just purely on the model experiments?"

"Excluding the correlation, I do not think there is anything, except possibly with water-jet propulsion instead of propellers, but KaMeWa, manufacturers of water jets, have no reservations about that at all."

"So any reasonable hull form with an L/B ratio (length/beam) around the ones you are interested in could be represented to get the characteristics of the speed/power curve. Our problem is solely to decide what is the (1+x) factor."[65]

Assuming, like David Jenkins, he referred to any variation of our semi-planing hull form, I answered: "I think so ..."

"Would you agree Mr Jenkins?" was the response from Dr Goodman ... He did.[66]

We had fallen into his ambush at the first shot. Unwittingly, we had apparently agreed that any 'reasonable' hull, with the same L/B ratio, would also have the same speed/power characteristics as ours, overlooking the importance of its having to be semi-planing and include hydrodynamic lift. That was misinterpreted by the MOD to claim that it was a normal hull with the same 'sinking' characteristics as any conventional hull of the same Length-to-Beam ratio, or 'L/B' ratio, but with a totally different correlation factor. It became the *pons asinorum* – the final verdict, based on semantics, not on science, by which our ideas would be condemned by the Inquiry, merely to support Rawson's seminal falsehood denying hydrodynamic lift.

There was no evidence that the Inquiry had considered a single page of the entire S90 Validation Programme. Thus Goodman, with all his verbal cunning, could never explain the huge difference between the speed-power estimates for the S90 made by the NMI and BHC tank estimates, and the Danish Technical University's full-scale trials on the one hand, and the YARD and MOD estimates on the other[67] – a difference of up to 36 per cent.

How could the Inquiry reach any valid conclusion without having accepted the evidence from TGA's Validation results, including YARD's 'Neapolitan' Speed/BHP chart? (See **Appendix 3**, Fig. 2.) Goodman was asking about the S90 hull form 'or any variation of it', which Meek, having

65 Also known as the propulsion factor. See **Appendix 3** and **Glossary**, p. 360
66 Extract from transcript of the meeting
67 See **Appendix 3**, Figs. 1–2

taken charge of NMI, on a blatant lie claimed 'Giles had rejected'. This loose wording allowed the earlier conclusion of the 1983 DSAC and the MOD to be repeated: "We fully accept the test data produced by BHC and NMI, but we have not accepted all the interpretations Mr Giles has placed upon them."[68] NMI provided our (1–x), not Giles.

In other words, the BHC and NMI tank tests were correct, although completely at odds with the computer predictions of the MOD and YARD for ordinary hulls of the same length, beam and displacement, that continued to sink with speed. The only difference was due to their claim that Mr Giles's exaggeration of the significance of (1–x), was his sole evidence of hydrodynamic lift, a phenomenon which – because they were unable or unwilling to accept, or measure it – they chose to deny. As explained by David Jenkins (**Appendix 3**, p. 304), the MOD/YARD propulsion factors using a (1+x) for conventional hulls was simply a ruse to support their argument: '*No Hydrodynamic Lift*', as shown in **Appendix 3**, Fig. 3.

*****

A week later, Dick Garwin had a long confrontation with the Inquiry. He described his concern about the 'Loose Minute' sent by the Chief Naval Architect of the Royal Navy to the Prime Minister's Policy Unit in 1985, which claimed that he, Garwin, had committed an error in his correspondence with Rawson:

> "I am bewildered as to whether those citing this correspondence had really read my part of it. It is this, too, that makes me wonder whether the Marine Technology Board of the DSAC in their reports of 1983 were properly aware of the ill-founded basis for any Royal Navy evaluation of the S90 concept."[69]

After devoting sixteen out of the twenty-nine pages of the transcript of this meeting to the semantics of the 1981 Garwin/Rawson correspondence, Chairman MacLeod could only come up with this bland statement: "The Hill-Norton Report itself ignored the fact that the end of the correspondence between yourself and Professor Rawson was that you agreed for whatever reason."[70]

68 Admiral Sir Lindsay Bryson: 'The Procurement of a Warship', TRINA, July 1984
69 Letter from RL Garwin to the Hull Design Inquiry, 17/3/87
70 Transcript of HDI Meeting, 20/3/87

The Fatal Flaw: from Inquiry to Inquisition

Garwin pointed out it was *he* who obtained universal agreement by all parties by the force of his argument, and that "his [Rawson's] recantation was couched in such unclear terms that the Royal Navy appears still to support his original error."

In other words it allowed the Inquiry's parties to agree that the benefit of hydrodynamic lift was confined, as Rawson claimed, to 'quite small craft': the opposite of the truth. It was a trick Garwin had often encountered in his long experience of bureaucratic battles. The Inquiry's version was not only false on theoretical grounds, as Dick Garwin had demonstrated in his letters to Rawson – but also supported by Sir James Lighthill, the Provost of UCL, a leading expert in fluid dynamics, of which Rawson was Visiting Professor. It was also shown to be false by the measurements carried out at three major testing-tanks and full-scale trials during the S90 Validation Programme, so insidiously ignored by the Inquiry.

Afterwards Garwin called me in an aggresive state, complaining of the way they were attempting to maintain the illusion that he and Lighthill had agreed to Rawson's initial argument, rather than the other way round. It was one of the worst reversals of the truth he had ever encountered in all his experience of dealing with bureaucracies at their most deceitful. (See context in pp. 86–87 above and facsimiles in **Appendix 5**, Item 11, pp. 318–319; and **Appendix 13**, Fig. 1, p. 337.)

CHAPTER 28

1987: A Whistle Blows from Within the MOD

Two weeks later, on March 26th 1987, a large brown envelope dropped through my letterbox containing a 40-page report, sent anonymously from one of our supporters, apparently within the Navy, MOD, or Lloyd's. The title page described its contents: "**Personal to Members** ... Ministry of Defence ... DSAC 36/86 ... 19 November 1986 ... Defence Scientific Advisory Council: DSAC Response to the Hill-Norton Committee on Hull Forms For Warships ... Loose Minute (Communicated by M Meek)."

As I knew from my 1984 visit to the Leander COs of the 5th Frigate Squadron in Devonport and from Lord Hill-Norton, we enjoyed much support within the serving Navy – particularly after the Falklands sinkings. A later contact within Lloyd's told a similar story – even to the extent of a certain young Lloyd's engineer resigning over the 'charade' of the Inquiry.

This latest 'Response' from the DSAC was in all important respects a repeat of its Report of March 1983. Although now much longer, it lacked any further examination of data other than the most preliminary concept drawings and the earliest tank tests for resistance provided by TGA almost four years previously, in February 1982. It ignored the later results of the S90 Validation Report – in particular the final tank tests for power and sea-keeping carried out by BHC and NMI on the S90 and Osprey, in compliance with the Danish Technical University full-scale measurements of the Osprey *Havørnen*. All the Validation data that we had provided to the MOD and the Inquiry on the S90, the Offshore Protection Vessel OPV-3, the 'Indian Summer' and *Prelude* projects, were ignored. The Sirius design family had come a long way since my rudimentary 15-page paper on its 'Design Philosophy' submitted to the DSAC in March 1983, yet the same cardinal points were maintained against the S90 throughout the Inquiry and within its final Conclusion – apart from Point 4:

1. The design is 'nothing new': it is a perfectly conventional hull form that continues to sink as speed is increased.

2. Excessive power is needed for a given speed and displacement, compared with traditional long-thin designs.

3. The short-wide hull form of the S90 has fundamental drawbacks as a concept for a modern class of frigate.

4. The S90 has inferior sea-keeping.

As it was classified 'Personal to Members' one assumed the Response was not intended for circulation to anyone outside the DSAC or Lloyd's Inquiry. The document was dated only two days after Lord Trefgarne announced the selection of Lloyd's Register, on 17th November 1986. Meek admits to having started work on the DSAC Response long before the Inquiry ever sat:

"We found ourselves yet again having to prepare a DSAC paper in the form of a response to Hill-Norton. I formed a Hill-Norton Working Party and together we got down to a critical examination of his report."[71] It was this Working Party that included Dr Goodman, Senior Naval Architect of Lloyd's, the most important Member of Lloyd's own Inquiry Board. Later he admits: "The DSAC submitted my formal report to the Hull Design Inquiry at the end of 1986."[72] The Lloyd's Inquiry first sat in March 1987.

Therefore, from Meek's account dated 19th December it is probable that MacLeod had read or heard of the DSAC's Response prior to the Hampshire dinner party on 24th November 1986 and that their conclusion had already been reached. When I next met Lord Hill-Norton, I made no mention of the report sent by the whistle-blower. Naively, perhaps, I still believed in the possibility of a fair assessment – and I knew it would upset him greatly. After the 'Dinner Party' issue, I had no wish to torment him any further. However, after the Inquiry had completed it's work, I thought it fair to acquaint Hill-Norton with this major aberration from the Terms of Reference. He was furious about it – as became apparent in his Foreword to this book: *"It was typical of the behaviour of such powerful vested interests before, during and after the [Lloyd's] Report was published."*

After some deliberation, in June 1987, the Inquiry proposed a design process for its new iteration of the S90 involving three 'Studies'.

71 M Meek: *There Go The Ships*, p. 207
72 Ibid, p. 208

To compare the S90, according to its final design, with the current Naval Staff Requirement (NSR) 7069.

- If deemed necessary, Lloyd's would develop a geosim (a geometrical enlargement) of the S90 hull, later known as the S-102 to meet the latest NSR.
- To compare this version with the existing Type 23 design in its ability to meet the NSR and its future significance for warship design, as per the Inquiry's Terms of Reference.

During an 'informal' session in late May 1987, Lloyd's showed us a preliminary sketch of the first version of their S-102, which they proposed to compare with the most recent Type 23 design. It was just as Peter Thornycroft had warned us: they would take our *worst* and compare it against the best and latest version of the Type 23, then approaching twice the size of S90. "Comparing apples with pineapples," he called it. We therefore insisted on being allowed to develop our own design proposal to satisfy the latest NSR, using all the more recent data at our disposal. This could then be compared with the Lloyd's version for the Inquiry's final decision.

This aroused strong objections from the Inquiry. MacLeod's response showed that we had raised a central issue upon which their decision had already been made: "It is difficult to define what the degree of latitude is without raising the question of when the S90 concept ceases to apply."[73] Lloyd's instead proposed to develop their own design for comparison over the next 4–5 weeks; there would be a series of Formal and Informal meetings with TGA and Per Sorensen for comment and advice; and then a period in which Lloyd's would produce their final iteration.[74]

Having proposed this plan, Lloyd's did not stick to it. There were no 'Formal' sessions for over four months, the next taking place on 9th September 1987. In the interval all sessions were 'Informal' and therefore not subject to record. Previous Formal transcripts had been released shortly after each session, and in the end we had to threaten legal action to obtain the transcript of the Formal session of 29th April 1987, in which we had tabled such strong objections to their obstruction of the new Terms of Reference. It was not released to TGA until January 12th 1988, nearly ten months later.[75]

73 Ibid, p. 16–17
74 DLG diary notes of criteria agreed at 29/4/87 HDI meeting
75 Letter, DLG to JG Beaumont, dated 11/1/88

The Inquiry's manoeuvre seemed to be intended to remove any official record of our objections to their S-102 design. Other conditions were designed to demoralise us and keep us guessing. Although the two Davids were able to obtain security clearance to inspect the latest Naval Staff Requirement, the vital engineering standards, the weight and materials specifications and the weapons fit for the larger 'gold-plated' Type 23 were denied us. We felt that we could demonstrate better compliance in some of these areas, if only we knew what those standards were.

Still, to match the capabilities of the latest Type 23, we proposed to produce a design study for a larger ship, the S-115, 115 metres in length, with reduced beam due to the narrower and lighter gas turbine engines than the heavy commercial marine diesels proposed by TGA for the S90 to save machinery and fuel cost.[76] Lloyd's informal notes stated: "agreement could not be reached on this matter."[77] My notes read differently: they *did* agree; moreover they offered £7,000 for the design work and paid TGA accordingly.

I had kept Peter Thornycroft informed of our progress and cheered him with the good news that, following the crucial April 29th meeting, we would be allowed to submit our own 'iteration' to answer the latest 'gold-plated' NSR, but only under the watchful eyes of the MOD. He seemed in good form, but reminded me that we should keep our fingers on the pulse of the Inquiry: "Maintain control of the design, otherwise Bath will wreck it for the Inquiry."

That was the last advice he ever gave me. Less than a week after that meeting, Arthur Mursell phoned early in the morning: "The Commander died yesterday, in his flat over the Pilot Boat Inn. They found him looking very peaceful, with a glass in one hand and a telephone in the other. It seems he just went out like a light." At least he had been cheered by the good news of doing our own enlarged S-115 design for the Inquiry.

What could one do or say? First, for some reason, I wrote a note in my diary: "PT: His soul goes marching on." Next, I sent a handsome donation from Osprey Ltd to the Coxswain and crew of the Bembridge Lifeboat, of which Peter was Chairman, for them to spend "in whatever way they felt Peter would have considered most appropriate". I sat down and wrote an

76 HDI Notes of informal afternoon session, 7/5/87, p. 273

77 Ibid. This contrasts with Meek's later claim that we accepted the Inquiry's decision on allowing no change to L/B, as in the Inquiry's 'S-102'. *There Go The Ships*, p. 209. Then why did they sanction and finance the design of S-115?

obituary for him and sent it to the *Times* – the newspaper that, above all others, had espoused our cause. I concentrated on the best things about him, emphasising his particular type of genius. It was published with few alterations.

With Frank Presland and Per Sorensen, Vanessa and I attended Peter's memorial service in Bembridge Church. It was a salty occasion, conducted in true Peter-style, with a large number of Nelson-owners, the lifeboat crew and numerous Isle of Wight and Royal Yacht Squadron characters. We sang the stirring old seafarers' hymns, such as 'For Those In Peril on the Sea', and heard the traditional lessons.

At the end of the service, the customary understated English memorial outpouring, 'Death is nothing at all ...' was read. Of course Death is usually a considerable 'something': certainly for me in this case. But the closing words, 'All is well', were a fitting reminder of the eternal optimism that typified Peter Thornycroft.

Afterwards we retired to his favourite haunt, the Bembridge Sailing Club, scene of so many happy occasions in the past, and the unofficial Board Room of the S90 Club. He might have made all the arrangements in advance, although it was obvious that his secretary, the redoubtable Jill Swallow, had contributed greatly to the organisation.

The *Daily Telegraph* published an obituary with a strange reference to the Osprey Case: "Later, as a partner in Osprey Ltd, Thornycroft designed a patrol boat which interested British Shipbuilders, who asked for a copy of the design drawing. A chance telephone call from an employee in a Ministry of Defence establishment asking for further details of Thornycroft's patrol boat design eventually led to a protracted legal action for breach of copyright." A curious distillation of the truth.

Peter's death affected me profoundly. Now I was on my own, although with the tireless and convinced support of David Jenkins and David Bridges to sustain the technical arguments and the design work. But Peter had so often been the helmsman who steered us through the worst storms, and we were now in the centre of another, perhaps the most menacing of all. I, for one, needed his reassurance and guidance – to nudge the tiller, to add a touch of his inimitable humour to deflate my indignation – to keep me on course. I could so easily play a hand too hastily, or utter a word too far. He was a great pilot and he knew the treacherous waters of the RCNC and the Navy, almost as well as those of his birthright, the Solent and Spithead.

His design genius was uncanny, his 'eye for a boat', his instinctive sense of the way that a certain shape would behave when transformed into a hull moving swiftly through the water. Without friends like Peter I would have lost my way long ago. My old mentors were fading away: Peter Salkeld had died the year before, now Peter Thornycroft – and I feared for Charles Adams, who was recovering from a second stroke.

Yet we retained a core of loyal supporters. The 'two Davids' were rising to the occasion, providing an enlightened combination of naval architecture and engineering expertise. Dick Garwin and Karl Kirkman would yield no ground to technical or semantic obfuscation. John Moore, Roy Corlett and Nigel Ling were resolute, while Per Sorensen would deliberate over the vagaries of ship construction with his usual solemn, very Danish, authority. Behind us stood Lord Hill-Norton: testy, irascible, difficult, some claimed; but inflexible in the pursuit of a just cause.

After Peter's death an episode of the popular detective series *C.A.T.S. Eyes* was shown on ITV, called 'A Naval Affair'. It concerned an obsessive ex-aircraft designer of a 'short, fat' frigate. His house was burned down, his plans stolen, his life endangered by a bomb, his sailing boat sabotaged and, finally, his marriage having been wrecked by an unscrupulous member of the MOD's design department, he had a nervous breakdown and took to drink. It seemed more than mere coincidence that much of the dialogue – about the Falklands and the need for higher radars and 'short, fat' frigates with greater stability – might have been taken from our own saga. The series was made by TV South, whose News and Current Affairs team had filmed our S90/Leander 1/10th-scale-model sea trials, so no doubt the story would have circulated at the television station. However, it may well have seemed to the MOD, that I had arranged it in order to curry favour with the public. It certainly did *not* improve our reputation within Whitehall.

In July an article by Andrew Marr, then Political Correspondent of the *Independent*, was published in which he stated, "Ministers now effectively consider the row dead."[78] I had lunch with Marr, who was charmingly dismissive of any hopes of the Inquiry finding in our favour: the MOD had already reached its own conclusion. His article aroused the wrath of Lord Hill-Norton, who wrote to Defence Secretary George Younger and Mrs Thatcher, complaining of the MOD's public rejection of the S90 while

78 The *Independent*, 21/7/87

the Inquiry was still in progress. This inspired a second article by Marr, in which he finally did trawl through the most egregious aspects of the saga, in particular the theft of our designs by BS and their secret testing of them, leading to the Settlement of the Osprey Case.[79]

With Hill-Norton's agreement, I complained to Defence Secretary George Younger and sought his assurance that Marr had been misled.[80] Younger's Private Secretary wrote to me on 4th August assuring me that Marr had been 'wrongly advised' by his contact within the MOD. This 'contact', Marr later told me, was none other than Lord Trefgarne, the Minister of State responsible for the conduct of the Inquiry.

79 The *Independent*, 28/7/87
80 DLG letter to Hon. G Younger, Defence Secretary, dated 31/7/87

CHAPTER 29

1987: Trials and Errors

After several turbulent sessions over niceties of hull design, the Inquiry finally agreed that the NMI/Danish calculations of the 0.97 (1-x) correlation factor should be put to the test. Using the Alan Bond yacht, *Southern Cross III*, which had just arrived in the South of France after a voyage from Australia, the Inquiry Committee proposed conducting a full-scale speed trial. This would be compared with tank tests, using an exact model, to be run a few months later in the MARIN tank in Holland.

Thus the tortured question of the correlation factor, upon which they claimed the 30–40% gulf in differing S90 powering predictions depended, would finally be settled and Giles's 'misinterpretation' revealed.

In early September 1987, Vanessa and I met David Bridges aboard *Southern Cross III* in Toulon. It was a perfect day: cloudless, flat calm and only a breath of wind as we stormed past Toulon towards the French Navy's trial course between the Île du Levant and Le Lavandou. We made a few test runs over the measured mile for calibration of the engine instruments. In excellent visibility the distance marks stood out clearly as we tested the various ways to determine the precise time over the course. We had the use of two radars, one short-range, one long-range, and the ship's Doppler Log as a double check on the radars. Finally there were three observers to check those measurements and obtain their own bearings as we made our transits of the measured-mile markers. This was all before the days of a GPS of sufficient accuracy.

According to the official Trials Report, "The French Marine Nationale gave a verbal assurance that the measured distance markers were visible and kept in good order, together with confirmation that adverse weather conditions were unlikely to occur."[81] The weather conditions for the critical high-speed runs, upon which everything depended, were: air temperature of approximately 25 degrees C., sea calm, barometric pressure constant at around 1014 to 1018 for the past six days, and a wind speed of 6 knots.[82]

81 *Southern Cross III* Speed Trials Report, p. 4 and see **Appendix 14**
82 Ibid. pp. 13 and 14, Tables 3 and 4

'Poor visibility': Taken as we left *Southern Cross III* to go ashore before the high-speed trial. David Bridges standing on the right. Le Lavandou is about a mile away in the background.

At about 3.00 pm, in the middle of the early trials, the Inquiry's representative suddenly ordered Vanessa and myself to leave the yacht using one of the motorboats carried aboard. It was, he said, a condition demanded by the MOD, which was paying for the trial. Had I been told that I would be prevented from witnessing the conduct of the full-power trial, I would have done my best to stop it, but it was too late. Lloyd's, however, agreed that David Bridges could remain aboard in my place. We were dropped ashore at Le Lavandou. Later I learnt that David Bridges was ordered to stay below in the engine control room to check the engine readings and thus was unable to observe the bridge readings or view the outside world during the full-power trials.

From Le Lavandou beach, Vanessa and I watched our fine vessel completing her trials. Although from that distance we could not see the course markers, there was no sign of any sudden reduction in visibility. When the trials were over we returned to our hotel.

David Bridge's phone call that evening came as a shock. The speed trials had been 'unsatisfactory'. The Lloyd's people had expressed doubts about their accuracy 'due to poor visibility over the course'. David couldn't contradict them, having been confined to the engine control room. The photographs we took from our vantage point on the beach show the conditions as clear. However, the official Report (which we were not allowed to see until after the Inquiry's findings were published) claimed, "The weather conditions under which the trials were conducted precluded the normal visual method of determining the distance run by the vessel. Therefore, in Lloyd's Register's opinion, little confidence can be placed in the ship speeds determined from these trials."[83] And later: "At 1515 hours the vessel proceeded along the measured distance in order to familiarise the time-keepers with the distance marks. During this time, a haze began to develop over the coast such that by around 1530 the distance marks and principal features on the land had become obscured."[84]

When David returned to the bridge there was no sign of any fog. Naturally Lloyd's took account of this in the Report: "The haze which had obscured the sighting of the distance marks during the speed trials cleared at dusk."[85] The effect of the 'haze' is shown in the photograph below. If there was a haze at all it was an intellectual fog conjured up by the Inquiry to obscure an inconvenient truth. The results of the *Southern Cross III* tank to full-scale Correlation tests and Commentary are given in **Appendix 14.** When it came to the tests at the MARIN tank in Holland, the credibility of the Inquiry was again undermined. Lloyd's, with Dr Ross Goodman in attendance, ran the tests on a model of *Southern Cross III* in precisely the same conditions. The results were even worse than 19.6 knots, the lowest of the two suggested trials speeds.

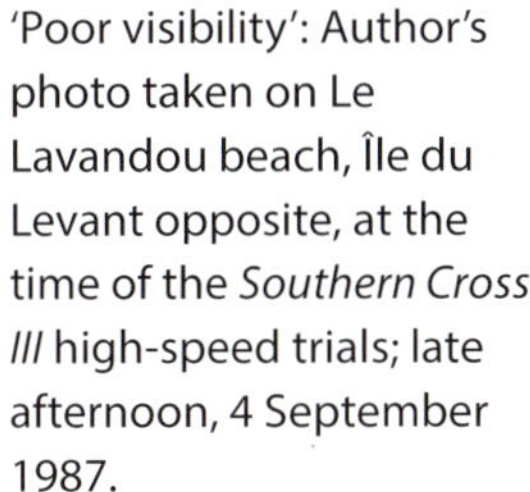
'Poor visibility': Author's photo taken on Le Lavandou beach, Île du Levant opposite, at the time of the *Southern Cross III* high-speed trials; late afternoon, 4 September 1987.

Later I attended some of the MARIN tests with David Jenkins who, with his long experience of working with BHC on semi-planing hulls, could not agree to the testing procedure. The results, he said, were from errors in the towing and calculation techniques arising from their overlooking the

83 Ibid, Introduction, p. 3
84 Ibid, p. 8
85 Ibid, p. 9

dynamic lift in the semi-planing hull form; a similar story to the Osprey tests run by David Moor at Vickers.

On his return to London, David Jenkins contacted Dr Goodman to protest vigorously at the apparent 'rigging' of the results. David was fuming about this altercation when I saw him at the Inquiry's February meeting, and was still upset thirty years later. The *Southern Cross* trials appeared to have been a complete waste of time and public money.

The S115: Developed 'Insofar as the Current State of Development of the S90 Permits'

The Inquiry's final act was played out in the New Year of 1988. True to its unspoken philosophy, the Committee arranged the last pieces of evidence to suit the verdict that it was determined to reach. The question before it was whether the S90 design would meet the latest NSR better than the Type 23 Frigate to which the MOD had given its blessing. As the reader may recall, we had designed the S90 according to specifications already out of date when the MOD gave them to TGA in January 1982. Since then the MOD's preferred supplier, YARD, had developed a design according to the latest specifications. We had not been allowed to update ours. When we asked to do so, the Inquiry stated that they would prefer to update our design themselves. The result, known as the S-102, was predictably lacking in various departments. We demanded therefore to produce our own enlarged design: the S115, for which the Inquiry agreed to provide £7,000. According to the revised Terms of Reference, MacLeod agreed to an Informal meeting, to take place on 27th January 1988, in order to resolve this issue. We saw it as a last chance to present our objections to their S-102 and present what we considered to be the most balanced answer to the NSR: the S115.

I phoned MacLeod to tell him that producing the S-102 for our inspection only days before sending it to the MOD as the 'definitive' interpretation of the S90 hull form was against the terms of reference agreed between the MOD and Lord Hill-Norton. If he were to proceed with his plan, TGA would resign from the Inquiry and go public over its conduct. He would not budge: there seemed an almost religious conviction in his voice as he told me it was his 'public duty' to follow his elected course of action.

The atmosphere at the subsequent Final Session was hostile. MacLeod was determined to maintain that while 'completely impartial in this matter' he would allow no further design refinement.

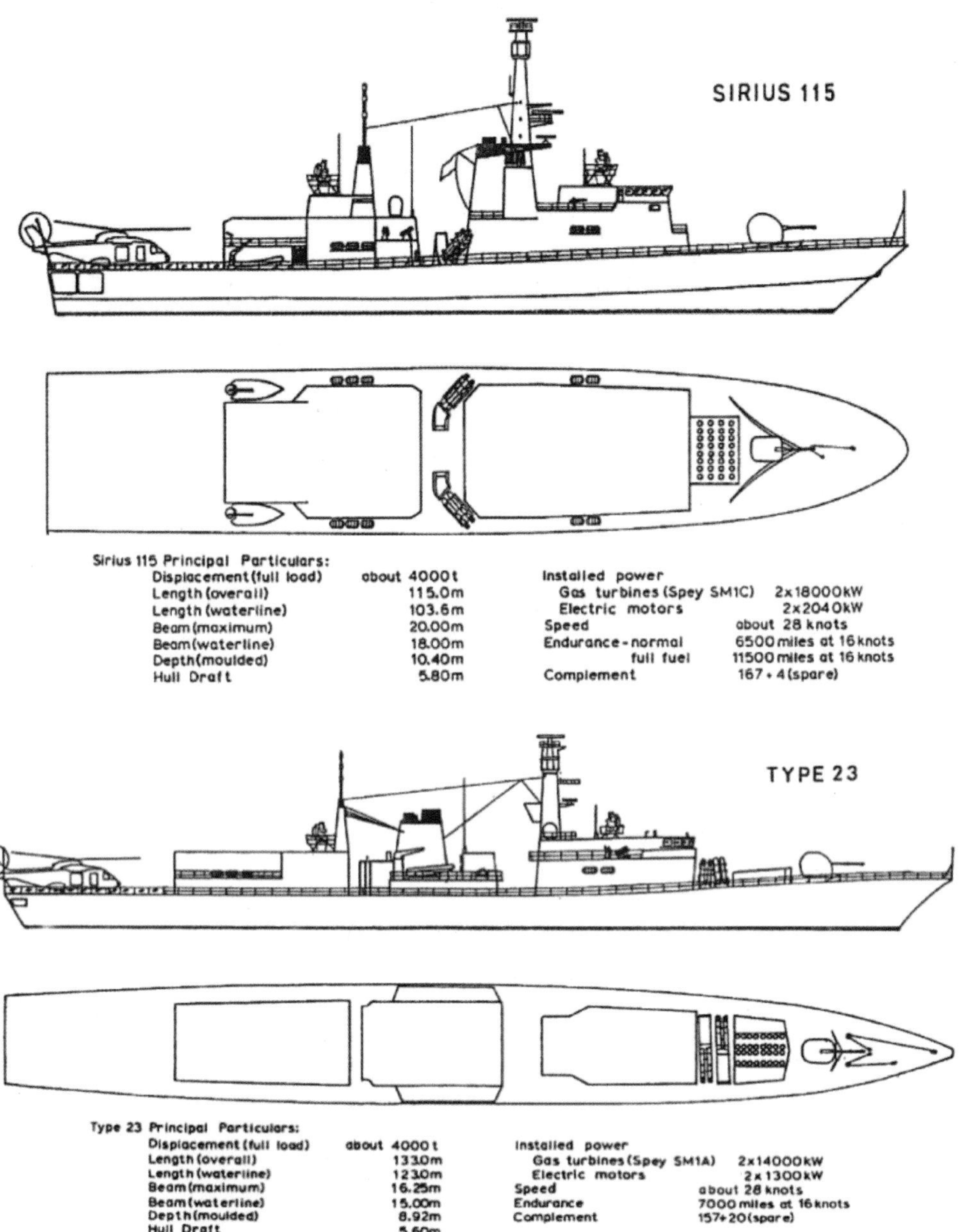

Comparative outline and details of the TGA S115 and Type 23 Frigates (to scale), showing their contrast in length and beam. Slightly more power is required by the S115, but its advantages in sea-keeping, increased speed potential, stability, reduced cost, useful internal volume and radar height are clear. The S115 bears more of a resemblance to the RN Type 45 destroyer than the preceding Type 23. (See **Appendix 11** and final Chapter 38 p. 284, on RN ship design following the Falklands War.)

MacLeod was in a difficult position. The technical discussion was beyond his expertise and his remarks became ever more ineffectual as he struggled to control an increasingly chaotic scene. As reproduced from the Meeting Transcript: "I would not accept for a moment that we are looking for reasons to dismiss anything. What we are setting out to do and what we have done from the start is to make an *impartial* technical assessment and judgement. That has not varied and it will not vary."[86] It sounded impressive enough, but the reality fell far short.

I asked the Inquiry whether the S115 would satisfy the latest Naval Staff Requirement. On every point except one they had to agree, in a grudging way, that it probably would. I asked the key question: "Does the S115 go further towards meeting the NSR than the S-102?" Then he gave his first answer: "In terms of speed and power it does. Yes."

"No," I replied. "In terms of the NSR?"

Beaumont was in a sticky position: "One has to go through each parameter again and make an overall judgement ... The overall judgement is that there is an improvement in terms of speed and power and therefore a marginal improvement in costs, but that has been done at the expense of those 'plus' attributes attributed to the S-102 as regards to shape; they have now fallen away to some extent and revert back towards ..."

I interrupted him: "... but those attributes are not specified in the NSR."

"I agree, but we have been balancing all the attributes which you have claimed for the S90 as being factors in the equation," he replied.

"But you were addressing the latest NSR, were you not; and the degree to which the S115 answers it or does not? I mean, in terms of sea-keeping, does the S115 now not meet the NSR?"

Beaumont: "It meets it."[87]

Having argued in circles, I came to the *dénouement*:

"Are you, notwithstanding that (i.e. the superiority of the S115 over the S-102 in meeting the NSR), still intending to go ahead using the S-102 as the basis for a comparison with the Type 23?"

Beaumont: "As a basis, yes."

Giles: "Would Mr MacLeod confirm that?"

MacLeod: "Yes."

86 Ibid, p. 8B
87 Ibid, p. 36 and 37

Giles: "And that the S115 comes closer to meeting the terms of the NSR, which I think we have agreed, have we not?"

Beaumont: "Yes."[88]

I made my final statement to the Inquiry: "Obviously we have come to an impasse at this stage and I see no benefit in going any further. It is perfectly clear to me that the S-102 will be the design that you intend to use as the basis for a comparison with the Type 23. That is contrary to the spirit of the Inquiry. I believe, or I did believe, that the whole purpose of the Inquiry was to find the best solution to the NSR for comparison with the Type 23 ... I believe that in spite of the small increase in fuel consumption and in power required (which is something we have always accepted) the S115 would provide a better ship. I think my two technical colleagues would agree with that. It would provide a better ship for the job; a better ship from the point of sea-keeping, from the manoeuvring point of view, and from the fighting point of view. I do not think the S-102 could do that because it would have too much stability and too little speed ... I don't really see any purpose in continuing this conversation any further."[89]

88 Ibid, p. 41D–G
89 Ibid, p. 47A–E

CHAPTER 30

1988: Breaking off the Engagement

How could we continue to co-operate with the so-called 'independent and impartial' Lloyd's Hull Design Inquiry when the verdict seemed a foregone conclusion? Hill-Norton urged me to take immediate action, as it was clear that their minds were set upon proposing an inferior version of our design for comparison with the Type 23. He suggested we tell them either to consider the S115 as the comparative study and do a detailed assessment of it; or, if this were unacceptable, to give good reasons why not: "If they do not, you may have to go to Court."

A big decision was needed: should we seek a compromise or 'shake off the dust from our feet' and walk away from the whole affair?

I was tempted to walk away. I had recently been in Washington for meetings at the Pentagon to present our 'SuperShip' concept to the US Navy, and there was enough encouragement there, and enough frustration in London, to make me think of establishing a company in the USA. But, largely out of respect for Lord Hill-Norton and a forlorn hope that the MOD might be persuaded to be reasonable, I decided to pursue the matter of comparative tank tests with the MOD. The Inquiry had already admitted that, on most material points, our S115 would satisfy the demands of NSR 7069 – in some cases better than the Type 23 and in every case better than its S-102. Perhaps it was worth slogging on to the end of this depressing business, if only to say that we had done everything possible, that we had not given up. Our final priority was therefore to establish whether or not our hull behaved like a traditional design over the tortured question of what the MOD called 'hydrodynamic lift'. All that was needed was to test a model of the S115 against a model of a Type 23 – or any other conventional hull – in the same tanks, under comparable conditions, to see to what extent the two models would sink or lift, precisely as suggested by Hill-Norton. In other words, was the S115 a conventional design or not? If such a solution was not forthcoming, we should dissociate ourselves from the Inquiry. Frank Presland assured me that, on the evidence, any Court of Law would decree that the S115 was a reasonable development of the S90 hull form, within the Inquiry's Terms of Reference. We dug in our heels, with a telex to the Inquiry:

> "TGA do not, repeat not, agree to any information concerning their design being passed to the MOD until the Inquiry has given to TGA valid reasons for their refusal to undertake a detailed assessment of the S115 or an equally compliant development of the S90 hull form for the purposes of meeting NSR 7069."[90]

A copy of this telex was sent to Percy Craddock, Chairman of the Joint Intelligence Committee in the Prime Minister's Office, with a request to halt any further action until the question of the S115 was resolved. We informed the Inquiry that, if we heard nothing by close of business on 29th March, we would call a press conference at 10.30 the following morning to bring the whole matter to public attention.[91]

By close of business on the 29th, '... no such undertaking had been received.'

I faxed my letter to MacLeod, formally withdrawing our co-operation from the Inquiry. I enclosed copies of our Press Release and a detailed statement that we had agreed with Lord Hill-Norton. Finally I asked for the return of all documents concerning confidential projects and data.

As arranged with the media, David Jenkins, David Bridges and I met at The Tower Hotel in St Katharine's Dock the following morning and handed out our Press Release together with pictures and data on the S115. I gave a short account of the reasons for our withdrawal. That evening and the following morning there were a number of articles in the newspapers that, in general, gave a balanced account of events.

Hill-Norton sent a further endorsement of our actions. "Thank you for the enclosures," he wrote. "I have no fault to find with them. It is most unfortunate that these circumstances have arisen in spite of our efforts – mainly yours – to prevent them."[92]

Our tactic succeeded. Younger's Private Secretary wrote, saying that we would be sent a copy of the Inquiry's Report shortly for comment – but only on the understanding that we would not release any of its contents. This was exactly what we had requested so many weeks ago, before incurring

90 Telex DLG to JG Beaumont, dated 28/3/88
91 Note by Frere Cholmeley on conversation with DLG, dated 28/3/88
92 Letter, PH-N to DLG, dated 6/4/88

the heavy legal costs involved in delaying its release. The Confidential Annex would also be sent to us on the same terms.

On 11th April 1988, I was informed that my copy of the Lloyd's Report had arrived at the Royal Hotel in Rosslyn, Scotland. I was taking a week's break with my family, in a small part of the Castle rented from The Landmark Trust. Due to its confidentiality the Report could only be delivered by hand to a destination where someone was always available to acknowledge its arrival. Thanks to all the CONFIDENTIAL stamps and MOD markings on the large brown envelope, the lady behind the reception desk thought it must be something of national importance.

Having signed the receipt, she handed it to me nervously – as if it were the plans for the latest nuclear submarine.

So there it was: a large brown envelope on the table in the huge vaulted kitchen of Rosslyn Castle. It was like the arrival of a school report that I knew would be a disaster. Why look? I knew it would echo, in detail, Meek's 1983 DSAC response. Since we had resigned from the Inquiry there was no need to take any action: whatever was said would be supported by the flimsiest of evidence, just like the original DSAC report. Why not just toss the whole thing into the roaring flames of the great fireplace? But to be fair to Lord Hill-Norton and those who had fought so hard for the Inquiry in the first place, I felt it my duty to take a deep breath and open the envelope. It contained a weighty document in a powder blue cover about the size of a telephone directory. I flipped through the 260-odd pages. There were only 6 or 7 pages of figures, largely theoretical, not one of them from any of the measurements by BHC, Lyngby or NMI that were fundamental to the S90 Validation Programme. It read more like a Manual of Naval Architecture or a much-expanded version of the DSAC Report. It had little to do with the agreed transcripts of discussions that took place between TGA and the Inquiry. It was no surprise that the design used for comparison with the Type 23 was Lloyd's own S-102. The summary claimed the S115 was also considered but, as confirmed at our final Meeting, it certainly was not.

SUMMARY OF PRINCIPAL CONCLUSIONS AND RECOMMENDATIONS

It is the Inquiry's firm conclusion that the Sirius design concept is unable to produce as good a solution to the NSR 7069 as the Type

23 and that the MOD were correct in their rejection of the S90 design proposal in 1983. The Type 23 solution is cheaper both to build and to operate and has a better overall performance than can be achieved by a vessel based on the Sirius concept. The Inquiry further concludes that the advice given to the MOD by YARD along with that of the Hull Committee of the DSAC in 1983 was also correct with regard to their overall conclusions.

The Inquiry concludes that the hydrodynamic performance of a Sirius hull form can be adequately assessed by conventional naval architectural techniques. The claim that a Sirius hull form does not obey conventional principles has been based on data being incorrectly interpreted. In particular the Inquiry finds that in regard to the propulsive performance of the S90 hull form, it requires substantially more power to achieve a given speed than that suggested by the Hill-Norton Committee.

The Inquiry's terms of reference require it, after considering the advantages and disadvantages of the S90 in relation to the NSR 7069, to identify any implications for the design of future destroyers and frigates for the Royal Navy. Based on its assessment of the claimed exceptional hydrodynamic performance of a Sirius hull form, along with its comparison of the S-102 and S115 against a specific frigate requirement, namely the NSR 7069, the Inquiry has concluded that there are no identifiable implications for the design of future destroyers and frigates for the Royal Navy arising from the Sirius design concept, as far as it is able to foresee the likely role and operating requirements for future warships of this type.

Accordingly, the Inquiry considers, and would recommend, that no further action is required to assess the applicability of the Sirius concept for future frigate and destroyer design.

*****

So there it was. Half a lifetime's work, recognised by many as pioneering and cost-effective, dismissed as a waste of time and money in four short paragraphs. As Meek claims, the Inquiry's conclusions were decided by him and his DSAC Committee and sent to the MOD at the end of 1986, before it had ever convened. All the subsequent detailed examinations, reports, tank tests, sea trials, meetings, and the reported expenditure of 'at least £1 million'[93] as he claims, might as well never have taken place.

I sent the TGA Response to the Secretary of State for Defence, George Younger, requesting a meeting before the Report was released to the public. We also sent a copy to Hill-Norton in preparation for a meeting he had arranged to have with MacLeod to express his concern at the way the Inquiry had gone. Hill-Norton had sent a copy to Sir Michael Quinlan, the Civil Servant newly-appointed as Permanent Under-Secretary at the Ministry of Defence, asking him "to take charge of the next steps" himself, pointing out that "in its present form the LR Report will not put an end to the whole controversy."[94]

For the final meetings, at the MOD in London, Whitehall was to field its Top Brass. This included Sir George Younger, Secretary of State for Defence; Sir Michael Quinlan, Permanent Secretary of the MOD; and Admiral Sir Derek Reffell, Third Sea Lord and Controller of the Navy, with assorted attendants. We prepared to discuss the central question of 'hydrolift', as Hill-Norton termed it, to establish whether the Sirius family was conventional or not. We took all the most egregious statements from the Report and analysed them word-by-word in a thirty-page Commentary to be presented to Mr Younger.[95] (A concentrated version is given in **Appendix 5**.)

Jon Boyes, a retired US Navy Admiral with whom I had been working on our US project, provided a lucid four-page statement on the importance of identifying the possible advantages of hydrolift in *future* warships of the US and Royal Navies, possibly for very different missions.[96] There were to be two meetings on June 30th. The first, in the morning, with Younger, Quinlan and Reffell; the second, a short afternoon session with Lord Trefgarne, the Procurement Minister responsible for the conduct of the Inquiry. Our two objectives were to propose comparative model tests to clarify the unanswered questions of 'hydrolift' and whether the Sirius hull was, or was not, a conventional form; second, to force a truce to delay

93 'There Go The Ships', p. 209,

94 Letter, PH-N to DLG dated 23/4/8

95 See: TGA Commentary on the Conclusions of the Lloyd's Hull Design Inquiry' **Appendix 5**, 5.

96 Jon Boyes' statement confronted the same prejudice as Alan Clark, MP, who replaced Lord Trefgarne as Defence Procurement Minister. Clark wrote in his 10/11/89 diary entry: 'I am in despair about the Navy or, rather, the sailors. This is the service which has to be the centrepiece of my plan – swift, flexible, hard-hitting. Yet the only thing they want is to be a forward ASW screen for the US Navy in northern waters. That's all over, I say: Forget it. The Soviet 'threat' no longer exists. Have not any of them read Mahan? Or Arthur Mader?' (p. 262, 'Mrs Thatcher's Minister').

the publication of the Report while the MOD considered our Proposal. Our plan for the tests was simple:

> The Inquiry, with our co-operation and that of the MOD, should undertake a short programme of bare hull resistance tests, using models at the same scale of the Type 23 and the S115 hulls, in the same tank and in comparable conditions of displacement, at scaled speeds of up to 50 knots: the only way to obtain a valid comparison. In addition to resistance, 'hydrolift' should be measured, for sink and rise, as employed by BHC and NMI, to identify any difference in these characteristics between the models. This very test was performed by the US Naval Academy 27 years later, between semi-planing and conventional hulls at speeds up to 40 knots, on YouTube. The results were as predicted by the S90 Validation: the conventional hull's stern sinking severely with speed, while the semi-planing hull rose slightly with level trim. (See Chapter 38, p. 290.)
>
> The Rawson/Garwin argument could likewise be settled by comparing the existing model of *Southern Cross III* or *Havørnen* with an existing model of Rawson's 'Ton' class minesweeper (which is of similar size and proportions) at the same displacement and at speeds up to 35 knots. This would not be conducted by the MARIN tank employed by Lloyd's, but by a truly independent tank such as Southampton University's Wolfson Unit. NMI was now clearly out of the question, Meek having been appointed Director and having himself rejected its S90 test results.

We considered the first test programme would cost about £40,000 – a fraction of the £1 million reportedly spent on the Inquiry to date. The second, whose cost would be much less, since models were readily available, would provide a second opinion.

Appealing Unto Caesar

I retain a blurred vision of our meeting: a crowded room with a green carpet, looking out over the Thames – it must have been Younger's office. We filed in to be introduced to a bland but welcoming George Younger with his bustling PA. The grim face of Sir Michael Quinlan contrasted with the cheerful seagoing demeanour of Reffell, while a posse of minor officials hovered in the background.

Our party included Dick Garwin, Karl Kirkman, who had not been able to change his suit since he fell out of the overnight flight from Washington,

Per Sorensen – equally generously proportioned and similarly ill-clad – and the two Davids, blinking in disbelief in these surroundings of unaccustomed grandeur.

Having introduced our team, I read out each of the Inquiry's objections, inviting our attendant experts to speak about the relevant slide on the screen before us. Our audience sat as impassively as bronze statues in a sculpture park. I stated that, in our unanimous view, the Hull Design Inquiry had produced a flawed Report. We suggest that, before its publication, our proposed short comparative hull test programme should be carried out. It was agreed that we should raise this Proposal at our session with Lord Trefgarne in the afternoon.

We then presented a number of measurements from the NMI S90 test series supporting our claims, as well as data from our response to the Lloyd's Inquiry. (See **Appendix 5**, Figs. 1–4.)

In the afternoon we met Lord Trefgarne. Jon Boyes did his best to pour oil on deeply troubled waters, while I made it plain that, if necessary, TGA would go to Court to protect its reputation against the fallacies (such as Rawson's) perpetuated by the Inquiry. Trefgarne listened to our request for further model tests with an expression of supreme boredom. All he could volunteer was that the MOD would consider our request.

It was obvious that no test results we could have analysed, no diagrams we could have shown, and nothing we could have said, would have altered the script the DSAC had written in 1983 and re-written at greater length in 1986. It seemed as if the public execution of the Sirius was our punishment for having persevered, financed and won the Osprey Case.

I was reminded of the statement of Lord Cohen, when we met, a quarter of a century earlier. As former Lord of Appeal and Chairman of the inconclusive and obfuscating Inquiry into the de Havilland Comet 1 disasters, he had said to me in hushed tones: "I was only the Judge – the Experts were the Jury" ... as they were on this occasion.

Would it have made any difference if, as Lord Hill-Norton advised, the Inquiry had been put under the authority of a 'learned judge'? On the authority of Lord Cohen, to say nothing of PD James and Detective Inspector Piers Tarrant: probably not. But, in the best tradition of those who 'Defend the Faith' in Her Majesty's Government, it worked out well for some: a few months later Roderick MacLeod received a Knighthood; and Marshall Meek a CBE.

As for Giles: ***"Little more was heard of Giles and the short fat ship, although I did see recently in the papers that he had lost a court case brought by a dissatisfied customer with heavy damages awarded against him. His company reportedly went bankrupt."*** (2006 Rawson memoir, p. 121.) A patently 'malicious falsehood' as the lawyers would say. Or was he confusing the *C.A.T.S. Eyes* TV programme (p. 219) with reality?

*****

On 27th July I was driving back to London from a Dartmouth sailing break. Somewhere around Bracknell on the M4, I heard a discussion on the radio between John Alcock, of LBC Radio, and Ken Purves, a former member of MOD Bath and designer of the 'Leander' frigate, who was being very rude about the Sirius concept: "It should never have caused such a furore if it hadn't been for ill-informed political backing," he said. I realised the meaning of the conversation, immediately confirmed by the BBC News. Younger had already announced his decision to go ahead with releasing the Inquiry Report in the House of Commons without the courtesy of letting me, or anyone else on the TGA side, know of his decision. My son Tom, in his car, also heard it – to his horror.

Younger's letter was there when I arrived home. My first reaction was 'Why the white paper?' Other official letters from Ministers and senior officials had been written on garish apple green stationery. Did they only keep white paper for Instruments of Execution by Secretaries of State?

Clearly they did: the only mention of the tank tests was a remark confirming that their decision was only made on grounds of cost. The question of any further implications for future warships was clearly subordinate to that.

> "In my view there was no new evidence in what was said to me that should cause me to dissent from the Inquiry's main conclusion – that the Sirius hull form would cost more to build and to operate than a conventional frigate designed to meet NSR 7069, and would offer no significant improvements in operational performance that would compensate for the increased cost. I have also concluded that there was no likelihood that this central conclusion could be invalidated by the further programme of tests which you have proposed."[97]

97 Letter, SofS, Rt. Hon. George Younger to DLG, dated 26/7/1988

The whole point of our suggestion for the additional model tests had been brushed off by amending one of the primary purposes of the Inquiry, which was to conduct an 'impartial' comparison between the S90 and the Type 23 as it stood in answer to the Operational Requirement ORC 2064 that we were given in April 1982. For such a comparison of power/speed it would be necessary to compare two hulls of similar size and dimensions but different hull form – as Moor had done at St Albans, illegally using Osprey model STA 2225 vs. 'Ton' class – to compare 'like with like'. What the Inquiry did was to base its conclusions on computer-derived simulations assuming the S90 was a conventional 'short, fat' hull of the same dimensions and proportions, but of a shape that sank, rather than lifted, with increasing speed. As someone humorously suggested, "the equivalent of a wind-tunnel test of an aircraft with its wings turned upside down".

Appendix 5, Figure 2 shows the NMI measurement of the S90 centre of gravity sinking up to 24 knots and rising continuously thereafter, with increasing speed.

I was tempted to use the outburst of publicity that would accompany Younger's decision to blow the whole thing to pieces with revelations about the Osprey Case. Gratifying as this might be, however, it would have broken the Terms of Settlement and might have backfired in the USA. Also I did not want to risk bringing even greater wrath down on the head of Hill-Norton, who was keen to maintain the peace.

That evening the Press Association phoned. I said that I had not been consulted about Younger's decision before it was announced. I continued: "The whole thing exploded in our faces – but we shall be proved correct ... Or someone else will benefit from British innovation. I am determined to find the conclusive answer to this controversy in a full-scale vessel. Younger's decision is the reason why people leave the country: the 'Brain Drain' – a path that I am now considering. We and many others are still convinced, notwithstanding anything said in the report, that an appropriate version of the S90 hull form could provide a cheaper and better solution to the requirements for the Type 23 frigate and we are confident future events will prove us right."[98] The following day it was all over the media: 'Plug Pulled on Short Fat Ship', was typical of headlines appearing

98 Press Release by Associated Press, 27/7/88

in the leading London papers. There was one more thoughtful leader in the *Evening Standard* under the headline 'MOD Irregularities':[99]

> "Defence procurement continues to give MPs grounds for concern. Yesterday the Public Accounts Committee published a report entitled 'Procurement Irregularities', recommending more investigators, and a confidential hotline for informants and whistle-blowers. At the same time the Defence Select Committee again insisted that the MOD, after three years of continuous refusal to co-operate, should publish details of the numerous senior procurement staff who leave to work for defence contractors ... Yet a third report, this time by Lloyd's Register, finds that the MOD were right not to have bought the controversial short fat warship rather than the Type 23 frigate. Here too there are irregularities: Lloyd's appear to have accepted uncritically that the Type 23, minus weapons, costs only £60 million, while Yarrows, who build it, say the figure is actually £120 million. Admiral of the Fleet Lord Hill-Norton, who had originally called for the Inquiry, says the key issues remain unresolved by the Report."[100]

Hill-Norton confirmed his disappointment. He was more concerned with the future interests of the Navy than with justification of the Sirius design family. He struck his colours with the words: "I am too tired and have lost too many friends." However, after hearing of our progress in the USA, he was to raise them again at the opening of the next century, in his stirring Foreword to this book. Our only recourse was to apply for a writ for Trade Libel, as recommended by Frere Cholmeley, in an effort to muzzle the shrieks of joy that would arise from the MOD, RINA and Lloyd's community. One meeting with David Pannick QC, confirmed that there were more than enough instances of malice on the part of the MOD, Lloyd's, Marshall Meek in the RINA journal *The Naval Architect*, and the Osprey Case, to support such an action. The writ was duly issued in the High Court on 3 August 1988.

It worked. Not another word was said, either in public or in print, until our three greatest antagonists, Marshall Meek, Kenneth Rawson and Jack

99 *Evening Standard*, 28/7/88

100 This perhaps explains why Meek is so extraordinarily rude about Hill-Norton in his 2003 memoir *There Go The Ships*. He refers to him as an 'SOA', or 'Silly Old Admiral' (p. 205) and, following the Inquiry: 'he toddled off muttering that he was glad that his main recommendation for an Inquiry had been met' (p. 209). Hardly – if one reads his Foreword for this book, written 15 years after the Inquiry ended.

Daniel, published their book or memoirs in 2006 and 2009, perhaps prompted by the MOD: *'de mortuis nisi nihil bonum'*? These were followed more recently by David Andrews on the 'secretprojects.co.uk' website. Thanks to these indiscretions breaking the 1987 High Court Settlement, I now feel permitted to respond with my version of events.

Eleven Words of Great Effect

In the years that followed my bruising encounter with the British defence establishment, the pain and indignation slowly faded. Yet questions remained. Why did these respected professionals, in public positions of the highest responsibility, behave so deviously? Lies, deception, manipulation of the truth, delay, destruction of evidence, defiance of court orders, cynical use of public funds, discouragement of enterprise and invention, and utter disregard for the lives and hopes of those unlucky enough to get in the way – and all officially sanctioned ... It was, and remains, an act of self-defence in the face of unquestionable guilt. Or was it wanton theft of an idea in which they were interested from the outset – hence the origin of 'Ospreygate'? Yes ... and their contradiction of the basic scaling laws of the world's leading testing tanks, technical institutions and physicists.

The 'Professionals' undoubtedly felt threatened by outsiders such as Peter Thornycroft and David Giles bringing in disruptive new ideas, especially when they could claim they appeared to contradict their contrived 'laws' of their trade. The shipyards tendering for new 'gold-plated' ships felt *financially* threatened; they stood to lose profits in the short term if they started building cheaper vessels, and it was an important part of the Thatcher Government's privatisation plan to make these shipyards attractively profitable before selling them off. If the yards had been open to new designs, however, who knows what export orders might have followed?

Most of the MOD and Royal Navy personnel were, of course, acting on orders from above; as loyal servants of the Crown they were perhaps able to square their actions with their professional conscience, although how they felt personally about it is another matter. If these senior figures were obeying orders, who was issuing them ...?

In 2013, twenty-six years later, an important clue came to light. In a file at The National Archives at Kew, under the 30 Year Rule, I found a scribbled memo dated 16th June 1982: "The Naval Staff recommend rejection of Giles' proposals from the outset."

This, we should note, was a month before July 14th 1982, when we first discussed our proposed S90 Validation Programme with senior political and MOD officials – a proposal which the MOD accepted. Thus the charade of testing our designs was set in motion, but with the preceding intention of rejecting them 'from the outset', as ordered by Admiral Sir Henry Leach – our most outspoken Naval opponent – then Chief of the Naval Staff.

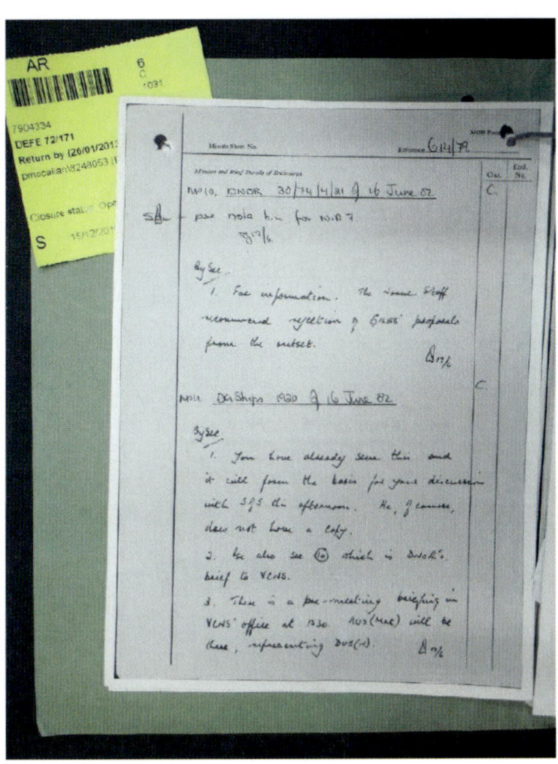

AR
7904334
DEFE 72/171

Minute Sheet No.

16 June 82

1. For information. The Naval Staff recommend rejection of G[illegible] proposals from the outset.

16 June 82

1. You have already seen this and it will form the basis for your discussion with SofS this afternoon. He, of course, does not have a copy.

The Memo reads like something out of the TV series *Yes, Minister*. Just as in those comedy scripts, the Minister (Pattie) was kept in the dark. Or, rather, partly in the dark: he was shown only those pieces of information that would lead him to make the preferred decision. He 'of course does not have a copy' – that 'of course' speaks volumes.

A further note adds a telling detail: "You have already seen this and it will form the basis for your discussion with SofS (i.e. Secretary of State for Defence, Sir John Nott) this afternoon. He, of course, does not have a copy."[101] (For the full-size memo, see **Appendix 6**.)

101 In the Parliamentary Defence Debate of June 25th 1981, John Nott had put down his marker: "First, if we want to build a reasonable number of new ships in the future, we must devise much cheaper and simpler designs than the Type 22 frigate. We must accelerate urgently, and I have provided funds in this programme for this, a new type of anti-submarine frigate, the Type 23, built with an eye to export ... I intend to pursue as well, the possibility of still more cost-effective, smaller, ships than the Type 23." The *Times*, 'Parliament' June 26th 1981. Nott resigned in January 1983, just as the S90 Validation Programme was starting. Thereafter it was a charade.

So too does the word 'recommend'. Nothing so crude as 'instruct', 'order', or 'demand'. Just a word in the ear, spoken softly but with immense authority from the man who had spoken of 'Giles's coracle'.

The reticence of Admiral Bryson, later the implacable resistance of the MOD and the Hull Design Inquiry, and the irrelevance of Defence Secretary Nott's position, discredited by the collapse of his 'Navy Cuts' after the Falklands War ... It was all explained in those eleven scribbled words, by the man who publicly claimed on the BBC five months later: "Even if the tank tests showed the traditional theories were wrong, the Navy would not accept them."

On December 22nd 1988, writing from his modest thatched cottage in the New Forest, Lord Hill-Norton sent me a sad, hand-written note on the aftermath of the Lloyd's Inquiry. It read:

> "After a long-ish and frustrating campaign with a deeply disappointing outcome, I should like to know from time to time how you progress. Nothing would please me more than to learn that some positive – & above all practical – results were coming to hand."

How could I do other than devote my future time and energy to repaying the magnificent support he had given me? Since that was bound to be further frustrated and wasted in Britain, I decided there was only one thing to do: focus on the USA, while keeping Hill-Norton informed of progress.

CHAPTER 31

1988: Hard Choices

Throughout the agonizing process of the Lloyd's Inquiry, after Peter's death design work on the high-speed project we called *Prelude* had started in the newly-established TGA drawing office, in Cowes, Isle of Wight. Manned by David Jenkins and Arthur Mursell, this was intended partly for Charles Hoste and his colleagues at Aberdeen Shipbuilders, and partly as a response to a request from US Navy's Admiral Metcalf for a new type of 40–50 knot frigate, the 'FF-X'. At BHC, just across the Medina River in East Cowes, we were conducting tank tests on *Prelude* at up to 2,500 tonnes and 50 knots. This was an unprecedented speed for any ocean-going warship, let alone of only 295-feet overall length: 'short and fat' ... and fast. It was only possible with water jets: an exciting project, and the best of a very difficult time.

On a boisterous day the previous April I had visited BHC and the TGA drawing office in Cowes. After arriving back on the ferry at Southampton, I had taken my favourite route home, via Charles Adams's wood-mill at Privett, in the pastoral surroundings of the Meon valley – the last anchorage of many an ancient mariner. Battered by the gale and driving rain, I slithered through small lanes across the Hampshire Downs and arrived to find Charles's hideaway ablaze with light and a log fire.

Here was warmth and kindness – a welcome contrast to the cold, evangelical world of the Inquiry, the DSAC, the MOD and the condemnation and contempt of so many Experts. While the wind roared outside, Charles cautioned me: "Get out of this place. There's nothing to be gained by hanging around hoping for someone to be nice to you. The Inquiry's rigged and you've blown it with the Establishment. Get out of England: go west, young man!"

My recent visits to Washington and the West Coast seemed to point me irresistibly in that direction.

In an amused, knowing way he went straight to the point: "What does Vanessa think about all this?"

I thought for a long time. I could only reply: "Good question."

"No hurry, old boy ..." he replied, taking a sip of his whisky. "We've got all night. You can bunk down here if you wish." He looked me straight in the eye. "I want an answer, you know. This is a decision you're going to have to face. Oriana could see this coming long before she died. You owe it to Vanessa, your family and yourself."

We had just moved from Wandsworth to a small house off Kensington High Street. It was exactly what Vanessa wanted: compact, charming, close to the shops, public transport and many of her friends and family, in a vibrant neighbourhood. She loved its village atmosphere in the middle of London. She had a good job working for the Antique Dealers' Association and the house was in her name. But I could see no future in England, other than capitulation, retirement to the country, sailing, digging the garden and going to the pub. The constant uphill struggle against the forces of reaction had affected my sense of proportion. I could only think of getting as far away from them as possible. That meant separating from Vanessa and my family. It was a horrible choice. She was a wonderful mother and companion. She had been extraordinarily patient over the highs and lows of the last few years of crashing between the MOD's agreement to the S90 Validation and its later rejection; and the collapse of the OPV-3 after, at much expense, we had developed a far more detailed design. Then, according to our S90 partners, the entire OPV-3 project was abandoned because we wouldn't settle the Osprey Case. Then came the sudden intervention of Lord Hill-Norton and his Committee and our success in the Osprey Case ... That was followed by the contemptuous rejection by the Lloyd's Inquiry and, finally – after the mass of detailed evidence obviously supporting our S90 convictions – that letter to me from Defence Secretary Younger. What did it all mean ...?

It had all put a terrible strain on our relationship. The choice was 'retiring to the country with my ill-gotten gains' – or following my instinct, knowing that we might be on the brink of a breakthrough in hull-design. That meant setting up shop – and home – in the USA and spending most of my time there. Understandably, she was not inclined to accept such an arrangement.

The Alternative

A beautifully converted old tide-cum-windmill was for sale on Langstone Harbour. Sailing, marshes, water, mud and wildlife – all the things I loved. Hayling Island just across the bridge, and Chichester Harbour round Thorney Island to the East. The pretty Royal Oak pub a few yards along the hard, and a good mooring available just off the Mill in Langstone Creek. Lots of friends from the Navy and my Hampshire youth not far away. Surprisingly, we had the funds to buy all of that, having sold the Wandsworth house in which we had lived since 1973, for the huge profit, such as was being made on London property at the time. But what about the Osprey settlement? I could hardly spend my share of all that on a jolly retirement in my early fifties.

Was that the easy way out? To give up the struggle?

"The new house in Abingdon Road will help," I said. "It's exactly what she wants, and its hers. With our youngest boy at university and the others graduated and out in the world, it'll be easier for her – and for me to set up shop in America. My boys agree I should follow my instincts – but not if that means splitting up."

Charles was not impressed: "Do you really think she wants the house more than you?" he retorted. "It sounds like a pay-off for all the support she's given you over the years."

I agreed that it did sound that way. We had lived for a year in America during my Pan Am venture. She had not enjoyed it and was not keen to go back. I would need to live there to set up a US Company to pursue such an opportunity as I foresaw – but it would require my concentrated and undivided attention, living in some small suburban apartment – a lonely, brutal environment, far removed from family and her many friends.

That was based on what I knew of our potential in the USA. I had the funds to give it a go for about a year; to continue to work with my superb team of Arthur Mursell and the two Davids while they stayed in England and I set up an office in the USA. There was no future for me over here – only 'over there'. Some version of our design was only likely to be taken up in the USA and that was my only future. In the UK I was anathema to the Naval architectural community, as to the Great and Good. One notable exception was Lord Hill-Norton; and I knew that he was losing both friends and credibility from his extraordinary commitment on my behalf. Hopefully, this book might redress some of that imbalance.

I explained all this to Charles. He was sympathetic: "Then make your peace with Vanessa and your sons – eschew the Lawyers – and then set sail for America." So saying, he led me to my car through the wind and rain, as he bade me farewell for the last time before his death a few months later. It was the best, if toughest, advice he ever gave me.

CHAPTER 32

1988: Westward Ho!

> And not by eastern windows only,
> When daylight comes, comes in the light;
> In front the sun climbs slow, how slowly,
> But westward, look, the land is bright.
>
> Arthur Hugh Clough: 'Say Not The Struggle Naught Availeth'

With the funds from the Osprey Case settlement now available, I could concentrate on our activities in the USA. At the current exchange rate, we had a little over one million dollars available to set up shop. In the spring of 1988, just as we were resigning from the Lloyd's Inquiry, I had been introduced to Dr Anthony Wells, a friend of John Moore's working for the US Office of Naval Research. Tony Wells was keen to help set up our American company. He found office space in Rosslyn, Virginia, just across the Potomac from the Lincoln Memorial, and brought in Vice-Admiral Jon Boyes, a submariner, he was a former CO of USS Albacore, of the experimental, high-speed, manoeuvrable, 'tear-drop' hull. He had already supported us at the final Younger meeting. As Chairman of the Armed Forces Communications & Electronics Agency, a lobbying group for the US electronics industry, he had retained his contacts with the US Navy's top-brass and was enthusiastic about prospects for the 'FF-X'. He agreed to act as President of our company, while another retired Admiral, 'Chuck' Horne, assisted him with other Navy contacts.

A series of meetings was arranged for us at the Pentagon. One of these was with Vice-Admiral Paul McCarthy, head of the US Naval Operations Command. With our two Admirals as escorts, we were graciously waved through security and navigated the labyrinthine passages and staircases of the then-world's biggest office building, ushered by a smart young Naval Lieutenant. Groups of visitors were being led past an array of flags framing the ever-open door of Defence Secretary Frank Carlucci. Under the watchful eye of a Marine walking backwards, they admired the many paintings of battles and war heroes that decorate the corridors. It was all in sharp contrast to our own MOD, where all is drab walls, with officials and service personnel moving about in civilian dress; certainly no parties of visitors – though not so in the gentrified offices of the First Sea Lord.

We arrived at a more 'Executive' office with views across the Potomac. Smiling Admirals sat in armchairs grasping huge mugs of coffee. We had been invited in to talk about SuperShip, our design for the new 'FF-X' frigate. Our proposal was based on the S115 – financed, but ignored, by the Lloyd's Inquiry – and our *Prelude* test programme for the suggested 40–50 knot speed, using water jets.

We also discussed our tentative ideas for a much larger vessel to be submitted under the 'Fast Sealift Initiative', the requirement proposed by Sam Nunn and the Senate Armed Services Committee for rapid return of heavy military equipment to US forces in Europe in the event that the currently encouraging Reagan/Gorbachev 'Cold Thaw' reverted back to 'Cold War':

NATIONAL DEFENSE AUTHORIZATION ACT FOR FISCAL YEAR 1989

REPORT

COMMITTEE ON ARMED SERVICES UNITED STATES SENATE

FAST SEALIFT DEVELOPMENT

For a number of years, the committee has been aware of, and concerned about, the significant gap in the Nation's strategic capability caused by the lack of fast sealift. Hearings this year revealed that solutions to this deficiency are no closer than when the Congress expressed its concern early in the decade by requiring the Congressionally Mandated Mobility Study.

Recognizing that a lack of progress in this area may be due to the nature of the bureaucracy and the internal ordering of priorities, the committee believes the time has arrived when fast sealift should enjoy a higher priority. Indeed, because of the timing an opportunity may exist to apply the benefits of new technologies to develop a fast sealift ship the design of which could prove of great benefit to the merchant marine and which may well serve to satisfy some of the requirements for the amphibious force which is fast approaching block obsolescence.

Accordingly, the committee will hold hearings attendant to the fiscal year 1990 budget request to determine the adequacy of the program to develop a fast sealift capability for the Nation. In preparation for that hearing the Secretary of the Navy will cause to have prepared a program for the development of a ship of suitable size and capability for the fast sealift mission and the amphibious mission. The ship should embody hull and machinery technology enabling speeds at least in the range of forty to fifty knots. To the greatest degree possible the hull and machinery technology should be common with that envisioned for the battle force capable surface combatant of the next century. While not a controlling parameter, the designers should be mindful of the potential for technology transfer to the merchant marine.

The Secretary of the Navy will form a design review panel to consist of the Commandant of the Marine Corps, the Commander in Chief of the United States Transportation Command and the Maritime Administrator. It should be anticipated that the program will be presented at the committee hearing and should detail characteristics, development costs, probable construction costs and best estimates regarding fleet size required for the sealift mission and the amphibious mission, as may be applicable. A report containing these particulars shall be made to the Committee on Armed Services of the Senate and House of Representatives by January 1, 1989.

The Secretary of Defense is invited to participate in the development of this program and to ensure appropriate funding is provided in the fiscal year 1990 budget request.

This document suggested that the Senate Armed Services Committee was in search of exactly the kind of ship that we could offer, on the basis of 'scalability' and was prepared to fund it. Boyes recommended that we invest in tank tests and preliminary design work for a more convincing Proposal. In order to qualify, however, this would have to be submitted by a US Company, controlled by US citizens. I would have to raise substantial additional capital, meaning, inevitably, forfeiture of my control.

Luckily a good friend, Gaby Grosvenor, a US citizen, was prepared to take up sufficient equity so that, between us, we could control fifty-one per cent of the company. With another American shareholder, an acquaintance of Gaby, who was fascinated by the US project and happened to own a yacht designed by my father, the majority of shareholders would still be US citizens.

Our US Company, Thornycroft, Giles & Co. Inc., was officially incorporated in the Commonwealth of Virginia in March 1988. I was able to pay $25,000 per month for the cost of the office and part-time work by Boyes and Wells, plus all expenses. With the addition of UK tank tests and design work by the two Davids and my own living and travel costs, I reckoned my funds would last about a year. But I was prepared to take the risk. There were enough opportunities, and we seemed to be talking to the right people about designs that only we could provide – contrary to the upholders of traditional wisdom.

At a second meeting with Admiral McCarthy, we were joined by a uniformed Lieutenant Commander from the Royal Navy, who sat in as a silent observer. I assumed he came from the large Ministry of Defence contingent at the British Embassy. Even in the comfortable surroundings of McCarthy's office I felt a pang of anxiety: were all our past tribulations to be visited upon us again, here in Washington? Surely senior US Navy officers would not be swayed by the views of their British counterparts – that our enlarged hull wouldn't 'lift' on its Captive Wave, it being confined to quite small craft – or would they? I couldn't be sure: But what about that 'Special Relationship' ...? Between them they would have all the data they needed to do a full investigation – from the Azteca, though the Osprey, S90, OPV-3, S115 – and now the forthcoming 'FF-X' SuperShip, based on our 50-knot *Prelude*. And there was the 25,000-ton Fast Sealift Initiative. If there was any merit in 'Giles's ideas', then certainly both Navies could have the evidence, if they needed and heeded it.

The next episode occurred when we met Rear-Admiral 'Mal' MacKinnon, the Head of the US Naval Sea Systems Command, generally known as 'NAVSEA', the US equivalent to 'Bath'. He seemed greatly taken with our SuperShip as FF-X. During April 1988 I gave him a paper on the subject and showed him a typical body plan for the vessel. MacKinnon was keenly interested. As Boyes told me later, however, he visited the UK and British Ministry of Defence in May 1988, where, no doubt, he was shown the Lloyd's Inquiry Report. Thereafter I had neither sight nor sound of him.

But, clearly, the idea was not forgotten – as proven by later events.

At the time I did not let these episodes worry me too much. In America there lived those I respected: eminent professionals with open, progressive minds, who supported my ideas: Dick Garwin, Karl Kirkman, Jon Boyes. They had looked beyond the past to new worlds, and they encouraged others to do likewise.

CHAPTER 33

1989: A Military Design with Commercial Potential

We had worked on the Fast Sealift proposals between the spring and autumn of 1988, creating lines plans and construction drawings, and running tank tests at BHC and Southampton University. The result of our efforts was the Monohull Fast Sealift Ship (MFS), a 25,000-tonne 775-foot Roll-on/Roll-off vessel, like a huge car ferry, propelled by General Electric gas turbines driving Riva Calzoni water jets. It was intended to carry 5,000 tonnes of military equipment or cargo over 5,000 nautical miles at 41 knots in North Atlantic conditions.

Through Jon Boyes's contacts we moved the TGC Inc. office to a much cheaper site in Fairfax, Virginia, where I rented an apartment in a newly built complex, the Hermitage – which it was indeed for me. Being on my own for many months at a time, I was able to concentrate on the Proposal for the MFS to the US Naval Sea Systems Command. The 'two Davids' visited me, to complete their extraordinary work on the MFS Project – all in contradiction to everything condemned by the British Naval Architectural fraternity, employing a design that was limited to 'quite small craft' – or 'an important contradiction of traditional theory', depending on the speaker and the audience, in this case the latter.

Back to Boyhood

During two summers of 1990 and 1991, I rented a house on the Chesapeake Bay. There, with my boys and, sometimes their girlfriends, wives – and later, grandchildren, we sailed and swam and indulged in the delicious Chesapeake Bay Blue Crab, *Calinectes Sapidus*: soft-shelled, boiled or as a crab-cake. It was a happy re-visitation of my own boyhood, as we sailed to some Bay-side crab-shack to crack a steaming pile of crabs, hammering away at their shells to enjoy the succulent meat.

Then rinsing our hands in the brackish water of the Bay, as we did in my boyhood.

From time to time members of my family and friends came out to explore the beautiful Virginia countryside, or to walk the Appalachian Trail. Afterwards the usual drive down winding country roads, with a spectacular sunset behind the Blue Ridge, to the Griffin Tavern in Flint Hill, Southern Virginia. There to enjoy a hearty bar-meal with 'Erster fritters', Wasmund's local Sperryville malt whiskey, cheerful company and Blue Grass and Country music – deep in the heart of the Piedmont countryside.

I also continued my weekly runs: four miles around the Bull Run Battlefield two or three times a week – beautiful paths leading through woods and fields, with deer, foxes, birds and other wildlife aroused by my passing. It continued for twenty-five years – in Virginia, or during visits to London, around Richmond Park or Clapham Common. To that, and the challenge of my chosen commitment to the 'Short, Fat (and Fast) Ship', I attributed my necessary good health to continue the battle.

When we submitted our four-volume, 800-page MFS Proposal to the US Navy in October 1988, Jon Boyes was confident of success. Our design met the specifications comfortably, and at reasonable cost. Yet somewhere

The 25,000 tonne/45-knot Mono-hull Fast Ship, or 'MFS', as proposed for the US Senate's Fast Sealift Initiative: the beginning of 'FastShip'. Yet our MOD still stuck to their guns: 'The semi-planing hull was restricted to quite small craft'; while the US Navy, after discussions at Bath, called it 'an extrapolation of the semi-planing hull to unprecedented size'.

behind the grey stucco walls of the Pentagon, rumour was stalking. Remarks that 'it won't scale' were being circulated, with other unsettling echoes of London, St Albans and Bath. In February 1989 NAVSEA informed us that our Fast Sealift proposals had been rejected. A conclusion that re-echoed all the dire warnings of Bath and Lloyd's.

Our ship, they said, was 'an unprecedented extrapolation' of small craft design. It would involve an eleven-year R&D programme costing over $1 billion, including construction of one or two interim test vessels. The design would obtain no benefit from hydrodynamic lift from the hull, or from the water jets themselves. These, they claimed, would suck the stern down, increasing drag and reducing sea-keeping capability – again the opposite of the calculated conclusions of the water-jet makers, the Swedish KaMeWa and the Italian Riva Calzoni.

This view, like that of the MOD and British Shipbuilders before it, was based on ignorance of the latest technology. A recent technical paper by Rolf Svensson, Chief Hydrodynamics Engineer of KaMeWa, had demonstrated the existence of high pressure at water-jet inlets under the stern and the consequent added lift from the water passing through the jets themselves. Svensson was a leader in the field. In 2003 he was to be placed in charge of the installation of similar KaMeWa jets in US Navy ships. But in 1989, as far as NAVSEA was concerned, he and his jets might just as well not have existed.

When NAVSEA returned our Proposal to us, they confirmed that they had retained neither copies of plans and documents, nor any of our proprietary data.

The Wolf at the Door

By October 1989 we had spent nearly a million dollars on our American venture, with no financial return. The collapse of TGC was imminent. Could we survive on a few thousand dollars, as we had done in the past? Swallowing our pride, Gaby and I arranged to see the old family friend with the Laurent Giles yacht – a possible benefactor – to give a full account of our situation. We had satisfied ourselves, and other more qualified minds, that we had the makings of a viable design. The only serious criticism seemed to come, as usual, from the MOD and its advice to the Pentagon, using those same old non-scaling arguments we had heard for twenty years. In May we had to close the Fairfax office. The two Admirals disappeared over the horizon with promises of thoughts and prayers, while Tony Wells was preoccupied with other projects. Again, I was alone on a hostile sea.

Our anonymous benefactor, without whose interest I would never have dared set up shop in America, asked if we needed money.

Lying, I said, "No: we need your advice."

He introduced me to his own corporate lawyers: the then-powerful lobbying and legal firm of Patton, Boggs and Blow, or PB&B, in Washington. A Senior Partner, David Dunn, took up our cause. David believed they might be able to re-mobilise the necessary political forces to propel us towards the Fast Sealift research funding that was due to become available in the Navy's 1990 budget.

Self-Defence

About this time, the autumn of 1989, I attended a Seminar by an MIT investment analyst. He expounded on the importance of having 'a Patent, a co-efficient of passion and a compelling idea', in raising funding for an innovative project such as ours. We already had a compelling idea; and, surely, the required 'co-efficient of passion'. What we needed to complete the trio was a Patent.

In June 1989 I had been introduced to Donald Stout, a leading young Washington Patent Attorney. He was encouraging, but he needed $5,000 to start writing the Patent with me. I immediately retrieved the last of my depleted funds – a deposit on a house in Maryland – and paid it into the account of Antonelli, Terry, Stout and Kraus. It precluded ownership of a house – but never have I invested more wisely.

The Lobbyist

The Lobbyist is an integral part of the political life of Washington. Ours, David Earl Dunn III, a swarthy fellow with a casual Southern manner and a Groucho Marx moustache, enjoyed a large corner office at the top of the Lego-style PB&B building between M Street and Pennsylvania Avenue. Represented by one of the great lobbying firms of Washington, we were no longer creeping into Congressional buildings as in the past, pleading with outer-office staffers for a few minutes of their Senator's or Congressman's time. Doors were quickly opened: doors of inner offices, resplendent with flags of State and Nation, the walls hung with framed certificates honouring the occupant's contributions to society or local causes.

In those days, before the advent of the mobile phone, David tirelessly worked the banks of pay-phones discreetly hidden around the stately halls of the Senate and House of Representatives Buildings. He spent hours bending the ears of the young staffers and clerks – cajoling, joking, sometimes gently threatening – as he sought to renew support for an item in the annual Defence Budget Authorization Bill that might provide us with the financial lifeline we so desperately needed: the much sought-after 'Line Item' ... Money.

David's knowledge of the corridors, elevators and underground passageways of Congress was dazzling. Dodging Mid-Western marching bands, Bible-Belt preachers, Civil Rights groups and environmental activists, I trotted behind him, panting like a spaniel. He would take swift evasive action, shooting up staircases or into an elevator, suddenly to emerge within a few doors of his target Senator's office. Contrary to Rawson's 1981 claim, Dr Garwin was surely no Lobbyist: based on his style of life.

He found support among some powerful names: Edward Kennedy, Ted Stevens and Daniel Inouye in the Senate; Jack Murtha, Charles Bennett and John Breaux in the House – all members of their respective Armed Services, Appropriations, Authorising or Maritime Sub-Committees. There was every chance that, with his contacts on 'the Hill', Dunn might get the right people to sponsor a resurrection of our 'line item' – finance for Fast Sealift in the Bill for the forthcoming Defence Budget negotiations. He agreed that the Patent on which I was working would be a big attraction to any potential investor and offered me the use of office space at PB&B. By November 1989 he had succeeded in getting $3.5 million incorporated in the 1990 Defence Budget as a much sought-after line-item, for a research programme into the 'Semi-Planing Mono-Hull Fast Sealift vessel'. There was a good chance that some of it might come to TGC.

Hope renewed

Early in 1990 Mark Bebar, the Head of Advanced Design at NAVSEA, phoned me out of the blue: apparently the funding for the Fast Sealift Program was likely to be included in the 1990 budget, thanks to Congressional support. He would be handing over the Program to a new team led by Captain Calvano, Director of Ship Design at NAVSEA, and he advised me to call him ... What a break!

Calvano claimed that, although nothing had yet been formally agreed, our Research and Development funding would probably go ahead. But there was a problem: the US Navy would insist they do their own tests on our hull at their David Taylor tank, to 'verify the numbers'. He would expect 'heavy involvement' of shipbuilders, but would not require TGC to be concerned in any way.

"But it's a proprietary design," I protested. "Every page of our Proposal was stamped to that effect, and we have just applied for a US Patent."

I told him that after their analysis of our Proposal, NAVSEA had written confirming they had returned every document to PB&B and had retained no copies for themselves.

"If you wish to do tests," I said, "you'll have to come back to us for the plans and hydrostatics: we must be involved in any test programme. Remembering the Osprey Case I told him how, in the past, people have done tank tests without our knowledge or guidance and have produced results that are misleading and inconsistent with their experience of the performance of full-scale ships."

At the other end of the line there was a long pause, followed by an exhalation of breath: "OK Dave, thanks for telling me ... I'll get back to you." He never did – and we heard no more of the idea, at least from the US Navy. We shall probably never know whether they ever repeated the performance of their 'Special Relations' across the Pond and undertook covert testing of our MFS design.

*****

Another promising lead came up soon afterwards. David Dunn introduced me to a former Navy officer, Ron Cornelison, who had good connections in the Pentagon and the Bush Administration. He thought our project interesting, decided to invest $135,000, and offered access to the Navy programme and funding.

To receive Cornelison's investment it was necessary to form a new Company, FastShip Atlantic Inc., into which Thoryncroft, Giles & Co. Inc., as future holder of the US Patent, and S90 copyright and Intellectual Property, was merged. No salaries would be paid until the Navy money came through. In the meantime I had to exist on selling some of my shares in TGC to friends and family who were prepared to back my ideas.

Unfortunately, the US invasion of Panama, which had occurred in December 1989, turned out to cost far more in time, lives and dollars than was expected. No funding would be forthcoming for Fast Sealift for the time being, and Ron Cornelison's friends from NAVSEA doubted that we would ever see any of it.

The Wisdom of Laura Ashley and Fred Smith

In parallel with the Fast Sealift design for the US Navy, we had been working on a commercial version of the ship: a 32,500-ton fast container carrier that we called 'FastShip'. Swifter than *Cutty Sark* or the SS *United States*, she would be the fastest ocean-going merchant ship ever. I remembered the wisdom of Laura Ashley, thirty years before: "What are you waiting for? Go for it!"

Thanks to David Dunn, I was re-introduced to Fred Smith, the Chairman of Federal Express. Like me he had been working with Pan Am in the early 1970s on a commercial use for the Dassault Falcon jet. By 1989, Fred's venture had blossomed into a highly successful worldwide service, against some prophets of doom within Pan Am and Yale University where he had first developed his ideas. They called him 'the Mad Marine'.

It was perhaps providential that, at this low point in our fortunes, I met Fred again.

I told him we were working on a commercial version of FastShip. To use his own word, he was 'thunderstruck' by its potential as the answer to the world's – and Laura Ashley's – logistical problems expediting High-Value, Time-Sensitive cargo. He coined the phrase the 'Middle Market' to describe the logistics gap between fast/expensive air freight and slow/cheap sea freight.

Following our meeting he wrote: "The parallel thinking about the opportunity for fast sealift was truly astounding. As I mentioned, for a number of years I have been working with Professor Aaron Gellman of the University of Pennsylvania regarding ultra-large freighter aircraft and/or fast, small ships to penetrate this 'middle market' between air and sea freight. Yesterday I met with him and described your project so that it can be included in a larger study we have underway with him."[102]

To have the support of such a luminary in the logistics field was encouraging. So, uncoupling from the US Navy, we headed for a possible commercial market. I was to work with Fred's adviser, Dr Aaron Gellman, for several years. That meant going through the statistics.

102 Letter, Fred Smith to David E Dunn, dated Oct. 29, 1989

'Office Machinery, Essential Oils, Ladies Hose, Non-Ferrous Metals ...'

The US Trade Statistics were recorded and stored on microfiches at the Department of the Census at Suitland, outside Washington. They read like some Global WalMart catalogue, where everything is obtainable somewhere: from computers to nappies to coffins. This was before the days of the laptop and the Internet, so the data was projected onto a glass screen to be written down in pencil. As I ploughed through hundreds of microfiches for the period 1982 to 1988, various patterns emerged. First, it was obvious that the higher value cargo – over $10,000 per tonne – was steadily increasing as a proportion of all cargo carried across the North Atlantic. Second, a growing amount of freight was being carried by air. Third, unlike the Pacific, there was a fair balance in the movement of high-value cargo in both directions on the Atlantic.

Airfreight across the Atlantic took between two and six days and charged between four and eight times the sea rate. Our service would take between five and eight days, costing between one half and an eighth of airfreight, depending upon market conditions. Compared with sea freight, we would be two or three times the price, but take two or three weeks off Atlantic door-to-door delivery times. We would operate between dedicated terminals, using an automatic 'roll-on/roll-off' loading system consistent with the Navy's sea-lift requirement; thus reducing port delays and with a turnaround time of only six hours.

For the time being, we lacked the Patent and the Business Plan to support our 'compelling idea', but had passion aplenty. Without those other two ingredients, all our lunches and dinners at clubs and restaurants all over Washington, the presentations before solemn committees and naval analysts, and the endless Board Meetings spent strategising, would largely have been a waste of time and money.

What was not wasted was the eternity spent writing, elaborating and defending the 'teaching' of our Patent: the descriptions, drawings and explanations of its key Original Features, or Claims. Valuable too were those lonely hours spent in the Department of Census, or the Department of Trade on Constitution Avenue, extracting quantities of data from the trade figures and the shipping contracts lodged with the Federal Maritime Commission. These were followed by yet more hours in my small Fairfax flat working out the FastShip operating costs and profit margins in comparison with almost every known type of air-freighter and cargo ship. It was back to de Havilland's and Sales Engineering.

I was applying for Patents in most major shipbuilding countries, including Japan, Korea, and Finland. My European Patent would protect our UK interests – and those in all existing and future EU countries. Prospects for a Patent were good, the economic studies would provide the foundation for a Business Plan, and we had limitless passion. Until ...

Patent Panic ...

A sudden cry of panic from our London Patent Agents: The MOD had issued a Secrecy Order to HM Patent Office. Our Patent Application was to be refused on the grounds that 'it might be of potential advantage to a hostile Power'. That also carried effect in all our other foreign Applications since registration in one's country of citizenship is an essential first step in obtaining any patent. Furthermore, the MOD claimed ownership of any such patent issued to me, as Patentee – or TGA Ltd as its owner – and we were prohibited from using it 'anywhere'. Of course such an action, following the rejection of our design principles by the very organisation that had issued the Order, was demonstrably absurd. Therefore our UK Agents merely had to send HM Patent Office a copy of the Conclusion of the Lloyd's Inquiry – and the personal letter of rejection to me from Defence Minister George Younger – and the Order was withdrawn. But it did show the determination of the MOD to frustrate any further progress with our technology ... and for what purpose? To lose it, or to use it? It was surely not the former.

As future events revealed, this action took place at precisely the same time as the MOD and the Marconi/Yarrow/BAE Systems alliance was resigning from the eight-year NATO NFR-90 multi-national Air Defence Destroyer project, to pursue its own independent version: that became the future Type 45 destroyer. This is all discussed in Chapter 38, '2021: Back to the Future – The Type 45 Destroyer'.

*****

Now that the outlook for both the Patent and the commercial market were positive, we thought it better to keep our distance from the US Navy. So where to turn for money? Who might benefit from FastShip? The ports, of course. So we spent much time and money courting both Baltimore, Maryland and Norfolk, Virginia. We courted the shipyards too, but they seemed caught in a time warp, mostly working and thinking as they had when the whistle blew at the end of World War II. They were largely concerned with building naval vessels for which they were paid on a 'cost plus' basis – convenient for them, and expensive for their clients.

We visited Newport News Shipbuilding, the premier source of nuclear aircraft carriers and submarines of the US Navy. There were several meetings with their Board and one with the President, Pat Phillips – a folksy ex-welder with a fine Virginian sense of humour. From his office window he gazed out at the latest nuclear carrier being completed. "I'd much rather be building big grey ships like that with antennas at the masthead like a bunch of broomsticks. But, if I can't do that, I suppose I'll look at anything, even some big speed-boat like yours."

Per Sorensen came out from Denmark and joined us for another meeting to provide Phillips with his estimate of the cost of building a series of FastShips. Having built twelve Ospreys and 42 large Ro-Ro ships at Danyard, he knew what he was talking about. He carried with him several huge files, all carefully tabbed, apparently the result of months of hard work. During the meeting, the Vice-President for Production asked Per for his estimate of man-hours required to build a FastShip. He chose a volume, selected a tab and, with some ceremony, spread the opened file on the Boardroom table. It was full of impressive spreadsheets, which he extended solemnly and scanned with his bifocals perched on the end of his nose. There was a hush around the room while Per took his time, finally intoning in his slow Danish drawl:

"I would say about nine hundred thousand man-hours: that's around 75 man-hours per ton of finished steel, or fourteen and a half million dollars at sixteen dollars per man-hour, for the steelwork."

Pat Phillips's jaw dropped. "What was that ... fourteen million dollars at 75 man-hours per ton? It can't be right ... I'd expect at least twice that."

"Oh, that's the right number. You see we keep very low overhead in Denmark. We do all the production planning on the computer and I have the software. We only employ ten production engineers."

"Only ten ...? We employ *a hundred*!"

"Well ... there you go!" replied Per with gleeful use of an all-American expression.

Later, over a beer at the Holiday Inn, we asked Per to show us his files. He would only open the one he had used at the meeting. At that tab it was indeed full of data.

"Why don't you show us the other pages? There are hundreds ... What's in them all?" asked Ron Cornelison.

"Oh ... paper," Per responded in his slow Danish way, flipping through dozens of blank sheets. "I thought they needed to be impressed. Ship-owners and navies – and big shipyards – they are all the same: more impressed with paper than anything else!"

Per had put on a fine performance, but there was no changing the basic economic facts: the costs for building at Newport News Shipbuilding would have been at least twice those of a commercial shipyard.

Another Logistics Luminary Supports Us: Rune Svensson of Volvo Transport

Glancing through my daily copy of the *Journal of Commerce*, I saw an advertisement for the annual International Intermodal Expo, to take place in Atlanta at the end of May 1991. It would include many of the leading international shippers of cargo: those who pay to move the cargo, not those who carry it. They would include car-makers, chemical companies and photographic equipment suppliers: the very people moving high value stuff from point to point around the world. They would also include the so-called Freight Forwarders: those who consolidate cargo and arrange its shipment. It was the biggest such conference in the USA. That was our market. I must go!

The Keynote Address of the conference, on the opening day, was given by Rune Svensson, President of Volvo Transport. He had the incentive for looking at all sorts of ways to improve Volvo's worldwide logistics. He was moving equipment, cars, trucks and parts, from brake pads to ten-speaker stereo systems, between 160 different companies all around the world. Everything had to go to and come from a remote North Eastern corner of Europe, many thousands of miles away from many of Volvo's suppliers and its main markets.

This enthusiastic 65-year-old Swede told his audience how the world was changing: "People are buying things they want rather than things they need. In Africa, a man ploughs the earth with a wooden plough drawn by oxen, but he is listening to a Walkman." And while the annual growth of world trade was at about 2%, the growth of the value of goods per tonne

'This ship is an absolutely wonderful idea': Rune Svensson expounds as the author explains the FastShip loading system at the Volvo HQ. (BBC TV: 'The Limit', October 1995)

shipped was about 4%. This was due to the huge increase in consumer goods being shipped from China to the rest of the world.

He went on to speak about the need for faster, more frequent, more dependable logistics. He encouraged ship-owners to look at new ideas to improve his supply chain. "We cannot go on saying that people will have their new car delivered 'in about a couple of months'. We must say 'It will be there next Thursday, at 2.00 pm'. By reducing warehousing and inventory costs we can cover the introductory investment of two new models of Volvo: the most important milestone in any car company's financial policy."

He argued that, if a fraction of the billions spent on improving productivity were spent on logistics, Volvo would greatly improve their profitability. But there still lurked the old paradigm: "freight doesn't need speed or reliability, only people do". Yet a growing amount of freight is far more valuable, in hard economic terms, than people. Volvo was wasting millions every day due to slack supply lines. "The supply chain is only as strong as its weakest link: and there are, today, too many weak links."

I was learning, from the very people who owned the freight being moved, that our ideas were not as crazy as the so-called 'steamship lines' would have us believe.

Our US Patent had been filed in May 1990. It was issued in May 1992. Further Patents had been, or were to be allowed in the EU, Japan and Korea over the next four years. This gave us the ability to promote our ideas worldwide without fear of misappropriation. At the same time, the Patent gave us a real asset with which to finance our entry into the commercial market.

In the summer of 1991 I met a market economist, Kate Chambers, at a weekend sailing party. She had abandoned a high-powered job at a leading Think Tank for a few months of release, and was just back from a double Atlantic crossing on a friend's yacht. Kate quizzed me on my project and, ignoring the social buzz around us, we were quickly deep in conversation.

A few days later she visited us at PB&B in Washington and started to make financial sense out of all the trade data I had collected over the past year. She expertly transformed my crude doodles on profitability and market penetration into a thorough business analysis. I knew we could not possibly afford to employ a person of her capability, but she volunteered to do some initial work for nothing. She started working on our Business Plan.

Without someone of Kate's talent, I realised, there could be no FastShip. Meeting her was another of those providential events in our protracted saga. She worked with us throughout the FastShip years, through fair weather and foul ... of which there was to be plenty of both.

A Philadelphia Story

The port of Baltimore, a few miles north of Washington DC, had built a vastly expensive new container terminal, but it just wasn't pulling the ships. We spent three years discussing FastShip with the City, but they were terrified of upsetting the few shipping lines calling there by supporting us. In Philadelphia there were no such scruples. The Port of Philadelphia was dead. Not one Atlantic container ship was calling there. But Pennsylvania had just spent hundreds of millions of dollars on raising the height of all its railway tunnels so that double-stacked container trains could follow the old inland route towards Chicago. Philadelphia was the old logistics centre of the East Coast, only being replaced by New York at the end of the 19th Century. It still had superb road and rail links inland, with three competing railroads leading north to New York and Canada, west to Chicago, and south to the new industries of Georgia and South Carolina. Seventy-five per cent of the US productive economy lay within 24 hours of Philadelphia by truck.

The Philadelphia Navy Yard, the largest employer in the area, was to be shut down in 1996. The Mayor, Ed Rendell, had a dream of turning it into the first up-to-date US shipyard dedicated to producing commercial ships. In early 1994, I visited the Navy Yard with Per Sorensen and Einar

Our first sight of the huge building dock at the Philadelphia Navy Yard.

Pedersen, President of TTS of Norway – a manufacturer of ship building and handling equipment used all over the world.

As these two experienced Viking shipbuilders gazed into the two huge dry-docks, one of them filled with the grey bulk of the aircraft carrier *John F Kennedy*, they agreed that it might be a good place to build FastShip, using the latest shipbuilding techniques. As employed by the Danes, these meant separate Hull Modules built in special units and joined up in the building dock, rather than from the keel up on a slipway. It was also a perfect opportunity to apply for a Title XI Federal Loan Guarantee, a financial incentive offered by the US Government to improve shipbuilding in American yards.

On a bitter afternoon in February 1994 the TGC team, accompanied by Rune Svensson from Volvo, trudged through the snow to Philadelphia's City Hall. Peter Hearn, a tall laid-back Philadelphia lawyer who had taken TGC under his wing, explained that he had gathered the big guns of the City together. In the Council Chamber were assembled Mayor Ed Rendell, the Heads of various City and State departments, plus Vince Fumo, a sort of *consigliere* for all matters Philadelphian. Nobody could do anything without his blessing. As Chairman of TGC, I stood up to introduce the team. I talked about FastShip and then handed over to Rune Svensson who could give the shipper's view of the world market. He held the audience spellbound.

"This ship," he said repeating his oft-repeated phrase, "is an absolutely wonderful idea. It will revolutionise the way Volvo does business in the global market, both for moving cars and components and spare parts. Logistics is becoming another part of the production line. We spend millions of dollars on productivity, but try to save pennies on transportation. It's crazy! The production line starts when the iron ore's taken out of the ground and it stops when the car's delivered to the owner. It all goes to the bottom line. We have hundreds of cars and thousands of containers, full of parts and gearboxes and axles and stereos, worth millions and millions of dollars, sitting in containers stacked in yards all over the world: all costing us a fortune in interest, security, insurance and warehousing."

"What do you do when you get a cheque for a million dollars? You put it in the bank! You don't leave it lying around on your desk for two or three weeks. Unsold cars, like unbanked cheques, cost a lot of money." One could immediately see the game moving TGC's way. After some minutes' whispered conversation, Vince Fumo spoke up: "What do you need to develop this thing?"

"Ten million dollars," I replied, grabbing a likely figure we had agreed earlier ...

After a few moments of huddled discussion with others behind him, he stood up: "Ya gottit!" he exclaimed.

Rune, running late for his plane, was given a police escort to Philadelphia airport.

A New Life

Of course it was not that simple, but the momentum was there and, within six months, we had the ten million dollars. We established an office in Alexandria, Virginia, as FastShip Atlantic Inc., or FSA. Terry Johnson, formerly Chairman of the Virginia Ports Authority, joined FSA as President. Ron Cornelison was Chairman, and David Dunn, Secretary. TGC Inc., under my technical direction, had responsibility for the research and development programme. I moved from Fairfax to a flat in Alexandria, across the Potomac from Washington.

CHAPTER 34

1990–2010: The FastShip Saga

Length: 265m/770ft. Beam: 40m/131ft. Max. Displacement: 36,300 tonnes. Max. Speed: 45 knots. 5x RR MT-30 gas turbines = 240,000 BHP driving 5x RR-KaMeWa water jets.

Meanwhile, TGA's UK Design Team was established, as Osprey (UK) Ltd, at the Old Coach House in Emsworth, outside Portsmouth. The offices were furnished with all the technical and electronic equipment necessary for a detailed design development programme. Five naval architects and draftsmen, led by David Bridges and David Jenkins, worked with the Business Manager, my ex-submariner friend Steve Williams.

Now we could go to the world's top technical institutions and University Ocean Engineering Departments, like those of the Massachusetts Institute of Technology (MIT), or the testing tanks at the Danish Technical University at Lyngby, for the latest in controlled tank-testing and computational fluid dynamics.

The Director of the MIT Department of Ocean Engineering, Professor Chryssostomos Chryssostomidis, invited us to give a complete explanation of our ideas before a large audience. After discussions with several technical teams, Professor Paul Sklavounos and his Ship Dynamics team, agreed to apply their latest computational simulations to a detailed analysis of our design. Having been involved in the design of America's Cup hulls – and normal commercial and military vessels – they had a particular interest

Professor Chrys Chryssostomidis, Director of the Department of Ocean Engineering at MIT. Of FastShip he said, "It's a great idea! Every now and then somebody comes along with a new idea and people all say 'it'll never work'. Then, when it does, they say, 'Of course, it's obvious!'" ('The Limit', BBC TV, October 1995)

in reducing all types of resistance. Their most *immediate* interest was in the flat wake flowing off the stern at higher speeds and the lack of a 'rooster-tail' as good evidence of a lifting stern, without recourse to tank tests.

Two years later, in July 1996, we had answers from the MIT Departments of Ocean Engineering and Transportation, and from large-scale model tests at the SSPA tank in Sweden: "Yes, FastShip will work. It scales up just as we would expect." There was never any question of it not scaling from any of the eminent people we consulted – test tanks, Universities or Classification Societies. It only existed in the fevered imagination of the MOD, YARD and Lloyd's Register: or as a monumental falsehood.

In the words of one MIT report, "The vessel is outstanding in all aspects considered of its hydrodynamic performance. Its remarkably low added resistance and modest ship responses, wave induced loads and relative motions, notwithstanding the extreme speed and high sea state, appear to be without precedence and point to the promise of the FastShip as a vessel capable of maintaining a speed nearing 40 knots in extreme North Atlantic sea states."[103]

Major shippers, such as General Motors, Dupont and leading Freight Forwarders agreed: "Yes, the market will pay the premium over sea freight for the service."

During the whole period of technical, commercial and financial validation, carried out over ten years by the best testing-tanks, University Engineering Departments, transport economists, financial planners and

103 *Evaluation of Hydrodynamic Performance of TG-770 using the SWAN Codes*, MIT, July 1996

shippers, the concept could not be seriously faulted – except by those *not* in possession of the facts.

FastShip, as a model being prepared for tank tests, and as finally classified to DNV Main Drawing Approval for a commercial container ship Class 100-A1+, showing layout of machinery, internal container decks and large internal volume available for Natural Gas or alternative fuels below the lower cargo deck. (For larger scale, see **Appendix 4**, Figure 6.)

Meanwhile the Philadelphia story was moving on. With the assistance of Tom Ridge, the new Pennsylvania Governor, and Tom Hagen, the Commerce Secretary, I helped interest the Norwegian shipbuilding group Kvaerner in reviving the Philadelphia Navy Yard. In October 1996, with Hagen, I flew to Finland, visiting the Kvaerner shipyards in Helsinki and Turku. We flew on to Oslo to meet their President, Diderick Schnitler, at a session attended by the Chairman of the Norwegian ship classification society, det Norske Veritas, who confirmed his confidence in our design. They were entirely supportive of the FastShip concept.

Using the possible construction of FastShip as a lure, we were able to ignite a spark of interest. Kvaerner went on to develop what today is known as the Philly Shipyard – the most modern commercial shipyard in the USA, in 2001 merging with the Aker and the French Alstom companies under the Aker name. By 2020 it had built twenty commercial vessels including the largest container ships from any US shipyard. Today it is the only commercially competitive shipbuilder of large container and other commercial vessels in the USA; and it only exists today thanks to FastShip's 1996 initial work persuading Kvaerner to take over the old Navy Yard.

The Philadelphia Port Authority was also financing the special FastShip terminal, incorporating the TTS-designed rapid load/unload system essential to the project. The TGC offices therefore moved to Philadelphia. TGC and FastShip Atlantic Inc. were combined into one Delaware Company: FastShip Inc. or 'FSI'. In June 1998, a new management team was selected, with Einar Pedersen, of TTS, Norway, as Chairman and Roland Bullard, a former First Union Bank Vice-Chairman and a well-known Philadelphia businessman, as President and Chief Operating Officer.

Kate Chambers came in as full-time Company Secretary. The management team continued almost unchanged, working on half-pay (and sometimes for nothing), while our UK technical team still comprised David Jenkins, David Bridges and their Osprey team in Emsworth. I was Founder and Technical Director, exclusively supervising the work of developing the FastShip design, overseeing its classification in Scandinavia and the USA.

Main Drawing Approval from Norske Veritas and The American Bureau of Shipping

Det Norske Veritas, or DNV, is the leading Classification Society for novel types of ship. Its standards for construction, design, safety and seaworthiness are the most rigorous in the world. To gain DNV classification for FastShip meant further testing of the hull. We carried this out in Norway, at the Trondheim University tank for normal sea states and by computational methods for extreme conditions – beam seas of up to 100 feet, for example, without power, fully loaded and at her most unstable condition.

By then we had used no less than four testing tanks and the computer analysis of five leading universities in the USA, UK, Norway, Sweden and Denmark. They all came up with the same answers. With the wealth of data on the 32,500-ton/770-foot FastShip vessel, the 3,000-ton/300-foot S90 and *Prelude* projects, it was quite obvious that our hull design would scale up to 42,000 tons and an operational speed of 40 knots, given the required power available from four or five water jets powered by RR MT-30 gas turbines. This was confirmed in 2001 when FastShip was granted Main Drawing Approval for DNV's highest Classification standard, 100-A1+, with construction starting immediately in any suitable yard.

Philadelphia and Pennsylvania remained keen to see FastShip happen – it could mean thousands of jobs in the shipyard and port, and throughout Pennsylvania. Under the US Title XI Loan Guarantee programme and a Strategic Partnership with Lockheed, we initially proposed building hull-modules at IZAR in Spain, where Lockheed were installing equipment in a new frigate of the Spanish Navy. These would be shipped to the new Kvaerner Philadelphia Shipyard for final assembly.

The sums of money involved were huge, as were the potential risks – and rewards. The illustration below, dated March 2003, gives an idea of the scale of the operation. We were within weeks of receiving the final agreement for the Maritime Administration (MARAD) Title XI Federal Loan Guarantees for $33.5 million as the foundation of our financial sources of $800 million, when it all blew up following the events of '9/11' – when a time of economic recession and financial insecurity was already established.

The Crippling of FastShip

The arrival of the George W Bush Administration, in January 2001, also changed everything. His newly-appointed Maritime Administrator, Captain Bill Schubert, was a Houston Shipping Agent who had worked closely for many years with our leading potential competitors, Maersk-Sealand, and

This company is ready to go

FastShip strategic partners are multinational and world-class:

Rolls Royce	Lockheed Martin	JP Morgan Chase
CP Ships	General Dynamics	BP
Republic of France	Delaware River Port Authority	Schneider Logistics
		IZAR

Current status

- FSI has already invested $37 million of private funding to create $2 billion project
- Ship technology developed, exhaustively tested, and approved by classification society Det Norske Veritas (DNV)
- Strategic partnerships arranged for technology, operations, marketing, project management and finance
- Supply contracts in place for engines, waterjets, ride control system, etc.
- Non-U.S. government commitments in place for $525 million of financing to complement MARAD Title XI funding
- Title XI account fully funded for FastShip at $33.5 million

Need MARAD approval for $800 million Title XI financing for two ships

Business Confidential 7 *FastShip, Inc.*

other major shipping companies in the container trade. The ghost of the Osprey Case had also returned to haunt us in the form of his Advisor, the British shipping consultant, Martin Stopford, who had been Director of Business Development of British Shipbuilders throughout the six years of the Osprey Case. He had also worked with Lloyd's List and Clarkson's, the world's largest firm of shipping consultants and was closely connected with the UK community of naval architects. On June 6th 2002, following a full-day presentation by FastShip, Schubert dismissed the FastShip programme as being 'too great a technical risk' (see **Appendix 7** for details of the commercial arguments supporting FastShip service). This was despite Classification by the American Bureau of Shipping and by det Norske Veritas to Main Drawing Approval and 100-A1+ standards. He also cited insufficient speed advantage over existing container services to justify the extra cost.

MARAD's decision only took account of the absolute comparative speed at sea, with no allowance for the 20–30 hours saved in port turnaround by the TTS rapid loading system, or the fact that container ships 'inter-ported' between several different ports during an Atlantic round trip taking a month or more, with frequent delays, while FastShip would operate entirely independently of other services – as applauded by leading logisticians.

A few weeks after this decision, there was a final meeting with Schubert and the US Navy Secretary, also attended by Representative Duncan Hunter and Senators Arlen Specter, John Warner and John McCain, the Senator from Arizona. Hunter, Specter and Warner were in favour of providing Title XI Federal Loan Guarantees, for which Congress had appropriated the required collateral of $40 million, but McCain, representing the landlocked State of Arizona, used his favourite expression: 'corporate welfare'. Faced with his formidable influence in the Bush administration, the others capitulated. It was the end of Title XI for large vessels and, as it later transpired, for anything except small conventional ships, tugs and ferries. It seemed like another intervention of the combined might of the Military-Industrial Complex, against which President Eisenhower had warned his contemporaries.

Despite the withdrawal of Title XI Federal Loan Guarantees, FSI struggled on – until two more unforeseen events made survival impossible. The first was the Madrid terrorist bombing in March 2004, which precipitated a change of Government in Spain and the cancellation of all shipbuilding subsidies. In May 2005, the Spanish shipyard IZAR, where we were about to start building hull-modules for FastShip, went bankrupt. Then, in 2008, came the world financial crisis, a crash in the freight market, and the collapse of our Lead Investor, Royal Bank of Scotland. This led to a chain reaction, with JP Morgan Chase and other banks and investment funds withdrawing their support.

National Defence Features

FastShip had not quite foundered, however. The US Military continued its interest in the availability of commercial vessels ideal for the rapid overseas transportation of military wheeled equipment, from heavy recovery vehicles, tanks and other armoured units to helicopters and Humvees. An MIT Class of US Naval Engineers had recommended FastShip as the most viable high-speed platform combining a 'Sealift Option with Commercial Viability' – the title of their thesis. This, however, was dependent on the existence of a commercial FastShip service. In the cautious years that followed the financial crash, when cargo ships took to steaming slowly across the world's oceans, our hopes of establishing such a service seemed ever more unlikely to be fulfilled.

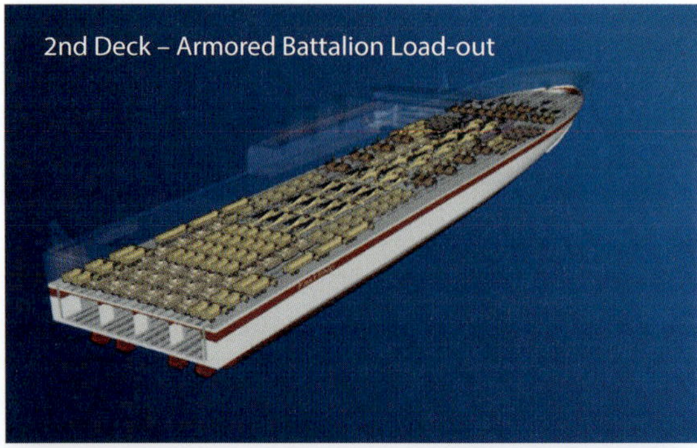

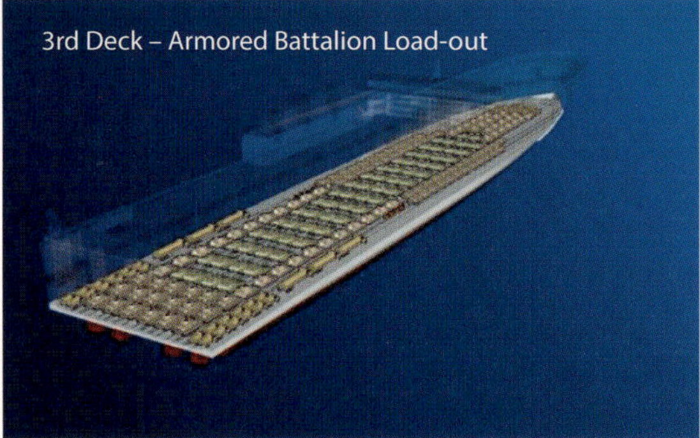

Fast Sealift: How FastShip could be loaded up with military or humanitarian equipment to carry 10,000 tons over 10,000 nautical miles in about 10 days. To carry the same load over the same distance in the same time would require 54 C-17's, the US Transportation Command's latest airlifter.

Prelude – TGA's Project for the Littoral Combat Ship, or LCS, as presented to Lockheed Martin and Gibbs & Cox, November 2003, 16 years after 'SuperShip' was proposed to the US Navy for FF-X.

CHAPTER 35

2003: The US Navy's Littoral Combat Ship

Our work on FastShip was not entirely wasted. Very slowly, between 1989 and 2003, thinking in the US Navy had evolved. The official attitude towards our designs changed from hostility to polite interest. On February 20th 2003 I was surprised to be invited by the Navy to an 'Industry Day' Seminar, at which they announced a requirement for a very unusual type of ship: the Littoral Combat Ship – the aftermath of the FF-X. This was to be a fast frigate of about 3,000–4,000 tons displacement, with a draft of under 20 feet, a preferred sprint speed of 45–50 knots at full load, and a range of 4,300 nautical miles at 20 knots (1,500 miles at 45–50 knots). The 'modular combat package' (i.e. weapon load) was to be 210 tonnes, with 'exchangeable' modules for mine-hunting, anti-submarine, anti-aircraft, missiles or torpedoes, based on the Danish Navy's 30-knot Stan-Flex 500, the GRP version of TGA's Osprey. With room for a crew of up to 75, the target cost was $250 million per vessel. It was to be the fastest ocean-going warship ever built.

The Requirement was chiefly in answer to the 42-foot Swedish-built 'Boghammer' 46-knot speed-boats firing missiles, high-calibre guns or torpedoes against tankers and US Navy ships in the Arabian Gulf during the Gulf Wars between 2003–2011; even suicide missions were mooted as potential threats ... An echo, perhaps of the Royal Navy's successful use of Thornycroft's 55-foot, 40-knot Coastal Motor Boats, one of which sank the Bolshevik cruiser *Oleg* at Kronstadt in June 1919, for which Lt. Augustus Agar was awarded the Victoria Cross.

The Industry Day audience was stunned. Many claimed that with existing technology the Navy's was an impossible Requirement. Its low cost demanded a relatively short, fat hull with unparalleled speed and long transoceanic endurance with a heavy combat load. It was unprecedented: a traditional hull would be twice the length and cost. There was talk of several types of design being selected. These included FSI's Semi-Planing

Mono-Hull – or SPMH as the US Navy termed it – and three versions of multi-hulls, all of them untried for this size and type of warship. Tom Ingram, head of the American Bureau of Shipping in Washington, responsible for the LCS classification, said: "It looks like they need one of yours – now's your chance!"

To me it seemed obvious: any such vessel would be a revival of *Prelude* or SuperShip, powered by gas turbines and water jets. The hull and propulsion system would be covered by my worldwide Patents. Lockheed Martin were our strategic partners on the FastShip project and would be one of the lead contractors. Six days later we were invited to visit the Lockheed Headquarters in Bethesda, Maryland, together with James Higney, President of Gibbs & Cox, one of the most respected names in US naval architecture, designers of the SS *United States* and builders of most of the frigates and destroyers for the US Navy since the beginning of World War II. We spent two hours going through the *Prelude* tank tests and data, which I had updated to the Navy's LCS Requirements. We discussed our US Patents and associated issues.

Afterwards, Higney questioned me about my presentation. I asked him if he thought the tests for this type of hull would scale to a full-size ship, contrary to the often-repeated views of both the Royal and US Navies.

"Of course," he replied, "but you were twenty years ahead of your time."

I thought of the Conclusions of the Lloyd's Hull Design Inquiry ordered by Mrs Thatcher in 1986, and MOD statements that "there are no identifiable implications for the design of future destroyers and frigates for the Royal Navy arising from the Sirius design concept". It was almost impossible to restrain myself from making some ironic remarks but, with the Lockheed people eyeing me savagely, I refrained.

A few days later we were invited to give a presentation to Rear Admiral Sullivan, in charge of the Navy's ship design programme. US NAVSEA had acquired a quantity of information and confidential technical data on FastShip over the past 14 years, so the claims of our design family were well known to them. But we also produced a project drawing and information on the *Prelude* design scaled to satisfy the basic Littoral Combat Ship Requirements. FastShip Inc. was immediately invited to join the Lockheed/Gibbs & Cox Design Team for the LCS. Formal Agreements for confidentiality and non-disclosure were signed and public announcements issued during February and March 2003. On June 23 2003, a press report was published on the LCS project: Arthur Cebrowski, Director of Force Transformation of the Department of Defense, stated there were "at least two designs that are very, very promising", without naming which contractor teams proposed them. In discussing one new type of hull, Cebrowski said: "It permits a ship to recapture its own wake." He predicted: "That changes all the arithmetic for ships ... and that's a real possibility. I'm so convinced it's a good idea that I'm going to buy a scaled-up version of that to prove scalability, and give it to Navy personnel for experimentation."[104]

Was he referring to the semi-planing hull, which recaptures its own wake (or captive wave), and was said by NAVSEA in 1989 'not to scale up'? It seems likely: only a month later, on 27 July 2003, the Lockheed LCS proposal was selected, along with two lightweight aluminium multi-hulls, for the second contract phase of the competitive tender.

Notwithstanding the enthusiasm expressed by Lockheed and Gibbs & Cox for our suggested solution to the taxing LCS Requirement, our subsequent participation in the design was minimal. In March 2003, we were sent a rough memo suggesting a budget of $15,000 for a performance comparison of alternative naval hull types, most of them unknown to us. I was on leave in London and it had to be ready within a week. For such a derisory sum, we risked forfeiting our Patent rights. Nevertheless, I did what I could in the time available, putting together a selection from our vast archive of data, and sent it off.

In September 2003, at a three-day meeting at the TGA Emsworth Office with Lockheed and IZAR, their Spanish shipbuilding partners, a large amount of FastShip tank-test, classification and construction data was passed to the Lockheed LCS Design Team. About the same time we were told that the hull design had progressed to the point that it was too late to incorporate any of the modifications we had suggested. They were clearly not going to heed our advice, although we showed a 25% reduction in power for 45 knots, based on FastShip data compared with our early FF-X estimates.

In May 2004, the Navy announced that Lockheed had been awarded the Final Construction Contract for its variant of the LCS. It was to be based on a 'revolutionary semi-planing monohull' powered by Rolls-Royce Trent MT-30 marine gas turbines driving water jets made by KaMeWa, now a

104 (71) *Navy News Week*, June 23 2003

Rolls-Royce subsidiary. It was almost the identical propulsion arrangement we had developed with Rolls for FastShip, with KaMeWa water jets, over eight years ago. Since Rolls then owned KaMeWa (in part thanks to TGA's work with their former owners, Vickers Ltd) over the course of sixteen ships – requiring at least 32 MT-30 gas turbines and 64 water jets – Rolls/KaMeWa made quite a killing. All largely due to TGA's FastShip initiative made in direct opposition to the conclusions of the MOD, Lloyd's Inquiry and the high priests of the UK's naval architectural community.

We immediately advised Lockheed that there were Patent issues concerning their design, as recognised in our joint Proprietary Information Exchange Agreement.

LCS USS *Fort Worth*, 336 ft/102.35m LWL. Beam: 59ft/18m. L/B ratio = 5.7; 3,500-tonnes Full Load Displacement. Max. Speed: 50 knots – FNo. = 0.79/SLR = 2.65.
The waterline shows the slight crest of the wave alongside the position of the water-jet inlets, indicating high pressure. Having sunk slightly up to about 28 knots, the onset of hydrodynamic lift occurs and, at 32 knots, she has risen back to her normal 'static', or stationary, waterline.

They rejected our claim on the grounds that theirs was now a 'double-chine hull', while ours was of 'round-bilge type', based on years – and generations – of experience of the semi-planing hull. That did not obscure the fact that it was clearly a large semi-planing hull within every possible variation of that hull-type allowed by the Patent. In our experience the use of the double-chine – meaning the hull comprises six flat surfaces as opposed to the four on the 'Italian shoe-box' type of fully planing hull – would degrade the performance and create undue stress at the joints, or chines, of those flat hull surfaces – particularly at speed in higher seas. IZAR made the same warning to Lockheed. And so it did, resulting in severe leaking. LCS also comprised a sixteen-foot loss of beam compared with our recommended hull-design for the same power, speed and displacement. The effect on low-speed stability, helicopter deck area, and useful interior volume was also detrimental. Finally, the water-jet outlets had to be installed against the advice of KaMeWa, squeezed onto the narrower stern. Very important also was the effect of extreme roll-angles on crew-comfort in ocean conditions in which the LCS is expected to spend some 70 per cent of its time at speeds of seven knots or below. It is in this area that Lockheed's Freedom class LCS has a poor reputation with its crews. The same mistake that was made in the Royal Navy's 'Osprey crib', HMS *Peacock*, during the previous century. The Freedom class of LCS therefore has not been a success and several were laid up after only a few years of service. But it did prove a point: that the benefit of hydrodynamic lift, enabling very high speeds, is not limited to 'quite small craft'.

A Farewell to our Champion

Over the years I had visited Lord Hill-Norton quite frequently and learned to relish his wisdom, gleaned from a lifetime in the Royal Navy and public service. His picturesque Cass Cottage, on the edge of the New Forest, became something of a place of pilgrimage for me, just as Charles Adams's wood mill had been until his death in 1998. We had many conversations sitting in his rose garden, while he deplored the iniquities of 'the abominable Bath men'. He had also suffered some distress – particularly from the deceit and contumely of the Lloyd's Inquiry, whose claimed 'independence' was initially endorsed by him.

Despite failing eyesight, Hill-Norton insisted on reading every page of an early draft of this book before writing his Foreword. I last met him in Dorset in April 2004, at a small beach hotel, where he was staying with some of his family, a few days before his sudden death. He was completely lucid and we spent a happy hour going over the 'short, fat' story in the hotel lobby. Typically, he raised sniggers from fellow-guests with his loud and salty turn of phrase. At long last, the conclusions of the 1986 Hill-Norton Committee had been vindicated by the work at MIT and by the SSPA tank tests – and, now most likely by the US Navy. In particular, in two important aspects of his Report: a significant increase in the operational speed of future destroyers and frigates in rough ocean conditions without a significant increase in power; and the ability to carry a much higher air-defence radar and associated close-in defensive weapons. His Committee had concluded: "If this is confirmed, it is a most important military advantage."

A few weeks beforehand, I had been able to tell him that the Lockheed version of our Patent was the front-runner for the LCS contract ... "We've won!" he had exclaimed joyfully down the phone from Cass Cottage.[105]

Lord Hill-Norton died suddenly on May 16th 2004. Only a week later, in an appropriate tribute the US Navy announced the award to Lockheed of the construction contract for the first Littoral Combat Ship. An important part of their success was their Contract Proposal's claim that FastShip Inc., "their design team partner, had provided access to $40 million-worth of Research & Development data". One could only wish they had used it more wisely.

I was honoured to be invited to Hill-Norton's Memorial Service at St Martin-in-the-Fields, traditionally the Navy's church, in London's Trafalgar Square. In his Address, the First Sea Lord, Sir Alan West, made a disparaging remark about Hill-Norton's interest in strange ideas, mentioning UFOs and the 'Short, Fat Ship'. How wrong he proved to be in the latter case, having claimed a few years later that the Type 45 was "the most capable destroyer ever". Afterwards, I lunched with a member of Hill-Norton's family, and there was no hint of regret for this great man's latter years, over the support that he had so courageously given us.

---

105 Kenneth Rawson writes: "It was sad to see in his obituary in 2004 the report that Lord Hill-Norton bitterly regretted his association with Giles and his short fat ship and the stain on his reputation that had been caused. I have to say that I feel rather less distress because Hill-Norton would have had call on numerous well qualified people who could have advised him." Ibid, p. 121.

DLG: Of course later events have shown that such advice would undoubtedly have been wrong. Before this statement, MOD Ship Department ('the abominable Bathmen' in his words) were not known to be participants in the Lloyd's Inquiry, unless, like the DSAC, they were employed covertly, to advise Lloyd's. In any event, that was contrary to the Inquiry's agreed Terms of Reference. In such circumstances the Inquiry could not have possibly have been considered 'independent'. Nor did I ever hear of any such obituary. KJR's statement is completely at odds with the obituaries I read (including the anti-S90 *Daily Telegraph*'s: 'He had an uncanny reputation for being right') and the company that I enjoyed with Lord Hill-Norton over twenty-one years until four days before his death.

CHAPTER 36

2008–2016: The Short, *Fast* Ship and an 8-Year Court Case

In October 2006 the first Littoral Combat Ship, LCS-1, USS *Freedom*, was launched. In August 2008 she went on trials, easily exceeding her contract speed of 40 knots, in later trials claiming a speed of 50 knots – albeit at a light displacement. Pre-launch photographs of the vessel on the slipway showed a slight forty-foot long 'hook' beneath the stern almost identical to that of TGA's *Prelude*, S90 and the Type 45 (see pp. 294–295, Figs. 2–5). Naturally, like a plane's wing, the length, depth and area of the 'hook' or lifting surface varies according to the hull's size, optimium lifting speed, proportions and displacement, to obtain the optimum lift coefficient (or C/L in aircraft). Later, in photos of the high-speed trials at over 40 knots (see LCS-3 *Fort Worth* in **Appendix 8**, Fig. 1) her full waterline was visible, with a slight wave crest at the point of the water-jet intakes, confirming there is greatest positive lift beneath the hull at that point. That is typical of our semi-planing designs. Having sunk slightly at lower speeds, the hull lifts at semi-planing speeds, back to its static level and progressively higher with increasing speed.

The USS *Freedom* was accepted into service by the Navy in September 2008. Thus the large semi-planing monohull, conceived in 1981 and roundly condemned by professional naval architects, was recognised 25 years later, in the form of USS *Freedom*, described by the US Navy as *'revolutionary'* or, to use Kenneth Rawson's memorable phrase, it represented *'an important contradiction of traditional theory'*. Although the hull of USS *Freedom* is not as we would have wished, the evidence of our principal design feature – its lifting quality – was complete.

On April 11th 2008, Don Stout, my Patent Agent since 1989, delivered an Administrative Claim to the Patent Counsel of the US Navy, for compensation in recognition of the breach of US Patents 5,080,032 and 5,231,946. Our Claim was put to one side by the US Navy Legal Counsel.

Our lawyers and some politicians representing FastShip had a meeting with Sean Stackley, the US Navy Under-Secretary for Appropriation, during 2008–2010, in an effort to find some form of accommodation. There was much correspondence, explaining in full the history of transfer of data to Lockheed and Gibbs & Cox.

However these overtures were abortive. Meanwhile, the launching of the second LCS, *Fort Worth*, was delayed by a year, to a few days after our Patent's expiry date, thus preventing her being subject to a further claim.

Two years passed without progress. The reason soon became obvious. On 28th April 2010, a few days before the expiry date of my US Patent, the US Navy finally acknowledged our Administrative Claim. It was curtly denied in a letter worthy of all those who had previously condemned the idea of the 'Short, Fat Ship' over the past thirty years. Finally acknowledging our Claim, the US Navy responded:

With respect to the rationale for the denial, each apparatus claim of the subject patents requires, among other things, "a hull having a non-stepped profile which produces a high pressure area at the bottom of the hull in a stern section of the hull," "at least one inlet located within the high pressure area," and "at least one waterjet coupled to the at least one inlet." Additionally, each method claim of the subject patents requires, among other things, "hydrodynamically lifting a stern section of a vessel hull at a threshold ship speed by virtue of a high pressure region at the bottom of the hull," and "propelling the hydrodynamically lifted hull via a waterjet system having water inlets in the high pressure region."

After a thorough analysis of the subject patents, including the corresponding prosecution file histories, and of the design of the LCS-1 U.S.S. Freedom ("LCS-1") by Navy personnel, including simulated testing of the hull, the Navy has determined that the hull of the LCS-1 does not include, among other things, the above-mentioned claim limitations. More specifically, the Navy analysis has determined that the hull of the LCS-1 does not include, among other things, a high pressure area or region, as claimed.

If you have any questions concerning this action, please contact the undersigned.

Sincerely,

Thomas P. Hilliard, Esq.
Section Head
Intellectual Property Law Section
Office of the General Counsel
Naval Sea Systems Command

Into Battle – Thanks to our Stout-Hearted Supporter

Due to the depleted financial state of FastShip Inc., it was first necessary to apply for Bankruptcy under the US status of 'Chapter XI'. This placed the company under the control of a Court-appointed Trust, but still able to employ its assets.

Our only viable asset was the TGC Patent, but we had to finance our Claim. This meant meetings with the sort of venture capitalists who specialise in financing court actions. These were dispiriting encounters. At one such meeting, the lawyer concerned announced his aversion to any sort of technical argument. The huge cost and consequences of depending on such a source to finance an unusually technical – though basically simple – Claim against such formidable defendants seemed prohibitive.

It was at this stage that Don Stout came to the rescue. Between our first meeting in July 1989 and 2010, he had taken a major stake in NTP, the company holding the Patent for the first mobile phone technology to connect with the Internet. After six years of haggling over the infringement of NTP's Patent by the BlackBerry, made by the Canadian company RIM, NTP obtained a Settlement of several hundred million dollars. This occurred after the US Government insisted on a Settlement because of the millions of BlackBerries in its service, use of which might otherwise be banned. Don then agreed to cover all of the costs of our Patent litigation.

How prophetic, I now realised, was the MIT Finance lecturer, in 1989, speaking of the need for "a business plan, a co-efficient of passion, and a patent ..." In our case, it also came with a source of finance to protect it. Without Don's help in writing my Patents – and without his and his wife Mary's generous financial support for our Court Case to defend them – we would never have had a hope.

Under Chapter XI rules a Trust for FSI was established in the New Jersey Court before we were able to file a Complaint for Breach of Patent. Setting up the Trust, including its financing, took two years, so we were not in a position to file our Claim at the Federal Claims Court until August 2012. Although weeks or even months can pass before a Judge agrees to take up such a case, ours was taken up by Judge Charles F Lettow in a matter of hours. It was perhaps our final turn of good fortune.

Judge Charles F Lettow was an exceptional man. He held degrees in Chemical Engineering and History, and had a special interest in 17th Century England – the time of Pepys's account of his work at the Admiralty and the birth of the Royal Navy. As a young Clerk to Chief Justice Warren Burger he had worked at the US Supreme Court. His favourite legal quotation was from Oliver Wendell Holmes, the former US Chief Justice: "A page of history is worth a pound of logic." His family history was also relevant to our case. His grandfather, General Paul Emil von Lettow-Vorbeck, commanded the highly effective, and respected, German guerrilla forces in East Africa in World War I, during which one of his patrol vessels on Lake Tanganyika was sunk by a small Thornycroft armed steamboat. This event became the origin of the film 'The African Queen', starring Humphrey Bogart and Katharine Hepburn. Later von Lettow became a close friend of the British C-in-C, General Smuts, and turned down the Ambassadorship to Britain in 1936 out of contempt for Adolf Hitler. He was replaced by Ribbentrop and suffered for his refusal, being reduced to poverty, though supported by his British admirers, led by his former adversary, General Smuts, and others.

We thought ourselves lucky to have his grandson as our Judge, considering his legal, historical and technical qualifications.

The Missing Link

The first documents were not released by the defendants until late 2013, five years after our initial Administrative Claim. During the following three years, until the full trial opened in October 2016, I examined over a million pages of evidence – most of it 'Protected' or 'Confidential', therefore unable to be published here: the 'Snowstorm Treatment'. Some limited information was made public by Lockheed and the Navy during the development of the LCS, but apart from that I am permitted to quote only from Judge Lettow's 'Redacted Trial Decision' of May 5th 2017, as published by the Court.

I am also free to reveal that those million pages of evidence were uncannily like the evidence meted out by British Shipbuilders in the Osprey Case: vast amounts of irrelevant technical detail flung into the air to blur the issues and confuse the inquirer. We had to understand that detail, and respond to it point by point with force and clarity: a laborious and enormously costly exercise designed to demoralise, and possibly bankrupt, the most determined opponent.

I was confident, however, that we had the ammunition to win this case on the three major issues. These were:

1. The presence of hydrodynamic lift from the 'hooked' stern on the LCS hull.
2. The additional lift imparted by water jets with inlets in the same high pressure area.
3. The LCS proportions, size and speed were within the limitations of my Patent.

The first two were the only two significant grounds for denial of our Claim given by the Navy. The other factual claims, being within the Patent, could not be denied.

The incorporation within the LCS of my Patents' Claims was a simple matter of fact, endorsed by an inspection of the hull of LCS-7 at the Marinette Shipyard some weeks before the Full Trial. I attended along with the Judge and Court officials and representatives of the Plaintiffs and Defendants. The gentle forty-foot long 'hook' beneath the stern was there for all to see.

The next day there was a similar pre-Trial visit by the same people to the Navy's huge testing tank, the high-security Naval Surface Warfare Center, at Carderock, Maryland. We were there to be given a presentation on the technique of tank-testing of LCS models by the staff; and to inspect the largest model of LCS-3, an adaptation of the original LCS-1 model, as used in the final tank tests.

Having examined all the model test evidence previously, it appeared to me that the LCS-3 model, and its original version as LCS-1, might *not* have the all-important hooked, or 'lifting' stern – the most visible contradiction of the USA's evidence. This was because the pressure-sensor readings in the disclosed Test Reports showed *negative* pressure in the area of the water-jet intakes. That would indicate that the stern was a convex, or 'rocker' stern, rather than concave, or 'hooked'. So in my car I took with me a yardstick straight-edge to define the contours on the model, if the opportunity occurred.

It did. So I asked if I could return to the car park to collect an item of equipment, accompanied for security reasons by a curious Drew Zager, the USA's Senior Counsel.

"Why do you need that thing?" he asked.

"You shall see," said I.

In silence we returned to the testing tank. All the personalities were assembled: the Court photographer, Judge Lettow, the Defendants' Counsels, Dr Garwin and our Counsels. Model 5640 – some 18 feet long – was set on trestles. As the assembled company looked on, I placed my yardstick along the base-line of the hull. Instead of a hollow space above the middle of the yardstick, showing a concave shape, the space occurred at the end – showing a gap of some three inches between the model's transom and the yardstick, as best shown by the shadow on the wall in the photo's background. There was an audible groan from the Defendants – and some expressions of satisfaction from the Plaintiffs. Judge Lettow was intrigued. The final Infringement Ruling of Lettow, published on May 5th 2017, tells the story:

> "The parties and the court observed Model 5640 in its current condition during their site visit to Carderock on September 7, 2016. Upon inspection, Model 5640 appeared as follows:
>
> 
>
> As this photo shows, Model 5640 appeared to have a convex 'rocker' stern rather than a hooked stern (Garwin). However, the lines and body plans of LCS-1 show a hooked stern rather than a convex stern. A convex stern would create low pressure at speeds over 40 knots, while a hooked stern would create high pressure at such speeds (McKesson). The court concludes that the stern of Model 5640, as inspected, does not accurately represent the stern of LCS-1; the totality of evidence shows that LCS-1 has a hooked stern rather than a convex stern, which generates high pressure and hydrodynamic lift." (Court Report)

In a very lame excuse for their error in exposing the model's false 'rocker', or sinking, stern the USA's contingent made a hurried, if laughable, explanation: It was due to the model's stern 'warping' during storage, 'having been stored upside-down.' Of course such an admission from, possibly, the world's premier testing-tank, where a model's accuracy must be presumed to be within a mini-millimetre of its full-scale progeny, the model's stern would have had to droop several inches. It was absurd to suggest that, at a time when the model was bound to be in constant use. As the growing fleet of LCS vessels entered service, with continual trials and tribulations to be investigated, the model would be much in demand.

With such evidence of intent to deceive, presented to Judge Lettow in all its garish detail – and recorded by the official Court photographer – we were optimistic of the outcome. Indeed, the famous photograph of my Bare Woods & Accessories yardstick disclosing the USA's intent to deceive, loomed large in Judge Lettow's final verdict.

CHAPTER 37

October 2016: The Trial Scene

Lafayette Square, Washington DC: St John's Church is behind the statue of President Jackson astride his rearing horse. The brick-fronted Federal Courts Building can be seen on the right.

Monday, October 3rd 2016 – Lafayette Square again ... over seventy years after my Sunday trips as a small boy – and forty years since Peter and I had established Thornycroft, Giles & Associates Ltd. The opening day of the Trial that would crown our efforts with success or failure. Would all the trials and tribulations over earlier years be brought to a satisfactory conclusion, or would they fail – to the delight of our detractors?

After contemplating my past associations with Lafayette Square, I strolled across to the red-brick Federal Courts Building, and through the courtyard with its silent stone fountain. In the lobby I met my team of lawyers and our Chief Expert Witness, the brilliant scientist Dr Richard Garwin. We all believed that certain Claims of my 1990 US Patent had been used, unlawfully, in the US Navy's new class of sixteen ships, claimed to be 'revolutionary' and 'game-changing'.

Entering the marble hallway of the Court, I was comforted by Lincoln's generous words inscribed overhead:

"It is as much the duty of Government to render prompt justice against itself, in favor of citizens, as it is to administer the same between private individuals."

*****

There followed the greetings of my own legal team and expert witnesses Chris McKesson and Dick Garwin, the suspicious eyeing of the five Defendants' lawyers from the US Department of Justice and their coterie of assistants; the inspection of bags and files, questions from the Security Detail; the checking of my Passport and personal data. Then, with the usual bland courtesies expected on such occasions, we all squeezed into an elevator.

We entered a large, high-ceilinged, oak-panelled chamber. A central Judge's throne faced us. Below it a polished barrier stretched either side, behind which were seated the Clerk of the Court and various assistants. The Defendants were allocated two large desks to our left as we entered, with our own Plaintiffs' tables to the right. Against the rear wall of the Court Room were two rows of seats for onlookers.

On top of the barrier, the Defendants had placed a fine model of the Littoral Combat Ship. To its right stood an accurate model of our 1988 Monohull Fast Sealift design that was the basis of my Patents. Mischievously, for good luck, and as a gesture to my father, I placed an exquisite model of a RN World War II 'Fairmile' patrol boat inside an Eno's Salts bottle, that he made for me as a child. He was a wonderful model-maker.

We unpacked laptops, iPads, files, exhibits and other legal paraphernalia, murmuring in hushed tones as we assembled our wares in high stacks on the tables. As the hands of the clock approached nine, we sat in silent expectation.

The Trial

US005080032A

United States Patent [19]
Giles

[11] **Patent Number: 5,080,032**
[45] **Date of Patent: Jan. 14, 1992**

[54] MONOHULL FAST SEALIFT OR SEMI-PLANING MONOHULL SHIP

[76] Inventor: David L. Giles, 4244 Hunt Club Cir., Apt. 1534, Fairfax, Va. 22033

[21] Appl. No.: 525,072

[22] Filed: May 18, 1990

[30] Foreign Application Priority Data

Oct. 11, 1989 [GB] United Kingdom 8922936

[51] Int. Cl.[5] B36B 1/04
[52] U.S. Cl. 114/56; 114/121; 440/3; 440/38
[58] Field of Search 440/40–43, 440/3, 4, 38, 6; 60/222; 114/56, 121, 125, 271

[56] References Cited

U.S. PATENT DOCUMENTS

1,270,134 6/1918 Emmet 440/6
2,185,430 1/1940 Burgess 114/56
2,185,431 1/1940 Burgess 114/56
2,342,707 2/1944 Troyer 114/56
3,122,121 2/1964 Krauth 440/40
3,826,218 7/1974 Hiersig et al. .... 440/3 X
4,004,542 1/1977 Holmes 115/14
4,079,688 3/1978 Diry 114/56
4,276,035 6/1981 Kobayashi 440/47
4,523,536 6/1985 Smoot .... 114/67 R
4,775,341 10/1988 Tyler et al. 440/38
4,843,993 7/1989 Martin 114/125

FOREIGN PATENT DOCUMENTS

409181 4/1934 United Kingdom .
739771 11/1955 United Kingdom .

Primary Examiner—Sherman Basinger
Attorney, Agent, or Firm—Antonelli, Terry Stout & Kraus

[57] ABSTRACT

A vessel (10) has a semi-displacement or semi-planing round bilge hull (11) characterized by low length-to-beam ratio (between about 5.0 to 7.0) and utilizing hydrodynamic lift. The bottom (15) of the hull (11) rises toward the stern (17) and flattens out at the transom (30). Four waterjet propulsion units (26, 27, 28, 29) are mounted at the transom (30) with inlets (31) arranged on the hull bottom (15) just forward of the transom (30) in a high pressure area. Water under high pressure is directed to the pumps (32) from the inlets (31). Eight marine gas turbines arranged in pairs (36/37, 38/39, 40/41, 42/43) power the waterjet propulsion units (26, 27, 28, 29) through combined gearboxes (44, 45, 46, 47) and cardan shafts (48, 49, 50, 51).

20 Claims, 13 Drawing Sheets

"All stand!" announced the Clerk of the Court in her ringing voice.

Judge Lettow appeared through a door in the panelling. He paused, looking down upon the Court. A tall, slim, distinguished figure in his seventies with a fine head of white hair, he sported a blue and white spotted bow tie, with the regulation black robe. I felt a surge of optimism: he looked exactly as I had hoped: a thoughtful, academic type, no extremes of appearance. I liked the bow tie in particular, a favourite sartorial touch of my father and his friendly rival designer, the great Olin Stephens.

Representing FastShip Trust LLC was our Lead Counsel, Mark Hogge. An urbane engineering type, with a good, methodical presentation. Out of Court, he was building his own flying replica of a World War I fighter plane, so was fully aware of the minor technicalities of the argument. Not that, in Court, there were many arguments to be settled apart from the 'lifting factor' – all the other Patent Claims were admitted by the USA.

This crucial matter to be decided was whether the LCS hull sank or lifted above a 'threshold speed'. And, if it did lift, was this due to high pressure exerted by the aft portion of the hull and further augmented by the flow of water through the water-jet inlets? Or was there no such high pressure?

This was the principal claim of the US Justice Department – as it had been of the British Ministry of Defence and the Lloyd's Hull Design Inquiry, the forces ranged against us thirty years before.

'The Proof of the Pudding ...'

Mark Hogge opened with a dissertation on the history of semi-planing monohulls in smaller vessels: how they had been proven during the twentieth century in patrol craft and workboats of up to about 120 feet in length and 350 tons displacement, with speeds of up to 30 knots. In the 1970s, a series of larger designs had been produced in Denmark, based on the 165-foot, 450-ton 'Osprey' design by Thornycroft, Giles & Associates. The Osprey was unusual in having a much greater beam for its length than traditionally associated with speeds of over 20 knots for its length and volume. It therefore became known as the 'Short, Fat Ship' – in contrast to the 'Long, Thin' shape traditionally used for fast vessels.

Variants of Osprey, he explained, were built for several Navies. Testing-tank and full-scale measurements showed that, up to a speed of about 22 knots, their hulls sank, as with traditional designs. Above that speed, however, they ran level and then rose significantly. This meant that the drag from the volume of water displaced by the hull was reduced in proportion to increasing speed compared with traditional designs of the same proportions, which sank as they went faster, thereby increasing their displacement and drag.

There was much opposition to the idea, claiming it was contrary to the teaching of Froude and other originators of the so-called 'Laws' of naval architecture, as well as most present-day naval architects in the Royal Navy and commercial shipping. Yet it had the endorsement of leading international testing tanks, Classification Societies and physicists, including the Plaintiffs' Chief Expert Witness, Dr Garwin.

*****

The Defendants' Lead Counsel, Andrew P Zager then stood up to present their case. A swarthy, shorter, if no less determined character, usually wearing black boots, he expounded the theory that the LCS was based on a variety of existing designs, unrelated to the FastShip hull form. He recited a family of lightweight 'V-bottomed' high-speed European car-ferries that were of very different proportions to LCS. He also claimed that the LCS was something of a 'hybrid', incorporating features of the 220-foot, 400-ton,

60-knot, fully-planing speedboat hull of the yacht *Destriero*. While both of these 'parent' hull forms only lifted at fully-planing speeds, Zager claimed, the LCS did not at all. This had been proved by the simulations of their key Expert Witness, Dr Stern of the University of Michigan, and by the depositions of the USA's previous Expert Witnesses: the *Destriero*'s designer and a hydrodynamics expert responsible for conducting the LCS tank-tests at the Naval Surface Warfare Center, which we had visited the previous week. These were their leading arguments. With objections from both sides and many interventions, their presentation took up the first three days. However the basic argument remained the same: did the LCS 'sink or lift' at semi-planing speeds?

Despite the tension, the high stakes, the gravity of the proceedings, there were occasional lighter moments. A friend who attended on the fourth day was coming up in the lift with a number of the Defendants' team. They were joking about "that funny old guy who's joining the FastShip people – is he one of their experts?"

Later there followed the first moment of drama as Dr Garwin was introduced to the Court. The Senior Counsel for the Defence complained that Garwin was not a naval architect. Judge Lettow quickly replied: "But he's one of the world's leading experts on fluid dynamics, and we're looking at the effects of an object moving through two fluids: air and water ... Objection over-ruled." This was greeted with a look of astonishment on the faces of the Defendants. So that explained the 'funny old guy'.

'... is in the lifting'

A few days before the opening of the Trial, Don Stout's wife, Mary, had suggested that a simple model of the LCS hull might be used to bring to life the lifting properties of the hull and water. I asked a few possible sources in the Washington area about such a model and was told it would take several weeks and cost $15,000. I therefore phoned David Jenkins in the Isle of Wight. He said that, in Cowes, there was a workshop that, from lines plans, could cut precise expanded polystyrene scale models of different hulls, from surf boards to racing dinghies, for tank testing. Using the 'Rhino' computer programme for the LCS-1 hull, obtained from the Defendants' disclosures, they would be able to provide such a model within four days, at a cost of £600. David ordered the 2′ 6″ model. Four days later, the model flew from London to Washington as hand-baggage and was delivered to the Court.

On the fourth day of the Trial, Dr Garwin was cross-examined. To demonstrate what he had already proved numerically, he placed the model in Court on a balance, with one of his laser-protractors directed to a scale on the wall. By moving a pile of US pennies backwards and forwards on the hull, he was able to reproduce precisely the difference in speed, trim angle and lifting force of the high pressure created by the hull and within the water-jet inlets. It was watched by everyone in the Court with expectation, awe, concern, or horror, depending on whose side one was on. Judge Lettow leaned forward attentively, fascinated by the demonstration. Dick Garwin takes up the story, as he described it to me – a marvellous account of a world-class physicist at work; like Faraday conducting his experiments on electricity and magnetism before an audience at the Royal Institution:

> "According to my calendar for October, 2016, I flew from New York to Washington early morning of Monday, October 3. You say that I was cross-examined on the fourth day of the trial, which by my reckoning is Thursday, October 6, and that is when I demonstrated lifting with the model of LCS-1. Each day, before trial at 10 AM, I would go to the Office of Science and Technology of the President in the Eisenhower Executive Office Building for a couple of hours to work there (uncompensated) for the US Government ...
>
> "When I heard from you, David, that the 3-D model had been created on the Isle of Wight, and hand-carried to you in Washington, I was wondering how I could make use of that, and late Sunday night, October 2, I decided that I could, somehow, prop up the model and move it slightly in order to show pitch, yaw, and roll. So I looked up the size of the model, and got a piece of wood about that same length and looked for something I could take to Washington to mount it. I decided that I could do no better than open-cell, lightweight packaging foam, which would give me no problems carrying it on the airplane, and I could even cut it with scissors to make a kind of trough for supporting the model, which I did.
>
> "How was I to exhibit pitch, in particular? So I retrieved the Zircon's laser ball from the Zircon's Home Decorating Kit, that I had bought, including a stud finder, online, six years earlier, and placed the laser ball on the plank. Then I found, experimentally, that I could easily depress the plank, compressing the open-cell black foam. Of course, with the foam blocks well separated, the resonant frequency of the plank in pitch

was very high, so you could hardly see it bobbing. But if I moved the two foam blocks until they were almost touching under the center of mass of the plank, the plank behaved much more like a ship and the horizontal line drawn by the laser beam on a distant wall bobbed up and down about its new location, if I placed a stack of pennies at one end or the other of the ship: stern or bow. All I needed to take to Washington on the flight Monday morning was the small laser ball and the two foam blocks.

"I didn't have time to photograph the setup in my Scarsdale home, nor in DC, when I first saw the model at the trial. But I had the plan well laid out, and I girded my loins for the opportunity, having scouted the low wall with its flat parapet between the witness box and the courtroom, and the distance to the panelled wall on which the horizontal laser scan would be visible.

"The actual demonstration was ad hoc, and, like many things in regard to your efforts over the decades, despite the adverse circumstances, it was crowned with success. When I finished, there was dead silence, and I feared that somehow the demo had fallen on deaf ears and closed eyes. Quite the contrary!"

*****

Of course the silence was one of respect for the Court – though the undoubted frustration at the Defendants' table was clear enough. We Plaintiffs were hard-pressed to restrain ourselves from applauding this simple experiment by the man who had also invented the same laser protractor he had so gainfully employed. It was the turning point of the thirty-five years of co-operation, friendship and trials of fact and fable, that Dick and I had experienced together.

Giles in the Dock ...

Later in the Trial the Defendants' Counsel chose to cross-examine me. First he questioned, in depth, my lack of qualifications in Naval Architecture. I was able to illustrate with slides the different attitudes, at speed, between the designs with which I had been associated, and those of traditional long, thin designs: my semi-planing hulls running flat and the others stern-down and bow-up. The most tedious questions took me, in detail, through every Claim in my two Patents. My cross-examination lasted over one and a half days. Despite the tenor of the questioning and its clear intent to make me look like a fantasist, I felt sure of my case and took the opportunity to prove it in as much detail as was required.

Having spent three years and thousands of hours going through every document released by both sides, I was able to assert with confidence that their arguments claiming the absence of 'hydrodynamic lift' in our hulls were unfounded. Furthermore, I had the previous experience of the Osprey Case, the MOD and S90 saga, the Hill-Norton Committee and the Lloyd's Inquiry. Finally, I had the US Navy's 1989 rejection of SuperShip and the MFS to assist me.

Perhaps my best moment was being challenged about having no qualifications in Naval Architecture, to which I replied that, had I been a naval architect, I would hardly have dared propose such an abomination as the 'Short, Fat – or *Fast* – Ship' ... And, of course, the primary basis of any Patent is that 'it should not be obvious to anyone skilled in the art', so heartily endorsed by all my 'skilled' naysayers, and raising a smile from Judge Lettow.

The Judgement *(All quotes from the final Judgement given in italics)*

As is customary for a non-jury trial, Lettow's final Judgement was deferred until May 5th 2017, after he had considered the transcripts of the trial and cleared up any confusion with the lawyers.

On the subject of lift from the 'hooked' stern, the Court concluded: *No. 12-484 C: "Using the data from the tank tests of Model 5623, Dr Garwin assessed the difference in trials at 39.6 to 39.7 knots between the hull with a straight buttock and the hull with the hooked buttock ultimately used on LCS-1. He then calculated that the difference in trim, 0.38 meters, corresponded to an additional high-pressure lifting force at the stern of 30.4 metric tons ... Thus 70 metric tons of lift corresponds to about 29% of the total thrust of the four water jets (600 MT per water jet, or 245 MT total) in the LCS-1."*

No. 12-484 C 1.116: "Dr Garwin later corrected this value to 80 MT for 1 meter of trim, or about 14%, due to the lifting force of the water jets adding 320 MT in total."

Judge Lettow's final verdict on the matter of the lifting of the hull at the stern was as follows: *"In sum, FastShip has shown by a preponderance of the evidence that the hull of LCS-1 generates high pressure under the stern, which in turn causes hydrodynamic lifting of the stern at a threshold speed, within the scope of these elements of the subject claims."*

In the same Court, in March 2019, during the Plaintiffs' later Application for Costs, Judge Lettow stated: *"One of the government's primary expert witnesses made repeated errors in his analysis that all happened to show it was less likely the LCS-1 infringed on the FastShip's patents."*

He later added in his Judgement on Attorney's Fees and Costs: *"Dr Garwin's testimony at trial was integral to FastShip proving infringement. His testimony directly contradicted the government's expert, Dr Stern, while exposing several significant errors in Dr Stern's analysis that ultimately led the court to 'accord little weight' to Dr Stern's testimony and expert report."*

The Judge's final verdict was as follows: *"FastShip has thus demonstrated that the acceleration of water into the water jets on LCS-1 contributes an additional lifting force at the stern of the ship, fulfilling this element of the representative claim."*

Thus, in the simplest terms, the two major areas of disagreement were found by the Judge to be in our favour. The other side could produce nothing that obscured the fact that their basic argument was seriously, and intentionally, flawed – their Expert miscalculated the need at *four times* the actual horsepower for a speed of 40 knots, assuming no lift from either the water jets or hull. Though USS *Freedom* benefitted from hydrodynamic lift generated by both, the Court concluded: *"USS Freedom generates high pressure under the stern, which in turn causes hydrodynamic lifting of 30.4 tonnes at the stern at 37.9 knots."* So this finally disproved the entire weight of the MOD and Lloyd's Register's false verdicts; also removing a huge weight from my shoulders and my conscience. I could only think of my gratitude to Dick Garwin who had stood by me with fortitude for half of my working life, undeterred by the deceit and insult of some of the world's most powerful governmental institutions on both sides of the Atlantic.

The Final Awards

In the final 2017 Judgement, the Court awarded FastShip Inc. $12.36 million in Settlement, plus $6,178,288.29 in Costs: a total of $18,538,288.29. With interest after four unsuccessful Appeals by the US Government up to the date of Final Settlement, in April 2021, the final Award to FSI was just over $20 million. After the lawyers, the witnesses, the FSI Trust, the priority shareholders and others had been paid, I received about two per cent of the Settlement, the Trust witholding the same amount due to a previous Agreement into which I was forced at its formation in 2012. Was it worth such a hazardous forty-year journey ...? My chief reward was the final confirmation of Charles Adams's famous dictum concerning the Bureaucracy:

> "There's often only one thing worse than being Wrong – it's being Right ..."

Inventers usually don't do as well as the lawyers. The eight per cent that Frank Whittle received from the British Air Ministry in 1944 for his shares in his struggling company, Power Jets Ltd, was quite generous. That was three years after his first jet-powered plane had flown. Later, he justly received an award in 1948 from the Royal Commission on Awards to Inventors. So much for the rewards of 'disruptive technology'. But what reward was there for Dr Garwin, without whose support over the previous thirty-five years, my good fortune could never have been realized?

On the last day of the Trial, Dick Garwin arrived late. It was due, he said, to "official business at the White House", but he did not say what the business was. Six weeks later, all was revealed ...

A Fitting Finale: Dr Garwin Awarded The Presidential Medal of Freedom

In his Citation, President Obama stated: "Dick's not only an architect of the atomic age. Ever since he was a Cleveland kid tinkering with his father's projectors, he's never met a problem he didn't want to solve. Reconnaissance satellites, the MRI, GPS technology, the touch-screen, all bear his fingerprints. He even patented a mussel washer for shellfish – that I haven't used, the other stuff I have. Dick has advised every President since Eisenhower, often rather bluntly. Enrico Fermi, inventor of the world's first nuclear reactor, and also a pretty smart guy, is said to have called Dick 'the only true genius he had ever met'."

Dick Garwin receives the Presidential Medal of Freedom, America's highest civil award, from President Obama on November 22nd 2016. Co-recipient Bill Gates applauds to the right.

CHAPTER 38

Back to the Future – The Origins of the Type 45 Destroyer

Returning to the UK in 2020 due to the Covid pandemic, I was advised by a Naval friend and ex-destroyer Commanding Officer that: "The Type 45 seems to be one of yours." Preoccupied with all our US Naval and commercial projects, I was unaware of the Type 45's extended development, or its being praised by a former antagonist and First Sea Lord as "The most capable destroyer ever." On the contrary, in the UK there still remained continuing adverse professional and institutional contempt for our ideas. Even in 2005, twenty years on, public comments on the 'short, fat ship' were circulating, verging on the poetic (see **Appendix 15**, p. 346): "It's vapourware", or "Fugazi, Fugazi. It's a wazy. It's a woozie. It's fairy dust. It's not real!"

Or was it ...? More recently, the unprompted reply of a serving Type 45 Chief Petty Officer, asked for an opinion of his ship (excluding its machinery), was beyond poetry: "Fantastic!" And that seems the gist of today's comments, apart from the problems of a highly innovative propulsion system.

The following photos and data of this chapter demonstrate key features of the Type 45 that repeat the very DNA of TGA's designs – and are alien to the classic destroter hull form, such as the Type 42 that preceded the Type 45. These could only have been obtained from detailed knowledge

The Type 45 Destroyer at about 32 knots, showing its high Sampson Air-Defence radar, high bow-wave, the wave-crest beside the stern indicating high pressure beneath; and the flat wake without the high 'Rooster-Tail' of long/thin hulls at speed. The wider beam allows for improved sea-keeping, manoeuvrability, power-for-speed, stability and crew comfort.

of copyright data obtained from the Osprey Case, S90 Validation, Lloyd's Inquiry and our Patent of 1981–89:

First: BAe's massive Sampson search radar. Due to the wider beam, it can be placed at over twice the height of previous air-defence destroyers like the Type 42. With a 350 square mile detection area, it can track up to six hundred targets at any one time. **Second**: The high bow-wave and the slight crest alongside the waterline at the stern indicates lifting pressure that maintains level trim of the hull, reducing 'squat' and thus power-for-speed. **Third**: The resulting down-wash of the wake, from the high pressure beneath the stern, into the propellers' thrust-line, improves propulsive efficiency. These features are exactly as described by a former senior RN Officer, previously CO of a traditional long/thin Type 42. His account of his son's experience as CO of a Type 45 describes the hull's advantages over his previous Type 42 Command, first in reducing the ill-effect of the WR-21 engine problems that have cast a shadow over the Type 45's entry into service: "The hull design is efficient so the scale of the problem is less than it might have been had the shape placed more demands on SHP for a given speed." He then quotes his son: "The hull performs better in head seas than in any of the other classes in which he has served. It is drier and far less prone to shipping heavy water over the bow. The stern hook reduces 'squat' and the 'rooster-tail' wake, which itself is wasting horsepower and thus fuel."

Below are shown key hull features incorporated in the Type 45 obtained from TGA's S90 Validation Project and later technical data:

Fig. 1: An Orca 6 CFD comparison between the waterlines of the Type 45 and TGA hull at equivalent speed FNo. = 0.44/32kts. Similar waterline with high bow-wave, amidships trough, aft crest, flat wake without stern 'squat' or 'rooster-tail':

Fig. 2: The aft lines of the TGA S90 hull form:

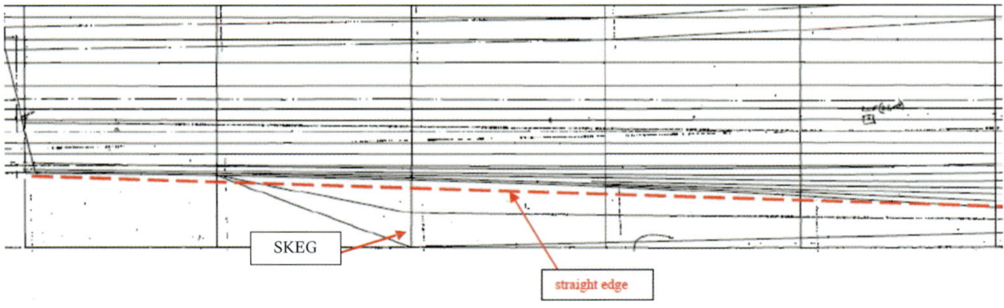

Fig. 3: Aft lines of the Type 45. At full scale the max. separation is 17"/43 cm:

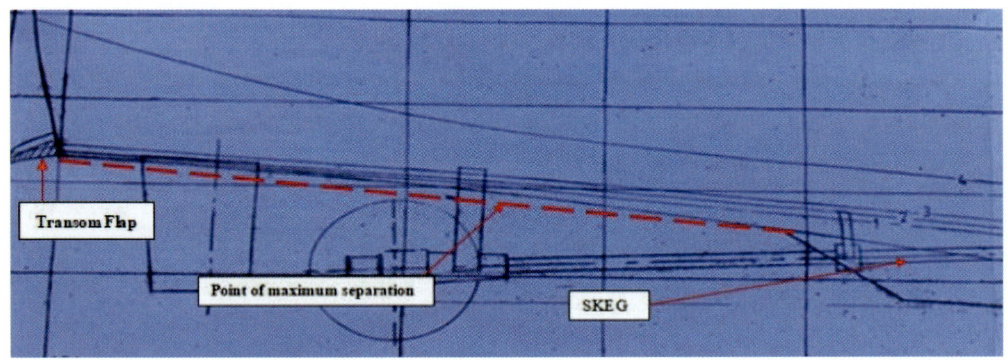

Fig. 4: Aft hull of a Type 45 out of the water showing the concave 'hook':

Fig. 5: Another view of the Type 45's concave stern 'hook', in dock:

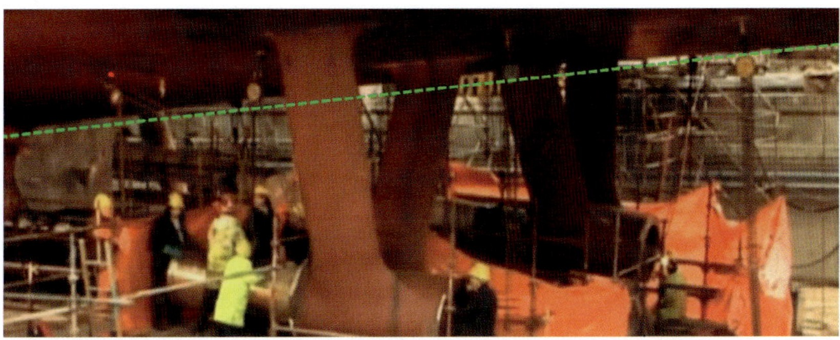

Fig. 6: 18-foot high 'Rooster-Tail' of a Type 42 Destroyer seen from below the helicopter deck at 30 knots (Froude No. = 0.42):

Fig. 7: Flat wake of TGA's semi-planing Azteca at 24 knots/Froude No. = 0.72 (70% higher than the relative speed of the Type 42 above):

Fig. 8: "She turns on a sixpence." The high manoeuvrability of the Type 45 due to widely-spaced rudders and increased stability:

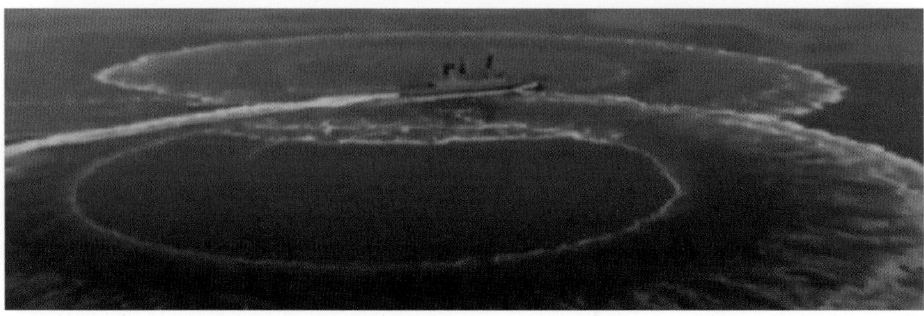

Chronology of Type 45 Development

After the 1984 collapse of the S90 Club and the later Lloyd's Inquiry, BAe Systems, with Marconi Marine, the new owners of Yarrow Shipbuilding, started on a twenty-year quest to produce a 'post-Falklands' Air-Defence Destroyer. It opened in 1985 with NATO's NFR-90 Project, which the UK abandoned in 1989 due to the RN's incompatibilities with other navies. This was at exactly the same time that the MOD placed its Secrecy Order on my UK Patent Application, claiming ownership of the technology and forbidding its use "anywhere, due to it's being of possible use to a potential enemy". Bizarrely, this contradicted the Lloyd's Inquiry and the MOD's previous public rejection of TGA's designs as being "of no future interest to the Royal Navy for future frigate and destroyer designs". After our Agents reminded the Patent Office of the MOD's S90 rejection, the Secrecy Order was immediately withdrawn.

Ten years later BAe/Marconi had better ideas, again resigning from the subsequent international UK/French/Italian 'Horizon' Project in April 1999. By then BAe had developed – with Plessey and Marconi Electronic Systems ('MES', now owners of Yarrow Shipbuilding) – the massive, highly effective, 'Sampson' early-warning Radar. This had to be placed as high as possible for maximum defensive horizon, requiring a hull with wider beam for improved stability, but without increased power-for-speed, as would apply to the more conventional hull adopted by the Franco-Italian design without the Sampson.

That alone illustrates the significance of the lifting hull feature employed by BAe in the Type 45 development that the MOD had tried to commandeer with its Secrecy Order. However, the S90 copyright had passed, via TGC Inc., to TGA's US Associate, FastShip Trust LLC, its present owner. (See S90 and Type 45 stern lines plans compared on p. 286 above.) In September 1999, MES obtained the contract for the Type 45. Within a week BAe bought MES, becoming the Type 45 Prime Contractor. Thus BAe became the major UK warship-builder, owning its former Yarrow and Marconi partners, building under Lloyd's Register's Naval Rules. It was a decisive moment in the development of a novel type of hull intended to resolve the least predictable of the Navy's future needs: Air Defence.

Previously much of the research and planning of the UK Government concerning Defence equipment and policy had been under the authority of the Defence Evaluation and Research Agency (DERA). In April 2001, much of DERA's assets, sites and employees were transferred to the private company, QuinetiQ. The MOD retained a share in the company, thus giving it an edge over its private-sector rivals. Among the previous MOD institutions transferred from the MOD to QuinetiQ, was the Royal Aircraft Establishment at Farnborough and Boscombe Down. It then acquired a US ingredient with the participation of the Washington-based Carlyle Group, leading to a special entry into the US Defence Community. During this period, under a twenty-five year agreement with the MOD, QuinetiQ also obtained control of the former Admiralty experimental testing-tank at Haslar.

There it could undertake a Research and Development project with BAe Systems using much measured design data from the 1982–8 S90 Validation Project and possibly that from the Osprey Case and Lloyd's Inquiry. This was incorporated in the Navy's new Type 45, with its unusual and controversial lifting hull. This lengthy experimental programme finally supported the MOD's alternative conclusion. As expressed by its Chief Naval Architect, it is reproduced below in the original typescript referring to TGA's 380-ton Osprey: "If there were any significant lift in the Osprey, it really would represent an an important contradiction of traditional theory."

After a twenty-year R&D exercise spent picking the brains of many other navies, and absorbing Yarrow and Vosper Thornycroft, BAe decided on its final Type 45 design. Hull modules were built at VT, then shipped to the old Yarrow, now BAe, yards at Scotstoun and Govan for final assembly. Those most conversant and involved with the 1980's S90 Validation – MOD, BAe/Marconi/Yarrow, their Consultants YARD and Lloyd's Register – all took part in the development of the Type 45. So TGA's most obvious and transitional hull feature, the concave lifting stern, or 'hook', was incorporated in the Type 45, virtually at the stroke of a draughtsman's pen.

According to the S90 Validation measurements of DTU and NMI, if scaled to equivalent size and proportions, the Type 45 power and fuel consumption at 30 knots has been reduced by some thirty per cent, compared with the MOD and YARD estimates based on proportionally similar conventional hulls. (See **Appendix 3**, Comparative Power for Speed and Displacement on p. 301 and D Jenkins's Commentary on p. 304.) Thus the Type 45, like the US Navy's LCS, has a large semi-planing, or 'lifting', hull. This was the basic principle rejected in one short type-written paragraph, by the MOD's Chief Naval Architect, in April 1981, perpetuated by the Royal and US Navies' Naval Architectural establishments over four decades, and at huge cost.

3. You will remember at our meeting on 27 February I said that if there were any significant dynamic lift on the OSPREY form, it really would represent an important contradiction of traditional theory. Unhappily, the information you now supply does not show such evidence and it must be presumed that conventional wisdom is not to be denied. Such dynamic advantage is confined to quite small craft and is lost at OSPREY displacements - not surprisingly because displacement varies as the cube of the dimension and dynamic lift as the square. We are safe in assuming that ships of SIRIUS size will certainly obey normal laws.

2021: The Final Judgement

The MOD's initial conclusion, though ill-conceived and erroneous, was endorsed covertly and publicly, over the same period, by leading naval architects, by the MOD and its 'independent' advisory bodies – and other eminent entities including the Royal Institution of Naval Architects. The opposing views of Lord Hill-Norton are succinctly summarized in his Foreword and his later Warship Hull Design Committee that reported to Prime Minister Thatcher. Again, to repeat Rawson's own words: if the hull of the Type 45 ... *"were to show signigicant dynamic lift, it really would represent an important contradiction of traditional theory"*. And so it did ...

Finally, a short personal note to me from a previously unknown, but highly experienced Type 42 and 45 CO, after a trip with members of a foreign navy aboard his ship: *"They were amazed by the ship as well as the rest of us. You should feel immensely proud of your work. They are truly world class ... It turns on a sixpence."*

He later added a verbal tribute: "In ocean seas we could leave a Type 42 standing, thanks to our level ride in the waves ..."

How could one ask for more? Other descriptions supported the conclusions of the 1986 Hill-Norton Committee. The first, in 1981, was the dissenting opinion of the two leading physicists, Dr Garwin and Sir James Lighthill, against the MOD's Chief Naval Architect. Their opinion – the subject of copious tank-tests, and full-scale ship trials – was upheld by the US Court, in April 2021, when it was finally established that the hydrodynamic properties of small craft scale up in direct proportion to equivalent size and speed for ships of any size. Also it appears that, with increasing size, the propulsive efficiency of the lifting hull improves, compared with tank propulsion measurements, as disclosed in the findings of the Osprey Case and S90 Validation. Mr Bill Crago, former Chairman of the British Towing Tank Panel and British representative at the International Towing Tank Conference, denounced the official 'non-scaling' theory, stating in a 1984 letter to TGA concerning the S90 Validation: *"It undermines the entire technique of model testing."*

This dichotomy was exposed by the US Naval Academy's own tank-tests in 2015. These were current at the same time as, in the US Federal Claims Court, the Justice Department was advancing the MOD and Lloyd's fallacy of the stern sinking with speed. The differing behaviour of the models is instructive.

The above USNA video 'still' shows the trim of bare-hull models of the semi-planing LCS, without the additional lift from its water jets, and a traditional hull, at about 40 knots. For a bow view, go to: https://www.youtube.com/watch?v=JGZcv_zCsq0.

The following Table shows how the approximate power-for-speed of the Type 45 compares with other classes of lesser and greater size (the Type 42 – Batch 3 and the Arleigh Burke Flt. II). Clearly, as claimed by the spokesman on p. 285 above, the Type 45 is also highly efficient.

Approximate Comparison: RN Type 45, USN Arleigh Burke Flt. II and Type 42 – Batch 3 classes

(Sources: Wikipedia, 'US Seaforces' and https://des.mod.uk/what-we-do/navy-procurement-support/type-45/)

Parameter	Type 45	Arleigh Burke Flt. II	Type 42 – Batch 3
Displacement – Tonnes	8,000	8,400	4,775
Length Overall – Ft.	500	505	463
Beam Overall – Ft.	69.7	66	49
O'all Length/Beam	7.17	7.65	9.45
Draft – Ft.	24.25	30.5	19
B.H.P.	2 × 28,800 = 57,600	4 × 26,250 = 105,000	2 × 25,000 = 50,000
Propulsion	2 × RR WR-21/IEP	4 × GE LM-2500	2 × RR Olympus TMB3
Approx. Speed – Knots	'30+'	'30+'	30
Displ./Length Ratio	64	65.2	48
Froude No. @ .95 × LOA	0.41	0.41	0.42

'Shortest & Fattest': Type 45 HMS *Daring*

'Shorter & Fatter': the USS *Arleigh Burke*

'Longest & Thinnest': RN Type 42 – Batch 3 HMS *Edinburgh*

Note: See **Appendix 1** for definition of above technical terms.

*****

In November 2003, some four years after the Royal Navy's Type 45 design and construction had begun, Lord Hill-Norton fired his final broadside in a personal letter concerning the drafting of this book. He starts by recalling his outspoken support for TGA's design principles during our first meeting on that memorable Trafalgar Night Dinner twenty years before, in October 1983:

> *"It all comes back to me now. I am fairly sure that a good many – particularly some of those more senior – very much resented and regretted what I had said so publicly. I am now even more glad that I did so ... There are a good many passages which make it luminously clear that you were faced with a carefully thought out and well-executed plan to do you down, and more infuriatingly, the S90 concept. The very care with which their disgraceful plan was conceived and executed does, I believe, make it clear how frightened and worried they were."*

Published trials, sea-keeping data, personal reports, media and periodical accounts of the Type 45 have realized all of the 1986 Hill-Norton Committee's conclusions. Although dismissed by the Lloyd's Inquiry, the Type 45's long construction time, partly due to its 'Block' production process in widely separated yards, alone contradicts the Committee's findings on hull-cost:

> **"We believe that certainly for ships up to destroyer size, the short/fat hull form offers enough advantages in the important elements of construction time, habitability, between-deck and weather-deck layout, stability and sea-keeping, and weapon-siting to merit much more serious consideration than it has so far been accorded ... We find that the short/fat hull form may offer a significant increase in top speed over the maximum which can be realized in a long/thin hull of similar size. If this is confirmed it is a most important military advantage."**

*****

In his farewell address to the American people, President Eisenhower warned of the new overpowering influence of the military-industrial (and governmental) complex – an 'iron triangle'. He spoke of the dangers to "the solitary inventor, tinkering in his shop ... where government contracts become a substitute for intellectual curiosity". Of a world where "we must guard against the acquisition of unwarranted influence, whether sought or unsought by the military-industrial complex ..." In other words, an inviolate community in which new ideas can be forcibly rejected, yet later adopted by those very Institutions and Experts that had denounced them. 'N.I.H.' or Not Invented Here ... But perhaps worth retaining for possible future use ...

Technical, Legal and Commercial Appendices 1–15*

Appendix 1

Introduction: The 'Laws', or DNA, of Hull Design

> 'The laws which govern the wave-making resistance of ships are not yet fully understood.'
>
> (Sir William White, Director of Naval Construction, in his *Manual of Naval Architecture*, 1910. That was 50 years after Froude proposed his principles, or 'Laws', as they became, in a Paper to the Institution of Naval Architects in 1861)

My father's old *Vade Mecum*, 'The Speed and Power of Ships', by Admiral David W. Taylor USN, provided a simple method to obtain Froude Number – or 'FNo.' – that traditionally fixed the maximum practical speed of any hull. First take the speed required and divide it by the square-root of the waterline length of a hull in feet (LWL) and that gives you the Speed-Length Ratio. This was the first of the three Ratios that governed the proportions, displacement and speed of a ship. So:

1. Speed-Length Ratio (SLR) = $\dfrac{\text{Speed in Knots}}{\text{Square-Root of Waterline Length in Feet}}$

   Or, writ small, Speed-Length Ratio (SLR) = $\dfrac{V}{\sqrt{LWL}}$

   To convert SLR to Froude Number (FNo.) simply multiply by 0.298

As an example, the famous old passenger liner *Queen Mary*, the fastest ship of her day, achieved a maximum practical speed of about 31 knots. On an LWL of 975 feet, the square root of which is 31.2. Then 31 knots ÷ 31.2 knots = an SLR of 0.99. Multiplied by 0.298, that gives a Froude Number of 0.296 or approximately 0.3. Thus the SLR is about one-third of the FNo. At that speed, of about 31 knots, the stern of the *Queen Mary* would

* As agreed by a leading independent naval architect.

begin to sink downwards, towards the trough of her self-generated 'captive wave', as my father called it. So, for any increase in speed the power would become near-exponential: the 'hockey-stick' curve (see Fig. 1). This 'Traditional Wisdom', was first broken by Thornycroft's *Gyrinus* in the Water Motorsports of the 1908 Olympic Games. *Gyrinus* won both Gold Medals: the only one of 17 competitors to finish in severe weather on the open sea course. It established the reputation of the semi-displacement, or semi-planing, hull for speed, seaworthiness and higher displacement than fully-planing hulls in future patrol craft.[1] The second Ratio is:

2. Displacement-Length Ratio (D/L) = Displacement in Tons ÷ (LWL/100^3)

Obviously, the volume of water the hull displaces has a major effect on both Frictional and Wave-Making Resistance, or 'Residuary Drag'. Therefore, to increase speed for a given displacement, a hull can be stretched like a piece of rubber, to give the same volume, or displacement, but lower relative to its length. It thus generates a longer wave and – as a natural wave's speed depends upon its length between crests – a higher practical SLR or FNo. Hence the origin of the 'Long, Thin' destroyer hulls designed by Thornycroft and others at the same time as Froude was working in his Torquay test-tank. So, back to the *Queen Mary*: On an LWL of 975ft, she had a displacement of 77,400 Tons or 'Long' or Imperial Tons. That gives a D/L of 83.9. The third Ratio is:

3. Length-to-Beam Ratio (L/B) = Waterline Length ÷ Waterline Beam

Again the *Queen Mary*: On an LWL of 975ft and a beam of 118ft, she had a Length-Beam ratio (L/B) of 8.2. So in terms of SLR, D/L and L/B, she was half-way between the 'Longest, Thinnest' ship ever – the French destroyer *Fantasque* with a D/L of about 30 and a L/B ratio of 11 – and a relatively 'Short, Fat' Battleship, such as HMS *King George V*, of 42,200 tons on 700 ft. LWL, with a speed of 28 knots (SLR = 1.06, FNo. = 0.3, a D/L of 123 and an L/B of 6.8).
N.B: Units given are in Long/Imperial Tons (2240 lb.), Feet and Knots unless stated otherwise.

1 Proceedings of the Society of Civil Engineers, March 12th 1909.

Taylor's Analysis

Displacement	DLR
ultralight	under 90
light	90 to 180
moderate	180 to 270
heavy	270 to 360
ultraheavy	360 and up

The 1930s French Destroyer *Fantasque*: An extreme example of the 'long, thin' hull, showing the sinking of the stern at a highest speed of 40 knots at a DLR of 30, defined by Taylor as 'Ultralight'.

Fantasque (L/B of 10.6, beam of 39 ft. on LWL of 412 ft.) was nearly half the LWL of *Queen Mary*. At 2,000 Long Tons (or 2.5% of *Queen Mary*) with 73,000 SHP she achieved a 10 knot higher speed (40 knots, V/√L = 1.97) with a D/L of about 30. The RN's Type 21, Leander and Type 42 – Batch 3 are in the traditional 'long, thin' Displacement category. The lower displacement-per-foot length of the longer and more highly-stressed hull reduces armament, fuel, etc., unlike the shorter, fatter semi-planing Type 45, Azteca and S90. The USN LCS achieves <47 knots on a much shorter, wider hull with 20% higher power than *Fantasque*, but at nearly 4 times the D/L (30 vs. 115). The blue and green plot-lines are intended to show a general tendency:

Figure 1: 

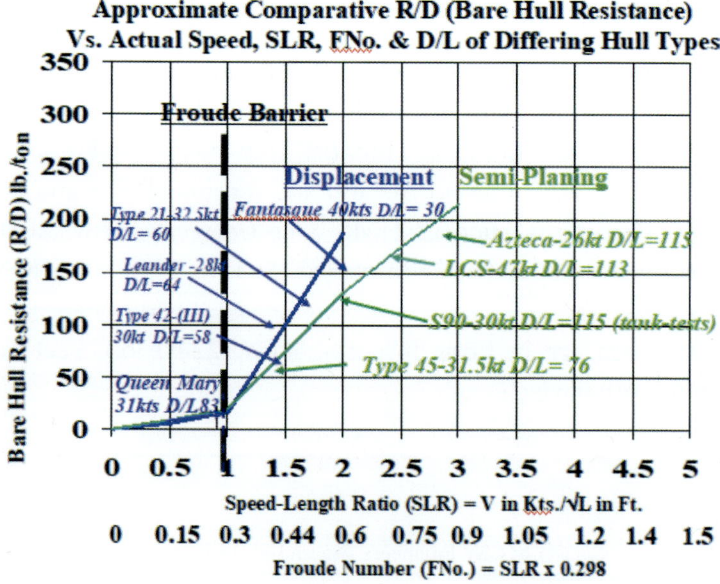

In the past, at this point the ship designer had hit the traditional backstop.[2] How to achieve the speed of the *Fantasque* or Type 21 – but without extreme L/B or massive power – on a shorter, wider hull with the same DLR as a Battleship? The extreme destroyer hull like *Fantasque*, or the shorter and wider, but higher-powered Type 21, could achieve higher speeds, but with a low DLR of 20–65 and reduced stability and seaworthiness. The Type 45, although 30% wider than the Type 42, benefits from a 30% higher DLR at almost the same SLR/FNo. and R/D – a significant increase in 'efficiency'. Extensive tank and computer testing (by four tanks and three Universities) of TGA's 770ft./31,500-Ton FastShip showed 47 knots (FNo. = 0.55, SLR = 1.84, DLR = 148) was practicable in a very large semi-planing hull.

Also the reduction in pitching and ability to maintain speed in high ocean waves, due to the lifting stern, was described by MIT's 'SWAN' seakeeping analysis as 'unprecedented'. Why?

A Similarity Between Aero and Hydro Dynamics:

Figures 2 and 3 below show the influence of Pressure Drag (Wave-Drag or Cd) on both aircraft and different hull-types. The technique of the Supercritical – or Rooftop RearLoaded – aerofoil, was based on reducing the sudden increase in drag as a wing passes from pressure-drag at Mach 1 (767.3 mph) and enters the more critical area of frictional drag. Due to the increasing relative density of air above supersonic speed, planes originally hit the 'Sound Barrier'. It caused a prohibitive drag-rise later reduced by the Supercritical wing that allows much higher speeds, similar to lightweight planing hulls (Fig. 3 below).

Likewise, when the speed of a traditional hull is increased to its theoretical limit, of $V/\sqrt{L} = 1.3$ or FNo. = 3.4 (see note 2 below), it creates wave-drag beyond which any increased speed requires prohibitive power. In a semi-planing hull the aircraft process is reversed and its lift reduces relative resistance as it transits from friction to wave-drag, allowing higher practical speeds of up to FNo. = 0.85 (see Fig. 1 above). This is similar to a Supercritical wing allowing aircraft to operate at high subsonic and supersonic speeds with reduced drag-rise.

2 "It is impossible to operate at a speed-to-length ratio higher than approximately $V/\sqrt{L}=1.3$. Beyond that realm even a trivial increase in speed requires a virtually infinite increase in power." *Encyclopedia Britannica*, 3/11/2023

Figure 2:

Comparison of Pressure Drag (Cd) Rise Between Aircraft and Ships:

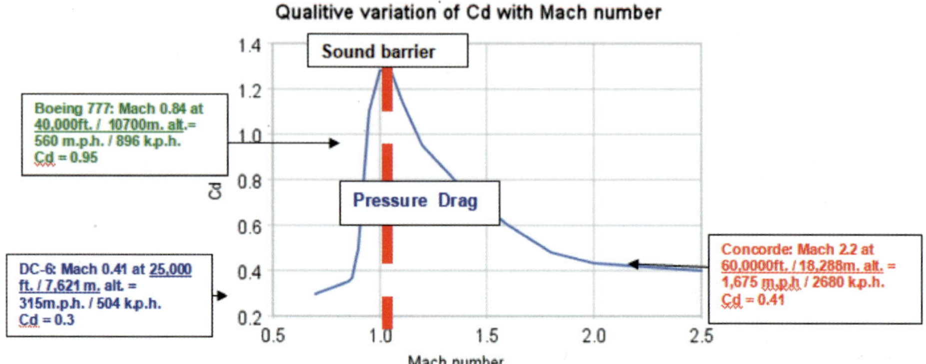

Aircraft - Cd (Co-efficient of pressure drag) Rise Vs. Mach Number (Mno):

As an analogy, the DC-6 (315 m.p.h./1,033 k.p.h. / Mno. 0.41 at 25,000 ft) equates to a Displacement hull ship. High Mach-Number subsonic aircraft, such as the Boeing 777 with a wing adapted to Mno. 0.84, equates to a Semi-Planing hull. Concorde (much-reduced pressure drag past Mach 1.0 & high friction drag (creating high structural heating at Mno= 2.2), equates to a Planing hull, such as *Destriero*.

Figure 3:

Ships – Total Resistance (Drag) Rise Vs. Speed:

Showing 'prohibitive' increase in Drag for Displacement hulls above Fno=0.3; constant increase in drag of Semi-Planing hulls above Fno=0.3; and higher increase in drag of Planing hulls above Fno.=0.3, but reducing at higher speeds above Fno. = 1.05:

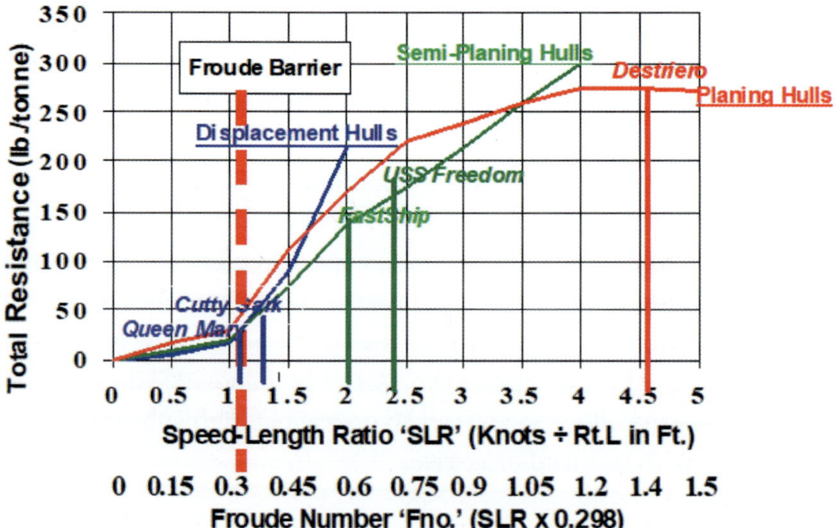

Appendix 2

Figure 1: (Pre-1970) Typical 20–25 knot Nelson Pilot or Service Launch. The most common worldwide variant:

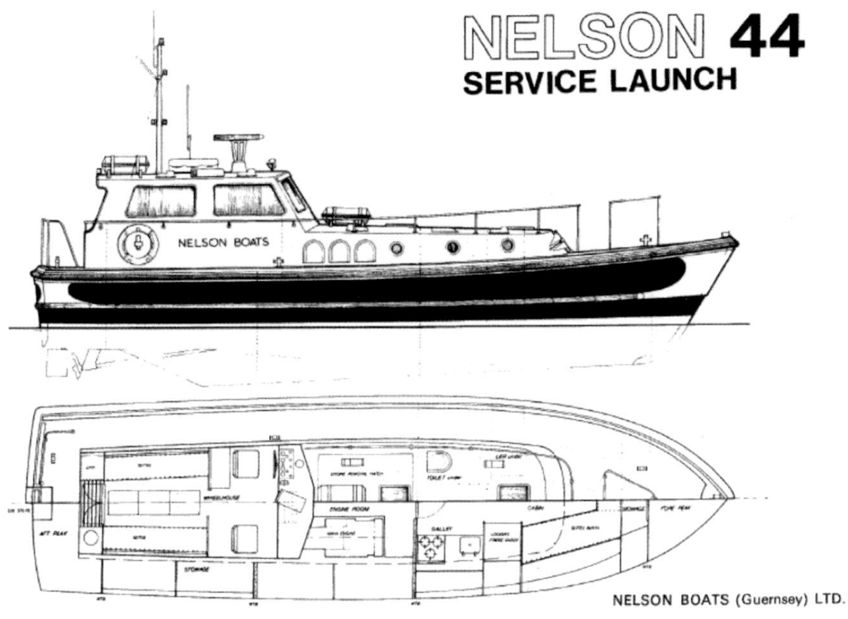

Figure 2: (1976) The 360-ton/36 knot Osprey Strike Craft developed with Westland's & Plessey for the Kuwait Navy, fitted with Exocet Missiles and a guidance-helicopter landing deck:

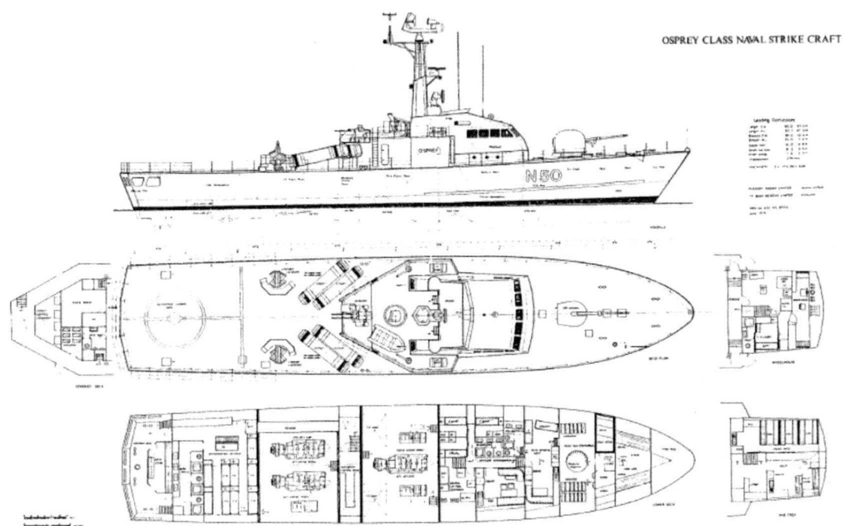

Figure 3: (1978) The 450-ton/22 knot Danish Fisheries Ministry Osprey *Havørnen* developed with Frederikshavn Vaerft. Helicopter and stern-launching boat shown in dotted lines:

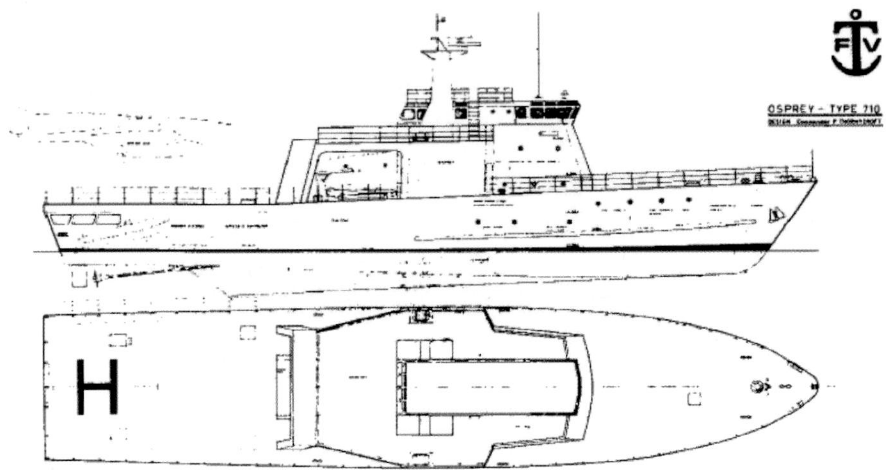

Figure 4: (1980) The first 360-ton/26-knot TGA design for the Hong Kong Patrol Craft:

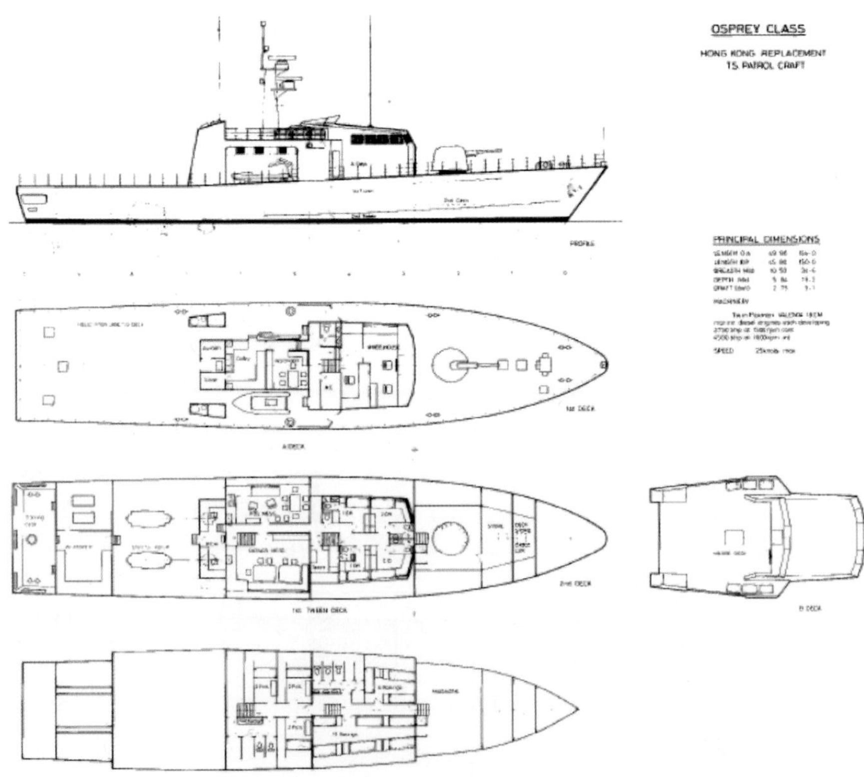

Appendix 3

Figure 1: Influence of 'X Factor', or (1+or–x) on comparative S90 SHP performance by NMI and MOD

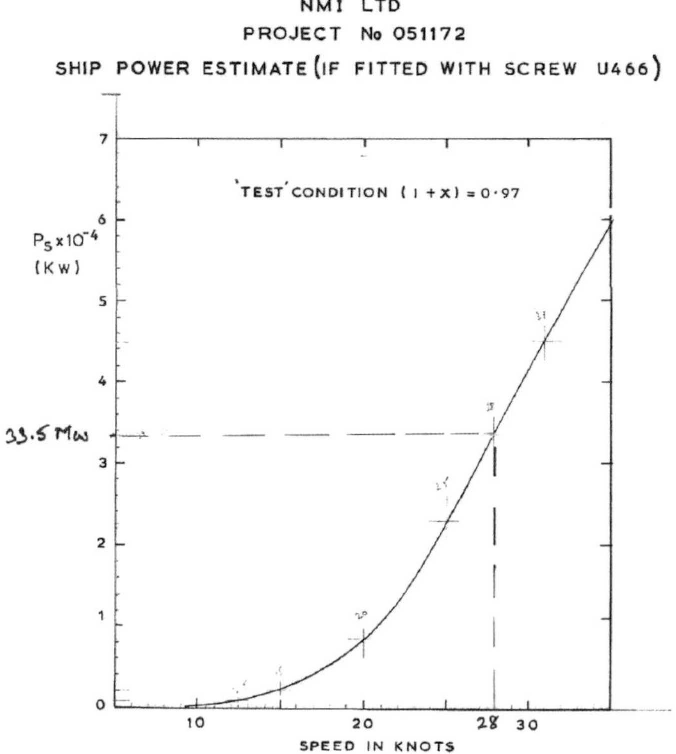

The MOD, Lloyd's Inquiry and DSAC arguments against the S90 propulsion all boiled down to two graphs including the (1+x) or 'X-Factor'. **Figure 1**, summarizes the NMI Speed-Power Tests at 2,600 tonnes; and **Figure 2**, in which the MOD's Advisors compare the NMI S90 results at 2,800 tonnes, with the YARD and MOD powering estimates, based on 'conventional' hulls of the same proportions and displacement as S90. This caused the confusion over the use of 'model-to-ship' correlation factor, usually (1+x) for displacement hulls; but in semi-planing hulls often a negative factor, or (1–x), as investigated in great detail by David Moor in his Vickers Osprey tests – later confirmed by the HMS *Peacock* Sea Trials that the Navy was so reluctant to release publicly. This was also the immediate source of 'Giles misinterpreting the correlation factor', as claimed by DSAC and the future Lloyd's Inquiry.

The NMI propulsion test estimate for the S90 at 2,600 tonnes displacement and 28 knots (without fouling) copied from its Report No. 051172 showing 33.5 Mw for 28 knots, at 2,600 tonnes displacement, as annotated in freehand by TGA. NMI's (1+x) – or negative (1–x) of 0.97 – is its precise calculation based on its own Osprey model tests, factored by the Osprey *Havørnen*'s speed trials, measured and supplied by the Danish Technical University and ignored by YARD and the MOD. Usually expressed as (1+x) this is the margin by which full-scale trials traditionally vary from the final tank propulsion tests. For conventional hulls it is usually *positive* and is due to the *frictional* drag of full scale propeller shafts, and brackets, rudders and other appendages and is usually up to x = 1.11. However, a semi-planing hull operates in an area of greater *pressure* drag, so the frictional effect of those appendages is reduced proportionately and a *negative* or (1–x) can apply. The NMI tests at a negative, or (1–x) = 0.97, show a 2.7 knot (+10%) improvement in speed at full power, compared with (1+x) = 1.11 (–13% at 200 tonnes higher displacement = –3% of speed) as applied by YARD/MOD. Likewise 'fouling' (i.e. weed, etc.) being a frictional factor, is proportionally less relevant at S90's higher operating FNo. This also reflects the downward flow of the wake of a semi-planing hull, giving greater propeller efficiency than the upward flow, or 'Rooster Tail' of a conventional hull at high speed.

Figure 2: YARD's diagram of conflicting S90 power predictions of NMI, YARD and MOD

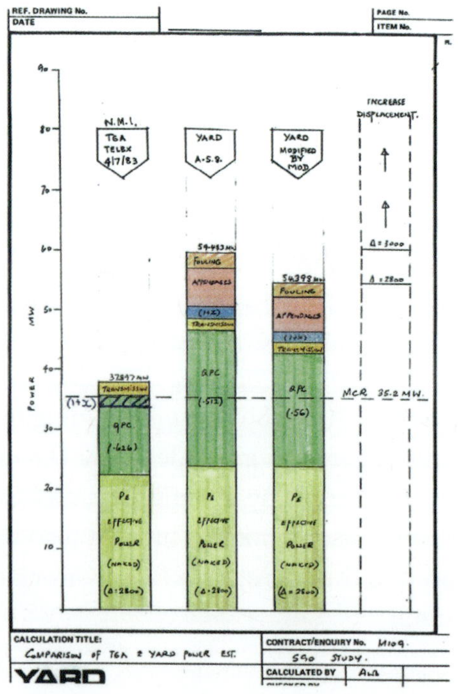

For the S90 at the higher displacement of 2,800 tonnes and 28 knots (with fouling): The NMI measured increase in S90 power for 28 knots is a reasonable 9%, compared with its previous measurement at 2,600 tonnes (Fig. 1 above). The power estimates of YARD and MOD are said to be based on computer data for a conventional vessel of the same length, proportions and displacement as S90, but employ much lower Quasi Propulsive Coefficients ('QPC' = .512 and .516) shown in green; higher Correlation Factor (1+x) and a very high Appendage Allowance, shown in red. This reduces their Effective Power ('Pe') shown in yellow, to near that of S90, thus concealing the advantage of hydrodynamic 'lift' in the S90. The YARD and MOD Powering estimates vary between, 54.4 Mw (MOD) and 59.5 Mw (YARD), a 37 percent increase in power over NMI's *measurements* of 37.9 Mw. ***TGA's graph in Figure 3 below shows the measured negative, or (1–x) of Vickers, Lyngby, NMI and MARIN***. 'Fouling' could not be accurately applied to S90, as also explained in the discussion on Appendages in Fig. 1 above. Thus both the S90 Appendage Allowance, and a negative (1+x) is included in NMI's QPC. This reduction is due to the propellers being in an area of high, rather than low pressure beneath the stern of a semi-planing hull at its higher FNo. This negative, or (1–x) was further confirmed by the Trials results of the semi-planing Hong Kong Patrol Craft (HKPC) HMS *Peacock*, which were only disclosed to the S90 Club after a High Court Judgement ordering their being released to the Osprey Plaintiffs. These showed a 2.4-knot increase in maximum speed compared with the Vickers propulsion tests of Model 2254B, the final as-built design of the HKPC, giving (1–x) = 0.95. Kirkman's Expert Witness Report for the Osprey Case and Moor's recorded remark to Crago, of BHC, on his testing the Osprey semi-planing hull ('this is something of which we know very little') could explain the confusion over the differing (1+x) and Appendage Allowance. The differing estimates and measurements for Osprey (1–x) are given in **Figure 3** below.

Also see David Jenkins's Note below on the inaccuracy of the YARD S90 powering estimates. David is one of the UK Naval Architects most acquainted with the testing and hydrodynamic principles of semi-planing hulls.

Statement by David Jenkins, C.Eng, FRINA, concerning inaccuracy of YARD's S90 Powering Predictions (Appendix 3, Figure 2) from their S90 Performance Analysis

I refer to the YARD coloured barchart designated Appendix 3, Figure 2, of your Appendices.

I remember looking at this chart some time ago; not being able to make sense of it, I dismissed it as not worthy of consideration. I now know why.

It is complete nonsense because appendage drag is allowed for in the QPC by NMI – although the NPL document is particularly for single screw ships, the principles are exactly the same for twin screw.

NMI would not have come up with a prediction for required power without allowing for appendage drag which is always significant. In any case the self-propulsion model in the NMI Report No. 051174 on Propulsion Tests, obviously had appendages.

In producing the chart, YARD have 'double dipped' or even 'treble dipped'. They have increased the naked ehp above the value obtained by NMI, increased the QPC, and then added appendage drag which should already be accounted for in the QPC (Quasi Propulsive Coefficient).

To add insult to injury they have then added a (1 + x) greater than 1 whereas NMI predicted this value to be less than 1, at 0.97 (see Fig. 1 above).

I cannot imagine that they have produced such misleading data by pure ignorance.

Ignoring fouling I have estimated (small scale chart so difficult to read off data accurately) the following OPCs (Overall Propulsive Coefficient) from the YARD data:

Yard ignoring (1 + x)	OPC 0.46 Yard ignoring (1 + x) adjusted by MOD	OPC – 0.50
Yard (1 + x) 1	OPC – 0.44	
Yard adjusted by MOD (1 + x) 1	OPC – 0.46	

Applying these OPCs to my analysis for the Type 45 dated 24/01/21 the max speeds predicted are as follows, at a displacement of 7350t., (1150t. below her Full Load Displacement of 8500t.):

Using the lowest figure of 0.44 predicted speed – 25.9 knots @ 7350t.
Using the highest figure of 0.50 predicted speed – 26.7 knots @ 7350t.

These are far lower than the widely quoted speeds of 30–32 knots for the Type 45.

Although the propulsive coefficients are due to the lower Froude No. for the Type 45 compared to the S90, it is not expected that corrections to the OPCs (could be rather time consuming) would yield significantly different results.

D Jenkins C.Eng. FRINA 26/01/2021

Figure 3: Comparison of Correlation Factors as (1+or–x) of three semi-planing hulls: HKPC (HMS *Peacock*), Osprey and *Southern Cross III*, by four I.T.T.C. testing tanks. Compared with MOD and YARD Simulations for 'conventional hulls'. Prepared by TGA for Lloyd's Inquiry:

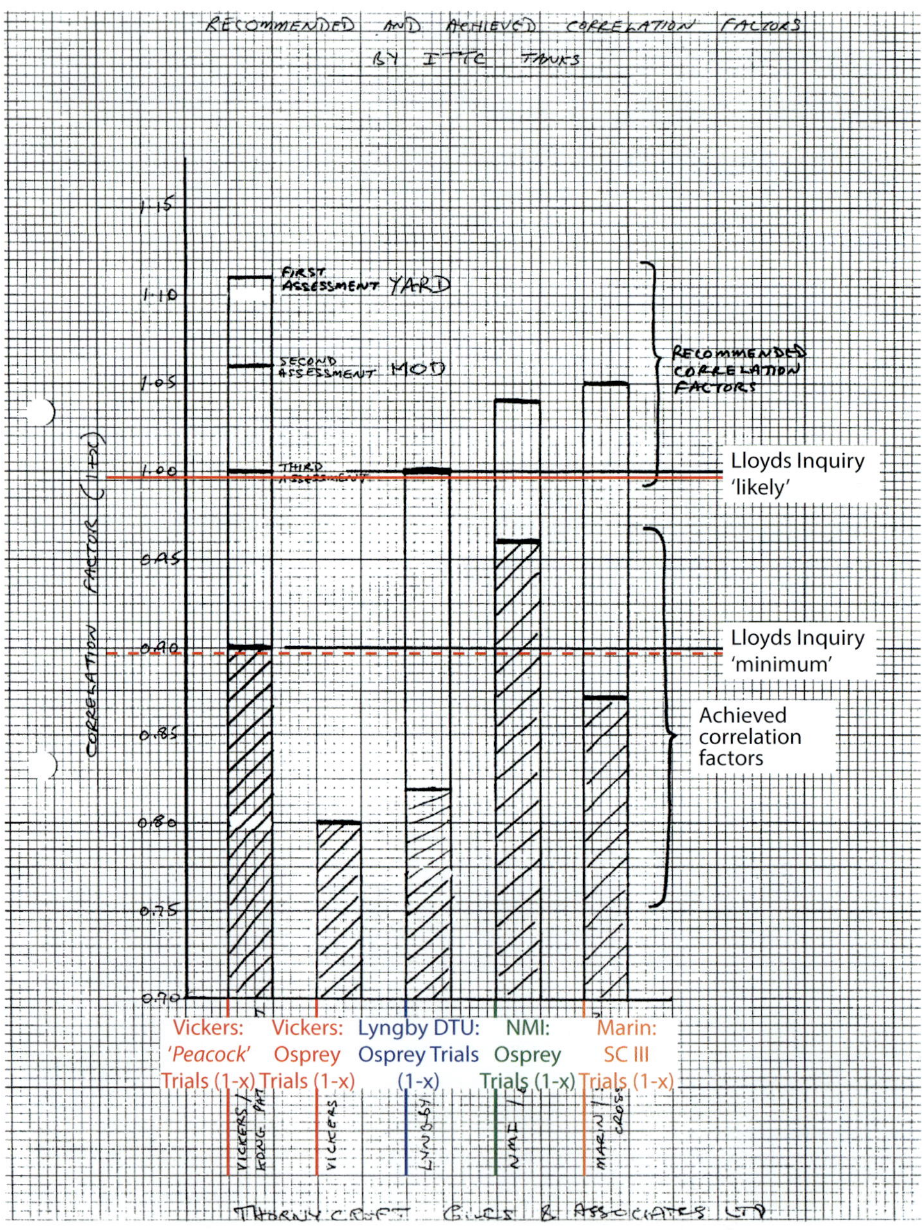

Appendix 4

Figure 1: (1984) TGA's OPV-3 Proposal to the MOD (Decks 01 to 05):

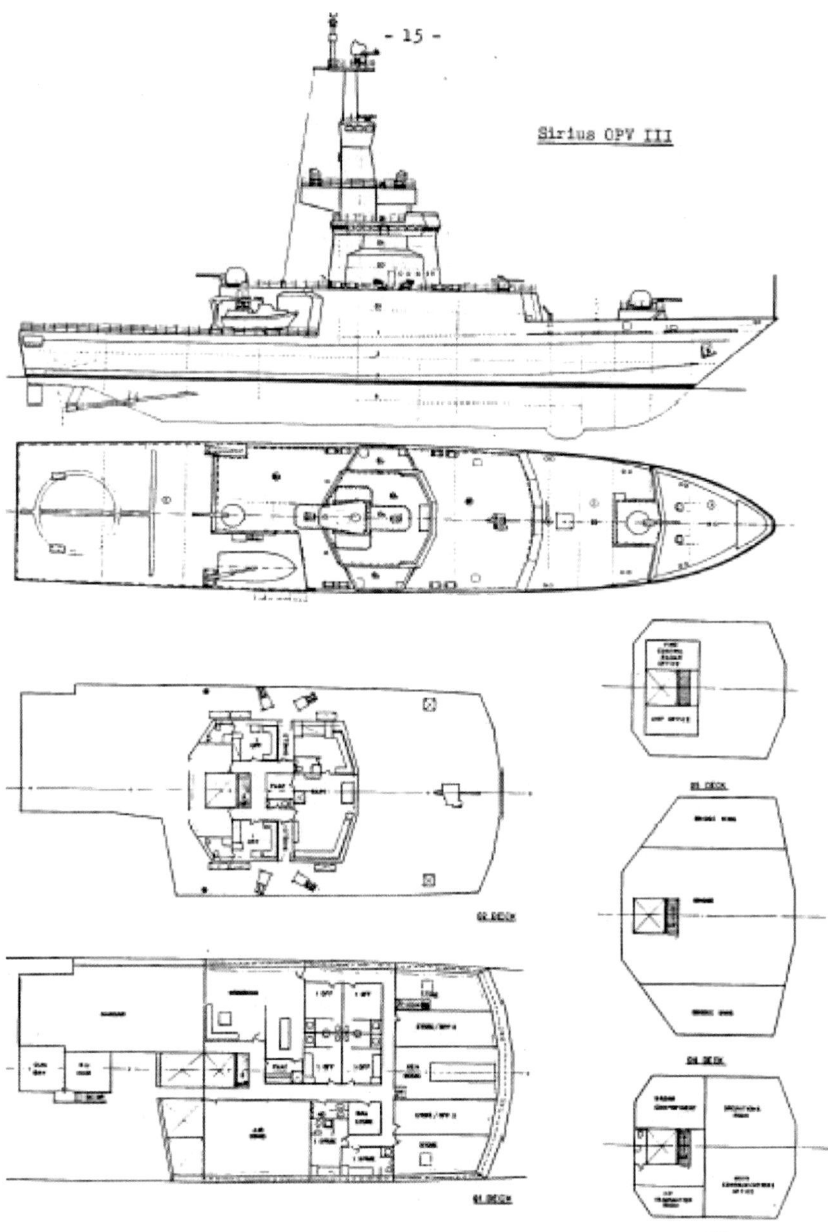

Figure 2: TGA's OPV-3 Proposal (Accommodation Plan by Conran Design Group Ltd):

- 16 -

Sirius OPV III

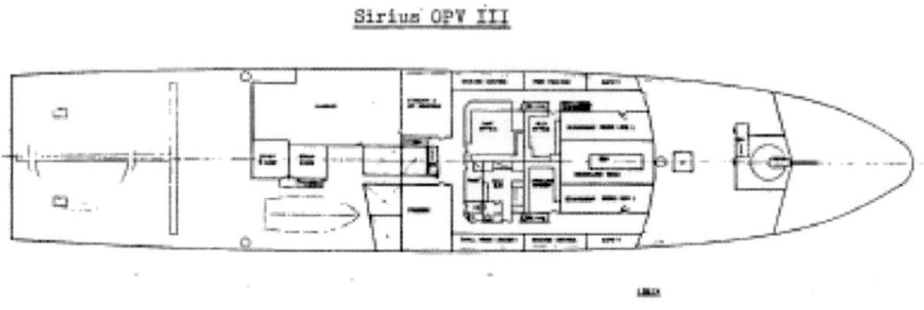

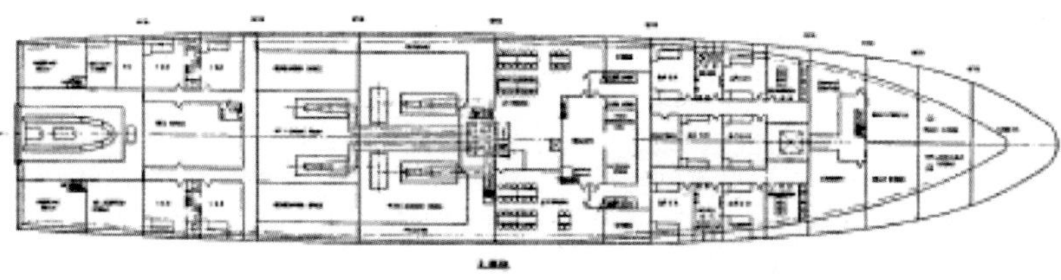

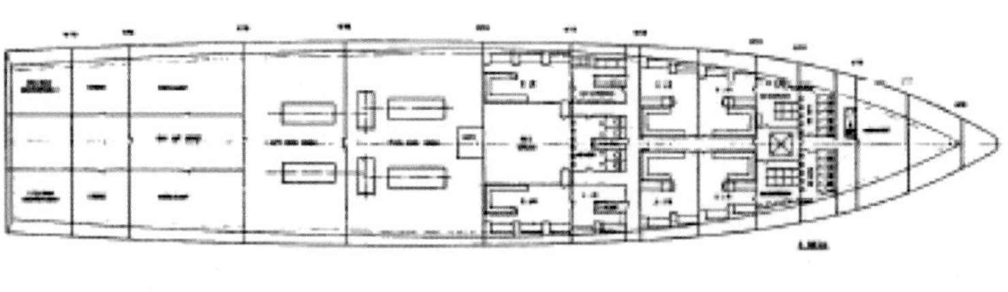

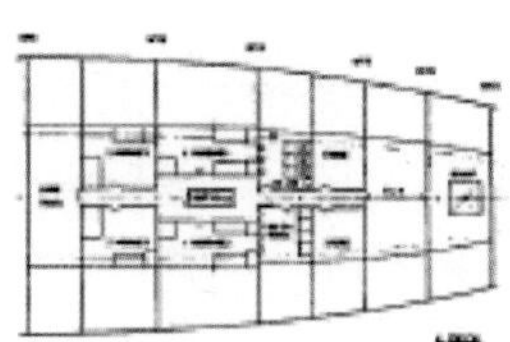

Figure 3: (1988) TGA's original S90 presented to the Lloyd's Hull Design Inquiry, compared to Lloyd's Inquiry's version, S-102 below (drawn to scale):

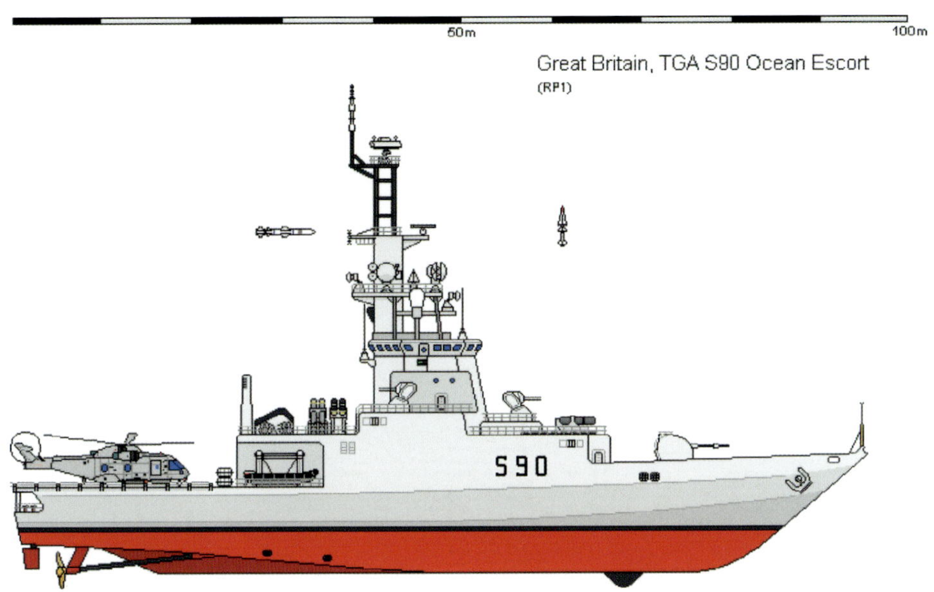

Figure 4: Lloyd's Inquiry's S-102 version of TGA's design to the final 1982 Naval Staff Requirement (NSR) 7069

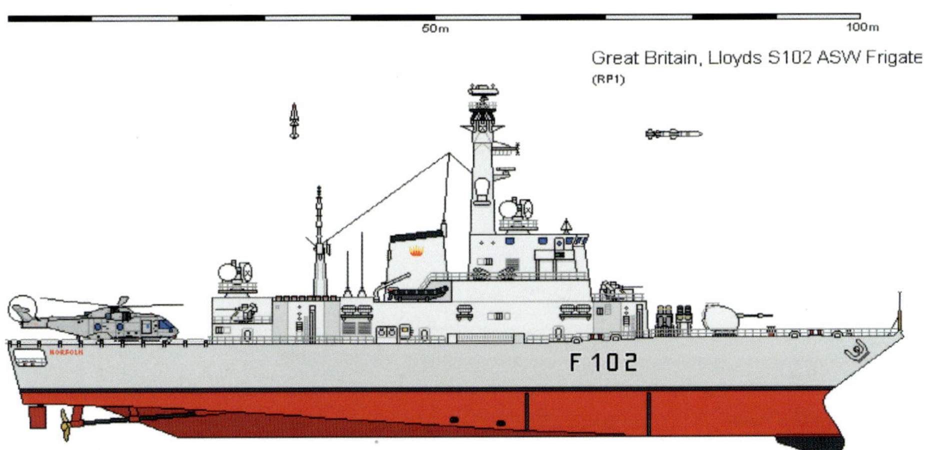

Figure 5: Comparisons between the three designs covered by this Summary, with approximate Length-to-Beam Ratios. This was defined by the Author's Patent as 'between about 5 and 7'. The S115 is S90 as enlarged to the Type 23 final size:

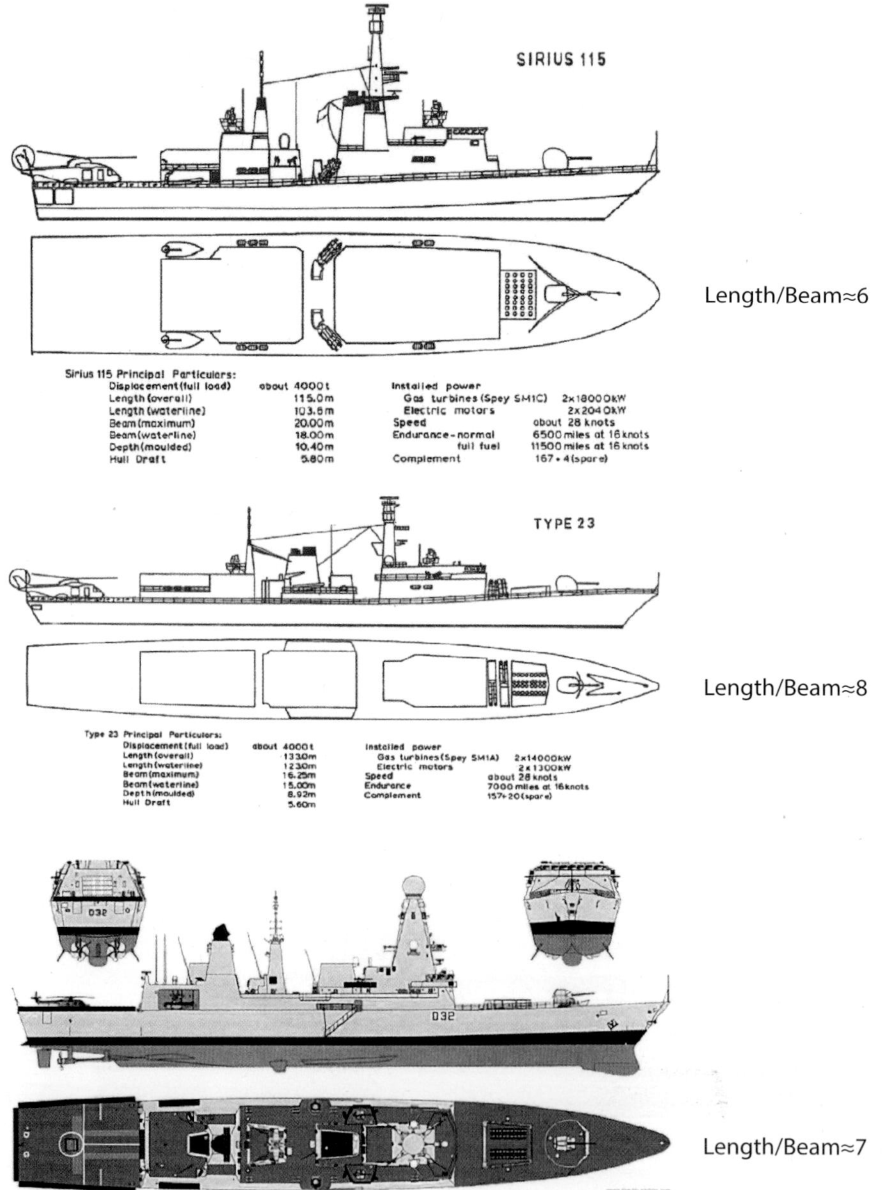

Type 45: LOA: 152.4m. LWL: 138m. BOA: 21.2m. BWL: 18.5m. Displacement: 8,500 tonnes.

Figure 6: The largest variant of the SPMH: The 2008 FastShip, after a 12-year/$40 million development, at the time her design had received Main Drawing Approval from the leading Classification Societies det Norske Veritas and American Bureau of Ships: LOA: 770ft, <42,000 tonnes, Displacement; 12,000t., Payload: 8,000 n.mls. @ 36 knots/6.000t. 4,000 n.mls. @ 40 knots; 5x RR MT-30 GTs run on LNG; 5 KaMeWa water jets

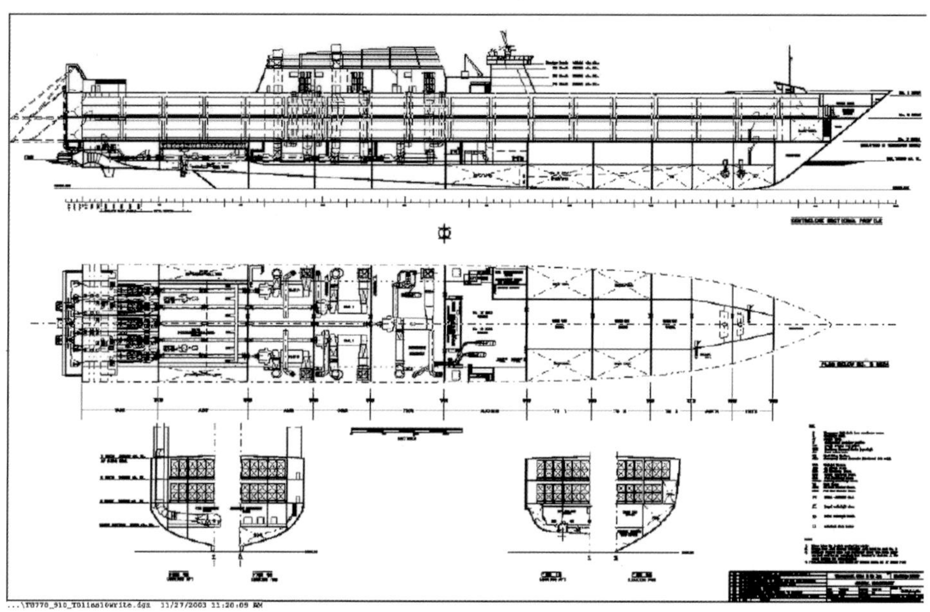

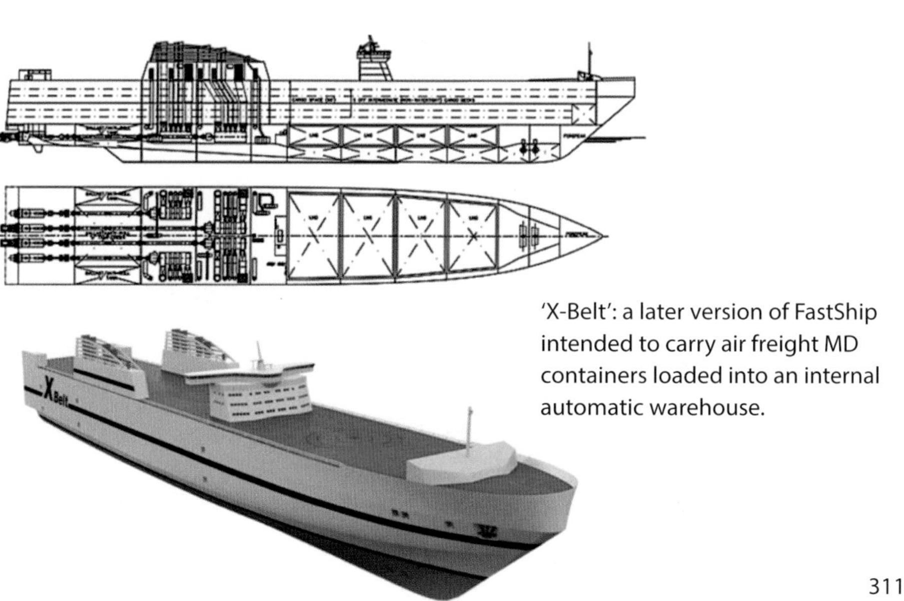

'X-Belt': a later version of FastShip intended to carry air freight MD containers loaded into an internal automatic warehouse.

Figure 7: The Type 45, first of the Royal Navy's new generation of 'Shorter, Fatter' Ships. Its scaled horse-power, or Resistance Co-efficient, for a 32-knot speed is almost exactly the same as the smaller S115 at 28 knots, as shown in Appendix 1. She has a similar Displacement/Length ratio and L/B of 7.17, 1/2 way between S115 and Type 42 – Batch 3:

'Shorter and Fatter': The Royal Navy's latest Air Defence Destroyer, the Type 45, ordered in 2003 and entering service to replace the Type 42 in 2009.

Figure 8: Last of the 'Long and Thin' Ships? The Royal Navy's Type 42 – Batch 3, two previous versions of which (Batch 2) were sunk in the 1982 Falklands Campaign. Replaced by the Type 45 Air Defence Destroyer eleven years after the 1988 rejection of the S115 by the Lloyd's Inquiry. The difference in beam-for-length is about 40% – similar to that between the S115 and Type 23. See Figure 5 above.

Appendix 5

Figure 1: TGA chart showing a constant increase in speed with power after the onset of dynamic lift, compared with exponential power increase at that point for a conventional hull ('Ton' class scaled to S115 size from Rawson's *coup de grâce*, Fig. 1). Onset of dynamic lift annotated by TGA:

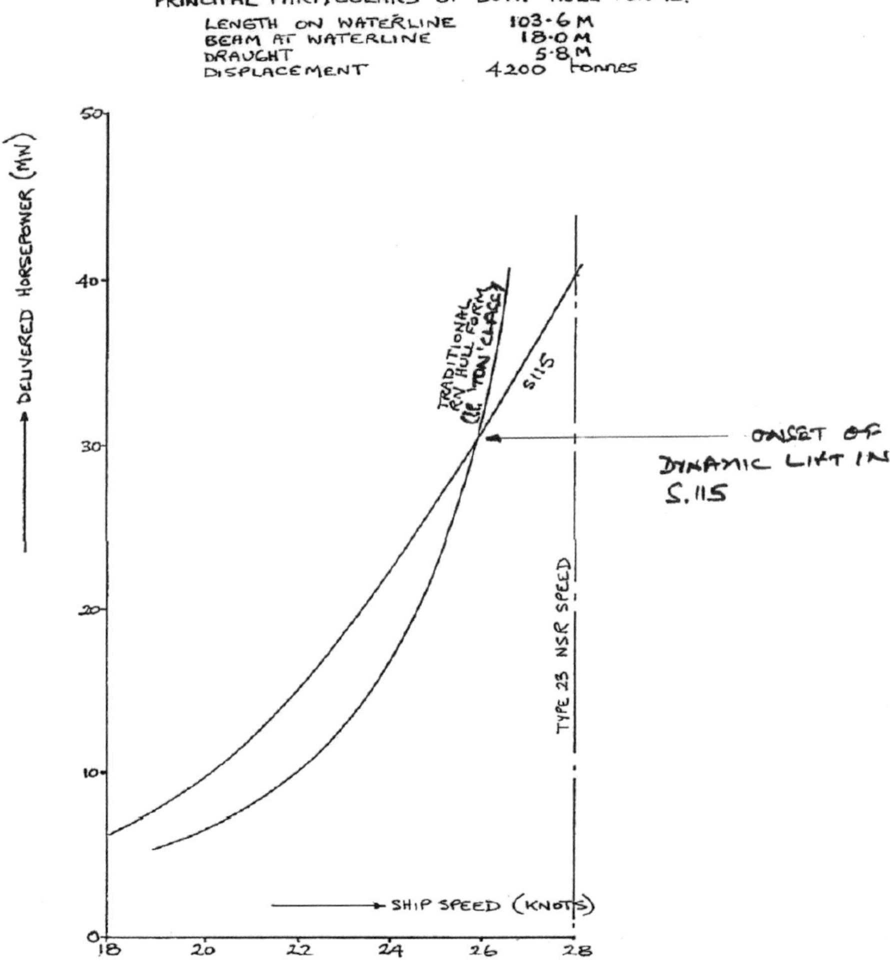

Figure 2: NMI measurements of S90 Trim and LCG rise versus Pe, EHP or Resistance with speed to 28 knots. Lower chart shows constant Pe, EHP or Resistance increase for S90 above 18 knots. This is exactly the data for the Type 23 the Lloyd's Inquiry and the MOD refused to supply for comparison. Onset of dynamic lift annotated by TGA, as in Figure 1 above.

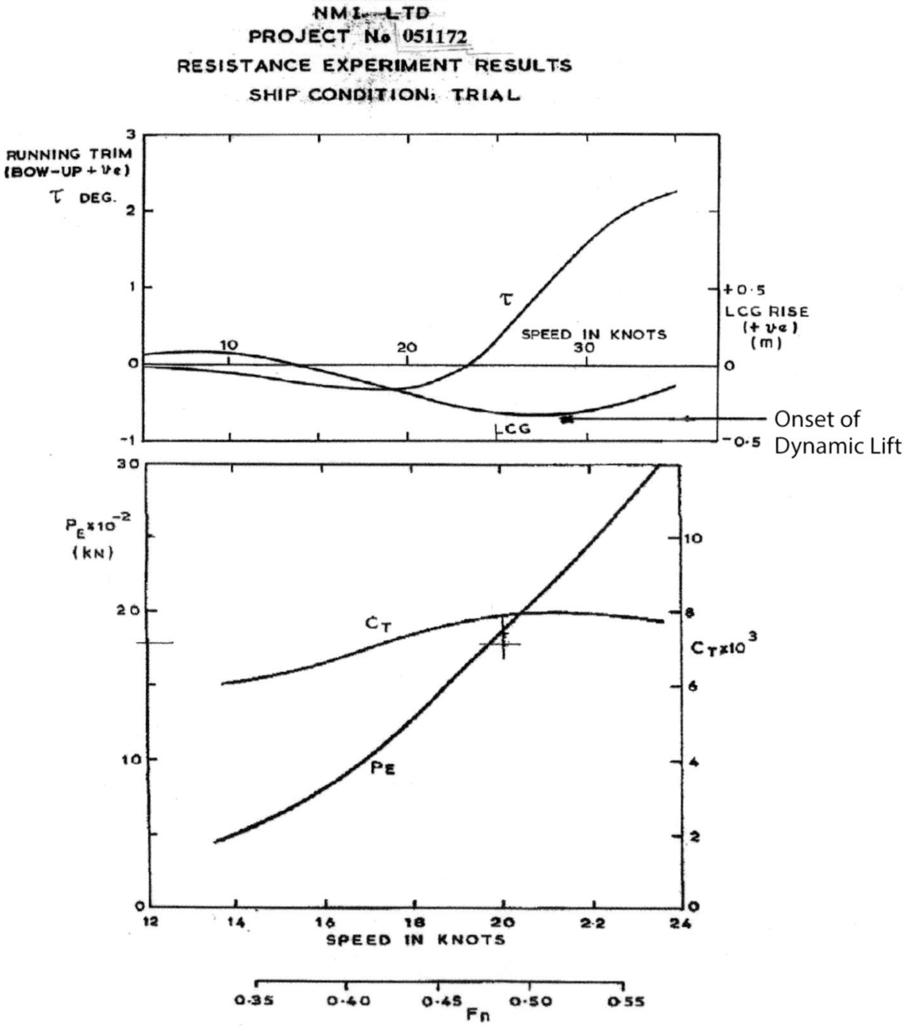

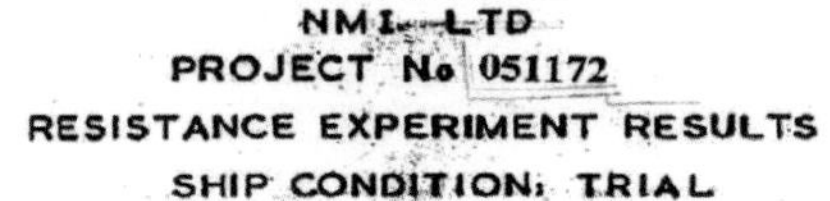

Figure 3: As shown to Defence Secretary George Younger Meeting in answer to the Inquiry's question: "Is the S115 short/ fat or long/thin?" Comparative Length/Beam ratios of TGA designs (red) and traditional frigates and destroyers (blue). Note that the low-speed 'Ton' class Minesweeper (green) is the same L/B as the S115. Though not shown, the Type 45 L/B (7.15) is exactly half-way between S115 (5.75) and Type 42 – Batch 3 (9.5).

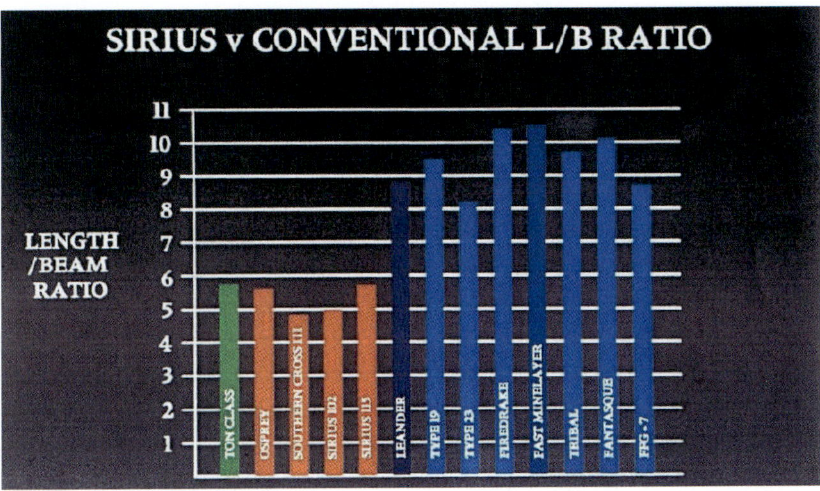

Figure 4: The effect of adjusting L/B ratio and Correlation Factor with the conflicting estimates using YARD, MOD and NMI powering factors for the S90:

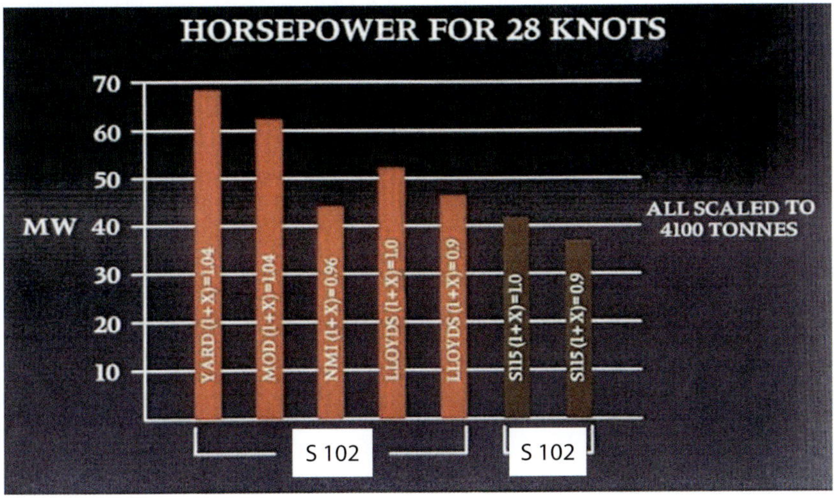

Figure 5: Summary of TGA's reasons for resigning from the Lloyd's Inquiry.

The Inquiry's Revised Terms of Reference for the S90 design to conform to the upgraded Type 23 NSR 7069. As agreed with the Lloyd's Inquiry, TGA and Lord Hill-Norton on 29th April 1987, one month after the Inquiry first convened:

'To consider the advantages and disadvantages of the S90 hull form for the purposes of meeting the Naval Staff Requirement (NSR 7069) for an anti-submarine frigate (insofar as the current state of development of the S90 permits), taking account of independent assessments by YARD and the Marine Technology Board of the Defence Scientific Advisory Council, and of the Hill-Norton Committee Report Hull Forms for Warships published in May 1986, and to identify any implications for the design of future destroyers and frigates for the RN.'

1. The 1983 S90 Validation Report was based on an early pre-Falklands version of the Type 23 produced by the Operational Requirements Committee, or ORC, which was already out-of-date when given to TGA. The later Naval Staff Requirement (NSR 7069) had been through two enlargements – from 2,500 tonnes to 2,800 tonnes and finally to 4,000 tonnes before the Lloyd's Inquiry. This meant that the criteria for the Inquiry's assessment called for a larger vessel than examined in 1982–83. The final versions of the S90 and S115, as presented to Lloyd's are compared in Figure 3 above – but, although Lloyd's financed the S115 design, and it was finally admitted by Lloyd's to satisfy the latest NSR – in most respects better than their own S-102 – it was rejected for comparison as being 'outside the authorised scope of the Inquiry'.

2. The Inquiry received the DSAC Response to Hill-Norton's Committee on November 29th, 1986, four months before it ever sat and contrary to the BS request that no evidence should be taken until 9 weeks after the settlement of the Osprey Case (Jan 20th, 1987) since it might be prejudicial to that case. It was, in fact, taken about 9 weeks before the settlement of the Osprey Case.

3. The DSAC Response claimed that it had not seen the final Sirius hull form data upon which the Hill-Norton had based its arguments in favour of the Sirius concept. However at least one member – Dr Goodman who, as the most important member of the Lloyd's Inquiry, should never have assisted the DSAC – received the DSAC

data in October 1986. Meek falsely claims to have seen 'voluminous documentation' on the S90. Such detail did not exist at the time of his February 1983 DSAC Report, and was only available in July 1983. Finally, the S90 Validation Report's most important ingredient, the NMI tank-test Report, he had previously rejected out of hand in July 1983, apparently without having read it.

4. Meek records that the decision against the Sirius family had already been made at the time of the settlement of the Osprey Case, on January 20th, 1987, ten weeks before the Inquiry first convened: "By that time the final conclusion that there was no value in the Giles design proposal was about to be announced."[3]

5. Dr Ross Goodman continued to act as lead naval architect of the Inquiry despite having been a member of Meek's DSAC Committee and assisting in writing their Response to Hill-Norton before the Inquiry sat. However he took no further part, after TGA had communicated to the MOD their knowledge of the DSAC Response being received by Lloyd's before the Inquiry sat.

6. The DSAC Response that was sent to TGA anonymously from within the MOD did not differ materially from their 1983 Report. Its major findings were used to demolish the final Sirius version (S115) developed at the Inquiry's cost, to be compared with the third and latest version of the Type 23 on the same grounds as were used against the S90 in 1983.

7. The Inquiry insisted on developing its S-102 version of the S90 hull form to be compared with the final Type 23. It refused to take account of the advantages available using normal naval architectural practice, suggested by Admiral Sir Lindsay Bryson in 1982, of adjusting Length-Beam ratios (L/B) within the Sirius design limits of 5 to 6 as prescribed by the Inquiry itself and sanctioned by Meek in his memoirs.

8. The *Southern Cross III* trials, intended to resolve the question of model/full-scale correlation factor were inconclusive and appear to have been adjusted to show a powering consistent with a conventional hull.

9. TGA were only able to obtain the Minutes of the Meeting recording TGA's objections to their proposal for an increase in L/B was not being allowed by the Inquiry, four months after the meeting took place. By that time the LR Study 1 (S-102) was complete.

3 M Meek: *There Go The Ships*, p. 194

10. In September 1987, the Inquiry introduced new Terms of Reference covering weight and space requirements from the Type 23 NSR to which TGA was not allowed access so they could not be incorporated in the S115.

11. The Inquiry represented Dr Garwin as agreeing with Rawson's initial statement that dynamic lift, measured in a tank model could not scale up to benefit a full-sized ship to the same degree. That was the opposite of the truth. At the end of that correspondence, in August 1981, all the leading Academics, including KJ Rawson (Visiting Professor of Naval Architecture at University College, London), Sir James Lighthill (Rawson's Provost of University College, London) all agreed with Dr Garwin's conclusion that the semi-planing hull's performance would scale up in proportion to increasing size, as shown by the concluding letters of each, given in chronological order below:

MINISTRY OF DEFENCE
Block G Foxhill Bath BA1 5AB

Telex Telephone 0225 61211 ext 3737

Professor Richard L Garwin
IBM Thomas J Watson Research Center
P O Box 218
Yorktown Heights
New York 10598
USA

Your reference
Our reference D/S/DDSD/312/81
Date 22 September 1981

Dear Dr Garwin

Your letter of 26 August addresses the specific variation of lift, per se, with size. My letter addresses the significance of lift in larger ships in reducing drag.

2. Like other dynamic forces the lift coefficient will be the same for geosims at the same Froude number. This, I believe, is what you are saying and, if so, we are in agreement.

3. Significance of lift depends on the attitude of the boat which will be perturbed to a degree depending on its weight and inertia that do not scale simply. Characteristic length, used for Froude number, may also change with attitude so that the significance of lift in affecting the power requirements is not directly scalable. As a generality, lift and planing are important in small craft operating at Froude numbers much higher than are usual with large ships. Saunders' 'Hydrodynamics' has much to say on the subject.

4. I regret that I have not the time and resources to continue this correspondence.

Yours truly

Richard L. Garwin
IBM Thomas J. Watson Research Center
P.O. Box 218
Yorktown Heights, NY 10598
(914) 945-2555

October 6, 1981

Mr. K.J. Rawson
Deputy Director, Ship Design
Procurement Executive
Ministry of Defense
Block G
Foxhill Bath BA1 5AB
ENGLAND

Dear Mr. Rawson:

Thank you very much for your letter of 09/22/81. I am glad that our correspondence comes to such an agreeable and harmonious end.

We agree that at the same Froude number (and attitude) the lift is as important in comparison with weight on a large ship as on a small ship. We agree that the significance of lift in affecting power requirements depends critically upon attitude.

I believe therefore that we must also agree that if lift can be used to reduce drag significantly in small ships, then by careful attention to trim and attitude in a similar large ship (or by good luck!) a similar beneficial effect may be obtained.

Therefore, the laws of physics (and in particular the "square-cube relation") in no way prohibit this happy result.

Very best regards,

Sincerely yours,

Richard L. Garwin

cc:
D.L. Giles

RLG:rsa:279'KJR:100681.KJR

University College London, Gower Street, WC1E 6BT 01 - 387 7050

PROVOST: SIR JAMES LIGHTHILL

9 November 1981

Dear Mr. Giles,

Over the last few months I have much enjoyed the technological argument that has been in progress between Mr. Rawson and Dr. Garwin. I assure you that I have been reading every letter in the correspondence carefully even though I have not hitherto joined in ..... Now, however, I will just make the comment to you that, in my own carefully considered opinion, Dr. Garwin's concluding letter of October 6 sums up the position perfectly correctly from the standpoint of hydrodynamic theory.

With kindest regards,

Yours sincerely,

James Lighthill

Sir James Lighthill,
Provost

12. The Inquiry ignored the MOD's July 1982 agreed programme of Validation for the S90, failing to take any account of the NMI controlled sea-keeping tests or the BHC comparative one-tenth scale model sea-keeping tests in the open sea between the Leander and S90 frigates using the MOD's recommended latest techniques. All were undertaken at considerable cost to TGA, BAE Dynamics and the UK Department of Industry.

13. The Inquiry took no account of the construction economies of building a shorter wider structure thus reducing the number of hull sections to be assembled, as advised by the Danish yard that had built several TGA Osprey-type vessels.

14. It refused to undertake tank measurements to establish differences in hydrodynamic performance between a Sirius hull and a conventional hull of the same size and proportions, thereby supporting its claim that the Sirius hull is conventional and therefore can be assessed using conventional computer predictions for existing hull forms generating negative hydrodynamic lift at all speeds.

15. The Inquiry refused to release data from Type 23 tank tests that might clarify the above issue due to the 'commercial confidentiality' covering those tests.

16. Based on prior advice from several sources we presume that this was because the Type 23 model tests did not include measurements of dynamic sink and lift with speed.

TGA resigned from the Inquiry in 30th March 1988, having received no undertaking that it would propose the S115 as the most compliant design for comparison with the Type 23 upgraded from 2,800 to 4,000 tonnes – despite having agreed to this at the final meeting with TGA on February 23rd 1988 and having financed the S115 design process itself.

D.L.G. **July, 1988**

Appendix 6

The MOD Memo: Twelve scribbled words that gave the game away.

Circulated two days after the end of the Falklands Campaign and one month before the S90 Validation Programme was agreed by the MOD: "The Naval Staff recommend rejection of GILES' proposals from the outset." And ... "You have already seen this and it will form the basis for your discussion with SofS (John Nott – Secretary of State for Defence) this afternoon. He, of course, does not have a copy." (Released in 2013 under the 30-year Rule):

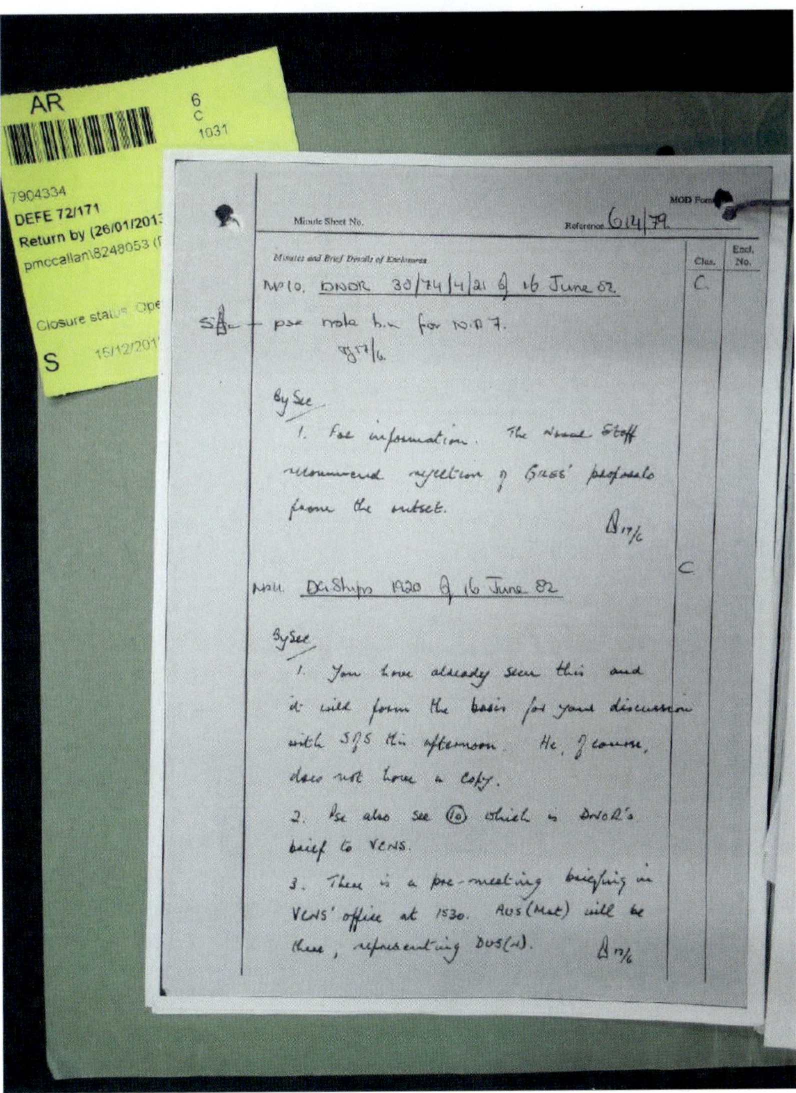

Minute Sheet No. Reference 614/79. MOD Form

Minutes and Brief Details of Enclosures | Clas. | Encl. No.

M° 10. DNOR 30/74/4/21 of 16 June 82. C

By Sec

1. For information. The Naval Staff recommend rejection of GILES' proposals from the outset.

17/6

C

M11. DG Ships 1920 of 16 June 82

By Sec

1. You have already seen this and it will form the basis for your discussion with SofS this afternoon. He, of course, does not have a copy.

2. Pse also see (10) which is DNOR's brief to VCNS.

3. There is a pre-meeting briefing in VCNS' office at 1530. AUS(Mat) will be there, representing DUS(...).

17/6

Appendix 7

Arguments for the FastShip Commercial Market:

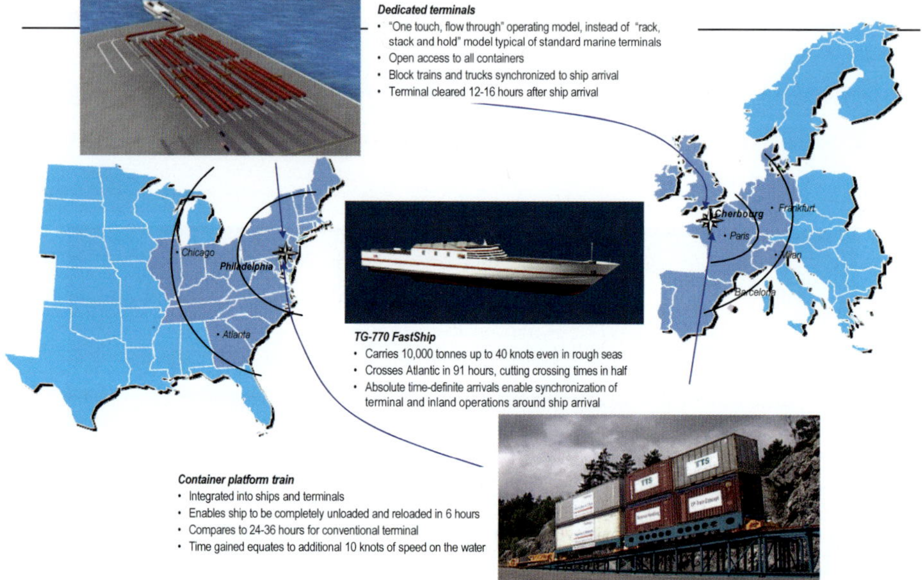

While the above slide gives the nature of the FastShip service, the following slide gives the comparative place of the FastShip service between Air and Sea Freight – and the amount of cargo required to meet the debt service agreed by the Banks:

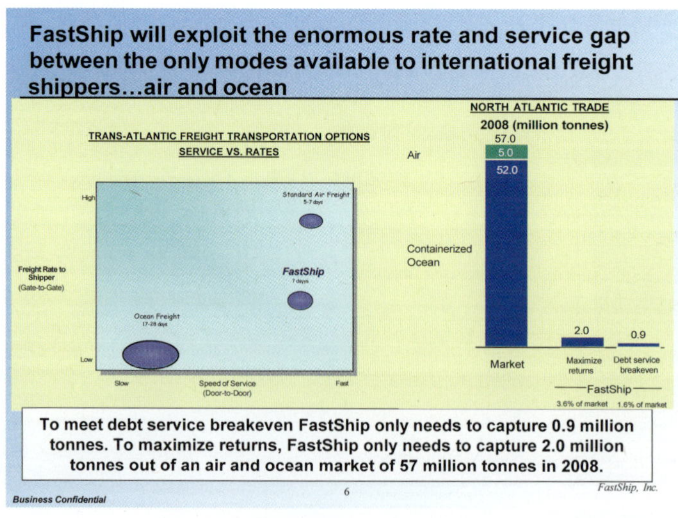

The following slide shows the gap between air freight costs and FastShip, as they stood in 2003, at the same time as the foregoing slides:

Air freighters cannot compete; high costs allow favorable pricing for FastShip

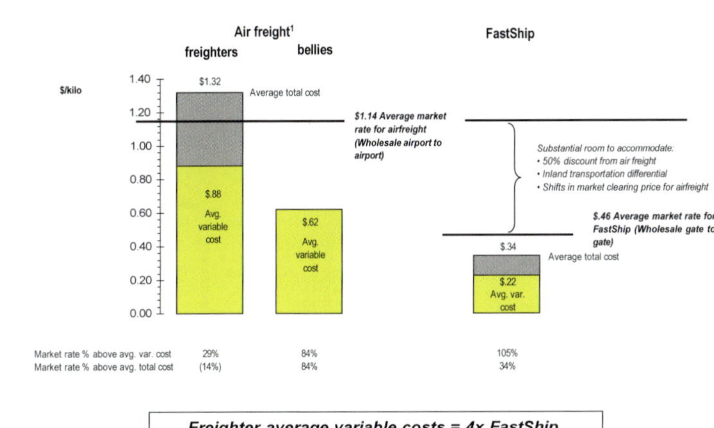

Freighter average variable costs = 4x FastShip

1. Source: MergeGlobal Inc.

Business Confidential

10

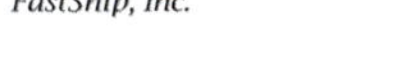

Proprietary & Confidential

Interviews and Market Studies Indicate Market Potential is Strong

- FastShip has conducted interviews with 51 major shippers to assess willingness to switch to FastShip
- Assumptions in interview questions aligned with FastShip's business case
- Very positive responses from shippers and freight forwarders alike
  - Freight Forwarders indicate a willingness to take an estimated 20-30% of FastShip's capacity on 3 year term take or pay contracts.
  - Shippers indicate an overwhelming support, with willingness to switch far in excess of the required upgrade/ downgrade rates
- Findings verified in additional market studies

Ocean Upgrade Diversion to FastShip*

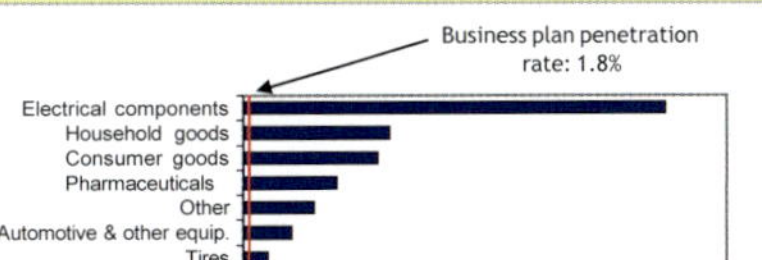

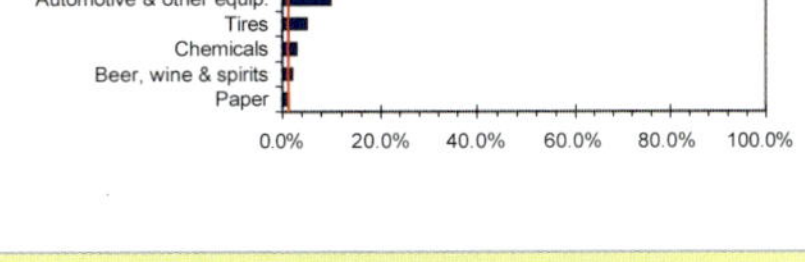

Air Downgrade Diversion to FastShip*

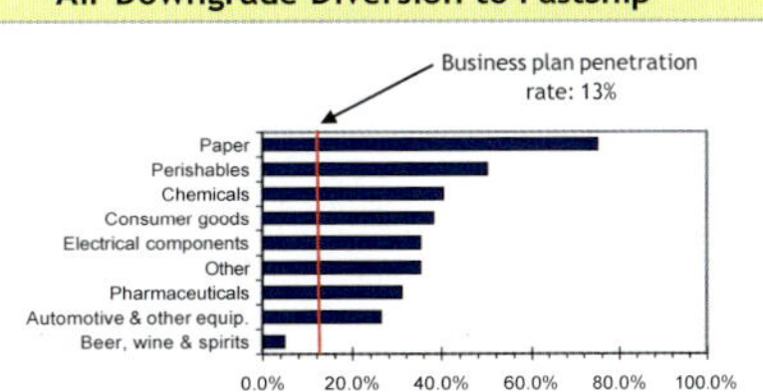

*Based on interviews with 51 major shippers

11

Selection from 87 Slides Discussed at 6.21.02 Presentation to Capt. Schubert (MARAD):

1. The FastShip financing situation when MARAD withdrew the Title XI Financial Loan Guarantee:

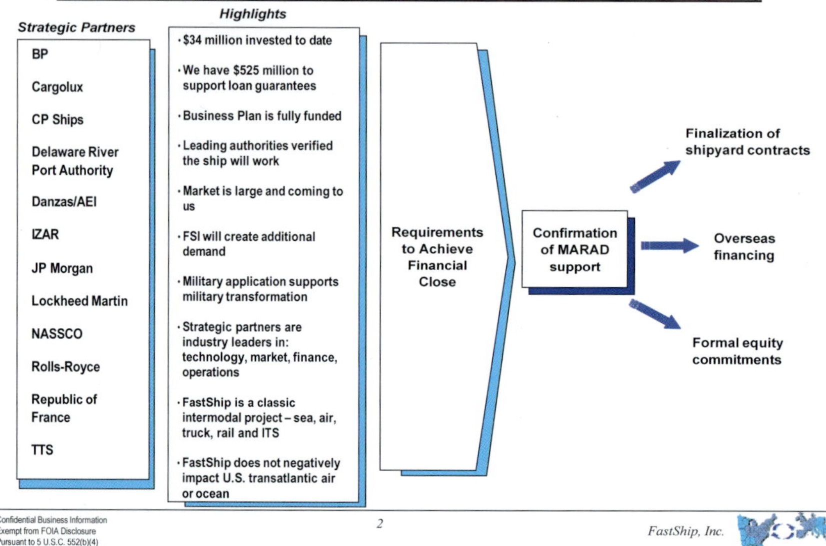

2. Eight-year Validation of FastShip Technology by Leading Institutions:

FastShip works -- validated by leading international technical authorities

- ✓ **MIT:** SWAN Codes Seakeeping Analysis
- ✓ **SSPA:** Resistance & Propulsion, Seakeeping Tank Tests
- ✓ **Danish Technical University:** QST, Computer Analysis
- ✓ **Trondheim University:** Extreme Sea Condition Analysis
- ✓ **Det Norske Veritas:** Classification to +100A1 CSA-2A for Unrestricted Worldwide Operation
- ✓ **U.S. NavSea:** Evaluation of National Defense Features (NDF)
- ✓ **J.J. McMullen Assocs:** Due Diligence for Investors
- ✓ **Rosenblatt:** Due Diligence for Investors

- **DNV** completed Main Drawing Approval (MDA) February 23, 2000
- **U.S. Coast Guard** accepted DNV's MDA, subject to construction verification, April 12, 2000

Confidential Business Information
Exempt from FOIA Disclosure
Pursuant to 5 U.S.C. 552(b)(4)

9

FastShip, Inc.

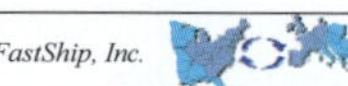

Our strategic partners proposed commitments

Rolls-Royce
- R&D investment on engines and waterjets: $15 million
- Vendor financing: $90 million
- Equity investment: $15 million

CP Ships
- Equity Investment: $5 million
- Equity investment: fund sales and marketing build-up: $15-20 million
- Performance risk on sales and marketing expenses: $20 million p.a.

Cargolux
- Equity investment: $5 million

Lockheed Martin
- Structuring financial commitment

JP Morgan
- No retainer for 3 years, value of resources approximately $3.5 million
- Equity investment: $15 million

BP
- Fund construction of tank farm and other infrastructure: $15 million

TTS
- Seed capital: $3.8 million
- Terminal development funding to date: $1 million
- Vendor financing: $4 million

Republic of France – Chamber of Commerce and Industry of Cherbourg
- Seed capital: $2.5 million
- Terminal financing: app. $90 million
- Infrastructure improvements: app. $100 million

Delaware River Port Authority
- Seed capital: $10.5 million
- Terminal financing: $75 million

Conrail (Norfolk Southern and CSX railroads)
- Terminal financing: $15 million

Danzas/AEI
- Equity investment: $10 million

Ship management company
- Equity investment: $5 million

Holt Cargo Systems
- Seed capital: $3.5 million

IZAR
- Participation in financing being structured

NASSCO
- Participation in vendor financing being structured

Total equity and long-term debt = $525 million

Confidential Business Information
Exempt from FOIA Disclosure
Pursuant to 5 U.S.C. 552(b)(4)

31

FastShip, Inc.

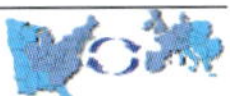

FastShip, Inc. - Summary of Capital Requirements

FastShip, Inc.
Summary capital requirements (through construction period)

Uses of funds	
Ships	
2 ships U.S.	792
2 ships Spain (after tax lease)	510
guarantee fees - Marad	71
guarantee fees - Spain	41
subtotal ships	1,414
Terminals	
Philadelphia	83
Cherbourg	90
subtotal terminals	172
other capex	5
working capital	146
Total	**1,737**

Sources of funds		
Debt		
Marad Title XI debt	762	
Spanish financing	441	
GIEK	21	Norwegian export financing
DRPA	75	Philadelphia terminal
Conrail	15	Philadelphia terminal
PIDC	5	Philadelphia terminal
French guarantees	82	Cherbourg terminal
Subtotal	1,401	
Equity		
Vendor preferred	256	
Total new common	80	
Subtotal	336	
Total	**1,737**	

Based on FSModel59 2plus2.NASSCO.Izar.revised.lease

Confidential Business Information
Exempt from FOIA Disclosure
Pursuant to 5 U.S.C. 552(b)(4)

29

FastShip, Inc.

FastShip Advantages to U.S.: Unparalleled Intermodal Opportunity

Economic

- Consumers get greater choice of goods, cheaper, more timely access
- American corporate productivity enhanced via streamlined supply chain
- 7,500 jobs created in Philadelphia Region
- Port of Philadelphia and South Jersey will be reconstituted into world-class transportation hub, leveraging existing rail, truck and air assets
- FastShip will pay guarantee fee double the Title XI appropriation, i.e., ($71 million vs. $32) million
- U.S. income tax projected at $160 million per annum upon ramp-up of first 4 ships

Military

- Rapid deployment for sealift and special operations; personnel and equipment
- 10,000 tons; 10,000 miles; 11 days: 50% better
- Integrates into "agile port" concept
- FastShip has DOD funding – no criticism possible that we take scarce resources from other projects

Security and Environment

- The FastShip network will be the standard for container security – new terminals, IT, closed loop network, customer base, 100% visibility
- New port capacity created to reduce overall terminal congestion; will feed existing regional infrastructure
- An ocean "I-95" is a future application
- Favorable emissions profile relative to both ocean and air: cleaner fuel and less fuel burn per ton

Confidential Business Information
Exempt from FOIA Disclosure
Pursuant to 5 U.S.C. 552(b)(4)

48

FastShip, Inc.

FastShip will provide increased sealift and rapid deployment capability for military contingencies or humanitarian emergencies at minimal cost

- Enhanced sealift capability and rapid deployment:
  - 272 helicopters to the Persian Gulf in less than 10 days
  - 12,000 tons of military cargo and personnel (e.g., two current armored battalions with heavy tanks or one light "Army Objective Brigade") over about 8,000 nautical miles at a speed of 36 knots without refueling
  - One FastShip vessel = airlift capacity of 45 C-17 military aircraft over 8,700 nautical miles, carrying a planned light 'Objective Brigade'
- Operation in austere (underdeveloped) ports without cranes
- Can be compatible with J-LOTS or future Theater Support Vessel (TSV)
- Reduced dependence on prepositioning of unit equipment
- Ability to carry personnel and equipment for Special Ops and force protection
- Elimination of bottlenecks through rapid logistics implementing DOD agile port concept
- No need for over-fly permission
- Speedy response to global humanitarian needs in time of emergency

Confidential Business Information
Exempt from FOIA Disclosure
Pursuant to 5 U.S.C. 552(b)(4)

42

FastShip, Inc.

Appendix 8

'Short, Fat and Fast' vs. 'Long, Thin' – Length, Displacement, Speed (Froude No. and Speed-Length Ratio) and Length-Beam Ratio Compared:

Figure 1: LCS USS *Fort Worth* 336 ft/102.35m LWL; Beam: 59ft/18m; L/B ratio = 5.7; 3,500-tonnes Full Load Displacement; Max speed: 50 knots – FNo. = 0.79 / SLR = 2.65.

The waterline shows the slight crest of the wave alongside the position of the water-jet inlets, indicating high pressure. Having sunk slightly up to about 28 knots, the onset of hydrodynamic lift occurs and, at 32 knots, she has risen back to her normal 'static', or stationary, waterline.

Figure 2: French 'conventional' destroyer *Fantasque* 440ft/134m LWL; Beam: 39.4ft/12m; L/B ratio = 11; 2,500-tonnes Full Load Displacement; at 40 knots – FNo. = 0.57 / SLR = 1.9.

Shows typical stern 'squat' at highest speed. An extreme example of the ultimate 'long thin' fast displacement hull of World War II – at 70% of the displacement; 25% greater length; 50% less beam – and 5 knots less speed, compared to *Fort Worth*.

Appendix 9

For sea-keeping videos of an LCS, visit: <https://www.youtube.com/watch?v=lHm-3eKo-ek> compared with an Arleigh Burke class Destroyer: <https://www.youtube.com/watch?v=Zvzld04Q5XI>. Stills shown below.

Figure 1: The semi-planing hull demonstrating what MIT described as its 'unprecedented' sea-keeping at speeds unattainable by a destroyer hull in approximate scaled seas:

The 3,450-tonne LCS-5, USS *Milwaukee* at 35 knots in estimated 6-foot seas, 'exploding' the waves, while maintaining level trim:

Figure 2: 9,000-ton DDG USS *Dewey* at approx. 25 knots in high seas: pitching of traditional hulls:

Appendix 10

The chronology of relationships between FastShip Inc. and the various participants in the LCS Program prior to the granting of the final Construction Contract to Lockheed and Gibbs & Cox. As presented by FastShip during its meeting in 2008 with Sean Stackley, the US Navy's Under-Secretary for Appropriation.

Year	US NAVY	ROLLS-ROYCE/KAMEWA	GIBBS & COX	LOCKHEED MARTIN
1988 1990	SuperShip & FastShip SPMH proposals presented to CNO, NAVSEA & Naval Staff	Kamewa water jets chosen for SuperShip & FastShip SPMH Projects.		
	5/18/90: U.S. PATENT ISSUED FOR SPMH, VALID TO 5/18/2010			
1992				
1994		Kamewa water jets developed for FastShip		
1996	FS/SPMH presented at US Navy SWC Carderock			
1998	FSI obtains 2 NDF R&D contracts from NAVSEA.	FSI select Rolls/Kamewa propulsion package for FS		
2000	FS NDF contracts completed. USN given full access to FS R&D program, tank tests etc.	Design & classification of Rolls/ Kamewa propulsion pkg for FS		2/01: Strategic partnership & CA acknowledging SPMH Patent signed between FSI & LM. FSI, LM & IZAR work on construction of FS. LM given full details on scaling SPMH to LCS & FS sizes.
2001 2002	FS/NDF presentations to USN, USMC, Army, TRANSCOM show SPMH feasibility.			
2003	FS give presentations to USN on SPMH for LCS, using SuperShip tank tests etc.	Rolls given details of FSI's LCS project based on SPMH design. Rolls/Kamewa selected for LCS	2/03: G&C & LM invite FSI to give confidential present -ation on its LCS project based on SPMH. Discussion on Patent issues. FSI invited to join LCS design team.	
2004	**5/04: LM TEAM WINS NAVY CONTRACT FOR LCS BASED ON SPMH TECHNOLOGY**			
2005				6/04: Challenged by FSI, LM denies infringement of SPMH Patent
2006	1/06: LM&FSI present SPMH Proposals for RSLS to US Navy			
	9/06: FIRST LOCKHEED SPMH LCS, USS *FREEDOM*, LAUNCHED			
2007				1/07: LM cancels Its strategic partnership with FSI
2008	**4/08: FSI ISSUES CLAIM AGAINST NAVY FOR INFRINGEMENT OF SPMH PATENT**			

Appendix 11

Figure 1: Technical Study – Was this the end of the 'Long, Thin' Ship?

It seems TGA's 1988 S90 and S115 designs and the Royal Navy's latest Type 45 Air Defence Destroyer HMS *Daring*, have comparable Total Resistance factors and Froude No. or Speed-Length Ratios, whilst having increased Beam and Displacement-for-Length. Appendix 12 below shows that the stern profile of the Type 45, of 8,500 tonnes, is 'hooked', and thus a scaled-up enlargement of the design of the original S90 of 2,800 tonnes or its larger derivative, the S115, of 4,000 tonnes. Between 1980 and 1988, TGA provided MOD, YARD, Lloyd's Register (Naval Standard Requirements of Type 45) and BAE (Builders) with a vast amount of measured data on the Osprey, S90 and S115 designs, some obtained legally and some illegally.

In October 1989 TGC Inc. applied for a UK Patent on the large Semi-Planing Mono-Hull. The most significant hull-feature was its 'hooked' stern profile, to obtain lift, thus reducing wave-drag to increase speed. The MOD immediately suppressed our Application with a Secrecy Order, on the grounds it might be useful to a foreign hostile power. That was swiftly cancelled by HM Patent Office on the strength of the very words used by the Lloyd's Inquiry in its rejection, that *ipso facto* made it 'not obvious to anyone skilled in the Art'. The Secrecy Order occurred at the time the MOD Type 23 and 45 design team started work on a Post Falklands Air Defence Destroyer in 1982 (NATO's joint NFR-90 Project) abandoned in 1989 at the time of the SO on TGA's Patent. The UK then joined the 'Horizon' Air Defence Frigate project, also abandoned in 1999, to proceed with its own Type 45 design, built by BAE Surface Ships, entering service in 2010. Had Type 45 been a conventional 'Shorter Fatter' design, as <u>estimated</u> by YARD and MOD (see Appendix 3, Fig. 2 colour-coded Power Estimates above) the Type 45 power for 32 knots (based on their QPCs of 0.39 and 0.42) would produce a PD of 76,600 to 73,150 BHP. That is 33 to 27 percent higher than the published Type 45 power of 57,600 BHP for 30+ knots.

Using the following chart, based on Hull Resistance of 53.5 lb./Ton, with FNo. = 1.4, the Type 45 with the S90's overall QPC of 0.684 taken from NMI's Osprey propulsion test (Appendix 3, Fig. 2 above) and 8,000 tonnes at 30 knots, shows an EHP of 39,078 ÷ 0.684 = 57,671 SHP, compared to 57,600 CHP as published:

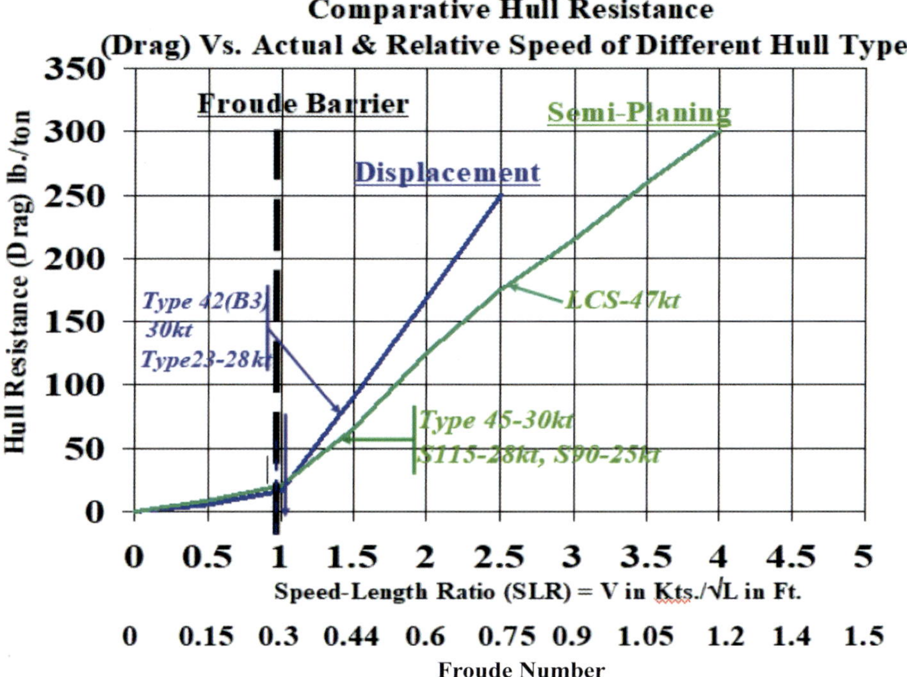

The above chart also compares the Specific Resistance, or Drag Coefficients, and relative SLR and FNo.s of different Displacement & Semi-Planing hulls. These include the RN's Type 42 and Type 23 classes of conventional Displacement category 'Long, Thin' Destroyers and Frigates with the RN's latest 'Shorter, Fatter' Type 45 and the Semi-Planing S90 and S115. Despite their increased beam, for a given SLR/FNo., they show a 25% reduction in Bare Hull Resistance. Allowing for their improved Propulsive Coefficient, this explains the c.36% reduction in Power/Speed shown in Appendix 3, Figs. 2 and 3 above, comparing NMI's S90 propulsion tests with YARD and MOD estimates for conventional hulls.

Figure 2:

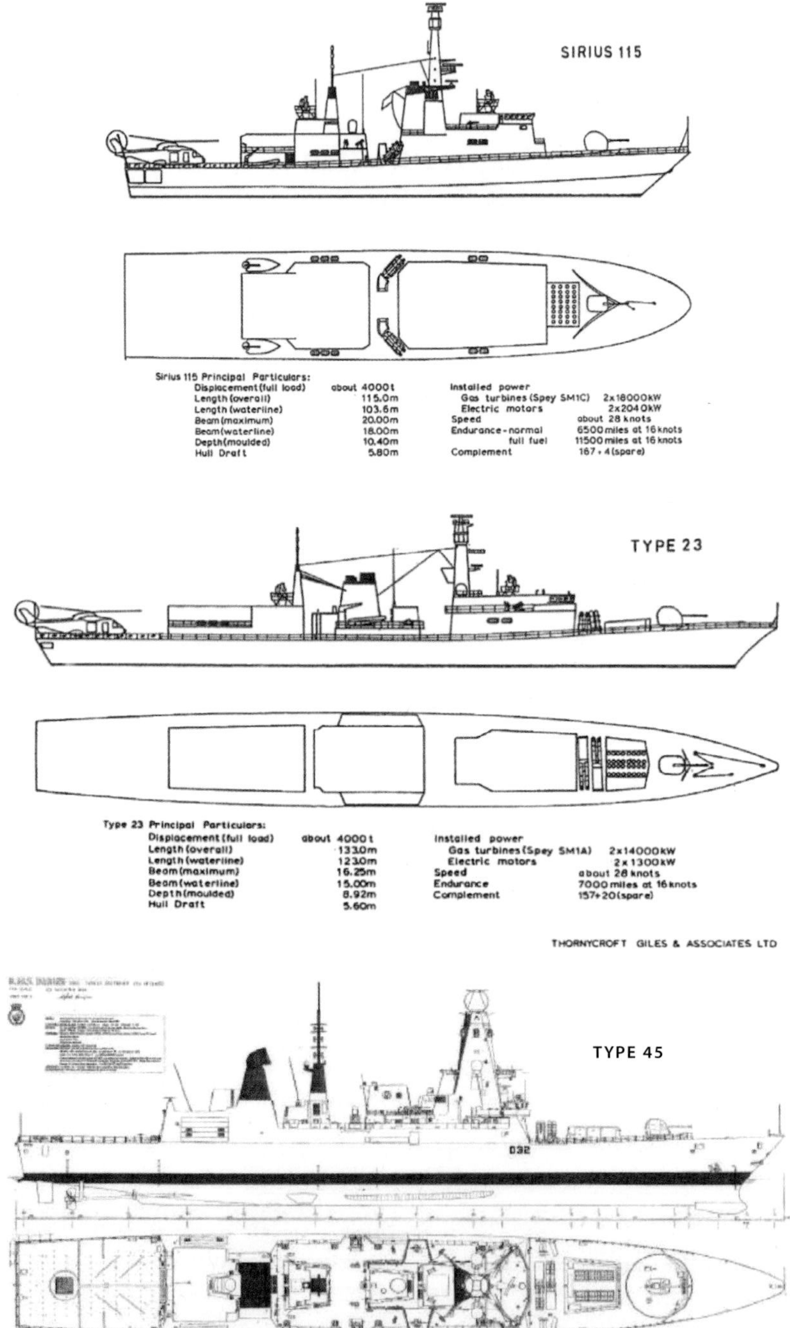

Length overall: 152.4 metres, Beam 21.2 metres, Displacement: 8,500 tonnes.

Appendix 12

Comparative stern designs of the Type 45, S90 and Type 23: Figure 1: Type 45 aft lines plan with red reference line showing a concave 'hook' similar to S90:

At full scale, based on this reduced original of a 1/72nd-scale drawing, max. separation of hull's hook from the red reference line is approximately 43 cm/17 in. (see Figs. 4 and 5).

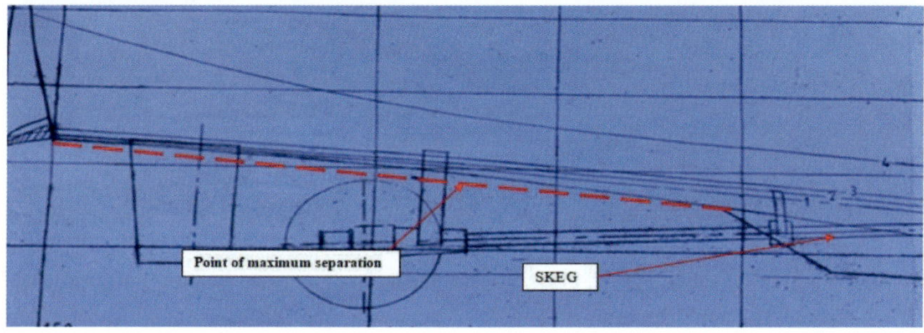

Figure 2: S90 aft hull lines with red reference line showing hook similar to Type 45:

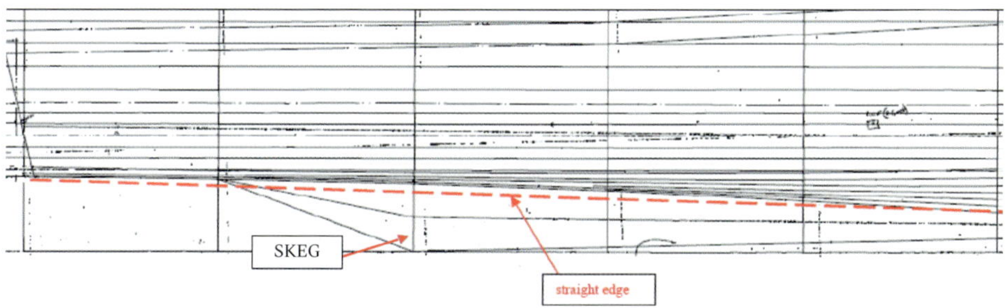

Figure 3: Type 23 (1980s pre-Type 45) aft hull lines with traditional straight profile:

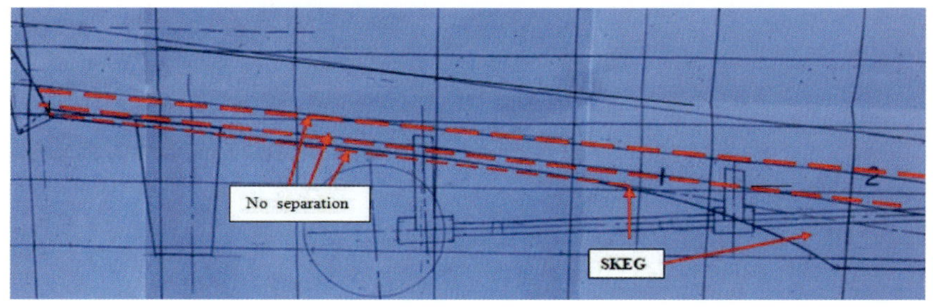

Proprietary & Confidential

Figure 4: Photo of concave stern and transom flap of Type 45 HMS *Daring* at launching:

Type 45 Transom Flap: The informative publication 'Seaforces Online' states (2021) that the Type 23 class 'ships are being fitted with a transom flap which can add up to 1 knot to the top speed and reduce fuel consumption by 13%.' That can only be due to hydrodynamic lift at the stern. With the added effect of the 'hooked' stern underbody, this must apply to the transom flap of the Type 45 with increased benefit: a further improvement in speed for less power and fuel consumption than for a conventional hull.

Figure 5: Concave stern of Type 45 HMS *Dauntless* in Dock:

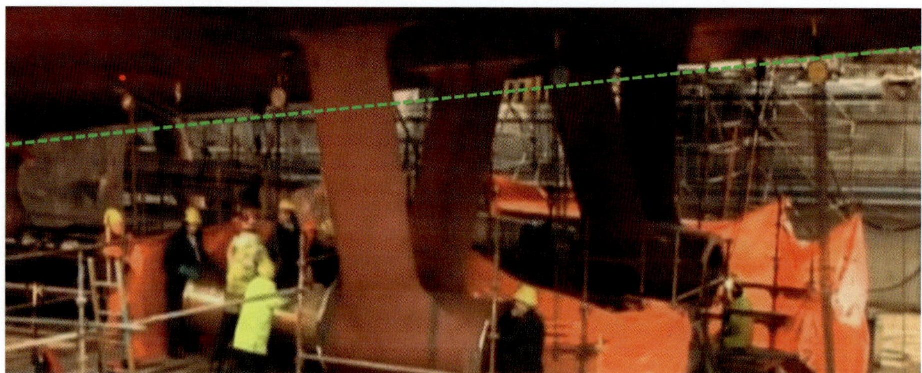

Figure 6: 18ft. 'Rooster Tail' wake of previous (1975) traditional 'long, thin' Type 42 Destroyer:

At about 30 knots (Froude No. = 0.42), with a traditional stern, similar to Type 23 (Fig. 3 above). Compare with wake of semi-planing Azteca below.

Figure 7: The flat wake of TGA's semi-planing Azteca at 24 knots/ Froude No. = 0.72 (70% higher than the relative speed of the Type 42 above):

Figure 8: The remarkable manoeuvrability of the Type 45 due to the more widely spaced rudders and increased stability:

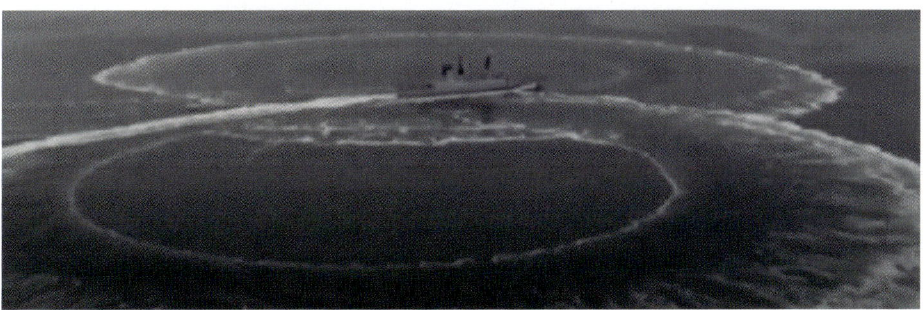

Figure 9: Two semi-planing hulls at similar Froude No. ≈ 0.45.

Type 45 at about 32 knots and TGA *Southern Cross III* at 20 knots. Waterlines show a low-pressure trough amidships and high-pressure crest at the stern, with high bow-wave and level trim.

Appendix 13

Figure 1: Rawson's 'Revenge', paragraph 3, opening the 'Short, Fat' saga:

DEPUTY DIRECTOR SHIP DESIGN

Procurement Executive Ministry of Defence
Block G Foxhill Bath BA1 5AB
Telephone 0225 61211 ext 3737

D L Giles Esq
Thornycroft Giles & Associates Ltd
24 Seymour Road
London SW18 5JH

Your reference
Our reference D/S/DDSD/146/81
Date 13 April 1981

Dear Mr Giles

PROPULSIVE EFFICIENCY OF OSPREY FAMILY

We have now had a look at some of the data contained in your letter of 17 March which followed our discussions on the 27 February. It did help to piece together some of the fragments of information on the OSPREY propulsive performance which you have supplied or published over the years. Some of the more tentative information, particularly that used in the unusual Telfer presentation of Appendix 8, has not been examined in depth because it appears to be based on conjecture which is not profitable to debate. There are two areas upon which detailed comment is worthwhile since they represent the essence of the problem viz the relative merits of short fat hulls such as the OSPREY family and long thin hulls which are the normal choice of naval architects, typified by the TONs or BLACKWOOD.

OSPREY SHP

2. In Fig. 1, I have brought together the trials data for both OSPREY and the TON Class minesweepers very slightly adjusted to be at the same displacement. The former accords with the model tank prediction at 15 knots which you have quoted in previous correspondence and I have therefore dotted the line down to this point. The TON line is the one which I have previously supplied to you (I observe that, in my letter of 25 March, the 16 knot figure was given as 1310 instead of 1810. As I said at our meeting on 27 February, the TON line (not the CASTLE line as you say in your letter of 17 March) climbs rapidly after 16 knots and would not be considered for speeds much above this. You, I imagine, would say the same in reverse for OSPREY viz that it is a high speed form not very efficient at speeds around 15 knots.

3. You will remember at our meeting on 27 February I said that if there were any significant dynamic lift on the OSPREY form, it really would represent an important contradiction of traditional theory. Unhappily, the information you now supply does not show such evidence and it must be presumed that conventional wisdom is not to be denied. Such dynamic advantage is confined to quite small craft and is lost at OSPREY displacements – not surprisingly because displacement varies as the cube of the dimension and dynamic lift as the square. We are safe in assuming that ships of SIRIUS size will certainly obey normal laws.

SIRIUS SHP

4. Your scaled-up version of OSPREY, SIRIUS, does provide us with an excellent opportunity for comparison with several well documented RN ships. We have chosen the Type 14, BLACKWOOD, at exactly the same displacement; the length/beam ratios of SIRIUS and BLACKWOOD are respectively 4.33 and 9.25.

COMMERCIAL IN CONFIDENCE

Figure 2: KJ Rawson's Figure I: Comparison of Osprey and Ton class at '373 tonnes'

The 'Smoking Gun': Rawson's hand-drawn graph, showing his use of 'Osrey tank data' that came from measurements at BS St Albans, using model 2230 and at the Dumbarton tank, using model 2225 on April 11th 1981. The 'tank data' did <u>not</u> come from TGA or BHC.

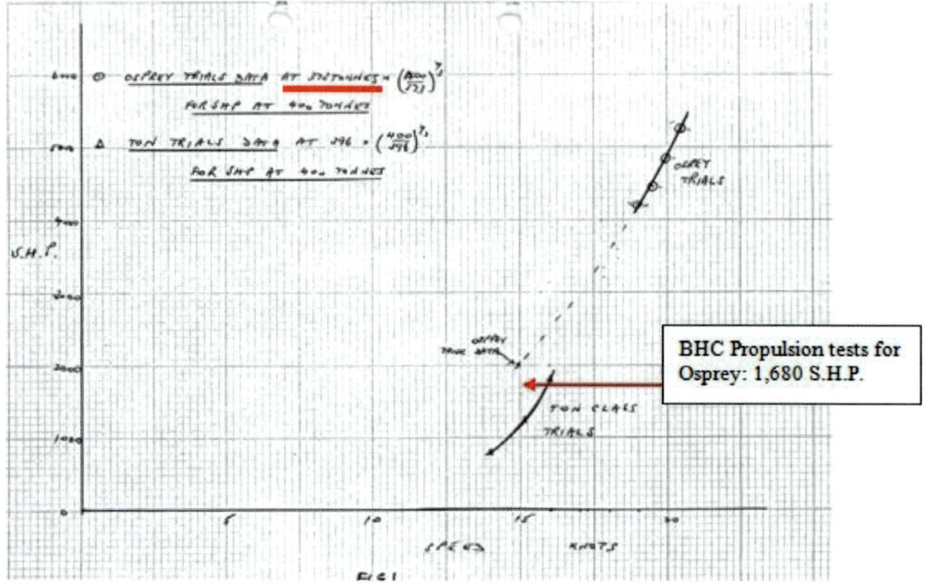

The data for the 'Ton' class Minesweeper, a traditional design of roughly the same size and displacement as the Osprey, was used to show the greater power required by the Osprey for 15 knots. The true difference was only 480 Shaft Horse Power (SHP), as given by BHC Tank data – not 800, as shown by Rawson, based on the tests ordered at BS Dumbarton for April 11th 1981. At higher speeds, the 'Ton' SHP/speed curve becomes almost vertical as its stern sank with speed, while the Osprey SHP/speed curve increases at a constant rate.

Appendix 14

Southern Cross III Speed Trials and Tank-Test Correlation:

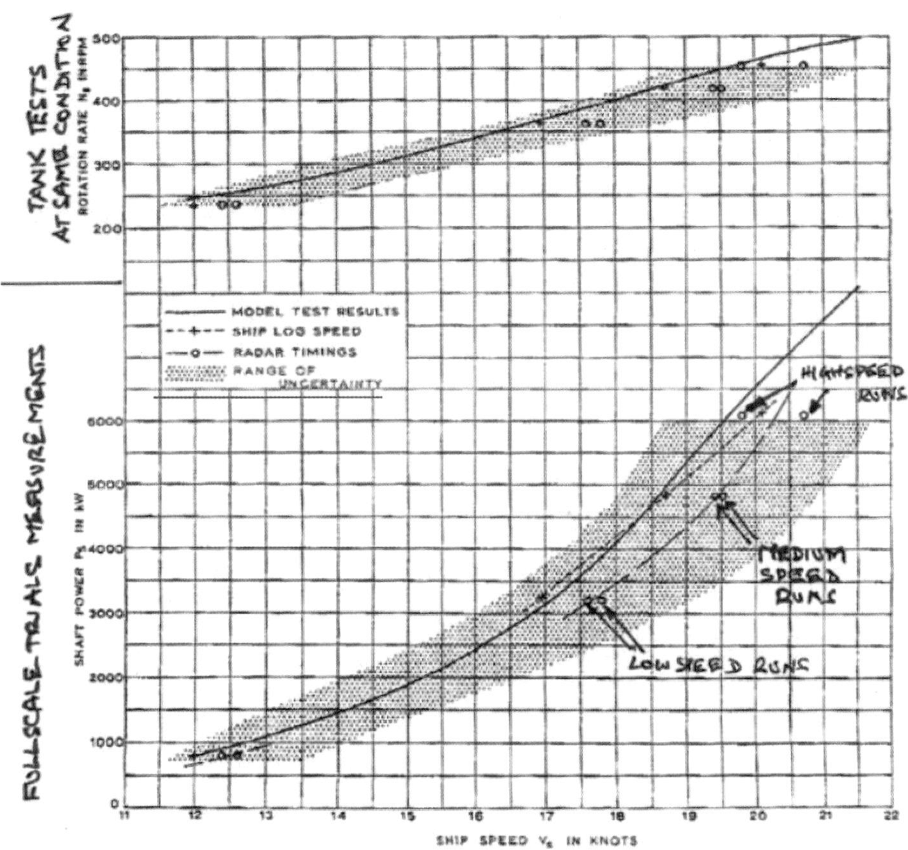

'Moving the Dots'

How the Lloyd's Inquiry created doubt over the two high-speed runs of the trials. Note the small difference between the two runs for low and medium speeds compared with those at high speed, their 1 knot difference being attributed to poor visibility, yet closely reproduced by the MARIN tank in later model tests. The shaded 'range of uncertainty' is strikingly similar to the upward curve of a conventional hull at its maximum hull speed, or Froude Number. Therefore the stern should be sinking, giving a bow-up trim. Yet all photos of *Southern Cross III* at maximum speed show her stern waterline is visible and her hull trim is level at her maximum speed. Having just arrived after a voyage from Australia, carrying a lot of extra equipment, her displacement was much higher than at initial trials, when she made 22 knots at 500 tonnes displacement and, after delivery to Australia from Japan, she was claimed to have made 25 knots at her lightest practical displacement. David Jenkins, who attended the MARIN tank tests was extremely upset about the way in which the model was towed.

When it came to the tests at the MARIN tank in Holland, the credibility of the Inquiry was again undermined. MARIN, with Lloyd's Dr Goodman in attendance, ran the tests on a model of *Southern Cross III* in precisely the same conditions. The MARIN results were all above those achieved at sea trials – particularly at the highest speed recorded on trials, which was 1.2 knots higher than the MARIN tests, showing a drag-rise more typical of a conventional hull, with a 'range of uncertainty' of 3.1 knots. I attended some of the MARIN tests with David Jenkins who, with his long experience of working with BHC and NMI on semi-planing hulls, could not agree to the testing procedure. The results, he said, were from errors in the towing and calculation techniques arising from their lack of experience in the semi-planing hull form; a similar view to that of our Expert Witness, Karl Kirkman, concerning the Osprey tests run by David Moor at Vickers.

On his return to London, David Jenkins contacted Dr Goodman to protest vigorously at the apparent 'rigging' of the results. David was fuming about this altercation when I saw him at the Inquiry's February meeting, and was still upset thirty years later. The *Southern Cross III* trials appeared to have been a complete waste of time and public money.

Appendix 15

Some Published Opposition to the Author and the 'Short, Fat Ship':

All of the following extracts, published by those who have served for the MOD and/or British Shipbuilders have broken the Terms of Settlement of the Osprey Case: that 'Each side recognises the professional integrity of the other.' So I am likewise entitled to discuss the lack of 'professional integrity' on the other side. But, of course I have also to record the published views of the Opposition to present a fair account of their arguments against the 'Short, Fat, Ship'. Many extracts from documents are given in footnote references; however these are the most striking published passages:

> **Item 1. Part 1, p. 39:** Marshall Meek: "I first became aware of the existence of David Giles through reference to him by Julian Taylor just about the time I was leaving Blue Funnel in 1978. I heard more about that link between Julian and Giles after I joined British Shipbuilders and my fears of some malign outcome increased ...
>
> "David Laurent Giles had an Oxford BA, had served in the submarine service, had an aeronautical background and although his business stationery carries the bold description of his company as Naval Architects, Consultants and Agents, he himself is clearly on record as saying that he was not a naval architect. His co-director in Thornycroft, Giles & Associates Ltd was Commander Peter Thornycroft, a famous name in the marine world. He was a naval architect, but he never seemed to play any very obvious active role in the short-fat controversy... Thornycroft had been closely involved in earlier days in the design of smaller craft such as pilot boats and was very successful. I think the trouble originated when Giles claimed that such successful small craft could be scaled up to the same size as frigates and still be as successful. There are several reasons why ships do not easily scale up and he should have known that the required horsepower would be very high. But he also claimed that the vessel would have the benefit from 'lift'. In other words it would start to plane on the surface of the sea. He did not seem to realise that to do this the propulsive power would be enormous, roughly double the power of a conventional frigate and so be quite uneconomic and unrealistic. There were many other features such as the ability to carry enough weapons, sea keeping and noise

reduction, that needed exploring. But as I have said, it took nearly ten years and vast cost to disprove his subtle arguments." (p. 39: Marshall Meek: *There Go The Ships*, The Memoir Club, published 2003, p. 192, with 3 further chapters devoted to the 'Short, Fat, Ship'.)

Item 2. Part 1, p. 45: As Rawson wrote in his memoir: "We rejected the design and thought that it would be the end of the matter. It would, in fact persist for another eleven years and cause no end of mischief." His views on the Osprey Case and S90 are revealing:

"In the mid-seventies, the Wilson government had nationalised the shipyards and created British Shipbuilders. There are many people with a deep respect for the political wishes of governments and do their utmost to make them work, irrespective of their own views. Civil Servants number highly among them. So too did those drafted into BS in the nationalising process, although this did not include most of the shipyards which deeply resented the intrusion into their affairs and offered only token cooperation. One of those to seize the opportunity was Jack Daniel, then the Director General Ships and Head of the RCNC, and there were other very capable individuals who did their best to make a go of the arrangements.

"Daniel decided to produce evidence of Giles' misrepresentations and built a model of *Osprey* to run in the ship tanks at St Albans and Dumbarton which were under the control of BS, aided by the hydrodynamicists there, Moor and Ferguson. They built models labelled prominently at the stern *Osprey* so that there was no attempt at subterfuge and their aim was to confront Giles and his team with clear proof that their claims were insupportable. However, Giles got to hear of this and his lawyers served a writ on BS for breach of confidentiality. It kept the lawyers busy for the next eight years in and out of the High Court.

"What his lawyers sought to do was show a conspiracy between the MoD and BS which would much enhance the damages payable and, although I was to leave the MoD in 1983, I was pursued by them and the MoD until the matter was resolved in 1987. There had been no such collaboration between BS and MoD but it is not easy to prove a negative and I was subject to severe cross-examination by the barristers in 1981 and in 1986. There were, of course, overtures to share information and I was to hear often the cry, 'come on, we all work for the same government' but I was extremely careful to avoid offending confidentiality. After all this time, matters were settled out of court in 1987 on terms that were not to be disclosed but which rumour had it were close to three quarters of a million pounds plus lawyers' fees that must have been considerable.

"I never understood why, when the models were disclosed, British Shipbuilders did not say 'Oops, sorry' and apologise but they persisted even after the writs had been served. What it did was cause great confusion when Giles made a bid for the Type 23 frigate contract. In order to prepare himself for this bid, Giles had gathered around him some very powerful friends. These were led by the Chief of Defence Staff, then retired, Admiral of the Fleet Lord Hill-Norton whose close supporters included the MP Edward du Cann, a Congressional lobbyist from the USA, Dr Garwin, Professor RV Jones, the distinguished electronics engineer who had pioneered radar during the war, and a string of journalists and newspapers who took the line inevitably that Giles' genius was being ignored by officialdom. It was a formidable team to face. How he managed to collect such support remains something of a mystery; Giles entertained them regularly at his club, so I understood, but what, if any pressures he was able to mount I have never known.

"In 1979, John Coates had retired early and I had been appointed the Chief Naval Architect. I decided that the disruption to the Type 23 frigate design, then underway must be avoided so I let myself be a shield for them and field all the problems that were to unfold. I also found myself nominated by the Minister for Defence to deal directly with press and television, which was very unusual without a shield from the public relations directorate. By that time, Geoffrey Pattie, a former adviser to Mrs Thatcher and a close friend of Giles, had been appointed a junior Minister for defence procurement. The Type 23 was evolving at that time with a small team at Bath and at Yarrow Shipbuilders in Glasgow to address the Naval Staff's requirement, and the Giles' supporting chorus were well aware of this. Numerous questions arose in Parliament and in the press as to why the Giles' design was not a front runner.

"There had been an outline proposal from Giles for an enlarged *Osprey* called the S90 or *Syrius* at around 1300 tonnes which, in the absence of any clear evidence we had found to be totally inadequate to meet the requirement. It claimed speed which we found was unlikely to be met and the idea that it would plane like a surfboard was, we believed absurd. Peter Thornycroft had by this time died and we were unable to discuss these claims with anybody capable from the firm but the pressures mounted. I actually introduced the term 'short fat ship' which I thought might deliver the right amount of ridicule but it stuck and the press loved it.

"I was instructed by Pattie to deliver the plans for the Type 23 as it then stood to Giles at his home. He did not have the requisite security clearances to receive SECRET documents and I had to delay the delivery for a week or two until the appropriate authorities had cleared him. When my representative arrived at the door, it was answered by Mrs Giles who said, 'Oh don't bother with those, we have already received them from Geoffrey.'", (KJ Rawson: *Ever the Apprentice*, The Memoir Club, 2006, pp. 115–116)

*****

DLG: I should clarify the fact neither my wife, nor I, ever saw – let alone received – the plans of the Type 23, nor did anyone approach or speak to Mrs Giles. All I saw was the Operational Requirements Document for the earliest version of the 2,500-tonne/25-knot Type 23, that finally grew to 4,300 tonnes and 28 knots.

Item 3. Part 1, p. 39, Note 3: Trawling through the Internet in July 2020, I suddenly happened upon a sinister website going by the Smiley-esque title 'secretprojects.co.uk' – with mention of the 'Short, Fat Ship'. It was initiated, appropriately, on November 5th 2005 – twenty-two years after the MOD saga was over – and it must have been written with certain lapses of memory. However, it shows attempts by a number of correspondents to re-write history without regard for the facts ...

Professor David Andrews RCNC, FRINA, FRSA, FIMechE, FRAE – also of Lighthill's UCL and holding numerous other Academic Awards and Honours, attended Admiral Bryson's final riposte (see 'Diet of Worms', Part 2, p. 3) at the RINA, on June 7th 1984. His reference to Marwood & Bailey's classic 1969 work 'Design Data for High Speed Displacement Hulls of Round Bilge Form' (NPL Round-Bilge Series) covered Semi-Displacement (or Semi-Planing) hulls up to 'Corvette size'. It had no connection with 'planing' hulls. I trust his confusion over the lifting characteristics of the large semi-planing hull, such as the US Navy's

16 'Freedom' class LCS, of 3,450 tonnes and recorded speeds of up to 50 knots, will by now have dispelled such arguments as he proposed in 2005. He wrote his polemic before the launching of the first LCS, although the argument against the 'Short, Fat Ship' has continued until November 2020. However it has now been rendered irrelevant by the US Navy's LCS and, I would claim, the contents of this book. But this is what Andrews wrote in defence of his – and the MOD's – Traditional Wisdom fifteen years ago:

"I was also asked to comment on the 'short fat ship' controversy of the mid-1980s, which I suspect many naval officers, without the benefit of naval architectural theory or subsequent practice in ship design, may feel shows a resistance to innovation by the constructor fraternity. Leaving aside the clear evidence that, historically, the British naval record is one of innovative technical adoption rather than resistance to it, any proposal has to be qualified by a clear demonstration that a given proposal is technically justified and good value for money. So any new ship design proposal needs to follow the scientific practice of being presented, preferably in an open forum, to a suitably broad range of informed professionals able to objectively assess its merits from the evidence. The proponents of the 'short fat frigate' proposal never presented this proposal to the naval architecture discipline, as can be seen from the written discussion to Admiral Sir Lindsay Bryson's paper on 'The Procurement of a Warship' in the 1984 RINA Transactions. While the public debate about 'short fat' turned on a legal issue, the private debate in the pages of The Naval Review was technically illiterate. In fact the hull form drew on small craft practice, namely the published National Physical Laboratory's guidance on such (Marwood Bailey) planing forms, which are valid up to 200 tons displacement but not for frigate-size vessels operating at typical frigate speeds when the vessel is over ten times heavier than the NPL limit. Elementary knowledge of the effect of Froude Number ($V/\sqrt{(gL)}$) on a ship's wave-making resistance would have ruled out such a form for frigate size and speed. Furthermore, the dynamic stiffness of frigate-sized vessels, given such a wide beam and a practical weight distribution, would have been such that the configuration would be highly likely to have too stiff a motion to operate in a seaway. The whole saga was more an example of the low status professional engineering is given in the UK when compared to our continental Ingenieur equivalents, plus the manner in which miracle cures are grasped by peacetime governments under fiscal pressure. In this case it was doubly ironic that the first scientifically (if not technically) educated Prime Minister, Margaret Thatcher, fostered a government attitude that positively distrusted professional advice, especially if the experts were in public service. That neither the Navy nor the MoD seemed to strongly support its own professionals, when (as I observed at the time from my NATO colleagues) no professional ship designer could understand the lack of official support for the professional engineering advice, says more than a little about the UK's decline from its former position of being the leading industrial power. But Matthew McConaughey puts it (admittedly while reading from a script) a little more poetically: 'Fugazi, Fugazi. It's a wazy. It's a woozie. It's fairy dust. It's not real.'" (Ref: 'RN Short, Fat Ship' saga, in 'secretprojects.co.uk', for similar comments on the controversy – in particular by Professor David Andrews, RCNC, written 5th November 2005)

Dramatis Personae

Adams, Commander Charles, DSC (1906–1988): Ex-Royal Navy wartime corvette and destroyer CO, later Cabinet Office adviser on Fisheries and Deputy Head of UK Coastguard. Trusted adviser to the author on the inner-workings of the UK Civil Service.

Ashley, Laura (1925–1985): Started life in a Baptist family in Merthyr Tydfil, Wales. In 1942 joined the Womens' Royal Naval Service, or 'Wrens'. Started making aprons, tea towels, napkins and floral headscarves in her home. Later she and her husband Bernard lived above a haberdashery shop she set up in Machynlleth, North Wales. This flourished and she began making frilly aprons and dresses in an old railway station in Carno, North Wales. She then moved her business to a shop by South Kensington tube station in London. In 1981 she bought a Nelson 40 motor yacht and forcefully persuaded the author to pursue the idea that became FastShip. The rest of her amazing career, her château, Welsh mansion, Bahamas villa and the worldwide expansion of her 'frilly frocks and tea-towels' business, is history.

Atkinson, Sir Robert (1917–2015): Wartime destroyer CO, Chairman of British Shipbuilders (BS) 1980–3, retired in 1983 during the period of the Osprey Case and the Sirius/S90/Short Fat Ship controversy.

Bach, Niels: Director of the Danish shipyard Frederikshavn Vaerft, partners with TGA in construction of the Osprey patrol vessel and the Osprey Case vs. British Shipbuilders.

Bannenberg, Jon (1929–2002): Designer of the exterior and interior of a new generation of super-yachts, including TGA's *Southern Cross III* and interiors for *Queen Elizabeth 2*, private planes and houses.

Bond, Alan (1938–2015): Australian businessman and brewer of Swan Lager. First Australian winner of the America's Cup (first raced in 1851), with his revolutionary winged keel on his 12-metre yacht *Australia II*. Built the TGA-designed 500-tonne mega-yacht *Southern Cross III* of carbon-reinforced glass fibre in Japan in 1985, the largest semi-planing hull and the largest such marine structure at the time.

Boyes, Vice-Admiral Jon, USN: Former Director of US Naval Communications, 1973–5. CO of the 33-knot experimental submarine, USS *Albacore* when Lord Mountbatten was given a demonstration aboard. Chairman of TGC Inc., 1988–9.

Bridges, David: Ex-Chief Engineer of Vosper Thornycroft who first worked for TGA on the Bond yacht in 1984 and has worked with TGA ever since. Highly qualified and, like David Jenkins, deeply experienced in the large semi-planing monohull.

Bryson, Admiral Sir Lindsay, GCB (1925–2005): Third Sea Lord and Controller of the Navy 1981–4. Later, President of the Institute of Electrical Engineers, Deputy Chairman of GEC-Marconi and Lord Lieutenant of E. Sussex.

Bullard, Roland: President and CEO of FastShip Inc. from 1998 to 2012. He held banking posts until he retired as Chairman of the Philadelphia Advisory Board of Directors of First Union National Bank, before joining FSI. Also Chairman of Marine Propulsion Systems of Miami, Florida.

Chambers, Katharine B: Head of Business Planning for FastShip 1991–8 and Company Secretary from 1998 to the present. Her skills with the Business Plan and presentation of the economic and logistics case for FastShip were essential to all progress made over the entire duration of the project.

Chryssostomidis, Professor Chryssostomos: Director of the Massachusetts Institute of Technology (MIT) Department of Ocean Engineering. In 1995 his Department undertook a detailed analysis of FastShip seakeeping and performance attributes. He was the first of several Academics in the US and Scandinavia to expound the advantages of the large semi-planing hull as being 'without precedent' in maintaining speed in high ocean waves. This was in direct contradiction to the previous conclusions reached in UK naval architectural circles.

Corlett, Roy: Ex-Royal Navy engineer officer who had worked for Vickers on advanced submarine technology. At the time of these events he was technical adviser to John Moore, editor of *Jane's Fighting Ships*.

Daniel, Reginald 'Jack' (1928–2016): Director General of the Ministry of Defence Ship Department in Bath 1964–79. Board Member for Warship Building, British Shipbuilders 1979–84. Designer of the first RN nuclear submarine *Dreadnought*. At BS he was the first public opponent of the Osprey in the November 1980 BS house magazine. Later involved in the illegal testing of the Osprey and Azteca class models by BS. Later founded Warship Design Services Ltd which is now part of BAE Systems.

Day, Sir Robin (1923–2000): Described by *The Guardian* as "The most outstanding television journalist of his day, he transformed the television interview." Abrasive presenter of BBC TV's *Panorama, Question Time* and BBC Radio's *World at One* from 1979–87, on which he interviewed the author on rejection of the S90 by the MOD in October 1983.

du Cann, Sir Edward, KBE, MP (1924–2017): Ex-Chairman of 1922 Committee that propelled Margaret Thatcher into power. Supporter of the S90 on BBC TV and in Parliament, recommending construction of a prototype after criticism of performance of HM ships in the Falklands.

Dunn, David E III: Political lobbyist, Senior Partner Patton Boggs & Blow, of Washington DC. Counsel to TGC and Secretary of FastShip Atlantic Inc. 1989–94. Active in pursuit of political support. Introduced Fred Smith of FedEx to FastShip.

Eberle, Admiral Sir James, KCB (1927–2018): C-in-C Fleet and C-in-C NATO Northeast Atlantic Fleet 1979–81, C-in-C Naval Home Command 1981–3. Later Director of the Royal Institute of International Affairs, Chatham House. The first Admiral to support and encourage the 'Short, Fat Ship' and the original concept of the Sirius Corvette and S90 Frigate.

Fieldhouse, Admiral of the Fleet, Sir John, GCB (1928–1992): Notable submariner, Third Sea Lord and Controller of the Navy 1978–81, C-in-C Fleet 1981–2, commanding Operation Corporate to re-capture the Falkland Islands, First Sea Lord 1982–5, Chief of UK Defence Staff 1985–8. His position on SFS and S90 was never clear, telling some he approved of such innovation – and others that 'he should not accept every new idea that came along'.

Garwin, Dr Richard L (1928-): Eminent American nuclear physicist, at age 24 was key member of the Los Alamos Project and thereafter Panel of the President's Science Advisory Committee to every US President since Truman. Head of the Naval Warfare Panel of the Science Advisory Committee for President Kennedy. Inventor of the first Spy-Satellites, MRI, Touch Screen, Laser Printer and 47 other inventions. Watson Professor of IBM. Holder of the US National Medal of Science and Presidential Medal of Freedom.

Gellman, Dr Aaron (1931–2016): Professor of The Transportation Center, Northwestern University, Evanston, IL. Advisor to FedEx on logistics and policy and to FastShip Atlantic Inc. and FSI, after introduction by Fred Smith, founder of FedEx.

Gow, Ian, MP (1937–1990): Old school friend of the author. MP for Eastbourne, Parliamentary Private Secretary to Prime Minister Thatcher 1979–83. Close friend of Capt. John Moore (q.v.). Assassinated by the IRA with a car-bomb at his Sussex cottage.

Griffin, Admiral Sir Anthony, KCB (1920–1996): Third Sea Lord and Controller of the Navy 1971–5, later Chairman British Shipbuilders 1977–80. In World War II escaped from sinking by lifeboat to the Cape Verde Islands. President Royal Institution of Naval Architects 1981–4.

Hennessy, Peter, now Lord Hennessy, FBA: *Times*, *Economist* and BBC journalist, historian, academic specializing in the history of government. Attlee Professor of Contemporary British History at Queen Mary College. Warned the author against the personal tactics used by Whitehall to discredit the reputation of those they wished to destroy.

Hill-Norton, Admiral of the Fleet, Lord Peter, GCB (1915–2004): World War II Norwegian Campaign, Arctic Convoys, Far East. CO of destroyers and aircraft carriers. RN First Sea Lord 1970–1 and Chief of the Defence Staff 1971–4. Chairman of the NATO Military Committee 1974–7. Supporter of the 'Short, Fat Ship' (SFS) during the S90 controversy. With his Committee on Warship Hull Design, was responsible for Prime Minister Thatcher appointing the Ministry of Defence to initiate the Lloyd's Hull Design Inquiry, with whose conclusions he strongly disagreed.

Ingram, Thomas: Director, American Bureau of Ships Associates. Responsible for American Bureau of Shipping Classification of FastShip – in association with det Norske Veritas – in 2000. Also responsible for ABS Classification of the US Navy Littoral Combat Ships to commercial standards 2003–6.

Jenkins, David: TGA's Chief Naval Architect, responsible for all hull design and performance estimates since 1982 to the present. Perhaps the most experienced technician on large semi-planing hulls and performance – as demonstrated by the failure of Lockheed and Gibbs & Cox to take full advantage of the hull form's potential in their LCS design.

Kirkman, Karl: US Hydrodynamicist, Director, Hydronautics Inc., Maryland. Assistant Director Ship Design, M. Rosenblatt & Sons Inc., Virginia 1982–9. Recommended by Olin Stephens as expert witness for Osprey Ltd and TGA in the Osprey Case 1986–7.

Leach, Admiral of the Fleet, Sir Henry, GCB (1923–2011): First Sea Lord and Chief of the Naval Staff 1979–82. Gunnery officer, World War II Far East, Battle of the North Cape, Korean War. Advised Mrs Thatcher in 1982 that the Falklands could be re-taken. Staunch opponent of the S90 and the author.

Lettow, Judge Charles F: Judge of the FSI Trust, LLC, Suit against the US Navy represented by the US Department of Justice. Apart from his Stanford Law degree, he holds degrees in Chemical Engineering and English and American Colonial History (16th and 17th Centuries). From 1968–9 he was Law Clerk to Chief Justice Warren Burger and later argued three cases in the Supreme Court. His paternal grandfather was the World War I German General Paul Emil von Lettow Vorbach, who commanded a force of African-led guerrillas in East Africa that caused confusion to the much larger Allied forces led by Field Marshall Smuts. One of his gunboats was sunk by the Thornycroft quick launch *TouTou* on Lake Tanganyika, the subject of the film *The African Queen.* Smuts became a close personal friend and admirer of von Lettow and received him with honour after the war in Pretoria. Von Lettow famously incurred Hitler's wrath by his rude refusal of the post of German Ambassador to London – taken up by von Ribbentrop – for which he endured house-arrest during World War II.

Lighthill, Sir James, FRS, FRAeS (1924–1998): Applied Mathematician and expert in Fluid Dynamics. Lucasian Professor of Mathematics, Cambridge University 1979–89. Succeeded by Stephen Hawking after being appointed Provost of University College, London. He finally endorsed Garwin's interpretation of Hydrodynamic Lift in the TGA Osprey Design, contrary to that of the MOD's Chief Naval Architect, KJ Rawson, his own Visiting Professor of Naval Architecture. The MOD and Lloyd's Inquiry both claimed the opposite: that Lighthill and Garwin had *agreed* with Rawson's interpretation. They had not – and how such an obvious reversal of the truth could have passed through all the Corridors of Power is typical of the conduct of the MOD and the Lloyd's Inquiry.

Ling, Nigel, FRINA: Marine Surveyor who advised Capt. John Moore, Editor of *Jane's Fighting Ships*. Recorded the August 1983 statement of Marshall Meek, at an MOD Meeting, that, as the newly-appointed Director of the National Maritime Institute (NMI), all of NMI's previous detailed tank tests and simulations of TGA's Osprey and S90 designs 'should be rejected' by the MOD in their assessment of the S90. Meek's advice was accepted by the MOD, and later, by the Lloyd's Hull Design Inquiry. As a result TGA refused to pay the £51,315 bill outstanding for the model test programme.

Lygo, Admiral Sir Raymond, KCB (1924–2012): Vice-Chief of Naval Staff 1975–8. Chief Executive and Chairman, BAE Dynamics and BAE Systems 1986–92. Supporter of S90 during the Validation Programme until promoted Chairman of BAE Systems, after they resigned from the S90 Club.

MacLeod, Sir Roderick (1929–1993): Chairman of Lloyd's Register of Shipping 1983–93. Chairman of the Lloyd's Hull Design Inquiry (March 1987 – August 1988), ordered by the MOD on behalf of Prime Minister Thatcher. The MOD Chief Scientist, Sir John Charnley, in 1982, said of the S90 Validation Programme: "It is important that the study be perceived from the outset as unbiased and independent." The same condition was requested by Lord Hill-Norton to Mrs Thatcher concerning his Committee's 1986 recommendation for an Independent Inquiry into the S90. However, the MOD appointed Lloyd's Register, of which MOD was the largest client, and allowed an earlier 1983 premature Marshall Meek-chaired Defence Scientific Advisory Council condemnation of the S90 to be used as a source of its conclusions, months before the Inquiry ever sat. His obituary in *The Independent* stated: "MacLeod's contribution to the National Interest was recognised by a knighthood in 1989. One particular activity he relished was the chairmanship of the committee appointed by the government to determine the future shape of Royal Navy Frigates."

Marr, Andrew: As Political Correspondent of *The Independent*, in July 1987, he published a news item in *The Independent* claiming that the Lloyd's Inquiry had reached its conclusion to reject the S90, a year in advance of publication of its final Report. He told the author he had received this information from Lord Trefgarne, as Under-Secretary for Defence Procurement, who was the Minister responsible for managing the Inquiry. Trefgarne denied it.

McKesson, Dr Chris: Consultant Naval Architect who has worked during thirty years for US Naval Sea System Command, as Professor of Naval Architecture at the Universities of New Orleans and British Columbia. As Chairman of the US Navy's 1997 Conference on Fast Sealift, he concluded that the FastShip project was 'State of the Art'. Later for Naval Architects JJ McMullen, he recommended the FastShip design as entirely feasible to potential investors KKR. As Assistant Expert Witness for FSI Trust in its Patent Suit against the US Navy over the LCS, he successfully argued against the claim of the USA's expert, Dr Stern, that the LCS-1 stern did not generate hydrodynamic high pressure, or 'lift', at speed.

Meek, Marshall, CBE (1925–2013): President of the Royal Institution of Naval Architects and the most zealous opponent of TGA's design ideas. Chief Naval Architect of the Blue Funnel Line and Ocean Fleets to 1980. Later Chairman of the 1983 MOD Defence Scientific Advisory Committee that condemned the large semi-planing monohull on the flimsiest of evidence before the final technical evaluation was half-complete. In 1983 appointed Director of the National Maritime Institute (NMI), having recommended the MOD's rejection of all NMI's previous tank tests for the Osprey and S90 in the 1982–3 Validation Programme previously agreed by the MOD. In 1986 he recommended the Lloyd's 'Independent' Hull Inquiry's false grounds for rejection of the S90 Frigate proposal, five months before it ever sat. Awarded the CBE for his work with the MOD in the dismissal of TGA's large semi-planing hull. A devout member of the evangelical Christian Brethren sect, he deployed all his religious zeal on discrediting the author's 'concepts', an expression he despised. His memoirs *There Go The Ships* devoted three whole chapters to the technical assassination of David Giles and all he stood for. Yet it proved a most helpful source of evidence of the powerful forces of reaction, and the many false arguments and calumnies raised against the Short, Fat Ship within MOD and naval architectural circles. His 1989 CBE honour was awarded at the same time as MacLeod's Knighthood (q.v.) for his work in the 'national interest' on the Lloyd's Hull Design Inquiry.

Monckton, Christopher (1952–): Lord Monckton of Brenchley, ex-Deputy Leader of UKIP. In 1984 was a journalist for the *Evening Standard* and member of Mrs Thatcher's Downing Street Policy Unit. Supported S90 in political circles. Later assisted with organising the Hill-Norton Committee and its Report 'Hull Forms for Warships'.

Moor, David: Superintendent of the Vickers Hydraulic Testing Tank, St Albans – later the British Shipbuilders Ship Model Experimental Tank (SMET) 1980–1. A long-time colleague of Marshall Meek (q.v.) and the man who carried out the secret and illegal testing of the Azteca and Osprey hulls in 1980–1, in developing the design of the Royal Navy's new 'Peacock' class of 600-ton semi-planing patrol vessels for its Hong Kong Squadron. After the High Court prevented the incorporation of all his work on the TGA designs, he was forced to produce a design that 'did the same, but looked different', but it did not. However, by attaining a higher speed than would have been possible from a conventional design, it helped establish the lifting properties of the semi-planing hull. As later proved by carriage log of the Vickers Dumbarton test tank, he was doing so on the initiative of the MOD, working with Prof. Kenneth Rawson, its Chief Naval Architect.

Moore, Captain John, RN (1921–2010): World War II submarine CO, formerly Royal Naval Intelligence Staff, specializing in Soviet Naval Intelligence. Editor, *Jane's Fighting Ships* 1973–88. A keen supporter of the S90 and with his former RN colleague, Lord Hill-Norton, First Sea Lord and Chief of Defence Staff, with whom he worked during the 1970s.

Morison, Sir Thomas, QC (1939–2022): Famed for his scurrilous sense of humour, he was a Recorder from 1987–93. Like Lord Younger and the author, he attended Winchester College and Oxford University. His *Times* obituary stated: "Thomas Morison 'snuffed it' on 19th March, aged 83. No funeral, no mourning, no flowers, no worries." In his Short Opening for the Osprey Case, he described the 1987 Defence of British Shipbuilders as: "Something you could tell to the Marines."

Morrall, Dr Tony: Chief of Hydrodynamics at the National Maritime Institute (NMI). He was in charge of the 1980 Osprey full-scale sea-keeping tests in the North Sea and the 1983 S90 Validation model sea-keeping and full-scale correlation testing for speed with the Danish Technical University Osprey trials measurements. In July 1983, he advised the author to remove all the experimental results from NMI before Marshall Meek took over as Director. As a result TGA refused to pay the £51,315 bill outstanding for the NMI model test programme.

Mursell, Arthur: Gifted Naval Architect working for TT Boat Designs from 1965 to 1988. Assisted in TGA designs such as Osprey, Sirius and S90. Drafted the Osprey and S90 lines plans.

Pattie, Sir Geoffrey, MP (1936–): Parliamentary Under-Secretary of State, later Minister of State for Defence Procurement 1981–3, Minister of State for Industry 1984–7. Director, Marconi Defence Systems. Fellow, International Institute for Strategic Studies. In April 1982, during the Falklands Crisis, Pattie was responsible, on behalf of the MOD, with Admiral Lygo, of BAE Dynamics, for inviting TGA to produce a proposal for the S90 Frigate, as an alternative to the projected Type 23. He strongly supported the S90, in the face of determined opposition from the Naval Staff, until removed from the MOD in June 1983, just as the MOD was passing judgement on the results of the S90 Validation Programme. Not a close personal friend, as claimed by Rawson.

Pedersen, Einar: Chairman of TTS Marine ASA of Bergen, Norway, associated with Monck Cranes, makers of freight handling equipment and concept designers of the FastShip Alicon and later, the Automatic Guided Vehicle (AGV) rapid Roll-on Roll-off (Ro-Ro) loading system for containers. Pedersen was Chairman of FSI from 1994 to 2006.

Presland, Frank (1944–): Copyright Lawyer of Frere Cholmeley. Solicitor for TGA, TT Boat Designs Ltd and Osprey Ltd, in the Osprey Case 1981–7. He advised Osprey Ltd to accept the settlement offered by the Treasury solicitors in the full trial of the Osprey Case. Later solicitor and manager of Elton John's management company, Rocket Entertainment.

Rawson, Kenneth J (1926–2010): Deputy Director Ship Design, MOD Ship Department. Visiting Professor of Naval Architecture, King's College, London, later Professor at Brunel University. The man who claimed the Osprey was a 'potential source of disaster', but also that, if the hydrodynamic lift of semi-planing hulls could be scaled-up from small craft to Osprey size: "it really would represent an important contradiction of traditional theory". They did scale-up and they did lift, as established in the US Navy's Freedom class of LCS. So *ipso facto*, they have proved exactly as Rawson suggested they might do.

Rendell, Edward (Ed) G (1944-): Charismatic Mayor of Philadelphia 1992–2000, "making one of the most stunning turnarounds in recent urban history" (*New York Times*) turning a $250 million deficit into a surplus. Elected Governor of Pennsylvania from 2002–11. Also Chairman of the Democratic National Committee (2000 presidential election). Fascinated with Rune Svensson's (q.v.) endorsement of the FastShip concept, he provided the initial financing to launch the FastShip commercial project in 1994.

Ridge, Thomas (Tom) J (1945–): Governor of Pennsylvania 1995–2001, later first Secretary of Homeland Security, who sent the author on a tour of the Kvaerner Masa Shipyards in Norway and Finland, with his Secretary of Finance, Tom Hagen. On the back of the endorsements of FastShip by the Kvaerner Chairman, Diderik Schnitler, and by the Chairman of det Norske Veritas Classification Society, we were able to bring in Kvaerner to take over the old Philadelphia Navy Yard, reviving a great Philadelphia shipbuilding tradition. Kvaerner was later bought by the Norwegian Aker Group and today, as Philly Shipyard, is the most competitive commercial shipyard in the USA.

Rix, Sir John, KBE (1917–2007): Chairman of Vosper Thornycroft, builders of destroyers, patrol craft and MTBs. Chairman of the British Shipbuilders Patrol Craft Sub-Committee. Was interested in acquiring sales rights on the Osprey in 1977, whether 'to use or lose it', we never learnt.

Salkeld, Peter: BAE Systems executive who arranged the author's first meeting with Admiral Sir Raymond Lygo, Chief Executive of BAE Dynamics. A determined supporter of the S90 and popular with serving RN personnel, he was deeply upset by BAE's terms of rejection of the S90 in 1983, shortly before his death.

Schubert, Captain William: Previously a Houston, Texas, shipping agent for the large container lines, including Maersk-Sealand, who were concerned about the competitive FastShip service. Appointed Maritime Administrator by President George W Bush in 2001. Advised by Dr Martin Stopford, a shipping economist and former British Shipbuilders Director of Business Development during the Osprey Case, Schubert was responsible for refusing the Title XI Federal Loan Guarantee of $300 million for FastShip construction in a US yard. His decision was made partly on grounds of 'technical risk'. He also refused to take account of the 75 per cent reduction in port turnaround time and distribution due to the FastShip AGV rapid load-unload system, removal of 'inter-porting' between destination/arrival ports, and direct port pick-up and delivery by major trucking firms, thus bypassing additional 'drayage' by small truckers to and from inland distribution centres. Effectively FastShip reduced trans-atlantic delivery times from about five weeks, to 10 days, with a big reduction in delays due to weather and port complications.

Silverleaf, Alexander: Expert witness for British Shipbuilders in the Osprey Case. Former colleague of David Moor and former member of National Maritime Institute.

Smith, Frederick W (1944–): Founder of Federal Express Corp., now FedEx. Associated with the author at Pan Am in 1971 when both were looking for commercial uses of the Dassault Falcon Business Jet, the basis of his initial delivery system. Since 1989, when 'thunderstruck' by the idea of FastShip for the 'Middle Market' fast container service, a supporter of the FastShip concept, though deeply committed to his support from Boeing and Airbus.

Sorensen, Per Holst (1923–1998): Naval Architect and Technical Director of Frederikshavn Vaerft, in Denmark. On the Technical Board of det Norske Veritas, the Scandinavian Classification Society. He first doubted and then enthusiastically supported the Osprey and all TGA's later designs. Assisted in the DNV Classification of FastShip in Denmark and Norway.

Stephens, Olin J II (1908–2008): Naval Architect and America's greatest yacht designer. Competitor and friend of the author's father, Stephens leant a sympathetic ear to the Osprey and Short, Fat Ship argument and recommended use of Karl Kirkman as expert witness for Osprey Ltd before the full trial in the London High Court. Kirkman's Report was decisive.

Stout, Donald E: Graduate in Law and Electrical Engineering, George Washington and Pennsylvania State Universities. Senior Partner, Antonelli, Terry, Stout & Kraus, of Arlington, VA, after working as an Assistant Examiner in the US Patent Office. He became patent attorney for Thornycroft, Giles & Co. Inc. in 1989, and has remained so ever since. Successful in managing his firm NTP's action that, in 2006, won $612.5 million against RIM of Canada over their infringement of the BlackBerry mobile phone patent. From those proceeds he financed FSI Trust, LLC's successful claim against the USA representing the US Navy, for breach of the author's patents in the LCS-1, the action lasting from 2008–21.

Strathcona, Lord (Euan) & Mountroyal (1923–2018): Laird of the Isle of Colonsay. Grandson of a Founder of the Canadian Pacific Railway. Former World War II MTB Commander and, later 'Nelson 32' owner. A keen supporter of PT's designs, he became Minister of State for Defence Procurement for the RN under Thatcher 1979–80. During his office, he took an interest in the Osprey, for which as her PPS, Ian Gow, significantly claimed, he was not sacked, but 'shot'.

Svensson, Rolf: Chief Engineer and Head of Hydrodynamics of Vickers/KaMeWa, who was one of the most influential people in the development of large marine water-jet propulsion systems. He provided TGA, FSI and Lockheed with vital data on the large water jets used in the near-identical Rolls-Royce Trent Marine Gas Turbine/KaMeWa water-jet propulsion systems for both FastShip and the Lockheed LCS.

Svensson, Rune: President, Volvo Transport 1984–95. In charge of all worldwide movements of Volvo cars and trucks – and their parts and sub-assemblies. A keen supporter of the FastShip commercial concept and crucial in winning the support of the Philadelphia City Council in 1994.

Taylor, Julian: Managing Director, Ocean Transport & Trading (formerly Blue Funnel Line), retired 1977. Adviser to the author, who, in 1978, introduced him to Marshall Meek, Chief Naval Architect of OT&T, later his staunchest and most public opponent. Assisted with organizing the S90 Club, but disappeared when TGA refused to be seduced into submission by a Settlement of the Osprey Case as required by himself and other members of the S90 Club.

'The two Davids': David Jenkins, naval architect, and David Bridges, marine engineer, former Chief Engineer of Vosper Thornycroft. First employed by TGA in 1985 as consultants during the construction of Alan Bond's yacht *Southern Cross III*. They continued to work for TGA for the next 36 years.

Thornycroft, Sir John I, FRS (1843–1928): Brother of Hamo Thornycroft, sculptor of Boadicea's statue on the Thames Embankment. Graduated with a Diploma in Engineering from Glasgow University. Starting as a boat builder of Thames 'quick' steam launches in Chiswick, he became the greatest innovator and builder of fast marine and naval shipping in the late 19th and early 20th centuries. His designs grew from his quick launches to the first torpedo-boat, later enlarged to the first destroyer HMS *Daring*, that he described as: "practically an enlarged torpedo boat": a similar enlargement as the 'Short, Fat Ship' from the Nelson semi-planing launches. With enlargement a similar improvement in power-for-weight was necessary, so he then invented the first lightweight water-tube, forced-draft, boilers; and, finally lightweight steam engines. This was a necessary process, as in the gas turbine and water-jet propulsion in TGA's later fast projects – and, as proven in the US Navy's Freedom class of LCS. Finally, Sir John turned his skills to steam and diesel lorries, and luxury cars, built in his Basingstoke factory. He could surely claim to have been the greatest of all English innovative marine engineers for fast seagoing vessels.

Thornycroft, Commander Peter T (1909–1988): RINA, Member of the Royal Corps of Naval Constructors, retired 1954. 1955 founded Keith Nelson & Co., builders of the Nelson class of semi-planing launches and pilot boats; followed by TT Boat Designs in 1961. Joined with David Giles in Thornycroft Giles & Associates Ltd in 1976. Without his design initiatives and reputation, the 'Short, Fat Ship' and large semi-planing monohull could never have happened.

Trefgarne, Lord David: British Conservative politician, Parliamentary Under-Secretary of State for the Armed Forces 1983–5, succeeding Geoffrey Pattie; Minister of State for Defence Support 1985–6, Minister of State for Defence Procurement 1986–9.

Trenchard, Thomas, 2nd Viscount Trenchard (1923–1987): Minister of State for Defence Procurement 1981–3. Son of Lord Trenchard, 'Father of the RAF'. Preceded by Lord Strathcona, who was 'shot' for his support of the Osprey and succeeded by Geoffrey Pattie, who was likewise promoted to Secretary of State for Industry following his support for the S90 'Short, Fat Ship'. His interests were chiefly aeronautical with his Air Freight Company, Treffield Aviation.

Wells, Dr Anthony: An ex-RN naval consultant with the US Office of Naval Research in Washington DC, a key member of the Thornycroft Giles team in the USA, 1988–90.

Younger, Lord George Kenneth, 4th Viscount Younger of Leckie, Baron Younger of Prestwick (1931–2003): Secretary of State for Defence 1986–9, who appointed Lloyd's Register to hold the Warship Hull Design Inquiry. A banker by trade, in 1988 he announced the final rejection of the S90 as being "of no interest to the Royal Navy for future frigates or destroyers". A few months later his MOD placed a Secrecy Order on the author's UK Patent Application as being "of possible use to a future enemy".

Glossary

(1+x): The percentage difference between the measured power for speed in tank propulsion tests and that measured in identical conditions in full-scale trials. The required power in conventional hulls is normally slightly higher – at (1+x) = 1.05–1.1. This is usually due to the reduced thrust of the propeller due to the wake flowing upwards from the low-pressure area beneath the stern, with consequent loss of thrust. It is most dramatically shown in the 'rooster tail' wake of a traditional destroyer at full speed. Due to the high pressure beneath the stern of a semi-planing hull, at high speed, the wake tends to be much reduced in height, as shown in the comparative photos of both stern wakes on p. 287, resulting in a reduced power for speed, or (1-x), see **Appendix 3**. This was a major argument rejected by the Lloyd's Inquiry – but supported by the NMI analysis as (1-x) = 0.97.

Amidships: Near the centre of the hull or ship.

Amplitude: Extreme value as opposed to average, or mean value.

Angle of attack: Angle of a moving body relative to its position when level.

Appendages: Any structure (propellers, brackets, rudders, stabiliser fins, bilge keels, etc.) not a part of the bare hull.

'The Auld Mug': Slang for the America's Cup. The term was first used by the tea magnate and yachtsman Sir Thomas Lipton.

BAE Dynamics: A missile and guided weapons division of British Aerospace (BAE), which was founded in 1977 as a state company, privatised in 1981, and reconstituted in 1999 as BAE Systems.

Baseline: The intersection of the lowest part of the hull with a horizontal line.

Bath: City in the west of England, location of the Ship Department of the UK Ministry of Defence, where the Royal Navy's warship design and procurement was carried out. Since 1985 the Ship Department has been renamed the Sea Systems Controllerate, and moved to Abbey Wood, outside Bristol.

Beam: Width of the hull. Usually expressed as overall beam or 'BOA' (the maximum beam at any point of the hull) or Waterline Beam or 'BWL' (the maximum beam at any point of the hull at its waterline).

BHC: British Hovercraft Corporation. The BHC test tank was established by The Supermarine Company in East Cowes, Isle of Wight, in 1932 to test seaplane and flying boat designs, with special interest in their waterborne lifting characteristics at increasing speeds. Supermarine became the leading experimental UK tank for planing and semi-planing hull designs. In 1947 it was taken over by Saunders-Roe for the design of their 'Princess' Flying Boat, and in 1966 by British Hovercraft Corporation, later renamed Westland Hovercraft. It was closed in 1984.

Bilge: The submerged transversally curved portion of a hull between the side and bottom.

Body Plan: The transverse cross-sections of a ship seen from in front, or behind.

Boundary Layer: The region of water closest to a moving hull which, due to viscosity, moves more slowly than the water outside. From bow to stern it may progressively be laminar, transitional or turbulent. During its transition, as laminar, it may add nothing to frictional drag, as transitional, add more to drag – or, as turbulent flow, nearing the stern – where it is widest – it adds its greatest amount to drag.

British Shipbuilders (BS): An experiment in nationalising the major shipyards of the United Kingdom, begun in 1977, ended in 1989 under Mrs Thatcher's privatisation programme. Nineteen major shipyards were involved. The ones that appear in this story were Ailsa (Troon), Appledore (Devon), Hall Russell (Aberdeen), Vickers (Barrow-in-Furness), Vosper Thornycroft (Southampton and Portsmouth), Yarrow Shipbuilders (Glasgow).

Chine: A sharp angled or rounded corner or knuckle in the hull form, continuous over a significant length of the ship, as in the junction of the side and bottom of a planing hull. There can be single, double as in the LCS, or even triple chines in a hull.

Correlation Allowance, or Factor: An addition that has to be made to the tank test for measured resistance and powering of a smooth hull to bring it into agreement with the actual ship performance as measured in, as far as possible, identical conditions on full-scale trials. Otherwise known as (1+x) because it is usually a positive factor to be added to the tank test results. However this can change to become negative (i.e. the full-scale hull performs better than the tank measurements, so (1–x)) depending on various factors, including the speed regime, or Froude Number, in which the hull operates: displacement, semi-planing or planing. (See **Appendix 1** and **Appendix 2** (Figs. 1–4).)

Deadrise: The angle in a hull's cross-section, from the keel upwards to the waterline.

DSAC: Defence Scientific Advisory Council – a council of independent technical experts advising the MOD on defence technology. Classified as 'Secret' until its conclusions on S90 were published by the *Financial Times* in May 1983, since when its terms and conditions have become more restrictive.

Directional Stability: A ship's ability to remain on a steady course in varying conditions of speed, heading to waves, wind, angle of heel and displacement.

Displacement: Generally, the weight of water displaced by a hull at rest – but this can change as the hull rises or sinks according to its speed. (For Displacement/Length Ratio, see **Appendix 1**, p. 295.)

Drag: Another expression for resistance – the fluid force acting in opposition to a moving body.

Drag Coefficient: A non-dimensional relationship between the Drag of a ship (or aircraft) and its dynamic pressure related to a specified area. (See **Appendix 1**, p. 297.)

Draught – or Draft: Depth of a hull from its waterline to the maximum depth of its bottom.

Dynamic Stability: The property of a hull to remain upright when disturbed by wind, waves, or manoeuvering, while underway.

EOPV: Enhanced Offshore Protection Vessel. An MOD project that became the OPV-3. (See **Appendix 3**, Figs. 1 and 2.)

Fin: A fixed vertical surface attached to the bottom or bilge of the hull to improve dynamic and roll stability. Also to provide a level surface for docking when attached to a contoured, rather than a flat-bottom hull. Also as an anti-roll fin: a surface angled out from the side of the underwater hull to reduce rolling – either underway, or when static.

FSI: FastShip Inc., a company formed by David Giles and colleagues in Delaware in June 1998, to pursue the FastShip project.

Frequency of wave encounter: The number of successive crests of a train of waves encountered at the bow of a ship over a fixed period of time.

Frictional resistance and Froude Number: See **Appendix 1**, p. 294.

Geosim: One of a series of models which are of differing sizes, but are geometrically similar.

Green water: Water shipped on the deck of a ship in high waves, as distinct from spray.

Hall Russell Ltd: Aberdeen shipyard that built the RN OPVs 1–3 and the five 'Peacock' class of Patrol Craft for the RN Hong Kong Squadron 1984–1988, after High Court enforced refinement from an Osprey-based design. All latter sold to Irish and Philippine Navies when Hong Kong returned to China in 1997. Sold in 1987 as privatisation of BS, to become Aberdeen Shipbuilders Ltd.

***Havørnen* (Danish for 'Sea Eagle')**: The first Osprey patrol boat, built for the Danish Fisheries Ministry by Frederikshavn Vaerft in 1979.

Heave: The up or down vertical movement of a specified point in a vessel, usually the longitudinal centre of gravity, that occurs at different speeds – in tank measurements – or in waves.

Heel: The steady inclination of a ship about a longitudinal axis – as opposed to roll, which is oscillatory.

Hull Tests: Bare, or naked, hull tests are undertaken as preliminary tank tests in which the hull form and its surface are represented without appendages or additions of any kind. Later, propellers, water jets or other propulsors are added, together with all major appendages such as shafts, brackets, rudders, stabilizer fins, etc., to obtain powering and, later controlled sea-keeping tests, for predictions for all significant performance factors.

Hydrodynamic Lift: The action of water on the concave underside of a ship's hull that, above a certain speed, depending chiefly on length and displacement, creates positive, or upward, pressure. This lessens the quantity of water displaced by the hull as speed increases, thus reducing the amount of power required for higher speeds, compared with a traditional hull of the same proportions.

Indaw: The first of three Ospreys built by Frederikshavn Vaerft for the Burmese Navy in 1980.

Jane's Publishing: Provider of defence studies and information, and publishers of books on military matters, especially *Jane's Fighting Ships* and *Jane's Defence Review*.

LCS: Freedom class Littoral Combat Ship, an agile, high-speed 3,450-ton/47-knot multi-role vessel built by Lockheed Martin and first commissioned by the US Navy in 2008. The USS *Freedom* was the subject of the FastShip Inc. Award for infringement of its Patents in the US Federal Claims Court, 2016. After Lockheed dropped FastShip from the design team, it suffered from major hull fatigue and propulsion problems that were predicted by FastShip and others, but ignored by Lockheed.

Leander class: A series of missile and helicopter-carrying anti-submarine frigates that served in the Royal Navy (and several others around the world) 1963–2012.

Lines plan: A drawing depicting the accurate form of a ship's hull pictured longitudinally. Usually includes a Body Plan (q.v.) showing transverse shape of the hull, from forward and aft.

Lloyd's Inquiry: A Public Inquiry set up by the British Government in 1986 to investigate the rejection by the Ministry of Defence of Thornycroft Giles's S90 design for a new Type 23 Frigate. The Inquiry was conducted by Lloyd's Register of Shipping and lasted from March 1987 to April 1988, when TGA Ltd resigned from the Inquiry over its refusal to consider its S115, upgraded to the current specification of the Type 23.

MFS: Monohull Fast Sealift Ship, a 25,000-tonne, 775-foot Roll-on/Roll-off vessel, designed to carry 5,000 tonnes of military equipment or cargo over 5,000 nautical miles at 41 knots in North Atlantic conditions. Basis design for the TGC Inc. Patents.

NavSea: United States Naval Sea Systems Command, which 'engineers, builds, buys and maintains the Navy's ships and submarines'.

Nelson: A 40–75 foot fast semi-planing power boat with exceptional stability in rough seas, designed by Peter Thornycroft in 1956 and sold initially to Trinity House, the UK Pilotage Authority, and throughout the world as police, pilot boats, the Royal Barge and yachts. Name attributed to the English naval hero Horatio Nelson, as well as to Arthur Nelson Compton, whose boatyard first built them. (See **Appendix 1**, Fig. 4.)

National Maritime Institute (NMI): At Feltham, Middlesex. Formerly part of the National Physical Laboratory, NMI had the largest and most advanced Experimental Testing Tank in the UK until it became British Maritime Technology, under the Directorship of Marshall Meek in 1984. Later its huge tank was transformed into a shopping mall under Mrs Thatcher's re-organisation of UK industry.

OPV: Offshore Protection Vessel. (See **Appendix 3**, Figs. 1 and 2.)

Osprey: A class of fast patrol vessel initially designed by TGA for the Kuwait Navy in 1976 and later built for the Danish Fisheries Ministry and as a fast corvette, the Danish Navy and several foreign navies. 165 feet long, with a speed of up to 36 knots. It had a wide aft deck and a hangar for a small helicopter. (See **Appendix 1**, Figs. 2 and 3.)

Osprey Ltd: Company established in Guernsey in 1978 on behalf of Frederikshavn Vaerft A/S of Denmark, to hold the Copyright and Sales Rights in the Osprey design. Therefore it was the lead Plaintiff in the Osprey Case.

***Peacock*, HMS**: Patrol vessel for the Royal Navy's Hong Kong Squadron, the reason for Vickers' illegal Osprey and Azteca tests in developing the Navy's first semi-planing monohull as a warship. However, due to the High Court injunction preventing the use of their successful Osprey-based model 2230, British Shipbuilders had to go through a whole new design process to provide a hull that 'did the same but looked different'. Although matching Osprey in power-speed, it failed in other areas.

Power-for-speed (or 'speed-power'): The power required for a vessel to move at a certain speed through the water.

'Prelude': A new type of fast frigate designed by TGA in 1987. 2,500 tonnes, 295-feet overall length, 50 knots top speed. It was intended partly for Aberdeen Shipbuilders, and partly for the US Navy's FF-X advanced frigate project. It became the first basis and proof-of-concept for the US Navy's Littoral Combat Ship, or LCS.

Propulsive Efficiency (PE): The percentage proportion of the Brake Horse Power output by the engine generated as Delivered Horse Power at the propeller or propulsor. This can vary widely, depending on speed and other factors – generally it is between 45–70 percent. (See **Appendix 3**.)

Quasi-Propulsive Coefficient (QPC): The ratio of Bare Model Effective Power (EHP) to the measured Shaft Horse Power (SHP).

QuinetiQ: A wide-ranging multinational defence technology company, formerly 'DERA', the UK MOD's Defence Evaluation & Research Agency, its HQ being the former Royal Aircraft Establishment at Farnborough. It has a 25-year Maritime Strategic Facilities Agreement with the MOD to provide maritime strategic facilities and capabilities including maritime hydro-mechanic facilities (tank-testing) at the former Admiralty Experimental Tank at Haslar, outside Portsmouth; and submarine structures, survivability and shock-testing at Rosyth, Scotland.

Royal Corps of Naval Constructors (RCNC): A Department of the Ministry of Defence for training and utilizing naval architects, founded in 1883. Originally based at MOD Ship Department Foxhill outside Bath. Now based at the Sea Systems Controllerate outside Bristol.

Royal Institution of Naval Architects (RINA): Founded in 1860, it is the oldest professional body of Naval Architects and has held the highest authority in such circles throughout the change from sail to steam – and the development of a mathematical approach to the design and performance of ships. It also was staunchly opposed to TGA's principal claim of the beneficial effect of hydrodynamic lift in larger hulls and published authoritative Papers to that effect, concluding that the performance of the S90 was no different to that of other hulls of a low length-to-beam ratio. In a response published by the RINA, Meek suggests the Author to be 'a quack'.

S90: An alternative design for the Royal Navy's Type 23 Frigate proposed by Thornycroft, Giles & Associates together with BAE and other companies. The name signified a 90-metre version of TGA's 'Sirius' 75-metre corvette. It was designed according to specifications provided by the Ministry of Defence which were several years out of date. The S90 was thus considered unsuitable.

S-102: An enlarged version of S90, designed by the Lloyd's Public Inquiry in 1987 to bring the TGA design into line with the current specifications for the Type 23. TGA objected to the S-102's obvious design faults and responded with the S115.

S115: An upgraded 115-metre version of the S90, proposed by TGA to meet the final 1988 specifications for the Type 23, which had increased in size from 2,500 tonnes in 1982, as was assumed for the original S80, to 2,800 tonnes as per the S90, and to 4,200 tonnes, as assumed for the S115. The Lloyd's Inquiry financed its design to compare it against the final Type 23 – but then refused to do so. Despite the final recorded meeting between TGA and the Inquiry admitting agreement of its superiority in certain aspects, it was ignored.

Sea-keeping: The behaviour of a ship at sea, in all weather conditions.

Semi-planing hull: A hull that lifts at the stern, reducing wave-making resistance and thus power-for-speed above the traditional critical speed-limiting Froude Number. Also improving pitching and therefore speed, by keeping the bow down and permitting a more level passage through rough seas, rather than excessive pitching.

Sirius: A design by TGA for a small warship of 1,750 tons, 75 metres/246 feet long, with gas-turbine engines and a 30-knot speed. As first requested by Admiral Sir James Eberle – and leading into the S80–S90–S115.

Skeg: A vertical fin, or keel-like attachment, extending downwards from the centre-line of the underwater portion of the hull. Intended to reduce roll and improve directional stability.

SuperShip: A 1988 TGA design for a fast new 40-knot frigate (FF-X) requirement for the US Navy, based on S115 and the 'Prelude' model test results.

Thornycroft, Giles & Associates Ltd: A ship design partnership between the author and Commander Peter Thornycroft. Established in 1976, with the chief purpose of exploring the possibility of enlarging the 'Nelson' semi-planing monohull designs of Thornycroft up to significantly larger size, using the improved power-for-weight of gas-turbine and, later, water-jet propulsion systems.

Type 23: A series of Royal Navy frigates first commissioned in 1987, and equipped with vertical-launched Sea Wolf rockets, a helicopter and anti-submarine torpedoes. They replaced the Leander class, originally intended as a 2,500-tonne anti-submarine frigate for the North Atlantic. However, following the Falklands, they grew to 4,900 tonnes as a General Purpose Frigate.

Type 42 Destroyer: A series of 14 guided missile and helicopter-carrying destroyers which served in the Royal Navy 1975–2013. Two of the class (*Sheffield* and *Coventry*) were sunk in the Falklands War (1982), the first by an Exocet missile and the second by a bomb dropped by a Korean War-era aircraft.

Type 45 Destroyer: Initiated to incorporate lessons from the Falklands, and classified as an Air-Defence Destroyer. The project started in 1981 as the NFR-90 NATO-multi-national air-defence project in 1989 before the destroyer losses of the Falklands Campaign. In 1989 the UK and USA resigned from the NFR-90 due to disagreement over design issues that made it unsuitable as the UK's Type 42 destroyer replacement. At the same time, the MOD also placed its Secrecy Order and claim of Ownership on the Author's Application for a UK Patent on his large semi-planing monohull, due to 'it's possible use by a future hostile power'. This was soon revoked due to the Lloyd's Inquiry and the MOD's recent rejection of the S90 concept. Britain and France then combined to work on the Horizon Project. Collaboration again proving unsuccessful, the UK again withdrew and Marconi Marine and BAE Systems joined forces to work on its 'in-house' design, the Type 45. In September 1999, a week after Marconi received the Type 45 development contract, BAE took over Marconi Marine, as BAE Systems – Surface Ships. The final stage in the development of a large semi-planing monohull that had started in 1984 with the collapse of the S90 Club.

Vickers Ltd: Engineering, shipbuilding and armaments company, founded in 1828, a supplier of ships, aircraft and guns to the British Government throughout the 20th century. Vickers Shipbuilding was nationalised and became part of British Shipbuilders in 1977. It was re-privatised in 1986. The Vickers Ship Model Experimental Tanks at St Albans, Herts, and Dumbarton, Scotland, undertook the secret Osprey and Azteca experimental test programmes in 1980–1.

Vosper Thornycroft: The 1957 amalgamation of Vosper Ltd and John I Thornycroft & Son Ltd, the family business of Peter Thornycroft's Grandfather Sir John I Thornycroft and his father, Thomas Thornycroft, later (2007) VT Ltd. Thornycroft's had specialized in fast naval designs ranging from the Royal Navy's first torpedo-boats and destroyers of the 1890s, to the 30-foot semi-planing monohull *Gyrinus*, winner of the 1908 Olympic Gold Medal for Powered Craft. After amalgamation with Vospers the firm always had a close relationship with Peter Thornycroft. This lasted until their disagreement over his design for the first significant enlargement of his semi-planing launches, 112-foot Azteca patrol craft, of which 30 were ordered by the Mexican Navy, built in three Clyde Shipyards and later in Mexico – but not at Vosper Thornycroft.

YARD: Yarrow–Admiralty Research and Development Ltd, wholly owned by Yarrow Shipbuilders. In 1983 the UK Ministry of Defence gave YARD the task of assessing the Thornycroft Giles S90 design for the Type 23 destroyer, despite a clear conflict of interest. YARD had proposed their own Type 23 design, and were rival bidders for the contract. Their power-for-speed arguments against the S90 were some 40 per cent above those co-ordinated by three individual testing tanks, NMI, BHC and the Danish Technical University, and full-scale trials of the Danish Osprey. This was because they regarded S90 as a 'conventional' hull without regard for the benefit of hydrodynamic lift.

Index

Page numbers in *italic* font refer to illustrations.

Anything but a *Still Life*

Behind the Scenes with a Hollywood Photographer

Carole Latimer

UNICORN

Published in 2026 by
Unicorn, an imprint of Unicorn Publishing Group
Charleston Studio
Meadow Business Centre
Lewes BN8 5RW
www.unicornpublishing.org

ISBN 978 1 917458 61 0
10 9 8 7 6 5 4 3 2 1

Design by newtonworks.uk
Printed by Publikum in Serbia

CONTENTS

FOREWORD

by Julian Fellowes

Anything but a Still Life, by Carole Latimer, is a book that comes from an unusual place. It is written by someone whose name is unlikely to be familiar to you, someone who has been successful enough in their chosen profession, but whose life has been comparatively normal except in one way. Socially, she has spent much of her time among the very rich and very famous of our age, knowing them intimately, judging their foibles, witnessing their high mountains of triumph, as well as their moments of deep despond, but from the viewpoint of an ordinary person. I do not mean to suggest that she has no sympathy for her subjects; sometimes she is full of sympathy, but, unlike many accounts of the troubles of the great, it is always mixed with common sense.

At the moment, there is an enduring fascination with people who seem to be living the magical lives of the rich and famous, and Carole Latimer is in the unique position of being part of many of those lives, and witnessing them at close quarters, but retaining the ability to look at them in a sensible way and without prejudice. Young people particularly like to think the lives of "celebrities" are endlessly rewarding and colourful and exciting, while our left-wing friends want to believe they are all unhappy and emotionally unfulfilled, but Carole is in a position to know that neither of these verdicts is invariably true, and that there is as much variety among the famous as there is among any other social group.

I feel, in a way, that I spent many years in a situation that was both. As a medium-successful, jobbing actor, I knew lots of these shining individuals but my position was much more normal than theirs, being neither famous nor especially well off, but then, after I struck it lucky with writing, I moved slightly into the position of being one of the envied that people wrote about and watched being interviewed on television. There is a terrible danger that you do start to think that your opinions on the merits of this side or that in a war are more important than those of real experts, that your understanding of the causes of child poverty are relevant to the discussion, when they usually remain the predictable ideas of a working actor or writer and only significant on that level. It is a sign of madness to think that fame brings a superior kind of understanding, which is easily fallen into, even at my modest level. I always hope that my years of a more ordinary existence stopped me going bonkers when things looked up. But I also acknowledge that my friends, including Carole, must be better judges than I am of whether or not that is true.

This is an interesting, detailed and informed view of the World of the Lucky, which most readers seem to want to belong to, but might not when they have read these words.

INTRODUCTION

> *"The photographer must possess and preserve the receptive faculties of a child who looks at the world for the first time."* – Bill Brandt

People frequently ask me how I became a "successful photographer" and my answer is quite simple: serendipity and hard work. I was in my early twenties, working as a unit publicist on John Huston's *A Walk with Love and Death*, when fate threw me together with the legendary photographer Eve Arnold. In 1951 Eve became the first female photographer at Magnum in New York, taking a place alongside Henri Cartier-Bresson and Robert Capa. Her body of work spanned more than fifty years, capturing people of all nationalities and in all walks of life and professions, who appeared in her illustrated books on America, China and Russia, and other subjects. Eve had been invited to take specials on the film set, and I was taking snaps for fun. Eve noticed what I had done and said that I could become a professional photographer. I have never tried to copy Eve, but she inspired me to do my best. We met up, intermittently, for more than thirty years. She would look through my portfolio and critique every shot, always adding that if she did not think I was any good, she would not bother to give me her time.

I have been fortunate to have worked consistently for many years in a male-dominated profession that has given me such a rewarding lifestyle. However, it has never been a cosy existence, and work and play blend into one. Being a photographer is a way of life, enabling me to make some of my closest friends, lovers and countless acquaintances from all over the world. As the daughter of two actors, I did not grow up in a nine-to-five environment. My father, Hugh Latimer, performed in the theatre, film and television, where sometimes he worked during the day and at other times at night. There were also periods when he was out of work and he would fill his days designing and making objects in his workshop. This was a truly enlightened upbringing, and we were taught not to be ageist, racist or sexist.

The people in this book are not all extremely well-known, but those who are not are exceptionally interesting in other ways. My perception of those about whom I have written is unique, and what I write is the truth only through my eyes; someone else's perception would be entirely different. I have tried to avoid the word "celebrity" as much as possible. This is a word that has become so devalued that it has become derogatory. Richard Gere, who came to my house while filming *Yanks*, referred to the "state of self-delusion that goes by the name of Celebrity". My memory is frequently jogged when I spot the names of people I have met in the past. They pop up like hardy perennials after a long winter's rest. Some years ago Jimi Hendrix's face stared at me out of the tabloids. My mind flashed back to the first time I set off professionally with a camera and flashgun when my friend Stanley Bielecki sent me to cover a reception for a pop group called Grapefruit. Two minutes of explanation as to how the flash worked were followed by mutterings of anxiety from me.

"I've never done this before ..."

"Szmondakowski!" Stanley cut in, in his endearing Polish accent. "How else do you learn?"

The venue was the Hyde Park Hilton in Knightsbridge, one of the grand, neoclassical London hotels. The usual gilt chairs, chandeliers and red carpets decorated a reception room filled with chattering guests, each one holding the statutory tulip-shaped wine glass and cigarette or canapé. At a glance, nothing special. Then slowly it dawned upon me that I was surrounded by a sea of famous faces mingling in a mass of people. Jimi Hendrix loomed up in front of my lens, his halo of frizzy hair filling the frame like candy floss at a fairground. We spoke briefly. He was so charming that I was left with an instant impression of a gentle, kind man. Thick and fast they now appeared in front of my camera: John Lennon, Donovan, Cilla Black, Paul McCartney and Jane Asher (Paul and Linda were to cross my path years later), and Ringo Starr. I kept praying that I had remembered my instructions correctly as I milled and clicked, pretending to be a dab hand as a party photographer. I snapped on in such a state of euphoria that I hardly noticed the weight of the power pack on

John Lennon, Hyde Park Hilton, London, 1967

Donovan, Hyde Park Hilton, London, 1967

Cilla Black, Hyde Park Hilton, London, 1967

Paul McCartney, Jane Asher, Cynthia and John Lennon, Hyde Park Hilton, London, 1967

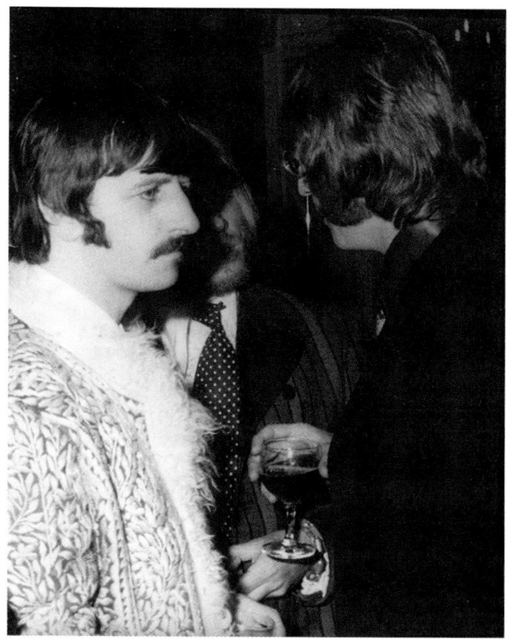

Ringo Starr (not quite so happy), Hyde Park Hilton, London, 1967

my shoulder. Suddenly the party was over, the room was empty, and the guests had disappeared into the night.

Quick as lightning, I jumped into my black Mini, whisked through Hyde Park, and dived into Stanley's basement in Holland Park. A team of Polish printers were working late into the night and soon we were buried away in the darkroom. Under the glow of the red light I watched magic in action, the blank pieces of paper slowly metamorphosing into the famous faces I had so recently captured. Into the fixer, into the wash, dried, and we came into the light to examine our work. Eureka, there they all were. It had not been a dream after all.

Photography gradually blossomed into a career for me. Though I struggled to make a living in such a highly competitive profession, I managed to create my own niche. The main body of my work has been commissioned portraiture, though I have occasionally taken a different path. One of these was to shoot a unique series of romantic pictures of men, semi-nude, lit solely by candlelight. Another was to photograph images for my exhibition *Flowers, Mirroring the Deep*. Then there are landscapes and gardens that I've tackled for the joy of being at peace with nature.

Portraiture, though, has been my trademark, and the days I've spent with fascinating people over my career could not have been manifested without my cameras: they have given me the entrance to a world beyond my wildest dreams. Out of thousands of clients, I have made many friends, some for a reason, some for a season, and some for life.

Whether photographing celebrities in their homes, in my studio or on location, I can honestly say I have learned legions of lessons since that evening shooting famous musicians and actors at the Hilton in Knightsbridge. Working with the actress Cherie Lunghi at Chiswick House was one of my favourite celebrity shoots and one of the most challenging. With its austere, neoclassical architecture, statues and grey walls, this house proved to be the perfect location for the ice-blue dresses designed by David Emanuel. David, with his wife Elizabeth, created Princess Diana's fairy-tale wedding dress. It was difficult to light the set-ups well and my two assistants and I worked nonstop all day to achieve six good pictures.

Shoots with the famous in their homes are often hazardous. There are guard dogs to woo and in the case of Alexandra Bastedo, well-known for

Cherie Lunghi, Chiswick House, London, 1990s

Alexandra Bastedo, West Sussex, England, 1990s

Alexandra Bastedo, West Sussex, England, 1990s

her animal rights work, there was a perilous guinea fowl called Georgina. This evil bird managed to get the better of my kneecap and prompted the stylist to threaten jokingly that Georgina would be served up in tarragon sauce. Alexandra's ferrets frolicking on her designer dress and her turkey fluffing up his feathers every time I mimicked him made up for my blooded knee.

Famous people who have agreed to be photographed are often genuinely busy, so I have had to wait days, weeks or months. In the case of Katharine Hepburn a year passed, plus a further wait of three weeks in New York. Patience is all part of the game. One of my family mottos is "Have patience and endure". These wise words are illustrated by a greyhound attached to a tree, which is the heraldic animal depicted on the Latimer family crest.

Over the many years of my career I have learned much about people and how to photograph them. My first and perhaps most important observation is that no two people are even remotely alike. It is hard to believe that there are so many billions of combinations of skin, hair, body types and personalities. Talented, talentless, famous, ugly, beautiful, vain, conceited, loveable, witty, charismatic – all have gazed into my lens. I have learned that everyone has two different sides to their face to a greater or lesser degree; the left side is said to be the spiritual artistic side and the right the pragmatic worldly side. I do find that the majority of my artistic clients are best on the left. A good example was Dirk Bogarde, a great actor who was also a wonderful writer and competent artist. Likewise, people have two types of eyes: spiritual and worldly. For spiritual, think of eyes like bush babies, lemurs, sloths, seminocturnal animals, eyes like deep bottomless pools, gentle, soulful, bulging, watering, beautiful, humorous. The worldly eyes are fast, twinkling, joyful, arrogant, hard, laughing, bright, restless, trusting to nothing but the control and planning of their lives. Spiritual or worldly, eyes are the most important feature in a face, as they are the only ones that reflect; they are mirrors of souls revealed to me through my camera lens. The eyes dictate my choice of lighting as much as the structure of the face, because light-coloured eyes are weaker, blue on the whole being the most sensitive. I sometimes have to compromise and use softer lighting or, if I am using natural light, I pull down a blind so that the eyes are not strained. Brown eyes can take far more light, and often with black people I break my rule and photograph in direct sunlight; the result can be stunning.

Another lesson I've learned is that clients need to be reassured. When they are feeling attractive, their eyes radiate confidence. Each session is like a performance, and I feel my way instinctively with every shoot, and no two sessions are ever the same. I try to enhance the best qualities of my sitters. Most are nervous when they come for their portraits, and many liken the experience to going to the dentist. It was quite a reversal of roles when my dentist came to my studio. He was as nervous as I had always been in his surgery. What an interesting experience for both of us.

Learning the tricks of the trade is a huge part of being a portrait photographer. After decades of portraiture, I have learned which lense and which format camera suits each subject, and that I must use brighter light and faster shutter speeds with children and fidgety, nervous people. Spare clothes, make-up, curlers and brushes need to be on hand, along with a fountain of knowledge to help subjects with their problems. I have learned, too, from professional clients who have tips to offer after years in front of the lens. Angie Dickinson protected herself from bad lighting by double checking where I had placed her by using a small hand mirror. This enabled me to capture a look of warmth and trust, although we had barely met. Her professionalism spoke volumes.

Another lesson I would pass on is that for the best portraits, you must be genuinely interested in your subject. The hardest part with portrait

Angie Dickinson, Los Angeles, 1990s

photography is being interesting and interested at the same time, while concentrating on the lighting, composition and subject. Everyone has a story, some more interesting than others, but there is always something new, and once in a blue moon I meet someone who is truly remarkable.

Finally, I quote Shakespeare for my final observations on how important it is to approach clients differently at different stages in their lives. In *As You Like It* the Bard writes, "All the world's a stage ... And one man in his time plays many parts/His acts being seven ages." This concept of our time on planet Earth could serve as a guide to working with clients.

"At first the infant/Mewling and puking in the nurse's arms." You have to genuinely like children, because they are even more instinctive than most adults. Also, it is important to remember that the very old and the very young have one thing in common: they must be photographed in a very short space of time, as they tire quickly. This does not mean that I necessarily only have a baby in the studio briefly, as it may take a long time to get him or her in the right mood. But once there, I have to move very fast, as often there is no second chance. I keep the studio as simple as possible and the lighting as even as I can. Then I add a few tried and trusted props, like bubbles, which are photogenic if caught in the shot. So often the plastic toys that come with the youngsters are too unsightly. The mothers invariably need more soothing than the children do, so it is a bonus when a nanny is in tow. It is too easy for the babes to pick up the mothers' anxiety and then all chaos can reign. Toys fly around, food and sticky drinks appear as bribes and end up on my backdrops, tears are shed, including mine, nappies are changed amid anxious cries of "Are you sure you've got Johnny's special smile?" I call a halt when I know I have achieved the shots. Clothes, hangers, toys, empty packets of sweets, bottles, teddies, Lego sets, all are scooped up and seldom do the sitters even stay for a cup of tea. Many of these pictures end up as Christmas cards to be sent around the globe by loving parents, or given to grandparents.

"Then the whining schoolboy, with his satchel/And shining morning face, creeping like snail/Unwillingly to school." Some youngsters invariably hate the whole idea of being photographed. Often the subject is a teenager at his or her most vulnerable age, decorated with the all too common braces on their teeth. I have never felt truly comfortable in these sessions. Again, proud parents push their children into being photographed to mark either their eighteenth or twenty-first birthdays and seldom are the subjects anything but reluctant. This is why I have done so little society photography, as I am uneasy about youngsters being forced into these conventional situations. However, I have enjoyed taking spontaneous pictures of my friends' children when the mood is right.

"And then, the lover/Sighing like furnace, with a woeful ballad/Made to his mistress' eyebrow." Drama students come to me for head shots. They are invariably involved with their first real love affairs, youthfully naive, secure in the knowledge that their love is like no one else's. They are in need of tender loving care and are invariably appreciative. I receive cards and presents and promises of being kept in the picture when they become famous.

"Then a soldier/Full of strange oaths, and bearded like the pard/Jealous in honour, sudden and quick in quarrel/Seeking the bubble reputation/even in the cannon's mouth." The adult, now a professional, has usually been photographed several times and is used to the whole procedure. Ambition and energy are at their peak, and the appreciation the subjects have is rewarding. They likely have many pictures of themselves as a point of reference, so the words "These are the best pictures I have ever had" are music to my ears.

"And then the justice/In fair round belly with good capon lin'd,/With eyes severe and beard of formal cut,/Full of wise saws and modern instances;/and so he plays his part." Middle age. In my work, many of these subjects are parents; mothers and fathers bravely going back into acting after the wilderness years of bringing up their children. They often bring shots of themselves in their prime and I know that they are hoping for a miracle, to look the same twenty years on. Others are business people at their most powerful who require a truly professional portrait, perfectly lit and achieved in the minimum amount of time. Time is money as far as they are concerned.

"The sixth age shifts/Into the lean and slipper'd pantaloon,/With spectacles on nose, and pouch on side;/His youthful hose, well sav'd, a world too wide/For his shrunk shank; and his big manly voice,/Turning again towards childish treble, pipes/And whistles in his sound." The old: the complaining and fighting against the ageing process now cruelly manifesting itself on the outside. Only those who have been growing and developing spiritually will achieve a higher level of consciousness that seems to shine through, creating a beauty of its own. They conserve their energy for what is important in life and invariably are truly grateful to be the lucky few in good health and able to keep doing good work. These are often artists who go on till they drop and retirement is not in their vocabulary.

"Last scene of all,/That ends this strange eventful history,/Is second childishness, and mere oblivion;/Sans teeth, sans eyes, sans taste, sans everything." This is where nature abruptly puts an end to any dreams of immortality and wins the day.

CHAPTER 1

BEGINNINGS

"The future belongs to those who believe in the beauty of their dreams." – Franklin D. Roosevelt

A bundle of square snaps in black and white shows friends about eight years old standing stock still, and in some cases they have lost the top of their heads. Not a promising start. From then on, not a shred of evidence of a budding photographer appears until my teens, when my father and I took over the top bathroom in Hampstead. Here we installed a second-hand enlarger, trays for the developer and an eerie red light. Tell-tale acid stains on my mother's white carpet lasted longer than the hobby. We counted out loud (we did not have a timer) as we exposed the negatives, then we whisked the paper into the fixer and into the wash, then lights on and, hey presto, a print. Hampstead Heath offered rich pickings, where I spent hours snapping landscapes, trees, fauna and flora, always with a keen eye for composition. Around this time I entered a competition in a magazine called *Girl*. A black-and-white picture of my sister Clare, Tommy the tortoise and Belinda, our black cocker spaniel, earned me the princely sum of £5.

My first school, North Bridge House, encouraged the arts and many artistic families chose it for their children. I was considered to be a good little actress, although, paradoxically, I was shy. Some roles were less demanding than others. My part as an oyster required no more than to join a chorus of protesting oysters from inside a cardboard shell from which I slipped out when eaten.

After those halcyon days, Channing School was joyless, apart from lessons in art, languages, history and tennis. But when it came to science and mathematics, I did not have the gene. My physique, light as a feather with long legs and a short body, resembled a newborn foal, and this spindly frame was ideal for running and jumping.

French with Mademoiselle was a joy, and history was made palatable by a cosy, pink-faced teacher whose name I do not recall. But English was a nightmare. Miss McRae taught through fear. A stout, androgynous woman with thick, bristling brown hair and a cavalry moustache, she enhanced her masculine appearance by dressing in tweeds and brogues. Due to her draconian teaching I cannot recite a word, I freeze. The school play gave Miss McRae her last chance to bully me, so she gave me the only singing part in Shakespeare's *The Merchant of Venice*. There, in front of my theatrical parents, I was expected to walk on singing "Hey, nonny, nonny no" with no accompaniment. So come the night, I simply did not appear. Upon her retirement Miss McRae took up flying.

Meanwhile, at home my upbringing was among actors, so I never dismissed my parents' friends as old or boring; in my eyes they were glamorous. Sparkling leading ladies who starred opposite my father were frequent guests at our home in Hampstead. As the theatre world is neither ageist, classist, sexist nor racist I had an enlightened upbringing. From an early age I was included in all my parents' parties, but as I was painfully shy these events were agonising. I would sit on the stairs outside our sitting room until I finally rose to my shaky feet and turned the china doorknob, which slipped beneath my sweaty palms. My fears were groundless as I was always met with a tidal wave of warmth and affection. Joyce Redman made a lasting impression. This petite, ultra-feminine woman oozed sensuality with her buxom bosom, sparkling blue eyes, glorious red hair and fruity laugh. Her sizzling sex scene with Albert Finney in the film *Tom Jones* often eclipsed her prodigious work in the theatre. When I first met Joyce she was starring opposite my father in *The Lionel Touch*, a play in the West End in which he had taken over the lead from Rex Harrison. That evening Joyce wore a black velvet dress with a white collar. The next day I bought some black velvet and copied the dress right down to the shiny black buttons.

I travelled to Oxford to spend the weekend with my father while he was touring in a play called *A Clean Kill*, directed by Alastair Sim and co-starring Rachel Roberts. Having been an actress, my mother knew the dangers that arise from propinquity when actors tour together for several months. She herself fell for my father and married him after starring opposite him in Terence Rattigan's *French Without Tears*. So, she sometimes had the wisdom to send me to stay with him for a weekend on tour. My presence did not deter Rachel, married to Alan

Dobie, from hitting on several of the actors. They included Peter Copley, who succumbed. When my mother went to stay, Rachel gave her a cool reception. She later remarked to my father that she had thought I was his mistress.

Mummy kept all the sirens at bay throughout their long marriage. The next time Rachel crossed our paths was when she starred opposite my father in *August for the People* at the Royal Court in London. In the 1950s the Royal Court had been the birthplace of the "angry young men", spearheaded by John Osborne. Rachel was by then married to Rex Harrison, and she hosted a small opening-night party at their grand apartment in Eaton Square. I was considered grown up enough to be included and I could bring my boyfriend, David Wills. I had a huge crush on David, although I would have died of embarrassment if he had so much as held my hand. Animated conversation filled the room until towards the end of the evening, when my cousin Peter Wilson, the chairman of Sotheby's, was talking eruditely about art. Rex appeared to be listening but then suddenly exploded, "Why don't you fuck off and go home?" He was out of his depth and did not appreciate being in that position. After all, he was the star. Chaos ... Rachel was in floods of tears, and Rex's son Noel was distraught until my father intervened. He ushered Rex aside and calmly talked him into the idea that we should all have one more drink before leaving. Soon Rex was in tears and apologising profusely and agonising over his rude behaviour. This was my baptism by fire into the world of film stars. With numerous films and marriages behind him, Rex was in a different league compared to the rest of that London set. Rachel had dropped her first husband, Alan Dobie, to be married to a star. I was learning fast the price of fame. Jeremy Irons once said to me that if the fire gets too hot, you can always get out, but many do not realise this until it is too late. Rachel became an alcoholic and even more outrageous than Rex, and eventually committed suicide. She was kind and warm towards me and I could sense the Celtic soul.

Meanwhile, there was the conventional side to my upbringing. My paternal grandmother in the country disapproved of my father working as an actor. "Not a job for a gentleman," she would remark. Holidays with Granny were steeped in tradition and routine. We rose at the same time every day and would do our exercises in her bedroom while my step-grandfather, Sir Alexander Anderson, nicknamed "Grandpups", went about his morning routine in his dressing room. Breakfast on time, always a stiff walk in the morning, prefaced by a glass of water from a decanter that had a glass stopper that made a resounding *ching* as it was replaced.

The days were filled with gentle pursuits, gardening, picking fruit, taking strolls along the river, where I would stop to play Pooh sticks. I have nothing but happy memories of those holidays spent with Granny. Lunch was at one o'clock, heralded by the sound of a gong, followed by the cook, Nan, in her crisp uniform, sweeping through the green baize door. This formidable character, tight-lipped due to a lack of teeth, sported an equally tight bun that forbad a single hair to be out of place. Children could only join the grown-ups at lunch if their manners were impeccable. Once I was reprimanded for mentioning the presence of a slug in my lettuce, so I scooped the leaves into my mouth as the little slimy creature slid around my plate. On another occasion I wanted to pee, but you were not expected to leave the table during lunch. I left it too late. At the last minute I rushed from the dining room, failed to make the bathroom in time and left a tell-tale stream across the kitchen floor. Needless to say I was mortified. Rest time followed, and I would lie on the chaise-longue in the bow window of Granny's bedroom, reading aloud to my teddies from *Ruthless Rhymes for Heartless Homes* and *Babes in the Wood*, cheerless Victorian children's books. I would gaze out at Granny's manicured lawn framed by herbaceous borders, with a beautiful pond and a sundial. Once, when I fell ill with flu, I was allowed to spend the days in Granny's four-poster bed working at a petit-point tapestry of a brightly coloured bird in a cherry tree. I could have been a child from a bygone era, though tapestry has come full circle and is now fashionable and I have the cushion to this day.

Guests often came for tea and croquet on the lawn, and as I grew older, suitable young men were invited. The vicar's sons were especially welcome, but I favoured the tear-away Irish lad, Michael Fogarty. This lanky, dark-haired boy would hit the balls as hard as possible to see how far they would ricochet off the wall. From then on he was not popular with my grandparents, and the vicar's son, with his pasty complexion and sprinkle of spots, was considered far more suitable. The early evenings were taken up with baths and changing for dinner. A drink in the drawing room was followed by dinner punctually at eight o'clock. If visitors were just invited for a predinner drink, it was always made patently clear when they should leave. Then Granny would sit down to dinner at one end of the highly polished mahogany table with Grandpups at the other end. Their individual tastes were catered for: unripe bananas for him, ripe for her, his butter salted, hers unsalted. Theirs was a harmonious marriage. Various members of the family sat on either side. After dinner, I would play backgammon with Grandpups, and he would offer me just two squares of Cadbury's Bournville chocolate and Granny would quietly do her tapestry. Only the sound of the logs crackling in the fire broke the silence. Bedtime was heralded by the nine o'clock news, at the end of which Grandpups would reach for his stick and rise to his feet for "God Save the Queen". The fender was then placed in front of the fire and we retired.

The only change in this daily pattern was church on Sunday. I hated our elevated position in the front pew. My grandparents contributed a great deal to the small community and at Christmas time Granny and I would go around the village to deliver the most modest of gifts, as though we were back in the Edwardian era. But the villagers understood that it was the thought that counted. Granny tried tirelessly to teach me what she had learnt as a girl, or "gal" as she would say. Family history was passed on, as Grandpups had visited Somerset House in London to obtain Granny's family tree. I took little interest at the time, but now I have become fascinated in the tree that goes back to Edward III and shows our descent from a veritable potpourri of interesting folk. Among them was George, Duke of Clarence, brother of the last Plantagenet king, Richard III. He was thought to have been exceptionally good looking, was considered a saint by some and a sinner by others, and was alleged to have been drowned in a butt of malmsey wine. On the Scots side were the Seaforths, a brave Highland clan said to have been as mad as a box of frogs. My granny's first cousin married a descendant of Charles Darwin, and rumour has it that Sir Isaac Newton is an ancestor, but as he bore no children this claim seems rather circumspect ... Mummy, who was born in Scotland, lost both her parents in her late teens, and she went to live with her aunt near Edinburgh, in a magnificent house called Johnstounburn.

As a child I can remember the most magical visits. To a child's eye the house was vast, and indeed, it was so big that after Colonel and Mrs Cruikshank died it became a hotel. The Colonel, who had famously been a pallbearer at King George V's funeral, used to ask guests the minute they arrived when they were leaving. Intimidating at first, but I soon became fond of this eccentric gentleman. Auntie Mamie was a match for her husband. Sophisticated and well-travelled, this grande dame had a passion for breeding toy Yorkshire Terriers. These pampered pooches took over an entire wing of the house, which had the distinctive sweet-and-sour smell peculiar to this breed. The grooming was time-consuming, as the dogs' fur was so long that they could not walk unless it was tied up in cotton curlers. The only time the fur was undone was when the dogs were placed on cushions at shows, where it would cascade onto the velvet. Mr Pim was the star at the international dog shows Auntie Mamie constantly entered.

I left school at fifteen and went to Paris to learn French. I attended a finishing school for *les jeunes filles de bonne famille*. We may have been *de bonne famille* but we behaved appallingly. Madame, who ran the school, was too old for the job and bitter about the loss of an eye during the Second World War. We soon learnt that if we kept very quiet she would fall asleep and we could escape the classes. The girl on her right would tip us off when Madame's one good eye closed, the other one being covered by a black patch. Three of us had to share a bedroom, and one night Madame's ugly little husband appeared at our door in his striped pyjamas and, without a word, exposed himself. We were disgusted. So much for the finishing school for young ladies!

We spent our afternoons wandering from one art gallery to another in the freezing cold. Too much of the Louvre and the Jeu de Paume dulled our senses. Our one treat each week was ice skating, when we would escape from our chaperone and chat up boys in the middle of the ice rink. This was all very well, but I really did want to learn French, as I needed to earn my living. My parents agreed that Paris was no longer appropriate, and L'Université des Étrangers (The University for Foreigners) in Tours was the next step. Tours is a gentle town of bicycles and students and lies in the valley of the beautiful, wide-meandering Loire river in central France. Here I spent the mornings in class and the afternoons sitting under a cherry tree eating its warm fruit and reading *Les Misérables*. The family with whom I stayed were teachers, so the lessons never really stopped. I was a diligent pupil and enjoyed learning French.

On my return to England, Granny remarked that I was "fluff-some and chub-some" (words of criticism were softened by adding "some"). A diet of too many baguettes had plumped me up, but I was tanned and fluent in French and my accent was good enough to fool most people, although some were puzzled as to which regional dialect I spoke.

Back home in London, I entered a dressmaking course in Knightsbridge, where I met several girls who were about to "come out". No, they were not gay. This was the expression used for debutantes when they were launched into Society. Again, I was thrown into a group who just wanted to have a laugh and to discuss future parties and dances. Soon I became the teacher's pet, as I loved designing and making dresses. My favourite classmate was Harriet Turton. She loved to play the fool and managed to cut a skirt exactly the right length with no room for a hem. I did attend Harriet's coming-out dance, but I lost touch with her until many years later, when my colour printers held a large party at London Zoo. Across the crowded room I spotted a familiar face. I remarked to the owner, Chris, "That's Harriet Turton!" "No, no," he replied, flustered, "that's Viscountess Bridgeman." He introduced us, and of course I was right. Harriet had married, and what is more, she has her own unique and highly successful picture library with offices in London, Paris and New York.

In spite of my being "teacher's pet", all the girls befriended me and I was swept up into the world of tea parties, drinks parties and, finally, the dances. My mantelpiece was festooned with smart invitations to balls in stately homes, grand hotels in London, and the countryside. As we were not rich, my dressmaking course served me well. I was able to make all my evening dresses with the help of our teacher, Mrs Bailey, and her tailor's dummy. Being shy made this a tortuous time that I only really enjoyed in retrospect. All too often I wanted to make my escape, which was easy in London. I could always gather up my skirts and rush out and grab a cab when some drunken young man's groping became too much. The country dances were another matter altogether. Invitations to dances included details about where I would stay and who would hold a dinner party for the house guests before the ball. These poor hosts would have to entertain several young people they did not know for the entire weekend. I seldom enjoyed those visits, because I spent the whole time worrying whether the maid would unpack my suitcase and if I had the right clothes for all eventualities. Then I would lie awake all night wondering what time I should appear for breakfast. The host would invariably think it amusing to say "Good afternoon."

The grim reality of a secretarial course followed this sybaritic period. My parents were anxious that I should train for a secure job to equip me for a life less turbulent than theirs. One disastrous job led to another and work was merely a means to an end, paying the bills. One week at Birds Eye frozen peas led to one day at the law courts, where dictation from a lawyer led to my flight at lunchtime: I could not read back the shorthand. So I went with my friend Virginia Ironside to the Spiritualists' Association, hoping for guidance. We sat in a circle on hard chairs in a shabby room with gloomy lighting while the clairvoyant turned to each of us in turn to pronounce his predictions. Suddenly, this spooky little man had

Virginia Ironside, London, 1991

Virginia in his sights and, without a pause, he predicted the imminent collapse of her marriage. Sitting next to her I could feel her palpable distress. The prediction did come true, but not before she gave birth to her son, Will. Many years later, she became a highly successful author. Soon I became the focus of his gaze. "I see negatives, pictures," he said, and added, "I'm sorry, that's all." This meant nothing to me, as my mind was focused on meeting my Prince Charming who would rescue me from the working world. It was not until years later that I remembered this man's prediction.

A fleeting glimpse of my enthusiasm for photography returned while I was running the ambulance office at University College Hospital in London. One of the ambulance men gave me a present of an ancient 6×6 camera, so I must have been efficient at sending the drivers in the right direction. I took my work at the hospital seriously, and each day I put on a dreary midcalf, blue overall, transforming the sexy young girl in a miniskirt into a dowdy spinster by the name of Miss Latimer. I visited terminally ill patients, arranged flowers, donated from funerals, in the wards and worked in an administrative capacity throughout the hospital. It was not long before I started thinking I had every symptom in the book. One stomach tweak and I was convinced I had appendicitis. Clearly this was not the job for me. Only the occasional crush on a doctor brightened my days, but I was as good as invisible in my synthetic blue overall. So I departed within a year, happy to be released from the daunting red-brick building, the proud owner of a 6×6 camera. With superior negatives from which to work, my father and I were back in the top bathroom, until the novelty wore off and the enlarger, chemicals and trays were banished to the box room.

Coincidentally, when the snaps went on hold, I got a temporary job at *Amateur Photographer* magazine, where I spent every day typing envelopes. This job was so deadly that I decided to spend every spare minute teaching myself Italian, in readiness for a six-week tour around Italy. Feeling that nothing could be worse than typing envelopes all day, I was quick to take on a bet with a friend. If I could make as much money as it cost to do a modelling course at Michael Whitaker's, I would win. I took on this wager in desperation, although I was not photogenic enough for photographic work and not tall enough for the catwalk.

Miraculously I got an in-house job modelling swimwear in Grosvenor Street in London. It was midwinter. I felt ridiculous setting off for the underground on those cold, grey mornings, heavily made up with false eyelashes and a false hair piece. Only when I arrived at the showroom, with its theatrical lighting, did I feel comfortable. Parading up and down a narrow catwalk, I modelled the complete range of bikinis, including mock leopard skin, while the other two girls got the one pieces. A live show for Lyle & Scott knitwear proved to be my valedictory modelling job, and this was more fun. I had plenty of clothes on, so I enjoyed strutting up and down the red carpet doing my perfect twirls as I caught the odd eye from the front row. I ignored the bitchiness backstage; after all, this was my last modelling job and not only had I won my bet but I had also learned to stand up straight, walk properly, descend a staircase without looking down, and get out of a car without flashing my underwear.

The shop girl replaced the model and I spent six days a week at Ning's, in Sloane Square, pretending to know everything about antique furniture. The shop also had a fast turnover of plastic flowers. With no formal training in flower-arranging I was thrown in at the deep end, but after two short lessons I was twisting, cutting, powdering to give a bloom, and doing anything else to make these plastic understudies as near real as possible. The highlight of my floristry days was a trip to Arundel Castle, my car piled high with plastic blooms in various shades of grey, white and green, with which we made huge arrangements that cascaded down pedestals set behind red ropes. The rooms were vast and dark, with small medieval windows more like slits, which meant that you could not tell that the flowers were artificial. The butler appeared as Rosemary, my assistant, and I struggled with the huge vases, twisting and bending the poker-straight flowers to make them look natural and burying them in plasticine to keep them in place. Would we care to lunch with the Duke and Duchess of Norfolk? I was delighted, but Rosemary was horrified and needed a great deal of persuading as we were covered in green plasticine. Her worries were unfounded, as one of their daughters joined us straight off her horse. I already knew that in England the more upper class people are, the more casual they are when they are in the country.

At this time, I was in love with a boy named Johnnie Minns. He co-owned a shop, Steam Age, with my boss at Ning's, Ivan Scott, so I was perfectly content in my dead-end, badly paid job, just living for the evenings when Johnnie was free to take me out. My aspirations in life went no further than the dream of a wedding ring. My family gave their silent approval, and even Granny was bewitched by Johnnie's good looks and charm, in spite of his fingernails, dirty from working with model steam engines. Aware of his assets, Johnnie was not without vanity, and he wore amazing jeans with lace-up flies and codpieces, all fitting to perfection. The ring was on offer, then Johnnie married another. I was heartbroken, and it was time for change.

Back to secretarial work, this time at Rediffusion Television, in the Schools department. I was still a fish out of water and could not see myself working my way up the ladder to become a personal assistant or a researcher. This job was brief, its only high spot descending to the recording studios to watch the first recordings of *Ready Steady Go!* where I saw both The Beatles and The Rolling Stones arrive for their first ever television appearances. The atmosphere was charged as we waited, then came the warning of their arrival as the fans screamed in the street, and into the studio they came, running up to the stage. So close were the Stones as they ran past me that I got a waft of warm armpits.

The next move was downwards. Promises of a job with a future with Hazel Adair and Peter Ling, who wrote a television series called *Compact*, led to typing scripts in the back room of an office in Finborough Road, West London. The only relief came every other Friday, when two of the writers, Ted Dicks and Myles Rudge, would take me out to lunch. I spent the other lunch hours alone, flicking through *Spotlight*, yearning to be a photographer. I never dreamt that this would become a reality. I had a crush on Ronald Allen, who was the star of *Compact*. Ronnie was reed slim, with chestnut-brown hair and an exquisite face, like that of a faun. Day after day I spent typing out his lines and longing for an introduction. Years later, in the late 80s on one of my long trips to Los Angeles, the telephone rang and Ronnie was asking me if I would photograph him. The shoot went well and Ronnie, handsome and well preserved, and his wife, Sue Lloyd, became new good friends.

In all of our lives there are chance meetings that can change our destinies. While still working for Hazel Adair and Peter Ling, I attended the anniversary party for their television programme *Crossroads*. As I struggled to be charming and cheerful, I struck up a conversation with Theo Cowan, who was the first press relations man in England to handle the stars, and I voiced my dream of working on feature films. Theo scribbled down his number, and soon I had swapped West London for a smarter area close to all the best hotels for Theo's client meetings. However, our offices at 46 Clarges Street were Dickensian. My desk stood beside a popping gas fire that leaked a thin trail of gas into the dark room I shared with Yvonne, on the switchboard; Jane, by the window; and Peter Thompson, who ran errands and collected clients from the airport. All three of them went on to be successful publicists. After the loneliness I had endured in my former job I loved the companionship, the humour and the shared love of movies. Our one treat was attending the film premieres, when we were expected to be on show and to help with the whole event. We would take turns to change, the girls into long evening dresses and the men into black tie, in the tiniest lavatory on the half landing. Nobody wanted to take long in there, as the room was unheated and spartan. Mimi, the office manager, who was loud and bossy, would sprinkle a thin film of talcum powder like snow over all of the surfaces, missing her armpits. I spent my days typing letters: "Dear Dirk" (Bogarde), "Dear Dickie" (Richard Attenborough); we never met any of these stars, as our offices were so shabby. Theo would hold meetings at various hotels, such as the Ritz and the Connaught. We worked ridiculously long hours, but Theo said that when I found what I really wanted to do, I could go with his blessing. The break came within months.

CHAPTER 2

PUBLICIST ON FILMS

"*Chance favours the prepared mind.*" – Louis Pasteur

It was a chilly spring day in Soho Square when the leaves on the immense plane trees were just beginning to bud and I too was facing a new beginning. The head office of 20th Century Fox with its impressive grey stone porticos, tucked away in the square's corner, loomed up in front of me. I pushed open the heavy door to be greeted by the doorman, smartly dressed in a black uniform and peaked hat, who directed me to the first floor. A worthy rather than glamorous personal assistant eyed me suspiciously, offered me a seat and continued typing. So here I was after months of muttering to everyone that I wanted to work on feature films. By chance, I had finally met someone who suggested that I contact Fred Hift, the formidable American head of production publicity for Europe. With no knowledge of what the job entailed, I hoped to become a unit publicist. Clammy palmed, I sat in silence listening to the tap, tap, tap of the typewriter, punctuated by the zing, thump of the carriage return at the end of each line. I longed to leap up to demand that this "perfect" assistant should lighten up and exchange a few words, but her eyes never strayed from her desk.

Eventually I was ushered in, but now it was Fred Hift's turn to keep me waiting for what seemed like an eternity as he took calls from all over Europe, switching from English to French to German, like a juggler with batons. Fred bore a striking resemblance to a raven, with his aquiline nose, slim face and thick, jet-black hair. He filled me with apprehension, as I thought he had something of the night about him. I had gone to great lengths to look my best in my new figure-hugging sage-green dress, cropped above the knees, with gold buttons running down the front. The right number of buttons were left undone, casual but not vulgar. My sheer stockings were just a shade darker than my flesh tone to flatter my legs, and a pair of black, fashionable patent-leather high heels completed the ensemble. My nails were neat and finished in a clear polish, and I was as groomed as a show horse, apart from my fine, shoulder-length blonde hair, which never could look structured. Yet I still felt painfully shy and nervous. The minutes ticked by; as I shifted nervously, I caught my heel in my stocking and a sensation like an army of ants shot up my leg, heralding disaster: not one but several ladders. As I glanced down to see the extent of the damage a button flew off my dress. With a glimpse of bosom and bra, dishevelled and blushing profusely, the image was shattered before I had opened my mouth.

The telephones fell silent and Fred stared down at me. He looked tall and formidable behind his executive desk (it was only later that I discovered his chair was purposely raised to give this appearance; he was only of medium height). He started to talk to me about the possibility of my becoming a unit publicist. How could I tell him I did not know what the title meant? So I had to bluff my way through the entire interview. Fred talked about writing stories and servicing the branches, at which point he could have been talking about trees as far as I knew; only later did I learn he meant sending out press releases to the various branches of 20th Century Fox. I felt sick at the idea of writing stories which would be sent all over the world as memories flooded in of struggling with English at school. I held my tongue and thought I would meet that problem if I got the job. In truth, I thought that would be a miracle.

The next day I was an assistant unit publicist on a feature film and I had not got a clue what I was meant to do. All I knew was that if I did not find out fast, I would be fired. The film was called *Joanna.* This was also Michael Sarne's debut in the film industry as a director, prior to which he had been a pop singer with a memorable song called "Come Outside". Mike believed his own publicity. He saw himself as a "Cool Dude"; he was witty, arrogant and the first person I ever heard use the word "fuck" as an adjective: everything was "fucking great" or "fucking awful"; now you cannot escape the word from all and sundry. Dressed in jeans, the perfect image for his fine physique, he drove a vintage, dark-red convertible Rolls-Royce and was dating the star of the film, Genevieve Waite. *Joanna* is of historical interest in the fashion world, since the clothes, designed by students at the Royal College of Art, were ground-breaking. This was the first time that we had ever seen skirts of all lengths at the

Michael Sarne, London, 2006

same time. Prior to this, only one length was acceptable in any one year. The students studied under Professor Janey Ironside, who happened to be the mother of my childhood friend Virginia. We had a feast of minis, midis and maxis in a rich variety of designs never seen before, enhanced by girls so pencil thin that they looked like they were on the brink of starvation.

I had a normal healthy figure with good skin and pink cheeks; not cool. However, I soon got the hang of writing biographies of the actors, choosing sets of stills for the press kits, and taking care of the visiting journalists and photographers. Each night I dined with the producer, Michael Laughlin, and Mike at the Aretusa on the King's Road or at Mr Chow in Knightsbridge. I lived the film from day one to the end and felt a kind of postnatal depression when it was all over, bar the editing.

During the editing period, I stayed on as personal assistant to Michael Laughlin, working out of a house in Regent's Park that he rented from the Churchill family. As Michael was single at the time, I ran his life like an efficient wife. I organised lunch parties at the house and dined with him regularly to make up numbers. We shared intimacies about our private lives, and eventually, when he was bowled over by Leslie Caron, I made the first call on his behalf to arrange an evening at the theatre. They were married within months. We entertained executives, went to endless rough cuts, and talked about nothing but *Joanna*. This was my whole life. Several newcomers started their careers in that film, but it was Donald Sutherland who stole the show with his sunset dying scene in Morocco ... stardom followed.

I only spent a couple of weeks on the next film, *The Touchables*, because the filming was nearly at an end when I joined. Bob Freeman was a fashion photographer, making his debut as a director, and David Anthony, also a photographer, played the male lead. The story revolved around four female models who kidnap a young man and hide him in a huge balloon in the countryside – a thin plot, but they were all the beautiful young things of their time. I got along well with the four girls, who were all roughly my age. Esther Anderson, who had been a girlfriend of Marlon Brando, hung out with me so much that a visiting photographer thought I was one of the cast. Soon after, German newspapers ran a picture of us captioned as two of "The Touchables".

With more confidence and experience, I returned to 20th Century Fox in Soho Square. The scene was all too familiar, with the relentless tapping of the secretary on her typewriter and Fred Hift juggling phone calls, but this time I did not panic, as I had seen the softer side of Fred and knew his height was an illusion. Now I could write a press release, knew that servicing the branches had nothing to do with cutting trees, and that colour had to be spelt without a "u" when working for an American company.

The telephone slammed down. Fearful Fred was in a rage and I almost expected him to squawk. He had to find a publicist who could speak French and time was not on his side. He needed one now. I let him rant on about the unfairness of being put in this situation at such short notice, then quietly put my head in the noose. "I can speak French," I piped up, without thinking about the consequences. *Un Soir, Un Train* would take up the next two months of my life.

Within forty-eight hours, I was packed and on a flight to Belgium, clutching my faithful Adler typewriter. The city of Antwerp, which is Belgium's major port, looked like a generously iced Christmas cake. Cars had packed the snow on the roads but at the edges it was loosely piled like billowing duvets. Arriving on the set at lunchtime, I was immediately struck by the difference of working with a French crew. A leisurely two hours for lunch was de rigueur in France, as food was given the respect it deserved: not for them the hasty sandwiches of their counterparts in England. I was greeted warmly, handshakes all around as I was introduced to the cast and crew; only the two stars were absent.

These stars were Anouk Aimée and Yves Montand. As I greeted all my new friends the following day *("Bonjour, comment allez-vous?"* with kisses all around on both cheeks) I was nudged by the first assistant, *"Je voudrais vous introduire à Monsieur Montand."* I swung around to find a tall, handsome man with broad shoulders, a huge smile and a wicked sense of humour. He was, as the French say and for which the British have no equivalent, *"jolie laide"*, a quality far more attractive than perfect features. I was bowled over, at a loss for words. This was one of the most charismatic men I had ever met. I had never been attracted to an older man, but this proved the point that age is irrelevant when it comes to

Me and Anouk Aimée on the set of *Un Soir, Un Train*, Antwerp, Belgium, 1968

true charisma. As it was freezing cold, I was dressed in my black Chinese mink (a smart name for weasel) fur coat, a black polo-neck sweater, and trousers tucked into long boots. We chatted on set a great deal, as after all, it was part of my job to get to know the actors. Yves frequently turned the conversation around to me, asking questions about my life. I felt shaken and stirred, as I was falling for this sophisticated man.

However, I had little time to dwell upon my feelings. There was so much to do. I did not even know the word for unit publicist in French, let alone any of the other film-making phrases, and I had daily press releases to write. The panic was rising, but I reminded myself that I had only just arrived, and at least I was settled into my hotel and had mastered driving to the set on my own. The next day, to my horror, I discovered that I had to pack all my belongings and move to another hotel. This was a little more complicated than it seemed. It was like moving an office, as I had already notified everyone of my whereabouts. I protested, but I had no choice in the matter. I was surprised to find myself in the smartest hotel in Antwerp, but the penny had still not dropped. It was not until the following morning, when I was standing with the entire crew in the foyer, that I finally twigged. There was Yves, at the top of the circular staircase with a huge grin on his face, his eyes unashamedly locked onto mine, oblivious of his audience. We both burst out laughing: what style.

I did not have to do much soul-searching to decide whether I thought it morally wrong to engage in an affair with this charismatic man. Yves's wife, Simone Signoret, was more than aware of my existence, so much so that when she called in the evenings, she would invariably ask to speak to "*la petite anglaise*". She was always warm and friendly. Whatever arrangement they had in their marriage, they were not living a lie.

Yves became my teacher, Professor Higgins to my Eliza Doolittle. He was emphatic that a successful career was as important for a woman as it was for a man. He was the antithesis of the young men back home who never boosted my self-confidence. Yves's knowledge of French politics, good wine, food and literature held me enthralled. He had such an enquiring mind and such a passion for life.

I lived and breathed the film in my every waking hour. Each day I awoke at six a.m. to write my stories, thin though they were, since how much could I say about two actors on a train? I then showered, dressed and hit the set. There truly were not enough hours in the day. I was always behind with my stories and receiving telegrams from London: "Unless receive stories by return you are fired." I needed to be on set all day to get the material for my meagre tales, and I had to keep myself as fresh as possible for my nights out with journalists and actors; then there was the fact that I was the girlfriend of the leading actor.

When the light had gone, I would return to the hotel, frozen to the bone, where only a hot bath could thaw me out. Pink as a freshly boiled lobster, I would collapse on my bed for a quick forty winks, make my notes for my releases to be written the following morning, and dress to go out to dinner. Most nights, Yves and I would dine at the best restaurants in this town renowned for its good and highly fattening food. I was growing visibly with the volume of delicacies placed before me and becoming more and more spoilt by the minute. Often we were joined by Anouk Aimée and her husband, Pierre Barouh, whom she had met when he wrote the music for her film *Un Homme et Une Femme*. Those evenings were fraught, because their marriage was taking a sharp decline and we were expected to perform like seals while they sat like thunderclouds, oblivious to the advantages of their privileged lives. I was mesmerised by the stark contrast of Anouk's ethereal beauty and her fingernails, which were bitten right down to the quick.

On the few evenings I had to myself, there was often a knock on my door just before I was about to go to sleep. It was always Pierre, on his way back from walking their dogs; he wanted a sympathetic ear. I never begrudged him the time as it was good to feel needed, but it was hard to deal with the atmosphere of gloom which filled my room, accompanied by the smell of warm dogs and manliness; Pierre did not use deodorant. To compound the situation, my windows were tightly sealed and the double glazing guaranteed a night of rich odours.

The entire experience of working on *Un Soir, Un Train* was like a dream which started when I landed in Belgium and ended as abruptly when I returned to England. Arriving home was a dreadful shock. I had worked extremely hard, but I had been spoilt beyond my wildest dreams, and the experience would stay with me all of my life.

I jump ahead to the summer of 2006. I was staying for a couple of nights with a friend in Paris, whose apartment was at the north end of

Père Lachaise cemetery. It was dusk when on a whim I decided to take a short stroll before the gates closed. The cemetery was vast and I felt I could easily get lost, then darkness would set in and I would be locked in for the night. However, I timidly ventured on, shrouded in the gloomy atmosphere of awesome tombstones. Quite abruptly I came upon a well-tended grave surrounded by fresh flowers. I read the inscription: Simone Signoret, and underneath, Yves Montand. Closing my eyes, I travelled back in time. I was full of laughter, thirsty for knowledge with the special naive enthusiasm of youth that can never be recaptured. I was carefree, immortal. Now, in the half light, tears silently coursed down my face. Had I made the most of the years between, had I become wiser, what did the future hold? (Little did I know then that my darling father would pass away a few weeks later.) A young Swedish couple stood next to me, one explaining to the other that these had been film stars; I was glad that at least one of them knew. Yves and Simone buried together, their love binding them together in life and in eternity. I remembered the brief note I still had from Yves in his tiny handwriting: *"J'ai frappé à ta porte à 8h moins 10 tu n'étais pas là – J'ai dîné tout seul – ensuite avec Anouk à demain – Yves"*. Now there would be no *demain*.

When I finished work on *Un Soir, Un Train*, the thought of cooking, shopping and standing over the kitchen sink appalled me, so the only escape would be to place myself on another film location before reality caught up with me. Soon I was flying to Vienna to work as a publicist on a film called *A Walk with Love and Death*, about a young couple who fall in love during the time of the student uprisings in France during the Middle Ages. Now, several hundred years later, we were being forced to move to Vienna because the well-documented student uprisings in Paris were in full swing in the summer of 1968.

For four months we were based in the Bristol Hotel, on the Königstrasse near the famous Sacher Hotel. From there, we travelled daily into the countryside to various castles. We were in a strange time warp, working six days a week, partying on the Saturday night, and sleeping most of the seventh. Everyone, apart from the crew, was dressed in medieval costume, from peasant girls to men in armour. The changing times were on hold as far as we were concerned until one day when we were shaken out of our world of make-believe. As we swept along the autobahn in luxurious

John Huston coaxing a horse into position for the camera

John Huston discusses a scene with
production designer Steven Grimes (left)
and first assistant Dick Overstreet

limousines, streams of refugees surged past us, weighed down by all the possessions they could carry. Their pleading eyes haunt me to this day. These were people just like us, the only difference being the roll of the dice, the fact they had been born in Czechoslovakia and the Soviets had just invaded their homeland.

A few people on the film panicked. What if the Soviets marched into Vienna? The majority stood firm, and we had the best director to lead us on. John Huston was renowned for his spectacular lack of fear. His loyal team had worked with him through thick and thin on many productions. Most were a generation older than me, a strong unit of professionals with decades of experience in the movie business. Gladys Hill, whom John nicknamed Glades, was the longest-standing member of the crew. She had started work in the film industry as a dialogue coach in 1946 and progressed to become a co-writer. But she was much more than a personal assistant/co-writer. She was at Huston's side through all his films, marriages and even holidays. I felt like a lamb among lions; Huston was the Lion King, and these were his courtiers.

John Huston studies the script while the producer, Carter DeHaven, feels the strain of being in charge of the purse strings. His wife sits behind him and the make-up man, Neville Smallwood, enjoys the opportunity to take a photograph.

My first encounter with Huston was on set when he ambled up to me to ask if I had read *Ulysses* by James Joyce. I replied that I had not. "Call yourself a writer?" he countered. "No," I retorted, "I call myself a publicist." He laughed and wandered off. The next day, John came up to me again. "Well, have you read *Ulysses*?" "Of course," I joked, my heart thumping and my cheeks burning with shyness. It worked, and from then on there was no more baiting of the new girl. Every night I would sit next to "The King" at dinner, listening to his hair-raising escapades, but he never bored us. He loved to talk and I loved to listen. I could never comprehend the quixotic nature of this man who appeared to share my love of animals and yet had made killing an elephant his bargaining point when he agreed to direct *The African Queen*.

The leading roles in *A Walk with Love and Death* were taken by two fledglings basking in the shadow of their famous fathers, Anjelica Huston and Assaf Dayan. Huston's daughter, nicknamed "Anjel" (the emphasis on the "el", as in "gazelle"), was a mere seventeen years old, fresh out of school and with just a little experience at modelling. She was truly thrown in at the deep end and what was more, she could not have felt

Anjelica Huston, Vienna, October 1968

more uncomfortable in the part. I could see that she was going to be a late bloomer but that she would grow into a beauty, with her classic features, large intense dark eyes, Roman nose and a bone structure that would stand up to the test of time. She loathed the hairstyle the role called for, swept off her face as she was sporting a heavy fringe. She thought little of the script and was not attracted to her leading man. Her eyes were on another actor, Anthony Corlan, who played the third lead and became known as Anthony Higgins later in his career. She complained that Assaf's armpits smelt when she had to do a love scene. The chemistry was at zero. This was a truly tough start for a budding actress.

Assaf Dayan claimed to be a poet, but he would not prove to be of the requisite Huston calibre, as acting was not in his genes. After the film he became known as Assi, as Huston had been calling him. His father, Moshe Dayan, was Israel's Defence Minister and a hero of the recent Six-Day War. Despite their differences, these children of famous fathers made a perfect combination on which to hinge our publicity campaign. The press took the bait and writers came in their droves from all over the world, along with photographers from a wide variety of magazines and newspapers.

Terry O'Neill, who carried an exceptionally long lens like a phallic symbol, was competition to any of the actors as a heart-throb. His talent and good looks sent him skyrocketing to the top of his profession. I bumped into Terry several times over the years; he was always warm and friendly, seemingly ageless, always with a beautiful woman, a veritable Peter Pan. Then there was Eve Arnold, a petite figure with beautiful thick grey hair caught up in a chignon. She moved quietly and unobtrusively around the set, her three Nikon cameras draped around her neck and shoulders, one for colour shots, one for transparencies and one for black and white. Everyone loved Eve, with her deep husky voice and bright twinkling eyes; a close personal friend of Huston, she quietly ruled the roost.

Somehow, Eve got to see some of my pictures. She was not over complimentary, that was not her style, but she told me in a matter-of-fact way that I should become a photographer. This was the turning point in my life that would take me from having a job to having a career, indeed a vocation. This greatest of female photographers had noticed my potential, and that was all it took to change the course of my life. Over the next thirty years and more, I often met up with Eve, and her constant encouragement and valid criticism spurred me on. She was my muse and my mentor, almost to her passing.

Huston, too, was a great influence. His style was unique. During the day he wore an olive-green corduroy suit with an inlaid belt of the same fabric and a stylish cream baker-boy-style hat. In the evenings, when smart, he wore a dark bottle-green velvet suit which enhanced his tall, lean body and distinctive craggy face. Although he was no spring chicken, his life was as full as ever. What an example he was of age just being a number. He still had the energy of a young man, though backed up by years of experience.

Huston had no scruples about propositioning women and girls: they were all part of the entertainment for him. I was no exception, and the day soon came when I became the focus of his attention, followed by an invitation to dine at his hotel, a converted castle called Schloss Laudon. I declined politely, as it went without saying that dinner was not all that was on offer. Huston took my reply in good heart. After all, he had not been short of wives and mistresses, nor would he be in the future. To him, seduction was a game. The following day he walked over to me on the set, grinning broadly, and said for all to hear, "Man's got a right to try." There were other times when the offer was repeated, but I never saw it as sexual harassment. Huston was a charismatic man, and I took it as a compliment. This was not the day and age for suing every man who found a girl attractive.

Each day I awoke at six a.m., wrote my press releases in bed, then ordered my breakfast while I dressed. Then I would pop into the publicity office to hand over my draft stories to my secretary and plan the rest of the day. There was nearly always a journalist or a photographer to escort to the set, and often the locations were far from Vienna. My chauffeur would appear just as the phones went mad with questions from the visiting press and people wishing to book a visit.

The story of *A Walk with Love and Death* was hardly a bundle of laughs, and some days were grim. On one occasion, Anjelica and Assaf had to convey to us that they were almost starving to death, so when they found a dead blackbird Huston insisted Anjelica should attempt to eat it.

The crew of *A Walk with Love and Death* arranges body parts for a quartering scene

Anjelica Huston, on the set of *A Walk with Love and Death*, October 1968

The bird was real, and she was literally retching as she forced herself to sink her teeth into the feathered corpse. When several of the horses were anaesthetised for an after-battle scene, one or two of the older ones never woke up. I found this hard to take and left the set in tears, but I was told in no uncertain terms not to show my disapproval to Huston. Another day, a group of knights on horseback came charging down a hill dressed in real armour, which weighed a ton. One of them fell to the ground, thrashing around like an upturned beetle, then stopped dead. And dead he was. Fiction was blending into fact in front of our very eyes.

Night shooting could also be a miserable, cold experience. We were spared one wet exterior of a castle by Huston simply rolling out of his seat in a haze of alcohol and exhaustion; it was a wrap.

The brighter days were magical, and on one occasion when we were shooting a scene with a group of gypsies, an enchanting honey bear was added. Why should a honey bear appear in that scene? Just because Huston wanted to have a bear on the set; I was delighted that he had the power to indulge his whims.

Finally, after four months, the film finished shooting and we said our farewells at the airport. I felt tearful and utterly lost, for these people had become my extended family.

Now, without a break, I was flying off to Tunisia to join a new film set, as a unit publicist for *Justine*. The film was an adaptation of *The*

Assaf Dayan walking away into the sunset; the wild boar just happened to walk across the take. "Cut and print," said John Huston, with extra conviction.

Alexandria Quartet by Lawrence Durrell. I had been looking forward to this assignment, as I had read and enjoyed the quartet enormously and I was to work with Anouk Aimée again. But most exciting of all was the prospect of meeting Dirk Bogarde, whom I had worshipped since my teens. I needed no reminder of his career. I had seen all his films and had kept a cuttings book largely devoted to Dirk, with pictures of him as Sydney Carton in *A Tale of Two Cities* and "the far, far better thing that I do" to shots of him in lesser films like *The Spanish Gardener*. These had all been carefully cut out and stuck in, with brown stains on the pages where the glue had seeped through.

I arrived at the Tunis Hilton in November and joined the unit publicist, Jack Hirshberg, and his secretary Spooky Stevens (apparently her father had taken one look at her when she was born and said "Spooky", and the name stuck for life). The film was already under way, so I was at a huge disadvantage, and in blissful ignorance of the fact that Jack and Spooky had worked together on several films, and she considered me a waste of space. Spooky terrified me; she was super-efficient and typed like a road-runner, but luckily we decided that since we were stuck with each other we might as well get along. Spooky, who is still one of the most popular unit publicists in Hollywood, has remained one of my dearest friends to this day.

The next problem, which was far bigger, was that the stars were unhappy. The script was an unsuccessful attempt to compress Lawrence Durrell's famous quartet into one film and, just to compound the miserable situation, the cast was unhappy with the director, Joe Strick. He was subsequently fired and replaced by the legendary George Cukor. Cukor filmed his scenes in a studio in Hollywood and cut in Strick's location shots or back-projected them from time to time. The end result was a disaster, with sex scenes including incest, homosexuality and nymphomania mixed in with political intrigues involving Coptic Christians and gunrunners in Palestine.

As I had never visited a desert, I was boiling over with excitement on my first day as I was driven to the set. The scorched brown landscape, with hardly a tree in sight apart from the odd date palm, was a far cry from Vienna. I felt like a pupil arriving late at a new school as I met the stars for the first time. To compound my unease, I already had a mild case of dysentery. All too soon I was desperately asking directions to the Portaloos and queuing anxiously in the heat. I did not need to turn around to recognise the distinctive voice behind me. "I wonder whose need is greater, yours or mine?" There was Dirk, my childhood icon. We were not alone with our problem; everyone seemed to be in the same sorry state. We all had to visit a local doctor recommended by the production manager. I took a taxi down the dusty hillside from the Hilton into Tunis, where I was dropped off on a narrow side street. I felt apprehensive as I asked the taxi to wait. The doctor greeted me and told me to lie on a couch, then asked me to remove my top. The red lights were flashing but before I could leap up, he was on top of me. I struggled with all my remaining strength, managed to break free and dashed from the building. The taxi was nowhere to be seen, so I walked unsteadily through the hot, dusty streets all the way back to the hotel. At lunch the next day with the director and the producer, I decided to denounce the doctor. Consequently, he was struck off the register, and it transpired that all the girls on the production had suffered the same fate but had been too embarrassed to complain.

Dirk, who played Pursewarden, managed to make something of his lines, in spite of his dislike of Joe Strick. He could truly make a silk purse out of a sow's ear. However, from a publicist's point of view, Dirk was not easy to work with: it was well documented how much he loathed the British press. With this foreknowledge, I was reluctant to ask him to agree to be interviewed by *Woman's Own*. So I approached him with great trepidation, but he agreed he would do the interview provided he was not asked about his private life. I was delighted, and I wined and dined the journalist at 20th Century Fox's expense and made her agree to the terms.

When I hit the set later that day, a huge crowd scene was being filmed. Suddenly, out of the mêlée, Dirk was shouting at me for all to hear, "Why didn't you tell her what we agreed?" He was too incandescent with rage to listen to my protestations, and I was too distraught to retaliate and fled without a backward glance. My driver took me back to the hotel, tears and mascara cascading down my face. The mild dysentery had left me fragile, and all I wanted to do was to hide away in my hotel. I grabbed my key, rushed to my room bathed in self-pity, and continued to weep and wallow. Once my eyes were as swollen as a couple of pink puff-balls, the telephone rang. I answered, knowing that I could not be seen, and managed a controlled, "Hello?" It was Dirk, mortified at what he had done. "I know that you are aware of how unhappy I am with the film but I know that is no excuse. Soon after you left the set, I found out that you had kept your word. How can I make it up to you?" He sounded so desperate to make amends, which completely threw me as I was not used to film stars ever apologising, let alone adding, "Would you meet me and Forwood [Tony, Dirk's good friend] for dinner in the mountains? Shall I send a car?" I needed no persuasion. My hero since school days and me still in my twenties. It all seemed surreal. I had no hope of finding any witch hazel to fix my resemblance to a pinfish, so I used ice from the minibar wrapped in a cloth. With clever make-up and evening light, I would pass for normal.

The car wriggled around hairpin bends up into the hills to a village called Sidi Bou Said, where we stopped outside a picturesque, traditional-looking restaurant, with whitewashed walls festooned with bougainvillea and trees draped with pretty fairy lights. Dirk was standing in the doorway, silhouetted against the light, hands in pockets, waiting to greet me. I was so touched by this gesture that I did not have time to wonder whether the three of us would have anything to talk about. Normally on location there is the film to discuss, but in this case that hardly seemed appropriate. As far as I knew, we had nothing in common except that I had seen most of Dirk's films, but there is something dreadful about meeting someone that well-known and listing all the performances you have enjoyed and why; they have heard it all a million times. As luck would have it, we found plenty of mutual territory: my father had tramped the same boards as Dirk at the theatre in Richmond. Also, I found out that, like me, Dirk had been brought up in Hampstead, so the conversation flowed. He chose for me from the menu, gave me advice on practically everything I could think of, and even advised me on what suited me in dress and hair-length. I was enchanted. We then discovered so many similarities in our childhoods, holidays in the English countryside, entertaining ourselves with nature and climbing trees. Both our mothers had given up the theatre for our fathers' sakes. Both were adored and wonderfully loving mothers and were quite capable of intermittently throwing huge theatrical wobblies. I had to remind myself that Dirk was a huge film star and matinee idol, but as we spoke, I could tell that was not his true self. I felt back on course to complete my month on *Justine* with far more achieved than just my job.

I have very few moments of remorse in my life, but one concerns a lovely friend I had on *A Walk with Love and Death*. Wolfgang Glattes, the second assistant producer, followed me to Tunisia and stayed with me for a couple of days until he too went on to another movie. I had to put my work first. He came sharply back into my orbit when I saw his name on the big screen: he had gone on to produce *All That Jazz*. I wondered whether we would have anything in common now or whether that film was all we shared.

So my days as a unit publicist came to an end. 20th Century Fox amalgamated with Rank, and Fred Hift set up on his own above the Carlton cinema in London. I freelanced for him on the distribution side, but I no longer wanted to live out of a suitcase for months on end on location.

CHAPTER 3

HOME STUDIO

"Some men see things as they are and say, why? I dream things that never were and say, why not?" – Robert F. Kennedy

Working as a unit publicist had been a glorious experience, but the gypsy lifestyle of living out of a suitcase was not what I would choose. I wanted to nest and to create a garden and a place called home, and above all, with Eve Arnold's encouragement, I now had a burning desire to be a photographer. My choice of Notting Hill was not a shot in the dark. Several people I had met while I was working in the film industry had already had the imagination and foresight to buy into this depressed and bohemian area of London. I too could see that behind the peeling paint and crumbling cornices lurked the remains of a genteel era. After months of searching I knew, before I opened the front door on that grey November day, that this was to be my home, and I have lived here ever since. As I entered the long, narrow hallway, my nostrils were singed by a pot-pourri of damp body odour and bad fat; not for the faint hearted. Built in 1840, this early Victorian house is beautiful. It is close to large expanses of greenery and is on the right side of the city for easy access to the country. My instincts were right; the area is now one of the most salubrious in London and yet still attracts an eclectic mix of residents.

My house was one of the many boarding houses owned by exploitive landlords who housed the first generation of Black people, known as the Windrush Generation, who came over from the Caribbean in the 50s. The well-proportioned rooms with their large windows were divided up into dreary bedsits, every one crammed with cookers and sinks. Large, ugly meters for gas and electricity filled the narrow hallway and a few bricks had been added on top of the back porch to make an apology for a bathroom. Rising damp and mushrooms decorated the rooms on the first floor. The roof was like a sieve, and the basement floorboards were rotten and smelly. The house was so dangerously in disrepair that the survey stated that it would not stand for twenty years and advised me against buying it; young and foolhardy, or brave, I went ahead.

It took nine months to convert the house from a wreck into the home I had visualised, the exact gestation period for a baby. I donned my workman's overalls and toiled from dawn to dusk with my team of Polish builders. This was a far cry from first-class hotels and the glamour of feature films, but I could not have been happier. In the morning the air would be filled with the smell of freshly brewed coffee, one of the young builders would play love songs on his harmonica, and the foreman, Mr Kalinowski, with his fine moustache, would arrive with cries of, "Where's my young lady?" before he gave me instructions on decorating. These men knew how to enjoy themselves, and working on my house was paradise compared to the time Mr Kalinowski had spent in a gulag in Siberia. In the evenings, we finished the day with a shot of neat vodka. I slept the sleep of the gods and my body was well-toned from painting walls, doors, windows and ceilings.

Finally, my home was restored to its former glory; the foundations were strengthened, bathrooms added, central heating installed, damp coursing and new roof all completed. I warmed my home with friends, filled my spare room with a tenant/friend, and my sister presented me with Emily. This beautiful, silky coated Cavalier King Charles spaniel became my shadow for sixteen years. Black and tan, with eyes the size of saucers and rimmed with charcoal, she was a challenge to any actress and feisty into the bargain. Emily's life was rich and varied, as her popularity ensured that she was invited everywhere, from private homes to film sets; the theatre and even restaurants were included, since an ancient law passed in the time of King Charles II had decreed that these were royal dogs and had the right to enter all public places.

The house had devoured all the money I had saved, so I scratched a living writing articles for magazines along with freelance publicity work on films at the distribution level. Even so, the money never seemed to come in fast enough, so I was always open to offers to make ends meet, and soon a big one came about. Based on my experience working as a publicist for American companies, I was asked to join a team obtaining a general distribution deal for a privately funded movie, *Chappaqua*; Conrad Rooks, heir to the Avon cosmetics fortune, had put his heart and soul, and money, into this semi-autobiographical film about drug addiction. The story was based on his painful journey through the Stygian fields of alcohol and drugs from his teens until his eventual sleep cure in Switzerland. The graphic depiction of the raw pain of withdrawal made it difficult to find a distributor, and our task was to invite top psychiatrists, pop stars, ex-drug addicts and anyone else who could further our cause to screenings. Out of the woodwork crept personalities who had survived their narcotic years, some of whom I found incomprehensible. William S. Burroughs, who had written the book *Naked Lunch*, totally floored me. Rumour had it that, under the influence of LSD, he thought he was William Tell and, having placed an apple on his wife's head, promptly missed it and killed her instead with the arrow. His altered state left me feeling as though I was on another planet.

Conrad, on the other hand, enthralled me with his magnetism, powerful energy and brilliant, quick mind. His piercing blue eyes could make me freeze like a hare in headlights, and his blond hair shone like silk; he generally glowed with good health. I was relieved when we moved on from the world of drugs to work on a distribution plan for Conrad's second film, *Siddhartha*, based on the novel by Hermann Hesse about a Brahmin monk's search for a meaningful life. The film had won a Silver Lion at the Venice Film Festival but it too was looking for distribution.

Conrad and his films and all who came with them soon took over my life; my portraits seemed pedestrian by comparison. One night I dined with the well-known fashion and beauty photographer Tony Moussoulides and his wife. I was thirsting for more technical knowledge of photography, so when Tony suggested that I could go to his studio the following day to assist him I jumped at the opportunity. I was thrilled that such a distinguished photographer was to share some of his precious secrets. I knocked on the door of a smart mews house just off Berkeley Square. Tony was expecting me and said he had this wonderful idea. "I can see it now," he enthused, sitting me in front of a mirror as his arms came around my waist and I froze. "Andy Warhol's head placed between your bosoms." His hands reached up and grasped the objects of his desire. I leapt to my feet, cheeks burning with indignation. I reminded him that I had come to learn about photography. With a skin as thick as rhinoceros hide, he grabbed at my shirt and I struggled, broke free and dashed for the door with tears streaming down my cheeks as I stumbled into the belly of Belgravia.

No wonder I found it so hard to believe I could have a successful career as a photographer when the men in the profession were still so determined to guard their territory. At that time, the only famous female photographers I could think of were American or German.

Conrad and I parted soon afterwards, partly because he was such a "free spirit". On a trip to England several years later, he told me I was the only girl he had really loved, and I pointed out that maybe that was because I was the only one who had broken up with him. Some things are just not meant to be forever. I still treasure a picture of Conrad on location in India with three Nikons around his neck; he had strengthened my resolve to carve out a career as a photographer. So another chapter came to an end, and I threw myself wholeheartedly into my work as a photographer; I would no longer be swayed off course.

Having worked with Brian Deacon on the film *The Triple Echo*, I decided that he would make an excellent subject for *19* magazine. So I took a portrait to accompany a piece, and in this way I became a published photographer. Brian then married Rula Lenska, which not only opened the door to photographing weddings, but after the appearance of my photo in *Spotlight*, the actors' directory, their friends from drama school appeared one after another. Soon the actors flowed in like a stream gathering momentum until it became a fast-running river. Now Emily had a full-time job: not for her the life of the lonely dog left at home. She flew up and down the stairs greeting clients, posing on their knees to put them at ease or simply politely observing. Pedigree Chum dog food at the end of the day was always well earned.

I had no studio at this time, so each time I took a portrait I had to clear my sitting room, which entailed taking a picture off the wall each day and putting it back at night. The sitters would then pose on my desktop with their heads obscuring the hook. It was Googie Withers, a grande dame of the theatre, who prompted me to do something dramatic about the situation. It seemed suddenly absurd to see such a grand lady perched on my desk, so it finally dawned on me that I had a career as a photographer and therefore should have a studio. So the builders returned to remove the roof and build a proper studio for me. The building work was supposed to take three months, but took a year. The builder ran out of money and left me high and dry while he acquired further contracts and advances. He was running, and the faster he ran the more he fell into debt. I remained locked inside the scaffolding, which he could not pay for and the company refused to take down. The tarpaulin over the unfinished roof flapped in high winds, making the house feel like a sailing boat in a storm. The bedroom ceilings leaked, and we had to put out buckets to catch the drops. Finally, one morning as as I was getting up, a large section of the ceiling collapsed onto my bed. Miraculously the builders finished the job, and when my studio in the sky was finally complete, I was like a horse that had been shut in its stable and was now galloping full speed ahead. Actors, models, writers, singers, dancers, businessmen and children all visited the studio. Like a snowball expanding as it rolls down a hill, my portraits of professional people were gathering momentum. Out of the thousands who have come, just a relative smattering have gone on to great fame.

Joyce Redman

When Joyce Redman called to say that her agent had recommended me to take her portrait, my mind flashed back to meeting her for the first time at my parents' house in Hampstead; since then we had lost touch. Decades later, I had the chance to tell her of my hero worship as a shy teenager. "Oh, don't be ridiculous, my darling," she retorted in dulcet tones, "you're as clever as paint." I described every detail of her dress; she no longer thought me "ridiculous". This time our friendship continued, and I spent happy times at Joyce's enchanting fifteenth-century country home in the heart of Kent, the county famous for hops and oast houses. I begged her to tell me stories of her fascinating life as we sat cosily by the warmth of the AGA cooker. She talked of working with Larry (Olivier) and Johnnie (Gielgud) and how, around the time of the release of the film *A Streetcar Named Desire*, Marlon Brando had appeared at her dressing-room door to ask her out to dinner. She had declined, as she had agreed to dine with the stage manager. "Coward," I said playfully. "You know that you would have had to sleep with him." We both laughed as we knew that with his reputation, dinner would be inclusive.

Joyce Redman, actress, 1980s

As I came down to breakfast the morning after this conversation, I announced, tongue in cheek, "I feel as if I am in a dream." "Why is that, darling?" Joyce called from the kitchen. "I just cannot believe that I am staying with a friend who has not only met Brando in his prime but been asked out to dinner by him." "Oh, don't be ridiculous," she giggled, beaming as she stirred the porridge.

My bedside table was always brimming with fresh flowers and books. I found a beautiful piece of writing by a monk who said that people needed constant noise throughout their lives because they were afraid of death. I remembered this when Joyce passed away peacefully at ninety-six; she was a strong believer in the spiritual world.

Dame Eileen Atkins

One of the earliest of these portraits was Dame Eileen Atkins. Eileen was memorable. Her personality was outstanding, so vibrant, so full of energy, and the forces seemed to agree: first the studio was bathed in brilliant sunshine, only to be dramatically plunged into darkness by a massive thunderstorm. Forked lightning sliced through the sky like Neptune's trident followed by claps of thunder which shook the house to its very foundations. We waited patiently, watching in awe at the power of nature. Then, just as suddenly as it had struck, the storm rolled on, the sky cleared and the sun broke through, shafting across the studio floor. Eileen's eyes sparkled, and the negative ions released from the storm lifted our spirits and I snapped on. The following evening, I took the contact sheets around to Eileen's home in Regent's Park, where she generously popped a bottle of champagne to celebrate the shoot. The sparkle in the champagne matched the mood of the pictures.

It is remarkable to see in a society obsessed by ageing that it is possible for any actress over the age of forty to achieve overnight stardom. Admittedly, Eileen had been well-known for many years, but the huge success of the television series *Cranford* propelled her into another league. And so she joined the elite group of Dames. I am always amused to see Eileen being so often typecast as a tight-lipped spinster, when in reality she has always been the exact opposite. For many years until his death, she was happily married to a younger thespian, Bill Shepherd. It is her wit, energy and sheer talent as an actress that is the key to her attraction.

Dame Eileen Atkins, actress, 1980s

Billy Connolly

Normally my clients telephone me, we have a chat, and if they like the sound of my voice, book an appointment. However, once in a blue moon, an agent or manager will make the decision and the client arrives cold. Billy Connolly, darling of the world of comedy, had been booked in. The stories abounded of his excesses. I loved his work, but I wondered about the man himself. I expected a large, grubby, scruffy bear of a man with doubtful manners. Now he was standing in my narrow hallway, where we were like two animals checking each other out. The boy from the tenements of Glasgow was confronting the girl whose mother came from Edinburgh: two Scots with opposite backgrounds and different accents. Before me stood a tall, attractive, red-blooded male bursting with energy and beautifully groomed. His hair and beard shone with good health and a well-cut jacket completed the look; only a pair of zebra-striped trousers hinted at his crazy side. Where was the rebel, the child sexually abused by his father, deserted by his mother and cruelly beaten by his aunt? Where

Billy Connolly, comedian, 1980s

was the anger and the scars that I had expected to be lurking behind the humour? The man was transformed, reborn, a chrysalis that had transformed into a butterfly. "My agent says you're very good," barked Billy in a challenging tone. "Oh! Please don't frighten me!" I whimpered. We both burst out laughing. The ice was broken, and the fun began.

The studio was like a playground with Billy coming up with all the ideas; he was fully charged, and he gave it his best and a bit more. It is rare that I witness that amount of unleashed energy, so I clicked and clicked, rolls of 35mm and 120 film for posters. Then, more 35mm as tea in the sitting room extended to more tea and Billy commented he was really enjoying himself. His wife Pamela Stevenson remarks in her book that Billy's humour is not dependent on a series of funny stories, but is more a way of seeing the humour in everyday life. How lucky for me to be experiencing my private show, but it was Billy's sensitivity, awareness and above all kindness that warmed my heart. A few weeks later, my picture of Billy was plastered on the side of every red London bus, reminding me of a magic time.

Rachel Weisz

It was the bleakest time of the year when winter seems to go on forever, and one murky February day, when the light was so poor that you could hardly distinguish day from night, Rachel Weisz came to my studio on the recommendation of her agent. She was tired, deeply tired in a very particular way that actors become when rehearsing for a show where they are giving their all. This was a testing role in Tennessee Williams's *Suddenly Last Summer*, in which her performance would be compared to others. I warmed to Rachel immediately and felt a great desire to do my best; she was interesting and interested, none of the all-too-common ego thinly veiling a sea of insecurity. Her beauty is mentioned all the time, but what is not discussed is her first-class brain. Rachel gained a degree at Cambridge University before becoming an actress. All too often in this celebrity-mad world, the public are not made aware of the fact that many big stars are not just pretty faces; retaining their status requires talent, brains, discipline and a strong personality. Rachel photographed like a dream. Her face had the plumpness of youth, which could hide any

Rachel Weisz, actress, 1999

degree of fatigue, and she had learnt the discipline of posing with confidence and a sense of calm, thereby making it easier for me to do my best.

Two days later, Rachel was back again after rehearsals. I offered her a drink and we retired to my sitting room, where she kicked off her shoes and curled up in one of the large armchairs, happy to relax after a tough day at the rehearsal room. She was appreciative of the results on the contact sheets, which she poured over with a magnifying mirror, circling her favourites to be enlarged. We giggled at the fact that her bra had been ill fitting and that I had been careful to disguise the fact. There is something so much more endearing and attractive about someone who is not always "perfect". But the ability to "scrub up well" when on duty as an actress is what separates the good from the stars. Years later, I was awed by a picture of Rachel in a Vera Wang dress walking down the red carpet to receive an Oscar, nothing short of perfection. There was the young girl who I had photographed with all before her, now a beautiful woman with a husband and child. I love the way a good photograph can capture a moment for eternity. Rachel's priorities are not shallow, and every time I read about her success, I feel a warm glow.

Lady Valerie and Claudia Solti

Within days of the death of the internationally famous conductor Sir Georg Solti, I took pictures of Claudia, his daughter, and Lady Valerie, his wife. Claudia had booked the appointment before his sudden death, but as she already knew me, she felt comfortable keeping the date and added that Papa would have wanted her to come. She radiated beauty in that state of grief before the reality kicks in, dewy eyed and in an other-worldly state. At this most challenging time for these two important people in Sir Georg's life, I felt honoured that Claudia and her mother trusted me to be of help in some small way.

Valerie met Sir Georg Solti while working for the BBC on the arts programme *Town and Around*, and they married three years later. She worked tirelessly in the arts and set up the Solti Foundation with her daughters to assist young professional musicians. She was also founder and chairperson of the Georg Solti Accademia and patron of the World Orchestra for Peace along with other charities. Valerie passed away in 2021.

Claudia has remained a special friend. She is an ageless, timeless soul who would make her papa very proud of her in every area of her life, both privately and in her work. She was but a fledgling when I knew her

Claudia Solti, actress and director, 1997; Lady Valerie Solti, 1997

well in London, and now, over twenty years on, Claudia has "arrived", with a good partner, twins and a huge body of work, largely as a director of films and opera.

Hugh Grant

I had the good fortune to meet Hugh before either he or my area of London became absurdly fashionable. He first appeared on my doorstep in Notting Hill as yet another actor needing new portraits; even then he stood out from the crowd. Rather like having a nose for a good horse, I can often guess an embryonic winner in the thespian world. Yes, he had good looks, charm and self-confidence, but I had photographed a humongous number of men and women with these assets. Few had that extra secret ingredient: "Star Quality". In those days, Hugh was not easy to photograph, because like many good actors, he disliked the stills camera. Actors are trained to portray others and they have always had my sympathy as they sit in my studio simply being themselves. Still, I found Hugh attractive: public school and Oxford had moulded him into someone all too familiar to me, with self-deprecating wit, typically British. That was his ace card. In true Virgoan style, he had a ground plan: "I shall give acting a chance," he told me, "but I must give it a time limit and if by then I have not succeeded, I shall do something else." Not for Hugh the bit parts for the older actor in a cast of younger actors touring the provinces and regaling them with tales of his one moment of glory in his youth. He had his safety net in place, ready for the worst.

Soon after I took my first black-and-white portraits of Hugh, he came to a gathering of my friends around Christmas. Then our lives went in different directions as happens all too frequently in London and he disappeared out of my life.

I thought no more about Hugh until I went to see Roman Polanski's film *Bitter Moon*. I was so impressed by his performance that I decided, on a whim, to write to congratulate him. The telephone rang within days. He seemed delighted to get my letter and, yes, he would love to do some pictures; "A chance," he said, "to kill two birds with one stone, to see you and get new shots." It was the same lovely Hugh; witty, sensitive and self-deprecating, but I sensed much happier about catching up on news with a friend than being photographed. We talked about the extremes of an acting life; one moment working on a film set being treated like royalty and the next battling against the crowds in the supermarket. It was February, the lowest time of the year and, endearingly, Hugh had bothered little with his appearance as he was doing radio work and loving it. With two films back to back in the pipeline, he was basking in the cosy atmosphere of the municipal green walls and the teas in polystyrene cups. He thought that the film *Sirens* would be the better of the two; the other one, *Four Weddings and a Funeral*, well, he was not so sure.

For the shoot, I was so eager to please that I raided my friend Marc Stanes's bedroom and borrowed his best black shirt. As Hugh was having a bad hair day, I experimented using a jewellery spotlight on a black background to isolate the light so that I could not see the outline of his hair. I covered myself by taking some more shots lit conventionally in case the others failed. They looked great and I so hoped Hugh would feel the same.

When I called for feedback, Elizabeth Hurley answered (they were living together at the time). "You don't know me," I started, but before I could continue, she interrupted: "Oh, but I do. You have just taken some wonderful pictures of Hugh. We both love them." She was warm and friendly and added that she, too, was keen to do a shoot with me. She then begged my forgiveness for being brief but she had to get a cake that she was baking for her godchild out of the oven. So this was my first impression of the girl who soon after became a celebrity.

Four Weddings and a Funeral opened and Hugh was swept off his feet to stardom, attached to Liz Hurley's safety pins. My pictures flew off to the magazines and a couple made the cover and an inside story of the *Telegraph Colour Supplement*; my hunch had paid off. I was over the moon. I thought, as far as I was concerned, Hugh had been launched into space never to return. But he did land back with a bump by making a very human error of judgement that the press fell upon like jackals after a kill. He had quite literally been caught in the headlights by the police in Los Angeles with a lady of the night. When I read about Hugh's encounter, my heart told me to write a letter, as I was all too well aware of the impact of the press, for better or worse. So I wrote, and Hugh replied, again within days:

> Darling old friend Carole, I am a heel and a half for not thanking you earlier for your letter, especially one so full of support and love. Isn't it amazing what a difference those letters make? I know you know. The dust is settling a bit but I'm not sure my lungs will ever be quite clear again … Anyway, pit pat and I do hope you are raking it in. Lots and lots of love, Hugh. Xxx

Hugh handled the situation brilliantly, not shying away from journalists and television interviews. The storm passed, and from then on his career went from strength to strength.

Once again, Hugh flew out of my orbit. Then one day in 1999 he came back into my life, but not in a way that I relished. The streets of my London neighbourhood were buzzing with gossip; the film *Notting Hill* was on everyone's lips as it was being filmed all over the area, especially around my favourite stalls in the Portobello Road. I spent weeks dodging the unit; I always feel desperately shy bumping into people I know when they are filming, as I am all too aware that they are trying to work and I never know whether or not to speak to them. However, others did not

Hugh Grant, my home, Notting Hill, London, 1999

have such sensitivities and everywhere I went I heard people's opinions of "Hugh" and "Julia" as if they were their new best friends.

On the whole, the stall holders loved Hugh, which did not surprise me. Then one bright, freezing cold spring morning, I came round a corner dressed in a fleece clutching my shopping bag stuffed with vegetables and walked right into the scene. Horrified by my mistake, I tried to beat a hasty retreat, but Hugh had spotted me and greeted me warmly. Opposite us stood a sea of faces watching the filming, probably wondering why Hugh could be greeting this flustered bag lady who was surely not part of the scene. Covered in confusion, I retreated with my groceries as fast as I could after we had both come out with banal remarks.

I hoped that one day when Hugh's frenzied rom-com fame died down that we would meet again, but there has always been another surprise. He metamorphosed into a campaigner against press intrusion and went on to become a serial dad. Whatever next? Well, next came the rebirth into an actor of great gravitas. No more romantic roles, and the next shoe seemed to fit more comfortably. At last Hugh was a man *bien dans sa peau* (comfortable in his own skin). Now in his sixties, no longer the rom-com pretty boy, Hugh has matured and grown along with the quality of the scripts he has chosen. I longed for him to get an award for his performance as Jeremy Thorpe, but so much depended upon the competition that year. However, a Golden Globe award, a BAFTA and an Honorary César are no mean feats.

Ed Shearmur

While Ed Shearmur was my neighbour in Ledbury Road, he composed the music for *The Wings of the Dove*. Still only in his thirties and looking younger, he had known that his destiny was in music since the age of seven, when he sang in the choir at Westminster Cathedral. This was followed by Eton, the Royal College of Music and a scholarship to Cambridge. None of this had gone to his head and he was *bien dans son peau*. So he appreciated my talent as a photographer and gave me all the time in the world to create a special picture. The lighting took time, but there was no need for him to boost his ego by saying how busy he was; the proof was already all too evident. Since then, commissions for film scores have rolled in, along with awards.

Ed Shearmur, composer, 1997

Craig McLachlan

I love the picture I took of Craig McLachlan during his time in London starring in the first cast of *Grease*. It truly expresses his naughty, sexy personality: a shameless flirt. He wooed ladies of all ages with his sensual good looks and talent as a singer, actor and dancer. When I took him to

Craig McLachlan, actor, 1993

the Halcyon Hotel in Kensington for lunch, his wicked Aussie sense of humour had me blushing to the roots. There he stripped down to a sleeveless T-shirt, leaving the businessmen at lunch in their city suits choking over their soup.

Kim Cattrall

There are days when nothing goes according to plan, and others when everything goes as smooth as clockwork. I was set to go to photograph Kim Cattrall, who would later be known for her work in *Sex and the City*, at her London hotel. I was not enthusiastic, as I had preferred the idea of Kim coming to my studio.

I arrived hot and bothered with my cameras and lighting equipment. The staff were polite and helpful, and I checked out a few suitable locations and set up lights ready to go. I waited and waited until I felt weary with waiting and just as I had almost given up Kim appeared, tired, tetchy and full of apologies that the filming had gone on so long. On days like this,

Kim Cattrall, actress, 1990s

all you can do is cut your losses and rearrange, and there is absolutely no point in battling against fate. So, a few days later, Kim was on my doorstep. I tried to take a few pictures in the garden because the weather was perfect, but Kim became nervous that with a slight wind she might get dust in her eyes. I did find her frisky and not easy, but as so often in these cases, I was 100 per cent sympathetic. In truth, I would feel just the same if I was working all hours of the night and day and having to do publicity shots in my time off. I think that because my understanding was not sycophantic, we found ourselves relaxed in my studio, where the light was stunning and everyone was happy.

Mike Myers

One of my early subjects was the very young and unknown Mike Myers. It was not until years later when he had become a huge star that I realised I had these pictures tucked away in my archives.

Neil Mullarkey and Mike Myers, comedians, 1980s

In my early photograph, Mike poses with Neil Mullarkey, with whom he had formed a comedy duet. Mike had just come over from Canada and was touring England for the first time, his fame all before him. Best known for his performances in *Shrek* and *Austin Powers*, his meteoric rise to fame has gained him a prestigious star on the Hollywood Walk of Fame, along with several awards.

Jason Connery

My picture of Jason Connery, son of Sean ("the name's Bond, James Bond") and Diane Cilento is a favourite of mine, capturing an actor devoid of any ego. He could so easily have been impossible, given his father's fame. I had enjoyed working with him on the production of *The First Olympics: Athens 1896* and was flattered when he agreed to this

Jason Connery, actor/chef, 1990s

portrait session in the studio and a spread for *OK!* magazine. Jason has faced the hard task of growing up in the shadow of a megastar, but he has found his own niche in life, working constantly as an actor.

Alistair McGowan

Many of the great British actors of our time return to me for pictures like hardy perennials, including Alistair McGowan, who is a great favourite of mine. These regulars are often the ones who are comfortable with themselves, so when they find a photographer who works for them they remain loyal. I have always loved Alistair's impersonations, but these are hard to sustain for too long at a time, so he has continued to enjoy a successful all-round acting career and has proved himself to possess a multifaceted talent.

Alistair McGowan, actor, 2000s

Toyah Willcox, musician and actress, 1980s

Lindsay Duncan, actress, 1980s

Anita Dobson, actress, 1980s

Celia Imrie, actress, 1990s

Jodhi May, actress, 1990s

Prunella Scales, actress, best known for her performance in *Fawlty Towers* with John Cleese, 2005

Ray Stevenson

Ray Stevenson came to acting late, having trained first at art school, after which he worked as a graphic designer until he studied acting. He then came under the wing of Jeremy Conway, one of a dying breed of old-school agents. Jeremy, a dapper dresser with razor-sharp wit and a tongue as smooth as silk, has always been much sought after and actors yearn to be part of his empire. "Darling," he said to me over the phone, "I have just taken on such a good-looking actor and he needs photographs. I know I can trust you, and only you, to do him justice."

So I looked forward to Ray's arrival with enthusiasm. He too must have been given the same build-up about my abilities, and the shoot was a huge success. Ray's career soon took off like a rocket, moving swiftly from television to feature films. We would meet from time to time for a catch-up, for I loved his kindness, wit and inquiring mind. I never failed to learn something new from our encounters. Not for Ray the endless talk

Ray Stevenson, actor, 2000s

about "the business" and himself; his interests spread far and wide and at heart he is a country man with a great love of the mountains. With the advent of the television series *Rome*, a co-production between the BBC and HBO, I began to think of him as Pullo the soldier, with his shorn head and rough manners. Then came his first leading film role, in *Outpost*, followed by the television series *Dexter*.

Amid all his travels for work, Ray invited me to visit him, his lovely lady Betta and their young son, Sebastiano, in Ibiza. My years of photographing this dear man were being generously rewarded with a magical week of sun and sea in a beautiful private villa up in the hills above the salt flats. I was delivered by taxi at the bottom of a dirt track leading up a steep hillside; there was no way that the driver would go any farther. At that moment, a jeep tore down in a cloud of dust and out jumped a handsome, tanned and bearded man dressed in a white shirt, sleeves rolled back and wearing long khaki shorts, looking more Indiana Jones than Ray Stevenson. But it was my friend, who greeted me with a bearlike hug and the enthusiasm of a Labrador puppy, full of bounce, warmth and excitement about all that he wanted to show me.

No sooner had I unpacked than Ray swept me off to the beach with promises of the best restaurant. El Chiringuito at Es Cavallet lived up to its reputation and, of course, being in the company of a film star always guarantees the best of everything. We were there for hours: drinks, seafood, catching up on the years gone by, and all the while a view of the beach with the crazy people creating the visual entertainment. Everything from the beauty to the beast, and invariably the nude ones were the beasts. At one point, a topless blonde with outrageous implants, arms and legs akimbo, dominated the sea view. As the sun started to set, we finally dived into the beach shop to choose a kaftan, Ray's idea again. I felt like a small child being indulged at half term, such a strange feeling after a lifetime of nurturing my actor clients. Back at the house, a swim in the pool, more chat, Ray played his guitar and sang as we gazed out at the sunset. Dinner under the stars and I slept the sleep of the gods.

The next day, Betta arrived with Sebastiano and all the younger members of her Italian family. As I lay by the pool, I watched the *jeunesse dorée* appearing one by one, dressed in white for a party. They all looked breathtakingly gorgeous with their perfect bodies and golden tans; no

English youth could match them with their olive skins. The golden evening light completed the scene; they were a joy to behold. Now I became one of the elite family, often dining with more than ten people. Languages jumped from Italian to English to French to Spanish, according to which was most appropriate; it was stimulating to try to keep up.

On my last night, Betta left Ray and me to have dinner on our own at Sa Trinxa, with promises of the best view of the sunset as we drank our aperitifs. The time of day induced a different tone of conversation. We shared our feelings over the loss of our fathers, both recently. "Everyone sympathises with the mother but what about the children?" Ray's remark hit such a chord and tears coursed silently down my cheeks, glistening in the flickering candlelight as I sprayed Autan around my ankles to keep the mosquitoes at bay. No more needed to be said. I no longer felt alone with that grief.

Since then, Ray's feet have never touched the ground, with back to back films and television series. The years flew by, punctuated by emails to each other with news of our lives and promises of many more walks in the mountains in Ibiza. A final email read, "I do look forward to seeing you soon, big love Ray." Then I turned on the news one evening to see that my friend had been admitted to hospital while filming *Cassino* in Ischia and had died the following day. Grief-stricken, I longed to turn to him for comfort on this loss of a loved one. He was the best friend for comfort in grief and he would have seen the irony that it was his passing that made me seek his comfort. He is with me in spirit.

CHAPTER 4

WEDDINGS

"Good luck is opportunity and preparedness coming together." – Deepak Chopra

The wedding: that most important day, when the photographer, and the photographer alone, is responsible for capturing all for eternity ... This is one of the reasons why wedding photography is an endurance test. Not only do you have to have nerves of steel, the strength of Hercules and a quick brain, but all these qualities must be wrapped in charm and accompanied by the ability to socialise with up to six hundred strangers. It is also imperative to look as smart as the guests; nobody looks worse than the seasoned wedding photographer in a suit shiny with food stains. I always dressed ready to go into battle, wearing a smart trouser suit and flat shoes, my hair tied back.

The day could often start at about nine a.m., and if there was an evening dance, it might end at three a.m. I photographed the day like a storyboard: bride getting ready, doing make-up and hair, bridesmaids fussing around, champagne being poured, cigarettes nervously lit, and anything else which could conceivably add to the narrative. I soon became adept at dealing with emergency situations. For panic attacks I packed Rescue Remedy in my camera case and I have added other tricks of the trade over the years. Once dressed, the brides have great difficulty with their dresses in the bathroom, and many a time I have held their skirts over their heads and heard small voices panicking from beneath the mounds of tulle. Stress would seize them up. Running the cold water tap is an old trick to make children pee in their pots and it works every time for adults, too.

My first wedding was a baptism of fire, as I had never even assisted at such an occasion and owned only one camera. In retrospect, I realise that the bride's mother, a smart lady from Hampstead, was delighted to get a photographer at an exceptionally reasonable price. As they only wanted black-and-white pictures, I thought that my one Pentax camera would be more than adequate. It was a standard type of wedding: pictures of the bride preparing, the arrival at the church, the ceremony, groups outside the church and at Fenton House up the road in Hampstead for the reception. The day went smoothly and the large pile of rolls of shot film reassured me that I had taken enough pictures. In blissful ignorance, I had failed to notice that one of the camera spools had remained static; the numbers were moving up but the film was slipping in the camera. The result: twenty rolls of blank film, so no images at all, as the film had never advanced out of the canisters.

Time passed, the memory of the disaster faded, and soon I was photographing weddings at weekends and portraits during the week. I started by luring a variety of friends and family to assist, but as I became more experienced the weddings grew bigger and grander and the assistants needed to be professional. It was not long after my disastrous experience that I met a charismatic priest, Father Kit Cunningham, who had a modern and lively outlook that drew many back to the faith. He worked tirelessly helping his parishioners at St Etheldreda's, tucked away behind the hurly-burly of High Holborn. This is one of the oldest churchs in England in current use by the Roman Catholics, and one of the only two buildings left in London from the time of Edward l. Father Kit also found time to visit Wormwood Scrubs and Brixton prisons, and he was never shy about coming forward with his opinion. Though not a Catholic, I was impressed by his sermons and his desire to appeal to all faiths. We struck up a friendship and soon he was asking me to photograph weddings in his church. The first of these occasions was the golden opportunity to photograph the wedding of Lord Kenneth Clark, best known for his iconic thirteen-part television series *Civilisation*. With quiet dignity, and without the ego of modern-day presenters, Lord Clark had opened up the history of art to the masses. This was a solemn occasion, with a small group of elite friends and relations, and the mood was religious and worthy of this ancient holy edifice. Lord Clark greeted me politely, and his bride was warm and friendly.

I knew that I had to be extremely discreet and that I must not use a flash, so I took pictures on a low shutter speed, and to avoid shake I held my camera against a pillar. In the candlelit church the daylight film turned

Lord Kenneth Clark wedding, St Etheldreda's Church, London, 1977

the pictures golden, adding a special quality to the moment that Lord Clark and his second wife, Nolwen de Janzé-Rice, exchanged their vows. Never had the words "till death us do part" sounded more poignant. Lord Clark seemed lost in the religious aspect of the occasion, while Nolwen, her hand resting on his, had a look of pure joy and a desire to nurture her new husband. Lord Clark's first marriage to a fellow Oxford student, Elizabeth Jane Martin, had lasted until she passed away in 1976. Now in the autumn of his life, Lord Clark was repeating the vows that he had honoured with Jane. In a day and age when such a high percentage of marriages fail, this was a treasured moment. Nolwen was by his side until he passed away in 1983.

The last time I tackled a large wedding without a professional assistant, I took my sister, Clare. Afterwards, she commented that she thought catering was easier and she would happily return to her own turf. The Dixons were friends of our parents and their daughter Tessa was to be married in Scotland to her childhood sweetheart, Nigel.

Two sleeper carriages on the train to Scotland were booked exclusively for the guests, thus ensuring that the party began from the moment we departed from King's Cross station. The chaps thought it was a wonderful opportunity to drink and smoke the night away, keeping everyone else awake and making sure that the sleeper compartments lost their meaning. My sister generously gave me the only peaceful berth outside the party block. We left London under a heavy grey blanket of cloud which created a mood as leaden as the skies. The gentle diddle-de-dee, diddle-de-dee rhythm of the train rocked me to sleep like a baby in a cradle until we arrived in Perthshire to clear blue skies and a thick blanket of snow. A procession of cars with snow chains swept up and drove us swiftly through the stunning countryside to Balgour, the family home. A huge breakfast of porridge, kippers, bacon, eggs and all the trimmings was a gift from heaven for the weary travellers.

The house was blissfully warm and we were looking forward to repairing to our bedrooms for hot showers and clean clothes. This was not to be, as my sister and I were to sleep in a cottage down the lane, where the conditions were quite the reverse. So cold was it that we had to sleep in sweaters and socks, and in the morning we had to chip our iced toothbrushes off the windowsill. We chatted most of the night as we were too cold to sleep; we could not wait for the day to begin. With the morning off we went to church for the rehearsal, part of the plan to create a pictorial record of the entire visit. The snow excited me, and I soon forgot my fatigue and found myself shooting everything in sight. Even when everyone else was having lunch, I took landscapes and anything else of interest, right down to the best man's boots left outside on the snowy steps; my enthusiasm saw no bounds.

The big day came and I snapped everything, from the bride getting ready to her father giving a last polish to his shoes, and soon we sped off to the cathedral, so vast that even six hundred guests barely filled half the pews. The service was running smoothly, hymns were sung and then a hushed silence for the marital vows. At that precise moment a thundering noise echoed around the cathedral. The vicar continued, desperately trying to drown out the noise, as several tardy and extremely inebriated guests burst through the giant doors. One glance to my right confirmed my worst fears: my sister's shoulders were shaking with suppressed giggles which spread to me as fast as the plague. By the time we had to appear in the vestry for pictures of the signing of the register we looked like a couple of panda bears. We had not anticipated the need for waterproof mascara.

With so many guests there is no guarantee that you will photograph them all, so you can only hope that you get all the ones that are most important to the family. A long crocodile of well-wishers, waiting to be presented, wound its way across the crisp, white, magical garden bathed in winter sunlight. After the reception, which was held in the marquee, everyone dispersed to bathe and change into evening dress for the dance. I never knew where all the guests not staying in the house went, but come evening they all reappeared in evening dress. Reels were danced and all made merry till the wee small hours as I snapped away until I could barely keep my eyes open. This time sleep won, and our bodies were too tired to argue with the cold.

The next day, lunch was followed by games and this was when my blood ran cold; I have a horror of house-party games and will do anything to avoid them, even if it means feigning illness. Somehow, I managed to escape playing Under the Blanket, where the person under the blanket has to take off an item of clothing each time he or she gets a question wrong. Neither my sister nor I escaped the train game. Here, one person at a time came into the room and joined the end of the train, not knowing what trick would be played on us. Each person in line turned to the one behind and kissed him or her on both cheeks and the procedure carried on until they reached me. I offered my cheek and was smartly whacked on both sides; the impact left my head spinning, and I had a strong urge either to hit back or burst into tears. I suppressed both as I witnessed with horror a row of laughing faces.

Not my idea of entertainment and certainly not the way to get the best out of your photographer. Reeling from the impact I snapped on, so tired that I was now on automatic pilot. That night, as we boarded the train, I took a parting shot of the bride's father in a top hat with a bottle of champagne sticking out of his pocket. I captioned it "The Last of the Big Spenders". Andrew Dixon had been generous to a fault. When we arrived back in London at six in the morning my mood matched the day, as it was still slate grey. My sister had a migraine from the smack around her head, and I felt depressed from exhaustion and a sense of guilt from what I had put her through. But nothing had prepared me for the scene which greeted me when I opened my front door to the ominous stench of alcohol and cigarettes. My lodger had held a party in my absence. In the sitting room the carpet had a huge red stain covered in salt, and I glanced up to see that the milk-glass shade of the Victorian lamp had been smashed. I ventured up to my bedroom and saw the door had a huge, ugly crack. A bedraggled, hungover creature in a dressing gown appeared to assure me that "it was not what it seemed". It was hard to understand what she meant.

Practically every wedding has some drama, and no two are ever the same. At one, the service was held up for half an hour because the bouquet had not arrived; at another the bride arrived at the church to find she had left her bouquet at home; fortunately, a member of the family managed to race back and collect it in time for her walk up the aisle. At an exceptionally grand wedding a page boy peed in his satin knicker-bockers during the group shots, thus ruining every picture as a large dark patch spread across the front of his trousers and onto the paving stones as tears splashed down his cheeks. I doubt he will ever have forgotten that day.

All these minor incidents paled into insignificance when I tackled a grand wedding in Gloucestershire for which guests had flown in from all four corners of the globe. It could not have started better. I liked the mature bride and I adored her mother, a real bonus. It was an exceptionally cold spring day but the sun shone and my only worry was that the bride would not be warm enough in the church. She, on the other hand, was far more concerned that she was being overly made-up, culminating in a battle of wills over whether she should wear lipstick or not. I refereed and a compromise was reached.

Soon we were gathered in the church porch along with the bridesmaids. The little girls clutched their pretty hoops garlanded with white flowers and shiny dark-green leaves; they looked divine. The bride bent to hug them and as the music struck up and we heard a shuffling noise as the congregation rose to its feet, I straightened out her train. But before I could rise to my feet or see inside the church the music came to an abrupt halt. The bride was in tears, the bridesmaids were hysterical, and the bridegroom appeared from nowhere with red-rimmed eyes. His mother had keeled over in the front pew; no, not fainted, but gone to Jesus, departed, passed on. The sound of an ambulance siren came so soon that I think time must have stood still, such was our state of shock. Within no time the ambulance tore off and the vicar was left to make an announcement. Amid many tears and much discussion it was decided the service must go on. I crouched down and clutched the bridesmaids who were still hysterical, made them do deep breathing exercises with me and instilled into them that they must be brave for the couple. The organ started up like the second take on a film set and the service was carried out in a surreal manner.

When we returned to the house my first thought was the arrangement of chairs for the group shot. The tell-tale chair for the mother had to be removed hastily just in case the newlyweds decided to have the pictures taken. However, I did think that they would prefer to do the photographs after their honeymoon.

Shock affects people in so many different ways, but I was glad the couple could get through the day; nature has a miraculous way of casting a veil of disbelief. Speeches were made and only when the bridegroom mentioned his mother did his eyes grow pink and dewy. Finally, late in the afternoon, the couple decided to do the formal groups. I was in the most appalling position: should I pretend to be jolly? After all, they would not want a wedding picture with everyone looking miserable. I hoped they understood when I cracked jokes to help them along. In desperation I threw a little dog into the groom's father's lap to relieve his haunted expression.

The next day, back in London, I sat wondering whether we should have done otherwise, and I felt uneasy about my jokey performance during the group shots. I felt so melancholy that I could not even call a friend to share the drama. Later that afternoon the telephone rang. "We are all sitting here in the country in a state of shock," said the bride's mother, "and the reason I'm calling is because we all agree that we don't know how we could have got through the day without your support." I thanked her profusely.

Another challenging nuptial came through a chance meeting with a journalist, Nina Prommer. We had met on a film set when I had taken some stills of the actor Anthony Andrews. We got along well, so she suggested that I should meet the boss of her syndication, Presse Impact, as they were about to cover the wedding of the grandson of the last queen of Italy, Marie José. Her husband, Umberto II, had reigned for a mere

Wedding of Prince Serge of Yugoslavia, grandson of the last queen of Italy, Geneva, Switzerland, 1990s

thirty-four days. I was delighted by the challenge, as I had never photographed a wedding on the continent, and soon I was busy discovering the logistics of gathering the equipment and finding a suitable assistant who would know how to behave in the presence of royalty. My knowledge of French would be invaluable, as the queen's generation still considered it to be the official diplomatic language. My assistant, Ashton, came through my friend Sir Geoffrey Shakerley, who assured me that not only could the young man do the job but, as he had been educated at Eton, he would be ideal for that milieu. The Broncolor lighting equipment would arrive in Geneva from Germany by train, which we would meet soon after our arrival.

We flew out on the Friday and arrived to blue skies and the magical sight of Lake Geneva. The winter sun shone brightly across the still water, which looked like a sheet of highly polished steel. We arrived at our hotel, cosy and charming but exceptionally small.

"Nous avons des reservations au nom de Presse Impact," I said with great aplomb. The concierge looked confused; I wondered if I should have been speaking in German. No, he had no such booking. Maybe it was in my name? I felt an icy chill run down my spine; if it was, I would be responsible for the bill. Knowing nothing about the owner of the agency, I had come purely on trust. I made a call to Nina in London, who said that she would sort out the matter immediately. We waited anxiously in the lobby. The train with our equipment was due, we were tired, and we had a long two days ahead of us. Finally, the bookings were sorted out and we sighed with relief and departed in a taxi for the station. I was feeling a little apprehensive about whether the rental firm might have neglected to include some vital part of the equipment, but as it turned out, that was not the problem. The nightmare was that the box was big enough to house a baby elephant but too large to put in the taxi. It was now after six p.m. on Friday night and all the van rentals were closed for the weekend. I decided that Ashton and I should split up and hit every car-rental office in the hope of finding a saviour, and miraculously my guardian angel produced a man whose brother was saving up to get married and would help at the weekend for an extra franc or two.

Eventually, tired and emotionally drained before the big day had even begun, I arrived back at the hotel. Ashton was talking obsessively about his girlfriend and how important it was for him to buy her a present. I rather acidly pointed out that the gift could wait until after D-Day. Transporting the "baby elephant" box into the hotel was the next hurdle. Yes, they would find some helping hands, but the box would not fit in the lift. I was sure they would not mind keeping it in the lobby just for one night. "Mais, non. Je suis désolé, mais non," said the concierge, working himself into a frenzy at the thought. It took several more francs over the counter to persuade him to allow the box into the kitchen, where it finally came to rest for the night. Silly jokes flew around about remembering to feed the baby elephant inside and we dived into the lift and went upwards to hot baths, dinner and a glass of wine to calm the nerves.

The next day I finally met the head of Presse Impact, who I nicknamed "The Wizard of Oz" as I seemed to have gone on a Dorothy-type adventure to find him. The Wizard was accompanied by a writer whom I subsequently discovered was a member of the queen's family. We lunched together and discussed the plans for the day, but only at this point did the Wizard inform me that the assignment would only go ahead if the queen liked me. I visualised the possibility of no photographs at all, as any press who were spotted up trees would be removed. I imagined them being shaken out of the trees like ripened plums.

The queen and I met in her garden. We shook hands and my first thought was that she reminded me of my grandmother. She had the air of someone who had been a beauty in her time, elegantly dressed with hair a subtle shade of ash to blonde and so thick and beautifully styled that I suspected a wig. We walked slightly ahead of the rest of our party and discussed gardening. Then, as luck would have it, her black cocker spaniel leapt out of the house and scampered across the lawn. I remarked that I had grown up with a black cocker spaniel called Belinda and the friendship was sealed. The queen returned to the house to get changed and the others hurried up to me, keen to discover how I had managed to chat in such an animated fashion.

The family arrived in a steady flow and we introduced ourselves. Everything seemed relaxed, but all was far more chaotic than I had imagined, for although we were in Switzerland, these were Italians with all their Latin passion. Furthermore, we were speaking French. If that was not confusing enough, the queen's grandson was Prince Serge of Yugoslavia. All was going smoothly until I saw Ashton marching up to the groom's father, Prince Alexander of Yugoslavia, with cries of "Snap!" as he pointed to their Etonian ties. Luckily, the prince seemed mildly amused and later offered to take me on a tour of the house, which in the style of many around Geneva was neither palatial nor particularly grand. Downstairs were adjoining drawing rooms off a wide hall with a staircase sweeping up to a large landing, where we were to photograph the groups. Prince Alexander even took me into the queen's bedroom, where he rather impishly pointed out her wigs on stands.

Ashton and I set up our lights, and all was readiness when the race began. Thick and fast the guests arrived and filed out to the private chapel in the garden. We snapped as many shots as we could and I ran across the lawn to establish the position I had chosen on my recce. To my horror, the chapel was jam packed. I could not get in and many of the guests were already standing outside. Manners to the wind, I pushed and shoved, and not realising how short the aisle was, I caught my foot in the bride's train, almost wrenching it off her head. Scarlet with embarrassment, I dived around the side of the tiny altar so I could get an excellent view of Prince Serge placing the ring on his bride's finger. My cheeks cooled and I followed them out, snapping all the while.

It took half an hour to round up the family for the group shot. The queen sat with great dignity while we rushed around trying to gather the family. Finally, all were in place, right down to the bride's miniature poodle on Prince Serge's knee. Just as I was about to shoot, the queen's pet ocelot strolled in to take its place. I was staggered by this beautiful creature, sleek with black spots and huge black eyes like pools of still dark water. As this majestic beast sat on the queen's knee, it never took its eyes off the poodle. I think it was ear-marking the little dog for dinner but had the good manners to wait.

I had no idea how to address such a group in French; I think I probably said something stupid like "Ready, steady, go." First shot, and a big bang: half the lights went out as we had fused half the queen's house. Now we were stuck with only one power head, as the only other one had exploded; we frantically changed the fuses and fiddled with plugs to no avail. Finally, I had to pretend that one power head was just as good as two, and to this day, I do not know why the pictures came out evenly lit; it was a miracle.

We fell into bed late after shooting party snaps way into the night and rose the following day to be driven into the countryside. This time, we were to play the role of guests for lunch at the bride's family home, where I took a few informal pictures to remind them of the day. I felt strangely at home because there, on the mantelpiece, was a china bowl decorated with roses and two cats, identical to one I had inherited from my grandmother. I had never seen one like it before nor have I since.

It is such a strange feeling to leave home and become completely involved with a family you have never met before, then soon returning home. I had a feeling that I was Alice in Wonderland who had fallen asleep, had this amazing dream then woken up back in Notting Hill. However, I soon had proof that it was real: tear sheets from all the European magazines were littered around my studio and there was the ocelot stealing the show with its eyes fixed on the poodle.

As the years went by, I became more aware of the pitfalls of photographing weddings. It is uncanny how often electrical equipment becomes faulty, lights explode, and assistants, normally calm and dexterous, become whirling dervishes, tripping over cables, making elementary errors and generally behaving out of character. I closed that chapter at the right stage of my life, and wedding marches, toast masters, three-tiered cakes and speeches were no longer a part of my weekends.

CHAPTER 5

CANNES

"We all have to live our lives and we do the best we can. Unless someone injures me personally, I'm not about to pass judgment." – Anjelica Huston

I have tackled the Cannes Film Festival three times, working on various occasions as an assistant, publicist and photographer. Each time I came away vowing never to go again.

I had always longed to go to the festival, with its promise of glitter and glamour. Everything I had read and the pictures I had seen looked so glorious, with those blue skies and stunning-looking celebrities in fairy-tale clothes. But invariably, as I learned over the years, the first week would be chilly and wet, and what is more dreary than outdoor gatherings and

Lydie Denier, actress, Cannes, 1988

parties on the beach in bad weather? When it rained I dodged through the dense throng of umbrellas in terror of losing an eye, at the same time avoiding the café awnings which poured water like jugs down my neck. All the time I was trying desperately hard to look groomed as I staggered into the hotels to attend parties from dawn to dusk. Whether I was sporting a publicist's or a photographer's hat it was never easy. It was harder as a photographer, as I had to produce good pictures.

The first time I flew into Nice airport to attend the festival, I was full of optimism and excitement, delightfully naive and keen as mustard to do well. I stayed in a little boarding house a few streets behind the seafront. Each day I walked down to the Croisette, the seaside promenade, which teems with suits all desperate to seal their deals through endless meetings in the various vast hotels that create an unbroken chain along the seafront. From dawn to dusk the terraces of these grand hotels bustle with business meetings, photo calls and parties. Starlets pose in outrageous outfits to

Jonas Rosenfeld, President of Advertising, Promotion and Research at 20th Century Fox, Cannes, 1988

attract the paparazzi, who descend like locusts upon a field of ripe corn. In the harbour the envied few observe the chaos from their luxury yachts.

The epicentre is the Carlton Hotel. This is where Fred Hift had a suite. I had worked for Fred as a unit publicist at 20th Century Fox, and now he was director of the festival and I was his assistant. I was fond of Fred, though we had crossed swords many times. This was fine, as I understood why he was so angry with life. One night over dinner in Vienna years before, when I was working as film publicist, Fred had let his guard down. He had been drawn to the piano in the restaurant and I had never before seen the look of joy on his face as he played. Afterwards, we sat late into the night as he told me the story of his childhood. I have never forgotten his description of his departure from Vienna by train, leaving his parents to the horrors of the concentration camps. How can anyone get over such an experience? I always had an enormous amount of compassion for him.

Fred's suite doubled as our office and his sleeping accommodation. Since he was a director of the festival, I thought this arrangement bizarre, but such is the demand for rooms that the grandest end up in the most unlikely situations. One night at nine o'clock we were bundling up press releases on the floor when I suddenly felt tearful, so tired from lack of sleep that I was feeling sorry for myself. Other people seemed to be having some fun and Cinderella did so long to be asked to the ball.

The next day, my prayer was answered. The telephone rang, and it was Stephen Friedman, producer of *The Last Picture Show*, asking me if I would like to go to a dinner party on Robert Altman's yacht. I was thrilled. Choosing an outfit was easy, as I had only packed one evening dress, just in case I metamorphosed from worker into guest: a long black number with a halter neck which looked elegant and tasteful and complemented my blonde hair and slim figure. No frills, no Cruella de Vil red nails. Simplicity was the name of the game. The finishing touches were a small gold locket inset with a tiny pearl inherited from my maternal grandmother and a family signet ring. The gold ring was set with black onyx bears and the Latimer family crest: a greyhound attached to a tree; the motto is "have patience and endure". I wear the ring to this day because I think this crest from my grandfather's side of the family suits me well. It was a treat to do my make-up with care, applying plenty of mascara, subtle shades of eye shadow and the finishing touch of a soft, translucent lipstick; I felt transformed from chrysalis to butterfly and set off in high spirits, determined to have a wonderful evening.

The harbour was brightly lit by twinkling fairy lights festooned across the bobbing boats. The inky darkness out to sea was broken by a shaft of silver light reflected across the still water from the waxing moon. Steve (Friedman) and I climbed the precariously narrow gangplank of Altman's impressive "gin palace" and were greeted by our host and hostess. A small group of guests were lounging around in wicker chairs with drinks in hand, looking "cool". Nobody made much of an impression on me that night, as all the women were self-consciously perfect and the men seemed to be hanging on Altman's every word. Altman, so large and imposing, like a mighty bear, held his audience spellbound, to the point where he totally eclipsed his guests.

Quite suddenly and without warning, Altman was shouting heatedly at Steve about *The Last Picture Show*, the low-budget movie that Peter Bogdanovich had directed and Steve produced. The film had been a surprising success, nominated for Oscars for best film and for two actors, including Cloris Leachman, who later became my friend and client. Whether Altman had turned the movie down and regretted the decision, I shall never know, but something made him react to Steve like a raging bull speared by picadors. Steve was desperately trying to back down and calm the situation, but Altman, with a few drinks under his belt, was having none of it. Finally, his wife managed to take him downstairs while everyone tried to continue their conversations, but all ears were trained apprehensively on the cabins below. Shortly afterwards Altman and his wife reappeared, and all seemed well until again he went on lambasting Steve over the same movie. This was all too much for Steve, and to my horror I saw tears were silently coursing down the face of this gentle, sensitive man. My dream was shattered as I witnessed the ugly side of the film industry. "Look at him," bellowed the by now extremely inebriated Altman, who towered over the poor producer, "just a snivelling little brat!" Everyone froze, but nobody rose to Friedman's defence, not wanting to offend "the King".

Without knowing quite how or why, I found myself, all five foot four of me, standing face to face, or face to navel more accurately, with this giant of a man. My rage overcame my inherent shyness, and my ridiculous need to please everyone evaporated into the darkness. I felt I was having an out-of-body experience. "Mr Altman," I started, my voice surprisingly strong, "you are the rudest host I have ever met and if I had known this before, I would not have set foot on your boat, let alone accepted your hospitality." Quite suddenly the mist of anger cleared, and I realised what I had done. Astonished faces were staring at me. No one spoke. There was a chill in the air and I winced at the sight of my fellow guests' disapproval. The women with their overly made-up faces stared coldly at me; they looked vulgar, hard and desperately over dressed with their gaudy jewellery glinting in the low light. The men remained equally still while blending into the half-light in their dark suits.

All I could think to do was to gather up my skirt and run away. I stumbled down the narrow, slippery gangplank. I just wanted to escape from the whole nightmare. Oh, those damned high heels! Off they came and as I ran barefoot along the darkened quay I heard voices shouting at me to stop and the clattering of running feet in hot pursuit. I soon outran them: in my childhood I had always been as fast as a leopard.

On reaching the end of the quay I turned left along the Croisette and just as I was wondering what I should do next, all dressed up and nowhere to go, I ran straight into Roman Polanski. I knew Roman through a friend. The three of us had dined together in London and I had also been to a party in Michael's magnificent house in Kensington. Roman wiped away my tears and took me off for a quiet drink on the softly lit terrace of the Carlton; his gentleness was almost too much. The whole festival had been such a tough experience, so when he held my hand and showed his concern, the tears flowed like mountain streams. It makes me sad to hear about the bad publicity Roman has received over the years. I always knew him as an attractive, gentle, highly talented Polish man with impeccable manners and a delightful sense of humour. We talked late into the night, and I returned to my hotel comforted and ready to face another day.

Robert Altman continued to be a highly respected director and, indeed, I am a great admirer of his work. I am enough of a professional to separate people's work from their private lives. However, I had no regrets. There was no way I could have remained a silent witness. Why should anyone get away with bad behaviour just because they are famous? It was not the drunkenness that bothered me but the baiting of a man who was his guest for the evening.

The incident did not go unnoticed. When I reached the office the following morning, Fred was furious. "I gather you are the talk of Cannes," he snapped, not even turning from his typewriter. Clearly he did not consider such behaviour to be part of my contract. I came home from that festival totally drained but I could live another day with the memory that to one man at least, I was a heroine.

A few years later when I was asked to return to Cannes, I should have reminded myself of my past experience and thought twice, but always the optimist, I believed that this time it would be different. By now I was well established in London as a portrait photographer and the commissions were rolling in at a steady rate. Theo Cowan suggested that I could combine my skills: I could cook breakfast – big English fry-ups, bacon, eggs and fried bread (after all I was a woman) – answer the telephones – because I spoke French – and meet and care for his clients, as I had been a publicist. Last but not least, I could take the photographs needed for the advertising campaigns.

Theo, his business partner Laurie Bellew and I shared an apartment which was also the office, another example of the ridiculous accommodation arrangements during the festival. We did have an enormous amount

Theo Cowan, Cannes, 1980s

of fun, as Theo had a knack for finding humour in everything, including himself. He was always kitted out in khaki from Laurence Corner, an army surplus store in London. He was not a dedicated follower of fashion but he was unique. Perhaps he set the trend for army clothes amongst civilians? Each year at the festival he could be seen in the same uniform: shorts, shirt and cotton hat to match, with flick-up shades on his glasses. His pockets bulged with emergency supplies. From tobacco to pens, string, boy scout's knife – you name it, he had it. I think he quite rightly saw Cannes as a battlefield. Theo was the dear friend and father figure to so many of us whom he had launched into the mad world of movies. He gathered affectionate nicknames, "Thumper" to Dirk Bogarde and "Threadbare" to another friend of mine, because he was bearlike and thin on top. He never drank alcohol but lived for food, and at press parties he stuffed his face with canapés.

Once during the night, while looking for the refrigerator, Theo sleepwalked into my bedroom. He was genuinely looking for food, as he had a habit of eating a comforting snack in the middle of the night. The evidence of his forays was always there in the morning, often in the form of piles of apple cores hidden under the bed. Laurie Bellew thought it was disgusting, but I found it hilarious. The more the tension of the festival built, the more Theo nibbled in the night.

The downside of sharing the apartment was that I had no time off at all. Each day I hit the ground running. Breakfast had to be cooked and telephones answered, and I never knew if the cameras were to come out or whether I had to meet a celebrity at the airport. Theo had a habit of giving me the tricky ones and none proved more so than Edy Williams, the ex-wife of Russ Meyer, who had produced *Beyond the Valley of the Dolls*. Edy made a grand entrance in a fur coat, and the minute my back was turned to pay off the cab driver she dropped the coat to reveal a see-through nightdress. She was no spring chick, and the fabric clung to her like cling film around a plucked boiling fowl, with her nipples protruding in the chill wind. I was mortified. Such near nudity was still unacceptable in those days even at Cannes, and I vowed that she should be Theo's for the rest of the festival.

The next time I was lured to Cannes was solely as a photographer. I was to be one of three snappers covering the festival for *Variety*, the

American show-business magazine. I was again working for Fred Hift, who had recovered from my previous performance in Cannes; time had proved a healer.

Each morning Fred held meetings to decide who should cover which function, from parties that went on from breakfast until late into the night. The three of us, one French boy, one French girl and myself, were responsible for producing all the pictures for two special editions for the festival. We were on our feet with cameras around our necks from dawn to dusk. I had walked into a trap. For two weeks I nibbled on canapés and ran from party to party, pounding up and down the crowded Croisette. I kept my eyes peeled for interesting shots until my legs ached and my feet, even though they were in trainers, swelled with fatigue.

As the days went by, I found it harder and harder to get out of bed in the morning. The pain as I put my feet to the ground and the stinging sensation of shin splints down the front of my legs made it almost impossible to continue, but still I soldiered on.

This was hardly my style of photography, chasing around after stars and producers; I was used to them asking me to take their pictures or being assigned to a shoot. However, stuck with this situation, I decided to make the best of it, so I learned fast how to get the shot when faced with a group of rival photographers. Being small, I darted down low and popped up just at the last moment right in the front. Admittedly I received several sharp digs in the back and bangs on the head, but the battles had to be won.

I held my own and even excelled, being published more than the other two photographers. I was even beginning to enjoy the situation and winning felt good. The assignments were hugely varied, from small parties on boats and private suites to huge gatherings in function rooms and castles in the countryside. Then came a lucky break. The Monty Python brigade was hosting an exclusive party from which the press was barred. As I knew Michael Palin and had worked with John Cleese, an invitation was extended to me, and that was the highlight of my trip. I was in my element chatting to the Pythons, who are certainly not known for taking themselves seriously, and I was laughing for the first time since I had arrived in Cannes.

On a particularly long day of snapping towards the end of the festival, when my physical batteries were nearly as worn out as those in my cameras, the last stop was a trip to the countryside to attend a party which started at ten p.m. It was pitch dark apart from the twinkling lights of the houses which dotted the mountainsides, blending with the stars in the sky. We drove for what seemed like an eternity until we arrived at a driveway lit by flaming torches, which threw the crowd into silhouette. We wandered around the courtyard, heaving with people, and tried to recognise the actors. A man before me in the half light, quite short with long, straggling, greasy hair, suddenly seemed familiar. I swiftly raised my camera and all was blackness.

"You've got your lens cap on," growled a disembodied voice. Too late. He had walked right on, grinning back at me like a naughty schoolboy. It was Michael Douglas; actors are sometimes hard to recognise when they have changed their usual appearance to play a role. I spent the rest of the evening trying to make up for my error, but we started to play a ridiculous game of cat and mouse. Every time I raised my camera he would spot me, grin and turn slightly. I never did get the shot. Clearly, I really was not cut out to be a press photographer.

Although a consummate theatre actor both in the States and the UK, Antonio Fargas is best known for his role as Huggy Bear in the 1970s television series *Starsky & Hutch*. The light was so perfect on the hotel terrace the afternoon I photographed him that I had what I wanted for *Variety* magazine in no time.

A multi-faceted talent who has starred in more than fifty films and on television, Peter Weller has also directed and lectured in history. Like many true talents, he was modest, warm and friendly. The light on the terrace of the Carlton Hotel was perfect the afternoon I photographed him for *Variety*. I wish that I had met him a few years later when he went on to appear in the film adaptation of *Naked Lunch* by William Burroughs. I had met Burroughs many years before when working on Conrad Rooks's film *Chappaqua*, which dealt graphically with the subject of drugs.

One day I was fighting to hold my own as a swarm of photographers cornered Bob Geldof on the beach, pushing him backwards into the sea. This was all a far cry from the care and attention I would normally devote to a subject while shooting a portrait.

That festival ended the way they all had. The locusts left, and the smooth sand of the beach, no longer dense with people seeing and being

Antonio Fargas, Cannes, 1988

Peter Weller, Cannes, 1988

Bob Geldof, Cannes, 1988

seen, was raked until it looked like virgin snow. Waiters, meanwhile, got busy changing the price lists. I had long ago become aware that the locals bleed the visitors unashamedly during the festival, doubling their prices at the beginning and changing them back on the last day. I love the French and enjoy being in a country where I speak the language, but I had observed over the years that this glamorous festival brings out the worst in everybody. The atmosphere of greed and fear in this glorified meat-and-money market inevitably left me drained and I was always glad to be up in the air heading back to London.

CHAPTER 6

LOS ANGELES, CITY OF ANGELS

> "*One of the diseases of the world is that we associate beauty with youth. We are wrong. The eyes and the face are the windows of the soul and these become more beautiful with age and pain that life brings. We are wrong. True ugliness comes from having a black heart.*" – Lobo Solo

I have always had a love/hate relationship with Los Angeles, where everything is more extreme than in London. I never quite fathomed out whether the nervous energy was due to the imminent and perpetual fear of the Big Earthquake or the burning desire for fame spawned by a city dominated by the film industry. When I first visited, there was the excitement of all that was new: the perfect climate, the beauty of the birds of paradise growing wild in the scrubland, palm trees, and above all bottlebrush trees, named for their brilliant red flowers. Then there is the promise of the city of opportunity, where anything is possible, stars are born, and the climate is perfect for shooting films.

In stark contrast is the ugly and endless sprawl of buildings covered in cheap advertising, the smog and the tangible presence of the broken dreams of the millions who are passed by. Taco Bells on every corner, the boy prostitutes on Santa Monica and the girls on Sunset, so blatantly out there – all this and more was a huge shock as I discovered Los Angeles.

I have been lucky: I have met stars, photographed them, and sometimes I have stayed in their homes and created firm friendships. Other good friends are successfully employed on the other side of the camera.

My first visit to L.A. was a short trip, which came about soon after I finished working with Michael Laughlin, the producer of *Joanna*. Michael had a script he wanted to sell and he did not want to go to Hollywood himself, so he asked me if I would go in his place. I was to stay two weeks, and he would provide me with a list of people to see at the studios. I had never been farther than Europe and loved the idea of having a reason to go to Los Angeles, but the trip turned out to be a baptism of fire. My suitcases went missing on the flight, so I arrived at the Beverly Wilshire Hotel with just my hand luggage. Dressed in a tracksuit well crumpled by the ten-hour flight, I looked suspiciously like a bag lady, and only my advance booking and credentials qualified me to stay.

The next two days I spent embarrassing the concierge, who longed for my bags to turn up as much as I did. It was so hard to buy anything I needed: I did not know where to go and I was totally unused to a city where you cannot just walk down the street. I felt imprisoned and lonely. It had never crossed my mind how desperately lonely the city could be, and I had only one acquaintance's address, a girl who I had only met once. As this was before the invention of the mobile phone, I had to wait endlessly in the hotel for calls from the studios to arrange my meetings.

Each day I would make my calls, then, clutching my reading book for company, I made my way to lunch by the rooftop pool. This routine continued until one day a waiter came over and told me a fellow guest would like me to join him for lunch. I looked across and, seeing a mature gentleman, decided to accept. We met daily for lunch as we were both there alone on business. He said that his name was Lane and that he was a producer; he seemed genuine, and in those days I tended to believe everything I was told. One night Lane suggested dinner. The conversation was as scintillating as ever and he behaved like the perfect gentleman, even escorting me to my bedroom door at the end of the evening. Suddenly, the bearlike man whom I'd mistaken for a teddy turned into a grizzly. Red lights flashed as he lunged towards me, and I shoved him backwards, slammed the door in his face and stood trembling on the other side. More alone than ever, I retired into my solitary life for the rest of the trip. All I had to look forward to were my car trips whisking me over the Hollywood Hills to meetings with various moguls in the studios. In my opinion, Michael Laughlin's script was not earth-shatteringly original, so I was delighted when the highly respected producer Anthea Sylbert echoed my thoughts. It gave me a certain amount of satisfaction to know that my finger was on the pulse. For the rest of the trip I was killing time, and I vowed never to return to Los Angeles until I had a better plan of action and an address book full of contacts.

Several years went by before I returned to California, and this time the decision just seemed like it was meant to be. Dan McBride had re-entered my orbit in London after three years of modelling work in Europe. I was thrilled to see him and the strange feeling of closeness remained. Some things are inexplicable, best expressed in *The Little Prince* by Antoine de Saint-Exupéry, "One sees clearly only with the heart. Anything else is invisible to the eyes." Dan was delighted to be back in England, but his joy was short-lived. The authorities discovered he did not have a work permit, so he was given two days to leave the country. I knew without a doubt that I had to go with him.

On our last night in England, Dan and I were up half the night, with pictures strewn all over my living-room floor, trying to create a suitable portfolio; I could have stayed up forever as my excitement had turned me into Superwoman. The plan was for us to spend the first few days with friends of Dan's, Jim and Ed, and then, for the first time in my life, I had no idea. As it turned out, six months would slip by with one opportunity leading to another.

We were met at Los Angeles Airport by Jim Daniels, a tall, bespectacled, willowy man with salt-and-pepper hair and a warm, friendly face. His soft melodious Southern drawl instantly put me at ease. It was dark as we drove along endless wide, straight roads then made a long, winding steep climb up the Hollywood Hills above Sunset Boulevard. I had arrived at the most incredible view of Los Angeles, a 180-degree panorama of lights twinkling in the city below and blending into a sky full of a myriad of stars. Jim and Ed lived right at the top of the hills in a high-tech house with huge floor-to-ceiling windows. Too tired to even unpack, we fell onto a futon bed and went straight to sleep.

The next thing I knew Jim was telling us to get up; it was dawn. Stunned and disorientated, we wrapped ourselves in a blanket like twins in the womb and followed him blindly onto the balcony. The most beautiful dawn breaking over the city greeted us. I had seen nothing like it in my entire life. The sky was the most breathtaking shade of pastel blue and pink stretching out forever. Never had I felt so comparatively minuscule in the universe, such a far cry from my limited view over the rooftops in London. We breakfasted extremely early and were joined at last by Ed, a lawyer. I soon realised that Jim was the househusband who spent his days at home with the dogs. This proved to be to my advantage as I enjoyed his company and he could help me find my way around town. After a couple of weeks, Dan moved down to Venice Beach and Jim invited me to stay on during the week so that I could try my luck in this land of opportunity, the City of Dreams. I spent the weekends with Dan in Venice Beach, where I ran on the sand before breakfast and chilled out ready for my next attack on the city.

I made endless appointments to meet up with useful people and friends of friends from back home. Jim would find my destinations on the map, then point from the balcony to where I was heading; this was an easy start to getting to know the layout of the city. He would then send me on my way in his huge pink Chevrolet truck. It appealed to my sense of humour to visit Beverly Hills in this absurd mode of transport, often with Jim's two large dogs in the back enjoying the ride.

One day I drove to visit my old friend Jon Bradshaw (better known by all close family and friends as just "Bradshaw"). Now living on Yoakum Drive off Benedict Canyon, this kind, witty, ruggedly good-looking man had invited me to lunch for a relaxed old friends' catch up. The truck roared and struggled up the hill until there was Jon, standing on his balcony grinning. "I knew it had to be you," he shouted against the roar of the engine. "No one else, apart from a workman, would be seen dead in a truck like that in Beverly Hills."

We had first met in Mallorca while I was holidaying with my parents and sister a decade before at the Sea Club, well-known for its notorious owner, Nora Cumberledge. An outrageous snob, Nora loved to tell the story of how she had swum out to an admiral's boat in the South of France, soon becoming his cook and swiftly thereafter his wife. Only those who were well born or famous were admitted, but Daddy qualified for both, which gained us endless free dinners and drinks at the bar.

I had no idea what a literary star my friend had become, with his brilliant brain and the looks of Indiana Jones. He was moving in the most elite circles, conquering all in London and New York, then moving on to gritty travel articles. Bradshaw was fascinated by people and far more interested in studying them in depth than in making money. So popular was he that people happily paid for his company.

So here I was enjoying a simple lunch *à deux* on his deck, gazing out over Beverly Hills, glass in hand, in the best company. I left glowing and waving and looking forward so much to the next visit, thinking of his words, "You know I am a great man, just look at who my friends are." I hoped a little of this stardust had rubbed off on me.

Two days later, I went to photograph Heide Lund, wife of the producer Lord Antony Rufus-Isaacs, whose films include the notorious *Nine and a Half Weeks*. Heide was beautiful and her husband was handsome, and they lived in a stunning house in Benedict Canyon with two beautiful little girls; everything was perfect. Just before we started shooting, Jon's name came up in conversation and I mentioned I had lunched with him two days previously. Heide paused, looked at me and asked in a fairly matter of fact way if I knew Jon had died the day after I had seen him. The news struck me like a thunderbolt, at first disbelief and then a dizzy sensation. I wanted to hide and be alone, away from these strangers. All I could do was ask to go to the bathroom where I threw up from the shock. I stared at myself in the mirror and saw that I had gone very pale.

Okay, deep breath. What would Jon have wanted me to do? The answer: go out there and kick ass, do those pictures and do them well; this was not the first and would not be the last time that the discipline of work has helped me through the hardest times. It was not until I got home that the tears flowed.

This setback aside, I was enjoying my time in L.A. I loved the climate and the crazy emphasis on a healthy lifestyle. I was young, carefree, "in shape" and fashionably tanned in no time. Three mornings a week, Georgina, the wife of actor Anthony Andrews, and I went to the Sports Connection on Santa Monica Boulevard. We struggled through absurd classes with names like "Stretch and Endurance" under the tutelage of a breathtakingly good-looking teacher, deeply tanned with golden shoulder-length hair and invariably wearing a lime-green sleeveless T-shirt to show off his muscles and tan to their best advantage. Soon we became the teacher's pets, as we were the only English girls in the class and he loved our accents. The other girls, a pushy crowd who would fight to get as close to the instructor as possible, would place lurid invitations in his suggestion box: "Come any closer and I'll give you a blow-job," to name but one. Needless to say, our classmates longed for us to return to England. Exhausted and sweating, we would go to Georgina's house where Anthony, still in his dressing gown, would browse through the newspaper while finishing his breakfast.

Anthony Andrews, Los Angeles, 1980s

Tony, who had just starred as Sebastian Flyte in *Brideshead Revisited*, would remark that we were completely mad and indeed, looking back, I think we were. Why did we want to go to an air-conditioned sports club when we could have been lounging by the pool at the most beautiful time of day? The Andrews were part of a big group of friends who embraced me into their fold, and several of them are still in my life.

As I acclimatised and started to understand the city, my confidence grew and I found it easier to pick up the telephone to perfect strangers. So the day came when I called Bill Bast, a script writer whose name had been given to me by Derek Granger, best known for producing the legendary television adaptation of *Brideshead Revisited*. Bill's voice was warm and friendly as he said, "If you are a friend of Derek's, you must come by the house." A date was set. By now I was confident enough of the layout of the main streets, as they all run in straight lines, either east–west or north–south, so all I had to do was look at the direction of the sun or the Hollywood Hills, which would be on my left when travelling east and on my right when travelling west.

The first trip to meet Bill was more of a challenge once I turned off Hollywood Boulevard and climbed steeply up narrow winding roads in the hills. Eventually I found the door, set into a white wall, and pressed the intercom. I stepped into a fairy-tale garden with mature blooming trees and plants and an arbour festooned with flowers. I walked up the path and there at the front door was a young man with blond hair who seemed familiar. "I thought it might be the same Carole Latimer," he said. It was Paul Huson, who had hardly changed since I last saw him at school when we were about thirteen. What divine force was at work that day? How else could I explain such a meeting in a foreign land? And so we picked up on a lifelong friendship which had merely taken a break while we were growing up and finding our places in the world.

I grew fond of Bill and Paul and saw them regularly in Los Angeles and London. We have even worked together on a couple of productions for which they wrote the scripts and I came in as a special photographer. On one production, a dalliance with the leading man gave Bill the ammunition to tease me, and I then felt that this gave me the opportunity to ask him about his friendship with James Dean. Bill had been the actor's closest friend during the last five years of his life. I would get an evasive answer along with a twinkling, amused look in his eyes. Only now do I and the entire world know the true story. Bill finally published his second book on their relationship, *Surviving James Dean*. The first, written soon after Dean's death in 1955, was only half the story, as in those days homosexuality had to be hidden. I am overjoyed that Bill published the second book before he left this planet, for theirs is the story of true love between two young men who supported each other through the early years of struggling in Hollywood.

Real love, in my view, is the most important thing in this life, whether it is between a male and female, two men, two women, whoever. Bill had a wonderful relationship with Paul, enriched by working together as writers and sharing their lives in every way. However, Bill admitted that not a day goes by without him thinking about Dean, wondering whether he could have trusted him with his love and how it might have worked out. None of us forget true love if we are lucky enough to have known it, but when that person is a legend and becomes public property, there is a different agenda.

My social life was rich and rewarding with all that a new environment brings, but this trip was not all fun and games. I had come all this way to try to get assignments as a photographer, and for the first time in my working life I was not heavily in demand. I soon found out that my portfolio with artistic snow scenes was getting me nowhere. Editors in L.A. wanted to see celebrities. I went on endless interviews in the big pink truck, and was seldom offered a glass of water or the chance to sit down. It was just, "Okay, kid, let's see your portfolio." And that was about it: nobody they recognised, so no interest. The lack of manners stunned me, but this was a useful crash course in how to handle the city, which would prepare me for future trips.

Looking back, Jim and Ed were incredibly hospitable as days turned into weeks and weeks into months. Just as I sensed that it really was time to go, their next-door neighbours, Frank Devine and his girlfriend Jan, insisted that I stay for another month in their guest room. It was not all doom and gloom on the work front, and I met and talked to a number of stars. A good start. One night the phone rang, and it was Natalie Wood returning my call; we talked for some time and she told me she would like to do a shoot with her children. She had heard from our mutual friend Delphine Mann that I did wonderful pictures of children. Of course I was over the moon and assured her I did not mind waiting. Tragically I will wait forever, as Natalie drowned on a much-publicised boating trip soon after we spoke.

Delphine, a realtor by profession, has remained one of my most constant friends over the years; her home on Benedict Canyon was like a sanctuary, somewhere I could retreat to when all about me was too crazy. When she entertained, the gatherings were small and intimate, so I enjoyed many good conversations. On one occasion, I had a passionate discussion about cameras with Robert Wagner, and we agreed wholeheartedly that the Hasselblad was far too unforgiving as the years go on and that the Nikon, with a 180 lens, was infinitely preferable. Delphine then moved to the most beautiful apartment block on Sunset Boulevard, which has been home to many a legendary Hollywood star. When I visited her new home, I was enveloped in the warm feeling of a calm oasis, well protected from the frenetic city. Of all my old friends, Delphine has always been one of the most constant.

Life took on a joyful routine; this time in L.A. was, in retrospect, one of the calmest periods in my life. I was living in a beautiful home with no responsibilities. I loved the routine, the sound of Frank and Jan getting up in the morning and going to work, followed by days to myself to arrange interviews and to meet friends. I always started the day gazing out across the city at the stunning sunrises as I ate my breakfast. In the evening, my hosts would come home and we often dined together.

The telephone was always full of surprises, as I never knew who would call back or not. It was on one of these happy evenings as we prepared dinner together that Frank answered the telephone for me yet again. I felt a little awkward that it rang so often for me but he never appeared to mind. "It's Joanne Woodward," he said, having learnt by now not to presume that someone was joking. We talked at such length that he was prompted to ask whether we were close friends. "No," I replied. "We have never met. We just have a mutual friend called Derek Granger." She was about to take a trip and the promise of a future meeting was mutually agreed upon.

Once again, when I decided it was time to go back to England, another opportunity occurred. This time, the request came from Anthony and Georgina Andrews. They asked me if I would house-sit their home on Blue Jay Way while Anthony was filming. I moved from one superlative view to another. Dan packed his bags at Venice Beach and joined me in the Hollywood Hills. In the morning, I would awake, press a button beside the bed, and the curtains would slowly roll back to reveal the swimming pool set in a pastel sunrise. Overlooked by nothing but the odd bird of prey hovering in search of gofers, I would jump straight out of bed and into the swimming pool to stretch my limbs. In the evenings it became quite chilly, so we lit the log fire, took another swim in the steaming pool and settled down to dinner, gazing out at the view. This was one of the most blissful months of my life. Time stood still; I was with a man I shall always love, in an idyllic climate, in a beautiful home with a host of friends nearby.

Now I felt I had the opportunity to hold a dinner party to thank my new friends for all their help and hospitality. As I was unfamiliar with the kitchen, I thought I would prepare something easy, and chicken with an imaginative salad seemed a safe option. About twenty people accepted; among them were Jack and Marian Neuman, Bill Bast and Paul Huson, Barbara Parkins (of *Peyton Place* fame) and Darren Ramirez, who had lived with Rachel Roberts up to the end of her life. It was Darren who was the most diplomatic when the chicken remained raw after two hours of cooking and made my salad sound exotic. It had never crossed my mind to check that the cooker worked. Everyone was remarkably kind and the evening went surprisingly well; the beautiful location certainly helped to sugar the pill.

My social life was rich and interesting. I dined with many fascinating Hollywood personalities, amongst whom was Marti Stevens, actress/singer and daughter of one of Hollywood's founder moguls, Nicholas Schenck. Marti lived in the Palisades, out towards the beach in one of the most desirable areas of Los Angeles. Christopher Cazenove, who I knew from years back when he acted in a play in London with my father, was there on one particular occasion, along with the singer Georgia Brown. Georgia was very drunk and became abusive. Apparently something I said helped to diffuse the situation, which caused my hostess to remark that I had "noblesse oblige". A distant memory of being in a similar situation in my teens when my father soothed a raging Rex Harrison sprang to mind.

I was not always so "cool". One day, Dan and I went to lunch with a couple who had rented a beautiful house in the Palisades, once lived in by Paul Newman and Joanne Woodward. I made a great effort to look casual but chic and dressed in a silk shirt and white trousers. I felt, fleetingly, self-confident. We had drinks by the pool, where I sat on a reclining chair under a blueberry tree, basking in the beauty of it all. When I arose, my butt looked like the rear end of a baboon. I had sat on a sprinkling of berries; I have never worn white trousers since. That was only the first act. Dan, who had been wonderful playing with the kids in the pool, somehow forgot I was an adult and threw me in. I emerged, mascara drizzling down my face like Coco the clown, the purple patch firmly in place. So much for trying too hard to look good. When we got back home, I lost my temper, drove straight off and went to stay with Steve Shellen, an actor friend who lived down the hill. I shall never forget Dan's stricken face when I returned the next day; I regretted my behaviour; it was cruel.

The time was coming to leave L.A. This had been a hedonistic time, a rich contrast to my working life in London; yet another experience in the quest to find balance in my life. I knew my L.A. life was a temporary phase as there was always an inner voice reminding me that this was just an interlude, not a lifestyle that I could adhere to forever. I would miss my many friends, and that perfect light that can be so perfect for film-making. Then there is the beach, which for me is like the icing on the cake. Even at the beginning of January, I can walk barefoot in hot sunshine along the wet sand by the water's edge. Little wonder I always feel a brief pang of sadness whenever I return to England.

The actress Elizabeth Mason dreaming of stardom, 1990s. She later created The Paper Bag Princess, a line of clothing based on vintage fashions.

CHAPTER 7

ON ASSIGNMENT IN L.A.

As it turned out, I would return to L.A. many times over the years. On one of those trips, I received an assignment to go to Monument Valley on the Arizona–Utah border to shoot for three days on *Winston Churchill: The Wilderness Years*, a television miniseries starring Robert Hardy and Nigel Havers. I flew to Phoenix with publicist Jenny Craven and writer Nancy Mills. Jenny drove the hired car across the Arizona desert, averaging eighty miles an hour. Her argument was that she was English and that the American speed limits were ridiculous. So Nancy and I spent the entire journey looking out the back, watching for police as we sped across the empty terrain. It was an exhausting journey, and I never had the chance to photograph the tumbleweed flying across the road in the high wind nor the staggering scenery which flashed past like an impressionist painting. After a couple of hours, our bladders were at bursting point, and it got worse and worse. There was not a bush to hide behind, and anyway, we were far too scared of snakes to crouch down. We had to wait and wait until we reached our destination near Monument Valley.

Any hotel at this point would have been welcome, and this one was characterless but comfortable. We dined with the actors and Nigel Havers and I bonded; both being Scorpios with similar backgrounds, we had plenty in common. Friendships had been established and we went to bed confident that all would go smoothly the following day.

We rose at dawn, still desperately tired from the long journey, and set off to a full day on the set, where the temperature would rise to over 120 degrees Fahrenheit. It was a daunting thought and the reality was tougher than I had imagined. I avoided sunstroke by wrapping a soft scarf around my large straw hat and tying it under my chin; I looked like a nineteenth-century English gentlewoman out for a stroll on a summer's day. The lack of humidity was so extreme that the fear of dehydration was very real, and we all had to drink water on the hour every hour and wear wet towels around our necks. I felt relaxed about taking the pictures, which

Nigel Havers, right, and producer Richard Broke on set, *Winston Churchill: The Wilderness Years*, Monument Valley, Arizona, 1981

The crew on set, Monument Valley, Arizona, 1981

I presumed would be exclusive: we were so far from anywhere that the chance of another photographer sharing my pitch was remote. To my horror, on day one I spotted not one, but two figures festooned with cameras. They were equally appalled to see me; we had all been promised an exclusive.

Luckily, we were all intelligent enough to realise that we were stuck with this situation and might as well make the best of it. Out of this predicament, I came to make a lifelong friendship with Peter Kredenser, who was shooting for *TV Guide*; his picture editor Cynthia Young accompanied him. Peter was so good looking that I was tempted to take more pictures of him than of the actors. Tall, slim and square shouldered, with a deep suntan and his long, reed-straight, raven-black hair tied back with a length of plaited leather, Peter looked quite the hero of the desert. The English crew looked a sorry sight with their lily-white skins and the odd alarmingly red neck where the sun had caught someone off guard. My hero turned out to be a real gentleman and has remained a close friend to this day.

For three days we battled with the heat and dust storms while the film crew shot one tiny scene in a vast landscape. In the film, Winston Churchill and his son Randolph, guests of William Hearst and Marion Davies, were watching the filming of a scene with cowboys and Indians; this was a film within a film. All the while the three photographers politely offered each other exclusives in a farcical fashion, doling out what little there was to shoot in equal portions like sharing a dime between three people.

Back in L.A., Cynthia, Peter and I dined at Morton's to celebrate the fact that we had all got along so well and the beginning of my relationship with *TV Guide*. I had never worked with a picture editor before, let alone in a foreign land; I was petrified and excited at the same time. Cynthia knew exactly what she wanted, right down to the film stock the photographer should use. This was a bonus at first, but soon became tedious when my taste differed. I hated the pinkness in the Kodachrome 64; my preference at that time was for Ektachrome with an 81a filter. On the second assignment, Cynthia allowed me to shoot some of the pictures my way and admitted she liked them better, so I continued to shoot as I wished.

My friendship with Peter would last over twenty years, and the promise of assignments for *TV Guide* and other magazines lured me back to Los Angeles many times during that period. I would often stay with

Peter and his wife Suzenna at their charming home, and my arrival was greeted with a wet kiss from Hank, their tiny Maltese terrier. No small part of the pleasure of these visits was spending time with Peter, looking at his impressive body of work.

Ted Danson

My first project for *TV Guide* was with Ted Danson, together with his wife and child. When the time came for me to leave, they all said how much they had enjoyed themselves and hoped they would see me again. This was a hollow wish as these were the early days of *Cheers*, the television show which propelled Ted to stardom. But I have my portrait shot on Ektachrome with an 81a filter, to remind me of that happy day.

Ted Danson, Los Angeles, 1981

The Woodwards and Dotrices

The Woodwards and Dotrices are theatrical dynasties, and few conjoined families can boast so many thespians. My friendship with Karen Dotrice gave me the opportunity to photograph the christening of her son, Garrick, for *Hello!* magazine. I had met Karen and her first husband, Alex Hyde-White, on the set of *The First Olympics: Athens 1896*. We bonded quickly, as we were all the children of actors and shared the same sense of humour. The friendship blossomed and on my many trips to L.A. Karen and I would spark off each other and Alex would remark that we would make a great double act, on top of which we all loved animals.

As the sun set in the canyons of Bel-Air, I gathered into my lens this thespian group that also included Karen's parents, the actors Roy and Kay Dotrice, and her sister Michele and her husband, the actor and singer Edward Woodward. Only one Dotrice sister had watered down the concentrated brew by marrying a businessman. These were not amateurs in front of the camera, so the task was easy and they were having a ball. Just as the sun tipped the top of the mountain, turning everything golden, Alex and his son embraced.

The Woodwards and Dotrices, Los Angeles, 1990s

Robert Culp

Robert Culp gave me a new challenge; a problem with his eyes meant that he could not stand bright lights, and even lighting him softly proved painful. He was utterly charming and as helpful as he could possibly be in the circumstances, and I discovered it was easier for him not to look into camera. I made up for my presence by taking a huge and genuine interest in a wooden windmill which he had built for his children's mice. The extraordinary coincidence was that my father had built me one that was almost identical for my white mouse Snowdrop. The only other person I knew of who had done the same thing was Sir Isaac Newton. How strange that two dads, both actors, living on opposite sides of the world, dreamt up the same toys for their children.

Robert Culp, Los Angeles, 1981

Harry Dean Stanton

Harry Dean Stanton presented another kind of predicament. He was a reclusive figure, best known for his superlative performance in Wim Wenders's 1984 film *Paris, Texas*. I arrived in the early afternoon to be greeted by Harry in his dressing gown, with a black cat in his arms. The

Japanese-style living room was gloomy, with dark wooden walls and half the curtains drawn, cigarette butts filling the ashtrays and books strewn on the sofa along with the telephone. There was something special, though, that in this day and age of homes looking perfect, here was an unapologetically lived-in space. Harry was polite and welcoming but wasted no energy in trying to impress; no fatuous smiles, only the occasional honest one. He excused himself to get dressed. In retrospect, I think he wanted to be sure that I would arrive, and that I was a good soul invading his space. I had time to reflect that I did not feel comfortable photographing this sensitive soul with the physiognomy of a poet. When I looked into his eyes, they were soulful, melancholy and reluctant to engage with the camera. Now, many years on, I can understand the desire to become a hermit and shun the world of stardom; it takes a certain type of toughness to survive in that furnace where so often the most sensitive, artistic souls perish. Harry made a choice that few are privileged enough to be able to make, for he has a great recognised talent. Buddhism had become his way of life, which seemed to suit the man I met.

Harry Dean Stanton, Los Angeles, 1980s

Cloris Leachman

One sunny morning, I received a telegram which turned my world upside down. My darling dog, Emily, had had to be put down. The bombshell hit on a day when I had an "at home" shoot with Cloris Leachman. Somehow I had to pull myself together and get through the day. Luckily, Cloris was magnificent and a real sport.

Cloris greeted me warmly as we entered the dark, low-ceilinged rooms. She could not have been more sympathetic, but when she said perhaps I should not have come, I explained that work is the greatest panacea and the assignment could not have been more welcome.

Soon we were both roaring with laughter at her crazy ideas. We worked well as a team as I snapped away in her study, holding a huge magnifying mirror to her eye, and in the kitchen putting mushrooms up to her ears and karate-chopping a poor vegetable; on and on the ideas flowed until the day was done and she invited me to stay for dinner. I knew that Cloris was vegetarian but at that time she was on a cabbage diet. I mentioned the problem of wind, which made her scream with laughter, and I made a mental note to buy charcoal on the way home.

Cloris Leachman, Los Angeles,1990s

Griffin O'Neal

I got no such sense of a good time from Griffin O'Neal, the son of actor Ryan O'Neal. He was like a wild horse recently corralled and determined not to be broken. While he was averse to the whole idea of doing a shoot, his wife, who was pregnant at the time, and his publicist saw this as a perfect opportunity for a story for *Hello!* magazine. You didn't need to be psychic to know that the marriage would only last five minutes. Make-up, hair, the manager, the publicist, all were there for the shoot. To my alarm, they were also there to watch my performance, but this did not bother me for long, as the grotesquely overweight manager never stopped raiding the refrigerator and cracking jokes about the lack of food on the shelves. The whole shoot was a struggle, and I felt uninspired; too much energy was going into placating Griffin and hearing about the manager's food requirements. The couple did as they were told, sitting side by side, but they looked bored.

Griffin O'Neal, Malibu, California, 1990s

Then suddenly I spotted a school of dolphins swimming along close to the shore and realised they had come to my rescue. I was so excited that I just stopped the shoot and dashed out of the door. The dolphins worked miracles. They are so magical, and I almost believed they had come to help. Now Griffin was cooperating, posing in a loving fashion with his wife, frolicking on the beach, handing large slices of bread to the seagulls and daring them to swoop low enough to snatch the morsels from his hand.

The shoot over, Griffin was hugging me as though we were new best friends, and the backup team was muttering about how much they appreciated the way I worked. My muscular assistant, who had spent an inordinate amount of time giving me relaxing bear hugs, probably because he knew next to nothing about lighting, now loaded up the car with the equipment to be returned to Sammy's rental. As we set off along Pacific Highway, the sun was setting over the ocean.

David Carradine

Another difficult assignment was my "at home" shoot with David Carradine. Tales of his wild hippie lifestyle fuelled by drugs and alcohol abounded. However, I was intrigued to meet the man who had shot to fame as Grasshopper in the 1970s television series *Kung Fu*. I arrived on time at a tiny house in the hills which was previously owned by Harrison Ford in his days as a carpenter. A secretary apologised for David's absence and explained that he and his fiancée had had to go to an urgent meeting with their lawyer and would be grateful if I could wait. Of course I could, and I did not care for the idea of an abortive trip, since getting the equipment together and finding the celebrities' homes is always 50 per cent of the work. The wait was long, I was beginning to wilt, and I was tired of drinking Diet Cokes and listening to endless stories of David's wild escapades from the hero-worshipping secretaries. Just as I was about to give up, the door burst open and David and Gail, his fiancée, bounded in, gushing their apologies. I instantly saw the attraction of this charismatic man, who must have been drop-dead gorgeous in his youth. No wonder there had been such a string of beauties in his life, including the lovely Barbara Hershey, who bore him a son, named, in keeping with their hippie lifestyle, Free. I was soon agreeing to come back another day, as they were hardly in the mood to do shots after a harrowing time with the lawyer. It was only later that I learnt that going to the lawyer was a way of life for David.

The next appointment was set up and this time I was given an even warmer welcome by Gail. "Honey, it's so great to see you," she enthused, then added, "I'm really sorry but David cannot be here today." Again, I feebly reassured her that this was perfectly okay; it would all happen when it was meant to in the true Californian way, or in other words, when Mr Carradine was in the mood. They liked me, so they wanted to make it up to me, and Gail had a wonderful idea: she would take me horse riding. I thought I had the perfect excuse that I was not dressed for the saddle, but this was brushed aside, hats and boots galore were offered, and soon I looked like an American cowgirl kitted out in a checked shirt, Stetson, jeans and cowboy boots. I muttered I had only sat on a horse twice in my life, but this too was brushed aside; we would only walk the horses and the experience would blow my mind. Well, it did not exactly blow my mind, but I certainly found it painful to walk for several days. An hour in an American saddle riding along the winding roads in the Hollywood Hills, with cars stopping in amazement and men cheering us on, and letting the horses "run" along the grass verge were experiences I was sure I did not want to repeat. I still had not taken my cameras out of their bags.

The third time his lordship was in the mood to be photographed. Having tested my nerves by swinging a chain around my head which would have decapitated me if he had made an error, he then posed with Gail in a more relaxed fashion and the friendship was sealed. Invitations flooded my answering machine. I never knew what to expect on my visits. An invitation to lunch seemed safe enough, but soon I was being bamboozled into taking a dip in the Jacuzzi. The swimsuit I was lent had little more than a thong up my butt which was unusual in those days, but I tried to be a good sport. On another occasion, David greeted me in the smallest pair of underpants, revealing an awesome display of tattoos. The sun's rays streaked down his thighs, suggesting that the sun did not so much shine out of his arse but right from his crotch. An eagle with its wings expanded was etched around one side of his sinewy chest. I

David Carradine, Los Angeles, 1980s

thought of the line from Hans Christian Andersen – "the King was in the altogether" – and asked whether he had forgotten to get dressed that day. As I knew it was all done for effect, it seemed a waste not to comment on the situation. His three-legged dog, who had been hit by a car, followed his owner devotedly before finally settling at the piano, where David proceeded to play and sing love songs. What a sight: a three-legged dog, a tattooed man at a piano, and the usual bunch of sycophants hanging on his every word. A picture opportunity, but not one I could use.

Joanna Cassidy

In the late 80s I met Mitch Kreindel, an extremely talented actor who supplemented his income with PR work at Freeman and Sutton Public Relations. He suggested I might work with one of his clients, Joanna Cassidy. I already knew her work, as Eve Arnold had recommended that I see *Under Fire*, as it was about a wartime photojournalist. Joanna starred

Joanna Cassidy, Los Angeles, 1980s

opposite Nick Nolte and won the Sant Jordi Award for best actress. She would later win an Emmy nomination for her performance in the television series *Six Feet Under.* However, Joanna's performance in *Blade Runner*, stripped to the waist with a boa constrictor snaked around her neck, has become iconic; *Blade Runner* has become a cult movie and every boy or man that I meet to this day is wide-eyed when I say that she and I are great friends. My stakes soar...

As we travelled west on Sunset towards the coast in Mitch's shiny black Beetle, the boulevard appeared to go on forever and each bend looked similar to the one before. Suddenly, we had arrived at a picturesque white house which looked more European than American. Wooden shutters framed the windows and little mansard windows peeped out of the grey-tiled roof. The house was enhanced by the most colourful front garden I have ever seen. Flowerbeds, neatly edged with box, were bulging with brightly coloured blooms: majestic dahlias, petunias, geraniums and a multitude of smaller plants of every hue. Weeping ficus stood proudly by the front door and a lush green lawn curved around the gravel driveway.

Joanna greeted us warmly and I knew instantly that I would enjoy the shoot. As we walked around the garden, I asked her what it meant to her. She pondered for quite a while and said: "I'll think about it."

Then came her response: "I am most comforted when I garden; the more dirt, fertiliser and water, the better. My day goes by in a great flurry of colour, perspiration and a lot of talking to myself, which I like a lot." Gardening has become one of our many shared interests.

Having explored the back garden, which was as well tended as the front, we wandered through the French windows into the cool, shady sitting room. Joanna changed outfits and posed in every one of her small, cosy rooms, and we treated the whole day as fun. Laughter was in abundance, drinks flowed, and rolls of exposed film gathered rapidly as I shot until the sun went down. I departed as flat as the batteries in my power pack, but overwhelmingly happy with the day's work. I promised to let Joanna see the pictures as soon as they were processed and she loved them. I was overjoyed, for now I had satisfied myself creatively and gained a new best friend, and the funny thing is that we have remained great friends through thick and thin over decades.

On my next trip to Los Angeles, Joanna invited me to stay at the same fairy-tale house on Sunset, in her enchanting little guest house. My view at the back was dominated by a weeping silver birch softening the lines of the curved-edged swimming pool and again the riot of brightly coloured blooms. On my first morning there was a little tap at my window and a smiling Joanna appeared with a cup of coffee, and I knew my stay would be joyful. I could not resist taking more pictures in her garden, where a bed of antirrhinums was sacrificed in the name of art as Joanna lay sprawled across them, with me up a ladder.

Then came another trip, and I arrived in torrential rain. When rain comes to L.A. it is so dramatic that roads through the canyons turn into rivers and people lock their doors and stay home. I drove hazardously through the storm all the way down Sunset towards the coast until I reached Joanna's house. When I arrived, I was surprised to see most of the curtains drawn, but thought no more of it. I parked the car around the back and ran to the front door. I had an awful foreboding that no one was home despite some lights being on. I was now frantically ringing the bell, as I was drenched. Truly in a dilemma, with no mobile and no immediate access to a phone, I returned to the car to wait a while just in case Joanna had popped out. After about twenty minutes, I climbed the stairs at the side of the house and caught sight of a pair of legs running past a window, so I dashed for the bell and hit it long and hard. "Can I help you?" asked a rather startled young woman. I was equally surprised that she did not know who I was. I explained I was looking for Joanna Cassidy. "Oh," she replied, "she moved several months ago." This bright young woman bothered to listen to my sorry tale and welcomed me in to make a telephone call. It then became clear that I was in Jamie Lee Curtis's house and this was her personal assistant.

Having driven twenty miles in the wrong direction, I finally swept up a short, steep drive overhung by dense, dripping trees and there stood a house which looked like a Venetian palace. A little stone balcony fit for Juliet shaded the solid wooden front door, which flew open, and there was Joanna, looking stunning, a vision in black; sharply tailored jacket and leggings, riding boots, a short elfin cut to her vibrant red hair and perfect make-up, enhancing her translucent skin. She was on her way to an interview; she was really sorry to have to leave me straight away but she would be back by five; I was to make myself at home and have some food.

My bedroom reminded me of my grandmother's style; piles of large tapestry cushions with velvet backings adorned the double bed. Clusters of flower paintings covered the walls even above the doors, an antique chest of drawers was decorated with china ornaments, and standing grandly in the corner was an English walnut tallboy; I felt very much at home. I began to unpack, filling the drawers of the handsome tallboy and jumping into the shower. Wrapped in white towelling from head to toe I collapsed onto the bed and spent the next couple of hours dozing and reading until I heard the bang of the front door, dogs barking and cries of welcome. Suddenly the place was alive, and I ran down to greet Joanna properly as her five dogs, who had been confined to the kitchen, leapt around us, all wanting to be the centre of attention.

So much to talk about, so much to do ... we did not stop until I left three weeks later. We would do the garden, hang pictures and move furniture, and I could not have been happier, although we looked forward to the end of the rain. These were halcyon days and everything seemed to amuse us, including one other member of the household. Kesch was from Argentina, twenty-seven, proud of his fine body and noisier than Joanna and I put together. Every morning, I would hear Kesch's stereo playing full blast while he took his shower. As soon as I smelt the fresh coffee, I would wander down to the kitchen, where a cacophony of sounds would hit me as I entered. Gleeful dogs, Joanna shouting "Good morning," Kesch's booming voice announcing that he was about to "maka de food". This hyperactive "Brazilian Nut", as Joanna nicknamed him, was there to help with everything from gardening to caring for the dogs to endless repairs; he usually spent his afternoons at the beach surfing. He would then make a habit of appearing with no shirt, flashing his pectorals whenever guests were present.

Then the phone call came from her manager, and Joanna looked serious for the first time. "He says I need to get back to work," she told me solemnly. Less than twenty-four hours later, she was called for an audition. We dashed to the clothes closet to choose an outfit. "No, too wild, too severe; mustn't frighten the suits," she commented, and then we were madly agreeing, "Yes! Smart but feminine." The true professional was transformed into the beautiful actress.

"They want me to go back and read for them. I don't read for people." Joanna was standing on the stairs, her blue eyes boring into me. She wanted my opinion. I tried to put myself in her position. "Right," I replied. "Do you like Dudley Moore? Do you like the script? Well, go and read, it's only your pride that's getting in the way."

The following morning, I awoke to find five dogs and a triumphant Joanna sitting on my bed. "I've got it!" she cried, as the dogs all leapt about as if they knew what was going on. Suddenly the atmosphere changed as rehearsals started that day, and coincidentally I had meetings to attend. The wine was replaced by water and we became disciplined. I would hear Joanna's lines sitting in our pyjamas by the fire, which would bring back childhood memories of doing the same thing with my father. The pilot show with a live audience would be recorded two days before I was due to return to London; the timing was perfect, as I would get to see it. The last week passed peacefully and happily, and Joanna worked all day and learnt her lines in the evenings. I, on the other hand, was out all the time, running in circles seeing all my friends and work contacts as my departure date was imminent. The day came all too soon and I felt so sad; it had been a great time for both of us. But it was time to leave, with Joanna's distinctive, melodious laugh ringing in my ears.

Joanna Cassidy, 2005

CHAPTER 8

PHOTOGRAPHER ON FILMS

"It is never long during the making of a film before the off-set atmosphere begins to reflect that of the story itself."
– Roman Polanski

"You should be a photographer ..." Eve Arnold's words changed the course of my life. Over the years, through good times and bad, her encouragement was a guiding light. Our treasured meetings over the decades kept me focused mentally and through the lens. For many years, I took my portfolio to our meetings and she would comb through the images and criticise them down to the smallest imperfection, once adding that she had recently made the same mistake herself. This was a great compliment, as she told me that when people showed no talent she simply said "lovely, lovely" and closed their portfolios, never wasting any time. However, after several years of visiting, I left my pictures at home and we simply met as friends.

Eve was always in my thoughts when I was assigned to shoot specials on film sets, because this was what she was doing when we first met, reminding me that diplomacy, charm and sensitivity are just as important as taking good pictures. Shooting specials entails taking photographs exclusively for publications and agencies. While the unit photographer covers every scene for the film company, a specials photographer needs to come up with something different that will ensure the photographs will appeal to magazines and newspapers.

My experience as a unit publicist and my friendship with Eve had trained me well for the unique, often claustrophobic atmosphere of the film set. There is this myth that everything to do with the film industry is glamorous, but the reality is just the opposite. The Venus on the red carpet is a far cry from the actress waiting between scenes in her trailer, often with curlers in her hair, her costume hanging in the corner and the ubiquitous polystyrene cup on hand. The locations are more often than not either freezing cold, boiling hot or just wet, and even when the sun shines, the hanging around requires infinite patience. Some of my experiences on set were, accordingly, quite challenging, while others I remember fondly as rewarding encounters with legendary stars.

Princess Daisy

One of my first forays into specials was also one of my more pleasant assignments. A television film of the bestselling novel *Princess Daisy* by Judith Krantz had all the ingredients for a fun day. The set was a glorious eighteenth-century, double-fronted house in the Boltons, one of the most salubrious squares in London. It was a perfect summer's day: balmy, azure blue skies and barely a breeze. For once I had an opportunity to dress in smart casual clothes, a welcome change from the sensible anorak in case of rain or the layers of woollies to keep out the cold. The director, Waris Hussein, and I had met socially, so it felt more like a cocktail party as he introduced me to Claudia Cardinale, Rupert Everett and the other

Claudia Cardinale, *Princess Daisy*, 1983

stars. I was enchanted by Claudia; here was one of Italy's biggest film stars with no apparent ego and good old-fashioned manners. It was a bonus to genuinely enjoy Claudia's company.

Terry O'Neill, a fellow photographer whom I had known since my days as a unit publicist, was visiting the set. He greeted me warmly. I had a great affection for Terry, as he was one of the few male photographers of his generation who accepted a female photographer.

"Meet Carole Latimer, she's taken great pictures of men in the nude," he said, as he introduced me to Ringo Starr and his wife, Barbara Bach, who were making a guest appearance in the film as husband and wife. I was flattered that Terry had noticed my exhibition and delighted to meet Ringo. Next I was chatting to charismatic Rupert Everett, who epitomises a rebellious member of the ruling class from days of yore. He introduced me to his mother, who was visiting the set, and whose strong personality and smart, tailored appearance was the antithesis of this theatrical crowd.

Finally, I was invited into the house to witness a scene I was hoping to cover, but I soon realised this was an intimate love scene, always a delicate moment in filming. The first assistant, chosen for his powerful voice, bellowed out to us to clear the set. As I walked towards the door,

Ringo Starr and his wife, Barbara Bach, *Princess Daisy*, 1983

Rupert Everett, *Princess Daisy*, 1983

I heard Rupert say, "No, not Carole. She's one of us." And so I found myself sitting next to Mummy watching her son, who was playing a character with the ludicrous name of Ram, rehearsing a scene where he abuses his half-sister, Princess Daisy. Stark naked, they looked vulnerable under the hot, probing lights. Every inch of their flesh was exposed to us apart from what looked like band-aids covering his tackle and her tits, which reminded me of the "no entry" banners stretched around the scene of a crime. I glanced sideways at Mummy as we watched this sea of glistening flesh, arms and legs entwined; she gave nothing away. Best not to ask why Rupert referred to me as "one of us".

Reunion at Fairborough

Another memorably pleasant assignment began inauspiciously as I set off for Cambridgeshire on a wet, blustery morning, hoping that by the time I arrived the weather would have improved. It did not. I had no intention of trying to persuade the two legendary stars of *Reunion at Fairborough*, Deborah Kerr and Robert Mitchum, both in the autumn of their lives, to pose for pictures in the pouring rain and a relentless, fierce wind. The publicist nervously introduced me to Robert Mitchum, who looked at me and silently hung his stick on my arm. I simply walked away with the stick. I could see, out of the corner of my eye, that he was mildly amused. Rather fun, I thought, but the publicist merely became more agitated.

Deborah Kerr proved a tremendous bonus, as we simply made the most of the time spent waiting around for the weather. She invited me into her caravan, where she offered me tea and we chatted like old friends. I was thrilled to be spending time with such a legend, and soon we were discussing friends in common. She had known my parents' great friend Jack Hawkins, so we reminisced about visits to Roehampton and his pet hare, who ran wild and came when called. The publicist kept nervously interrupting like the chorus in a play, mentioning famous scenes from Deborah's films, even adding the one on the beach with Burt Lancaster. I think he felt I was being disrespectful for not mentioning her work. I, on the other hand, felt that she might prefer a change of subject from her films. So, although Deborah Kerr barely removed her headscarf because of the howling gale, I had a memorable afternoon with a woman who was everything she had been cracked up to be: utterly charming.

Judi Trott, Deborah Kerr and Robert Mitchum, on set, *Reunion at Fairborough*, Cambridgeshire, UK, 1985

Foreign Affairs

A decade later, I would spend time with another legendary actress when I drove down to the English countryside to shoot specials on a production called *Foreign Affairs*, in which Joanne Woodward starred opposite Brian Dennehy. As Joanne knitted to while away the time between her takes, we talked about our shared passion for gardens. We sat under a vast cedar tree in a beautiful garden belonging to a grand estate. The late afternoon sun created that soft golden light so typical of England at its best as it moved lower in the sky, casting long shadows. I was enchanted by Joanne's warmth and down-to-earth attitude; stars of her calibre remind me of why I enjoy being around the film industry. Maybe her long and successful marriage to Paul Newman partially accounts for her air of inner peace, but on the other hand maybe that very quality was the secret to their union.

I temporarily broke off our conversation to take some pictures of Eric Stoltz, who looked stunning in a dinner jacket and black tie. As I peered

Joanne Woodward, on set, *Foreign Affairs*, Buckinghamshire, UK, 1993

Eric Stolz, on set, *Foreign Affairs*, Buckinghamshire, UK, 1993

through the lens, I studied his finely structured face, the perfectly shaped mouth and intelligent hazel eyes. But above all, I shall never forget the colour of his hair: a stunning deep shade of auburn, as deep as an autumn chestnut; with the added enhancement of the afternoon light, I could not believe my eyes. This was an occasion when I was truly happy to have shot the pictures on transparency and to have got the exposure spot on.

Steaming

The day on the set of *Foreign Affairs* had been a gentle reminder of how rewarding my work on films could be when the conditions were harmonious. On some films, not surprisingly, this is not always the case, though it's often possible to find some blessings even in the most difficult circumstances. *Steaming*, based on the book by Nell Dunn, was an unhappy picture, as both the director, Joe Losey, and one of the lead actresses, Diana Dors, were terminally ill, and this would prove to be their last film. A director would rather die than give up his baby mid-filming, so Losey was directing from his chair with an oxygen mask by his side. Nobody can take over your part halfway through shooting, so Diana struggled

Director Joe Losey, *Steaming*, Pinewood Studios, UK, 1984

Vanessa Redgrave, *Steaming*, 1984

through in great pain, which she did her level best to hide, but I caught glimpses of her discomfort out of the corner of my eye when she thought no one was looking. It must have been agony for Vanessa Redgrave and Sarah Miles to watch their pain, but both are such professionals that they could not have handled the situation better.

The highlight of that week was Eve Arnold's arrival to shoot specials. The producer was deeply in awe of the great lady and before she arrived, he kept nervously telling me that I must keep a low profile. I said nothing. The time came and Eve was escorted onto the set by the fawning producer. I stood quietly on the other side of the set and observed the scene. Quite suddenly, Eve spotted me and without a pause flew across the floor, arms wide, and greeted me with joy. In the end, this turned out to be a day when there were no photo opportunities, "one of those days when it would have been better not to have got up", Eve said with a wry smile. Eve's patience paid off, for the next day she was given her exclusive pictures and I stood well back with due respect and fascination as I watched my mentor at work.

Photographer Eve Arnold, on set, *Steaming*, 1984

The First Olympics: Athens 1896

Another assignment that proved challenging was my work on a television miniseries, *The First Olympics: Athens 1896*, the true story of the American team who entered the first modern games in Greece. I jetted off to Athens with the mostly male cast. I was happy to be escaping part of the English winter; December in Greece would be a treat, and what a wonderful excuse to forget about the Christmas duties. Alas, it was not so perfect. Day after day we were soaked to the skin in the vast stadium, then dried out by the sun. I was deeply disappointed in Athens, as the Parthenon was then out-of-bounds amid the heavy traffic, noise and, above all, pollution. Our hotel, the Athenaeum, was vast and the foyer was like an airport departure lounge. A white Flokati rug in my suite was the only ethnic furnishing; otherwise I could have been in any international hotel anywhere in the world.

The schedule was so tight that the director could never wait for the light to match, and the lighting cameraman grew more morose by the day. "They should call me mushroom. They keep me in the dark and feed me bullshit," he muttered as I plied him for information as to what we were going to do next, wondering what film stock I should use. The lads became visibly more tired as their trainer forced them on and on, racing around the stadium all day and training back at the hotel at night. Then, when they were actually filming, Ed Wiley, who played the part of the coach, was still yelling, "Let's go, let's go!" I'm sure the sound of his voice must have been ringing in their ears for weeks afterwards.

Team training for *The First Olympics: Athens 1896*, Pinewood Studios, UK, 1984

The First Olympics: Athens 1896, Panathletic Stadium, Athens, 1984

Hunt block jumping, *The First Olympics: Athens 1896*, Panathletic Stadium, Athens, 1984

There were some bright spots. The shortage of women in the film did my self-esteem the power of good, as I was thoroughly spoilt with invitations and, on one occasion, a single red rose appeared in my pigeon hole at the hotel. When we had the energy to go out in the evenings, the stars, crew and I enjoyed traditional dishes complemented by the local retsina in tiny restaurants.

On the very day we flew back to London, a double-page spread of my pictures of the film appeared in the *Mail on Sunday*. The timing was perfect and made the trip seem worthwhile, and I eagerly passed the spread around to all the cast and crew on the plane. One of the air hostesses became aware of what was going on; when she promptly told the pilot, word came swiftly that I was to be invited into the cockpit. Apparently the pilot was intrigued to meet a female photographer who had been published in a newspaper.

Once again my camera was opening doors for me, to the cockpit to be precise, and soon I was wearing headphones, listening to our flight information. I barely asked any questions, because I did not want to be a nuisance in case they sent me back to my seat. I was duly rewarded when the captain asked me if I would like to stay for the landing. What a question. I was overjoyed. Being in the cockpit was like sitting in a giant

Honor Blackman, *The First Olympics: Athens 1896*

Louis Jourdan, *The First Olympics: Athens 1896*

Angela Lansbury, *The First Olympics: Athens 1896*, Pinewood Studios, UK, 1984

Virginia McKenna and Bill Travers, *The First Olympics: Athens 1896*, Athens, 1984

fishbowl, the glass folding right down to our feet. This did not seem so alarming with the sky and clouds around us, but as we started to land, warnings of dense fog came through on the headphones and I wondered if ignorance would have been bliss. The pilots made jokes about landing on automatic pilot, and all too soon the ground loomed up from nowhere, speeding faster and faster. Then, quite suddenly, two straight rows of lights were racing towards us and in a blink we had landed.

The Sun Also Rises

My work on *The Sun Also Rises* turned into a nightmare but began in a promising way, with a day trip to Bath to photograph one of the stars, Jane Seymour, at home in a Tudor mansion which she had bought after working there on a film for television. We snapped all around the garden with her daughter, Katie. They held hands by the orangery, ran down the bank, played with the kittens, sat on a pony and so on until we had exhausted the great outdoors. We broke for lunch under a sun brolly and established that we knew many thespians in common, which is usually reassuring to an actor I have never worked with before; it is like a club. The bond was formed, and Jane said she looked forward to working

Jane Seymour at home, near Bath, England, 1984

Jane Seymour's Tudor estate, near Bath, England, 1984

with me. After lunch, we continued to work enthusiastically, with Jane changing into different dresses to suit each interior. She was utterly professional and a joy to photograph.

The film was to be shot in Paris and Segovia, and in Paris I stayed with a friend, Monique, on the Champs-Élysées, which gave me a wonderful opportunity to brush up on my French in the evenings. Despite this friendly connection, here I was again, though, arriving on a strange set, and to my consternation the publicist was not there to introduce me, so I had to fend for myself. It is always hard at first before the crew knows they can trust you not to snap on a take.

Just as I introduced myself to the director, James Goldstone, a booming voice rang out, "If it ain't the callipygian Venus!" I swung on my heel. It was Bob Joseph, a producer I had met at a dinner in Los Angeles. ("Callipygian", he explained, was derived from the Greek, meaning perfectly shaped buttocks.) "Show the director your arse," he continued. I tried to conceal my horror as I studied this man who looked distinctly like a toad, and decided to make light of it. I did a mock twirl, smiled at the director and asked if I would be suitable for the job as a photographer. Although embarrassed, I still felt far too professional to be thrown by Bob's behaviour.

Day after day, indoors and out, I followed the stars from one set to another. Each day I had to field Bob's chauvinistic remarks as I was a guest on the set. I simply concentrated on my pictures and developed a good relationship with the stars, all of whom were totally cooperative, but they too were having their difficulties with Bob. Hart Bochner had a nervous stomach, as every time he missed one word Bob would yell "Word!", and even Jane was reduced to tears. Only Leonard Nimoy escaped, as Bob worshipped the ground he walked on. In the evenings, I would return to Monique's apartment; she was astounded at the bad behaviour, which was hard to comprehend for someone who had not worked on a film set. Monique always suggested that we should *boire un coup*, a nightcap of whisky, in her study, where we would soon be relaxed enough to laugh at it all.

The hours spent hanging around film sets are long and tiring, so before filming moved to Segovia I took a week off in Mallorca. Sadly, instead of arriving tanned and fit, I joined the film looking like one of the victims

Jane Seymour, Segovia, Spain, 1984

Leonard Nimoy, Segovia, Spain, 1984

of the Spanish Civil War. My arm was bandaged as I had been severely stung by jellyfish. The poison had seeped up to my armpit, so a cortisone injection had been inevitable. I was a sorry sight with my open wounds, but the make-up artist was thrilled and wondered if I would mind if he took shots as reference pictures for the gruesome wounds he had to create for the shooting.

And so I carried out the rest of my assignment with one strong arm and the other too weak to lift a feather. Days were spent filming in the streets of Segovia and around the bullring, where the heat rose and Bob became even more abusive as his alcohol intake increased. The baiting did not stop until finally on my second to last day, as the temperature soared, he yelled, "You're full of shit and a complete phony." I walked calmly away, but when one of the sound men put a comforting arm around me, tears flowed from beneath my sunglasses and I fled from the set.

The last night in Segovia I dined with the cast, relaxed at last, knowing that I had finished my assignment. However, as Hart Bochner and I left the restaurant a lone, drunken figure, propping up the bar, bellowed

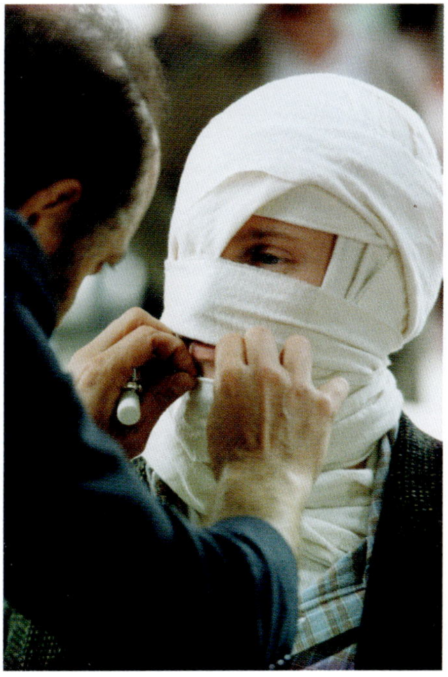

Hutton Cobb (in bandages) and Hart Bochner, on set, *The Sun Also Rises*, Paris, 1984

across the foyer, "Come here." Hart told me not to go, but I decided not to run away. I walked purposefully towards the bar and asked Bob what he wanted. "You're full of shit," he repeated. "I've heard that

Robert Carradine and Leonard Nimoy, *The Sun Also Rises*, Paris, 1984

From left rear: Robert Carradine, Željko Ivanek, Leonard Nimoy, Ian Charleson, Jane Seymour and Hart Bochner, *The Sun Also Rises*, Segovia, Spain, 1984

already," I replied, and calmly wished him goodnight. A few years later, I heard that Bob survived major surgery only to die when he fell down the steps of his garage and cracked his head. I was sad that we had ended

Director James Goldstone (left) and producer John Furla Jr and crew, on set, *The Sun Also Rises*, near Segovia, Spain, 1984

Left to right: Robert Joseph, Robert Carradine and Željko Ivanek, *The Sun Also Rises*, Segovia, Spain, 1984

our acquaintanceship on such a bad note, as he could be so entertaining socially. Sometimes people metamorphose into monsters in a work situation and, in this case, I think he cared too much about the project: it was too close to some personal experience of his. I vowed never to go on location again unless I had vetted the men in power.

Hart Bochner, Jane Seymour and Leonard Nimoy, Segovia, Spain, 1984

Gulag

A couple of assignments have had all the earmarks of disaster but have eventually evolved into satisfying and enjoyable work. Things certainly did not seem promising as I drove down to Surrey in the slashing summer rain to photograph David Keith on the set of the television film *Gulag*. Tales of his fiery redhead's temper had gathered momentum with each telling and spread like wildfire. I struggled to find the abandoned Victorian lunatic asylum buried in dense woodland and realised I might as well accept that this was going to be a horrible day. Just when I had almost given up finding the location, tall turrets rose out of the mist and a red-brick Gothic monster appeared; I hated it on sight.

Wet cables, like slimy black eels, guided me up the grimy, grey stone steps where, in the mist and gloom, I found the unit huddled around a low-lit scene. The publicist, Doreen Landry, appeared out of the dark, and as we hugged each other, she warned me that David was in a foul mood, so I had better wait patiently. The hours passed. People I knew from other film sets came up and greeted me while all the time I noticed, out of the corner of my eye, that this exceptionally tall, toned, fiery, sexy hunk was observing me from afar. We were stalking each other like feral creatures, gathering information without a word spoken; I made no attempt to approach. At five o'clock in the afternoon, when I had all but given up, I was shaken out of my reverie by a booming voice, "My name is David Keith. I gather that you're a very good photographer." I swung around to be confronted by a large expanse of chest. I felt like David confronting Goliath.

At last, the mountain had come to Mohammed. As it turned out, we got along famously, and though David is American by birth, we have

David Keith, *Gulag*, Surrey, UK, 1985

Scottish blood in common. His red hair and the dimple in his chin gave me clues to his fiery nature, but my mother was also a red-headed Scot. He could not have been more helpful and said that if I came back the next day, he would be at my disposal, provided he was not needed for a take. It is not uncommon to spend a whole day without getting one good shot; patience is all.

The next day the soft, English sunlight transformed the exterior of the faded Victorian institution, even though the interior remained grim. I now had a feast of wonderful locations: stone arches, broken stained-glass windows with plants woven into the cracks and sunlight shafting through the gaps. I was spoilt for choice and David was keen to work with me when the light was at its best. He was fun, with a great sense of humour, which showed through the lens. A few days later, he invited me for a drink at the Athenaeum Hotel in London, where we downed a whisky or two to celebrate. He loved the pictures and I thoroughly enjoyed his company.

Wild Geese II

My ability to tame a difficult actor must have impressed the publicist, Doreen, because she called a couple of years later with another formidable assignment. I was cleaning out the grate in the drawing room, which was filled with grey ashes to match the dark winter day, when the telephone rang. "Hello," I said briskly, cursing inwardly as I covered the receiver with cinders. "What are you doing?" Doreen asked. "Right now," I replied, "I'm cleaning out my fireplace, if that's of any interest." Doreen quickly got to the point. "Could you catch a five o'clock flight to Berlin? I desperately need a specials photographer on *Wild Geese II*. Two photographers have walked off the set, both male, unable to get along with Scott Glenn." She added hesitantly, "I thought perhaps a woman might be a good choice at this point." I loved a challenge.

Berlin was still divided into western and eastern sectors, and as we flew over the city I looked down at the floodlit Wall, which was nine feet high and stretched for one hundred miles. On the eastern side it was painted white, while on the west it was covered with garish graffiti. Maybe my schoolgirl German would, at last, be of some use; at least I knew how to order a boiled egg (*ein gekochtes ei*).

The next morning, after a leisurely breakfast, we were swept off in a chauffeur-driven car to a military academy which was the German equivalent of the British SAS training centre. Doreen and I sat in the back, while Scott Glenn fidgeted around in the front like an excited child being taken out for a school treat. His nickname, "Bendy Toy", suited him well. He was tall, slim and wiry, with muscles standing out like whipcords, and he had a craggy face and bright, naughty eyes. He certainly had a strange kind of attraction. He fired cheeky questions at me, testing to see if I could stand the heat. I gathered that throughout the filming he had made it clear that he wanted to have a shooting lesson at the academy, and his dream was about to come true.

We drew up at the gates, where two armed guards stepped forward to check our papers, leaning in through the windows, their eyes scrutinising us closely. They took their time, then abruptly waved us through, closing the gates firmly behind us. We entered the shooting area, a vast hall with three cardboard figures as targets, and each of us was given a set of earphones to deaden the noise. Eager not to miss a photo opportunity, I put mine to one side for a minute and rapidly opened my camera case.

Scott Glenn, Berlin, Germany, 1985

Before I had time to load a roll of film, Scott fired the first shot. I was deafened for a good ten minutes; a bad start. I gathered my composure and started shooting as Scott fired something a great deal more dangerous than my camera. "I wonder," I said to Scott, "whether you could fake a shot facing me with the gun angled slightly to one side? But do not point the gun at the camera."

I had been fiercely trained by my father never to point a gun directly at anyone. I raised the camera; the gun barrel was directed straight into

Scott Glenn, Berlin, Germany, 1985

my lens. Click ... I took one shot, looked up and said, as calmly as I could, "I knew you were going to do that," before turning on my heels and walking away. I was seething. I did not care if I was on the next flight back to England. I had already had enough; obviously a woman photographer was going to get just as much grief as the men.

To my great surprise, I received an apology, so I felt we could continue and hoped there would not be any more "tests". On the journey back to the city, Scott regaled us with lurid tales, including a recounting of the time he delivered his own baby in the sitting room with several close friends. I remarked casually that it sounded like a Tupperware party; he laughed, and I realised we would muddle along just fine. He could go on trying to shock me and I could go on acting like a crispy, crunchy Norland Nanny.

I knew we were "new best friends" when Scott offered to carry my camera case. Doreen jumped at the opportunity to arrange another photography session the following day. This time we were in a paddock and Scott was galloping around on a feisty chestnut stallion. As I focused on the horse, I realised Scott was galloping straight towards the camera; I remained calm and took the picture; after all, horses are not dumb and do not crash into people for the sake of it. I asked whether Scott could repeat the course again. No, he could not. "My balls are killing me," he growled, and added that he was not wearing underwear. I doubted the great equestrian skills of which he boasted, and Doreen looked embarrassed. The photo opportunity was limited and I wanted good strong portraits, so I made a rash promise that I would take some of the best portraits ever. Scott agreed to do the course again. It was not until a year or two later that his manager saw the results and raved about them. I had kept my side of the bargain.

I thought that the experience at the training school had been daunting enough until Doreen came up with her next plan. We were to take a day trip into the eastern sector with the British actor Edward Fox. Our driver could not emphasise enough the need to be vigilant and the dangers of not returning to Checkpoint Charlie by sunset. I sensed he was enjoying the drama.

Once inside the eastern sector, we were transported into another reality. There were few signs of modern life anywhere, except the monotone little

Edward Fox, Checkpoint Charlie, Berlin, Germany, 1985

East German Trabant cars, which looked like grey biscuit tins, unaesthetic and functional. No billboards or neon signs, the people were visibly poor and the lack of colour was weird and depressing. We looked glaringly out of place in our big black Mercedes.

Every interesting building we had time to visit became a backdrop for pictures of Edward until all too soon the light began to die. With pounding hearts, we hastily returned to Checkpoint Charlie. We arrived just in time, but the car was searched, the boot scoured, seats removed, not a mat left unturned, and once again my heart was in my stomach and I prayed they would not destroy my film. God was on my side and we returned safely to the brightly lit streets of the West, festooned with neon billboards and brightly lit shop windows.

Now I was due to photograph Barbara Carrera, who had starred as Fatima Blush in the Bond movie *Never Say Never Again*. After that, I would be returning to London. I was taken aback when Barbara said she would have to have her hair washed, which threw off the day's schedule and meant that I would miss my flight. Finally, we were being whisked off to the Charlottenburg Palace, together with Barbara's boyfriend Nicholas.

Barbara hated Berlin, and everything Doreen suggested went down like a lead balloon, but when we arrived at the palace, she became totally professional and a joy to photograph. Draped in a wonderful mixture of fuchsia and purple, she posed around statues and a vast fountain, her glossy dark hair highlighted by the late winter sunlight cascading down her back. I complemented her incredible complexion with a soft golden reflector. She was a photographer's dream.

I never got my *gekochtes ei*. That would have to wait, but Doreen knew Berlin well and we enjoyed some of the best food I have ever eaten from all parts of the world. Memorable meals were often among the great perks of visiting a film set as a special photographer. Mind, body and spirit greatly satisfied, I returned to London a fulfilled traveller.

Barbara Carrera, Charlottenburg Palace, Berlin, Germany, 1985

CHAPTER 9

ON LOCATION

"The world of reality has its limits. The world of imagination is boundless." – Jean-Jacques Rousseau

In 1973, I had only just stopped working as a unit publicist on films, and I was still feeling like an imposter in my role as a photographer. Then I met an actor, Gilles Millinaire, who was the stepson of the Duke of Bedford, and charming and divine-looking, with high cheek bones, a classic nose and wavy chestnut hair. How could I not be flattered that he should praise my photographs and invite me to take exclusive pictures of his friend Lindsay Kemp's dress rehearsal of *Flowers*? This was to be the first of several opportunities to photograph actors working on theatre and television projects.

Lindsay Kemp, *Flowers*, London, 1974

The title conjured up an image of romance and beauty; I was in for a big surprise. *Flowers* was Lindsay's interpretation of Jean Genet's *Our Lady of the Flowers*, a story about pimps, sailors, murderers and drag queens. This gave the highly talented, overtly gay Lindsay the perfect vehicle for his style of mime and dance, and a wonderful excuse to harvest beautiful young men from the parks.

Kemp's mime technique was brilliant and unique, and he has influenced David Bowie and other pop stars. Even so, he was hardly a cosy household name in 1974; more of a decadent, camp performer who shocked his audiences with shows full of glitter, blood, pansexual orgies and beautiful naked youths. I found the show deeply disturbing, but it was a feast of visuals for a photographer. I shot enthusiastically, only stopping briefly to change the rolls of film in the half-lit auditorium. The play was beautifully lit for my needs, something that I would not always experience as a new world opened up.

Next I was assigned to photograph a tour of *No Sex, Please, We're British* in the provinces, which led to a string of fringe plays, then *Exclusive Yarns* at the Comedy Theatre in London. The highlight of this period was seeing my pictures of the musical *The Matchmaker* in the huge boxes outside Her Majesty's Theatre, Haymarket. But no play since *Flowers* has shaken and stirred me to such a degree; the sheer brilliance of Kemp's mime and direction and my utter revulsion at the subject matter left me with a boiling cauldron of emotions.

Jump forward twenty years and I was encountering a very different sort of actor, none other than Bond, James Bond, at that time being portrayed by Roger Moore. Dusk was falling as I tried to keep calm for this tricky assignment. Roger had been filming all day; he would be tired and I was to be given a very short session.

Clutching all my bags of equipment, I was greeted politely by the doorman at the Connaught Hotel and followed by staff leaping out from nowhere to guide me smoothly to the correct suite. Roger's wife, Luisa, answered the door and beckoned me into the sitting room. She was polite but far from friendly, and I felt a certain wariness in her, like a lioness guarding its cub. I inwardly sympathised, as I could imagine what it was like to be married to a James Bond. The door from the bedroom flew open and Roger appeared, breaking the ice with his warmth and good manners.

"How lovely to meet you," he said as he offered his hand. "I have always admired your father's work."

"How kind," I replied, deeply flattered that he had taken the trouble to research me. I was also aware of his intelligence in checking the credentials of the press before agreeing to pictures or articles. I felt Luisa melting

Roger Moore and Luisa Mattioli, London, 1990s

like an ice cube in a warm drink. I was one of "them", not a "civilian", so the session became a total joy. We chatted on as though we had known each other forever, which only helped to get the pictures I needed for this one-off assignment for the *Daily Mirror*. I knew that Don Short, the well-known show-business journalist, had chosen me simply because I was the right person to work with the Moores.

Later that evening, Don appeared on my doorstep to collect the pictures, and I was delighted to see that he was carrying a bottle of champagne. I opened the bottle while Don adjourned to my drawing room, but my joy turned to horror when I saw Don lounging on my beautiful, velvet-button-backed sofa, trousers at half mast. I brushed his advances aside, adding that I had a friend studying upstairs who might appear at any moment. I was furious that he should think that I owed him anything! I was already a well-established photographer and the right one for the assignment. However, back then, we did not take such men to court; we just moved on. That night I shut my eyes with an image of James Bond, 007, the quintessential English gentleman who had lived up to my expectations. I always try to look on the bright side ...

Another assignment brought me face to face with not one but two great actors. Alone in the tiny, claustrophobic bar of the Almeida Theatre, awaiting Mandy Patinkin's one-man show, I found myself sitting opposite a pair of piercing blue eyes that have put many bums on cinema seats. It was Ralph Fiennes, and I thought how ridiculous it was that I had been talking to his agent that very morning. So I introduced myself and explained that I had come to see Mandy's show as research for an imminent photo shoot with the great man. I shifted uncomfortably under the intense gaze of this seriously good-looking man. Ralph quizzed me about my connection with his agent, and I was immediately struck by his dedication to his craft and all that it involved.

Our mutual admiration for Mandy eased the rest of the encounter and soon I was buried in the womb-like darkness of the tiny auditorium. When Mandy walked on, something about his huge presence, as powerful as a gladiator, his striking dark features, shining thick dark hair and lack of artifice instantly captivated his audience. There was no fumbling for sweeties and last-minute mutterings, just silence. Then the voice ... such rich tones, such range. I was moved and taken out of myself in a way that I have seldom felt in the theatre. I was not alone, and far too soon the audience rose to their feet to give Mandy a standing ovation.

So it was with great humility and excitement that I tapped on Mandy's dressing-room door. Still bursting with energy, the adrenalin racing through his great frame, Mandy greeted me with warmth. No British reserve here; this was a New Yorker with fire in his belly. We exchanged a mutual admiration for each other's work and, of course, he had taken the trouble to look at my photography before our meeting.

I have observed that the more successful my clients are, the more likely they are to have checked out my work before we meet; they do not want to waste time meeting someone with whom they do not want to work. We agreed to meet at the Comedy Theatre the following day.

The Comedy is the only theatre in London with the original ropes high up above the back of the stage, used to lower the huge backdrops. To use this space, my assistant Robert Simpson and I had to negotiate a ladder with all our lighting and camera equipment. Luckily, I do not have a fear of heights, and anyway, I was just grateful to have been given access to the location due to my father starring in a play there many

Mandy Patinkin, Comedy Theatre, London, 1990s

moons ago. Mandy arrived still bursting with vim and vigour, but in spite of the charm I could sense that with him I had to be efficient, swift and do a good professional job and all would be well. Performers of Mandy's calibre, understandably, become bored rigid hanging around for photo shoots, but he had had the good sense to bring his script to learn while we did our setting up for a second shot in the auditorium.

A couple of days later, I was rewarded for my good work with a lesson in how to prepare a mango. I had gone to meet Mandy at his hotel in Victoria and a friend had just sent him a box of the succulent fruits. Not only did he gift me several to take home, but I watched this multifaceted talent dexterously tackling the slippery fruit. Two slices released the two halves from the stone, followed by vertical and horizontal cuts, a quick flip, and the half section was inside out like a yellow hedgehog. As we tucked into the exotic fruit, Mandy talked about his family, who are never eclipsed by his success. His wife, Kathryn Grody, is his anchor, and he knows that she and their children are more important than all the stardust. Typical of this man is how he turned a brush with prostate cancer into a positive; he now raises money for the charity. I am reminded of Mandy's powerful enthusiasm when I see him act or listen to his music, and whenever I peel a mango, I give a thought to Mandy.

My first foray into the world of television was quite memorable, shooting stills for a 1973 pilot show in which Diana Dors hosted well-known guests who brought unusual animals to the studio. The point of a pilot show is to see whether an idea will work as a series, and this one apparently did not.

We arrived at a theatre in the West End of London where the auditorium was full of guests, including some who had also brought their animals. The day began well, with a cowboy arriving on horseback. The show started with Dudley Moore at the piano accompanied by a parrot, then Zsa Zsa Gabor and Diana Dors received the other guests. Lord Bath appeared with a lion cub and soon an alligator materialised, much to

Dudley Moore with Zsa Zsa Gabor (left) and Diana Dors (right), London, 1973

A cowboy with his horse, London, 1973

Alan Price and a fox, London 1973

Diana Dors with Lord Bath and a lion cub, London, 1973

Diana Dors and Richard Wattis with a baby alligator, London, 1973

Dudley Moore and parrot, London, 1973

Diana's alarm. I spotted a fox in the audience eyeing a goat in the front row; it was all becoming quite surreal.

David Cassidy and Long John Baldry, London, 1973

As the day wore on it got hotter and hotter and more and more humid, and the mixture of smells became overwhelming, what with hot people, hot animals and the occasional slip-up, which was inevitable. I cannot imagine how anyone could have thought that all the animals would restrain their bodily functions, but this was probably the deciding factor in not continuing with the series. I shall never forget how sporting the celebrities were under these conditions, especially Dudley, who managed to continue to see the funny side.

On a later television assignment, I was able to spend time with one of Dudley's early collaborators, Jonathan Miller. Through my connections in Los Angeles, I was hired to shoot specials for HBO on a 1981 co-production of *A Midsummer Night's Dream* at the BBC. Jonathan was the producer, and I could not wait to meet him. I was in awe of his enormous talent and intellectual qualifications, juxtaposing acting, directing, writing and above all, his life as a doctor. When I mentioned I had just spent six months in Los Angeles, he became extremely animated and said that the minute he got off the plane there he started to laugh and loved every bit of the absurdity of the place. I sighed with relief and knew that the shoot would be fun, even though I was not popular with the BBC

Helen Mirren, in *A Midsummer Night's Dream*, London, 1981

staff photographers, who were extremely territorial. The American side of the production was, in some ways, threatening to them, and it was still slightly frowned upon for a woman to be doing the job. Helen Mirren played the fairy queen Titania, an omen of things to come. Who would have guessed then that her portrayal of the real Queen of England would later gain her an Oscar?

My long association with another great and multifaceted actress, Maureen Lipman, began inauspiciously in 1995 with an interview for the assignment as official photographer for the *Agony Again* publicity shots. Maureen was to reprise the role that she had played in the television series *Agony* in 1979. I felt a little irked that, after all these years as a professional, I would have to go before a panel like a university student applying for my first job, or in truth, was it more a fear of failure? Whatever, I duly turned up at a venue in Kentish Town and appeared before a panel of four with my portfolio; they were polite and an old television producer friend, Humphrey Barclay, was businesslike. As I drove away, I felt a cloud of depression descending. I was so sure that I would

Maureen Lipman and cast, *Agony Again*, London, 1995

not be chosen and that I had wasted a lovely, sunny day; the list of male photographers they were seeing was awesome. The early years of battling my way into a male-dominated profession had taken their toll.

The telephone call to say that the assignment was mine came as something of a shock. I hired my favourite studio, where my assistant Bill Burlington and I were greeted by the entire cast, along with Humphrey. It was a long day's work, with group shots and individual portraits of every cast member, which were to be used on the programme credits. But I have never felt happier or more fulfilled doing what I love most with the kindest, funniest team, and it certainly showed in the pictures.

Maureen did not forget me and a few years later she asked me to shoot new portraits for *Spotlight*, the actors' directory, followed by pictures for a DVD cover and then a book jacket. I have several loyal clients who have come back to me over the years but few are more talented than Maureen, a prolific television and theatre actress and a writer of several extremely witty books. Yet a sustained successful career does not come through luck alone, nor just talent, and it is here where Maureen truly shines. Over the years of working with her, she has never been one minute late and her hair and make-up are always flawless, as she brings her trusted stylist. There is no negative conversation about private problems that any human being is bound to have from time to time, and the energy is 100 per cent professional. Only when we've had lunch or tea together have we opened up about ourselves. Then I see the compassionate side and the loving mother who was such a devoted wife to the playwright Jack Rosenthal. This is the art of being a true professional.

One of the few big crises of my career as a photographer occurred through no fault of my own. The assignment was to photograph Leslie Ash, star of the British television show *Men Behaving Badly*, at Le Mas Candille in Mougins, in the South of France. All started well when on the flight from London I was offered the chance to sit up front in the cockpit, enjoying breathtaking views of the snowcapped peaks of the Alps then the oyster beds along the French coast.

I then spent a delightful week with my friends Sarah and Anthony Beerbohm, at their new house in Provence. For a week I saw no one except my hosts, the gardener and the delightful Comte Philippe d'Argence, a guest for Sunday lunch. My senses were filled with the smells of the yellow broom, lavender, rosemary and thyme. Each evening from dusk onwards, Sambo the dog sat gazing across to the woods; he was on "hog alert" for the wild boar who would charge through the garden at night, grunting and snuffling their way around the flower beds, leaving a trail of hoof prints but surprisingly little damage to the plants. The cats placed dead mice and rats under the kitchen table as though it were an altar, under the misapprehension that these were gifts we wished to receive. Tree frogs and giant toads strolled across the terrace and tree rats, with eyes like saucers, peered out of the creeper as we sipped our wine at dusk. The nightingale sang at night and the fluffy cat, Boy, had taken to relaxing on the chopping board on the kitchen table. This was the animal kingdom and we humans were here under sufferance.

Then it was time to go to work. Le Mas Candille is named after three cypress trees planted here more than 300 years ago ("*candille*" is Provençal for candle) to mark the route from the coast through the valley into the mountains and up to Grasse. It's said that Napoleon camped beneath them on his escape from Elba. After settling in and enjoying a long swim, I went to find my celebrity and received a warm welcome. Dressed in a tiny denim skirt and pink gingham shirt, Leslie looked as pretty as I could have hoped. I felt excitement mounting at the thought of working with her and the mound of colourful outfits we had chosen back in England. She then cheerfully announced that she had trodden on a hot coal which had fallen off the

Leslie Ash, Provence, France, 1996

barbecue back home, creating an open wound on the heel of her foot. I saw my lively shots disappearing before my very eyes and the problems did not stop there. Suddenly I was under attack from Lee, her husband, a massively fit footballer who had played for Nottingham Forest and Leeds United.

"How do we know we can trust you?" he growled.

"Lee, we've been over that," protested Leslie.

I thought back to the meeting with her agent at Soho House in London, and how drinks had led on to dinner because we had got on so well. The couple battled like two tennis players, with me stuck in the middle.

Drained before we had even started, I retired to my room to unpack and change for dinner; I yearned for the tree frogs and Sambo on "hog alert". By the time I came down for dinner, all had been resolved and Lee was hugging me and begging my forgiveness. For the next three days, we were part of a big happy house party. Even so ... I shall never know if it was Lee's paranoia over being shafted or some sort of premonition, but although I was in no way to blame, there was a black cloud on the horizon.

Leslie was always up for the next idea, even though we were trying to shoot in the early morning and late afternoon when the light was best. Each day, the sun shone in the morning before the storm clouds would build over the mountains by lunchtime, destroying any hope of golden shots at sunset. One lunchtime, as we sat roasting on the terrace in a heat which normally comes in August, not June, we decided to make use of the black clouds. Leslie rushed to change into a silver evening dress and we set up lights to shoot her against the blackening sky; the result was sensational, and no sooner had we finished than the heavens opened.

We rose early on the day of our departure because we still had work to do, as the bad weather had held us back. The sun was shining and Leslie, who now had an infected and swollen foot, posed wonderfully, leaping like a spring lamb. I had to keep reminding myself that she was working with one leg, not two! At last we had time to visit the village of Mougins, where we tucked into a huge farewell lunch before piling into two taxis bulging at the seams with cameras, lights and costumes, for Nice airport.

Back in London, Lee and Leslie arrived for dinner at my house, armed with several bottles of the most delicious wine. They looked at the photos from the trip and were delighted, so we celebrated. My agent and close friend, David Maskrey from Frank Spooner Pictures, was there to meet

Leslie Ash, Provence, France, 1996

them and all seemed too good to be true. The final step was to deliver the material by hand to *OK!* magazine to be absolutely sure nothing went wrong at the eleventh hour. Anyway, I was keen to see the art editor, Andy Becker, because he really showed his appreciation if he liked a shoot; he did, and confirmed six pages.

Then the axe fell, in the form of a double-page spread in the *Daily Mirror.* One glance and my spirits sank to an all-time low. There were my pictures of Leslie under a lewd headline, and they had even used a shot for the cover. Had I not given Leslie my word that the pictures would not appear anywhere without her blessing? Had I not sworn that they could trust me? What would they say? How had this happened? It had never crossed my mind that someone at *OK!* would give the pictures to another publication, thus breaking the rules of copyright. I had never in all my working years as a photographer experienced such a blatant steal.

The next day, my phone went crazy. The art editor was full of apologies, the picture editor was appalled, Leslie's agent was furious, my agent was equally enraged. I sat like a soldier on a battlefield, dodging the bullets from all sides. First Leslie's agent thought she should sue, then my agency wondered if I should do the same, then compensation was discussed. Lee and Leslie had such great plans for us, but I could kiss goodbye to that idea; that it was not my fault was irrelevant. I chose not to sue, but I agreed, on the advice of my agent, to a small amount of compensation.

I also encountered some glitches on another assignment but overcame them. I was at Cliveden, the stately Italianate house famously owned at one time by the Astor family, the location of my shoot for *OK!* magazine with Nigel Havers and Gina McKee. The actors were not a couple in real life but were co-starring in the television production of *Element of Doubt*.

Nigel Havers and Gina McKee, Cliveden, Buckinghamshire, UK, late 1990s

As the day began and I drove down the long winding driveway along which giant rhododendrons nestled under the trees like verdant Norman arches, I was in blissful ignorance that there was just one snag – Gina McKee had a reputation for hating photo shoots. When faced with this hurdle, I naively thought that with such sympathetic company and such a stunning location all would go smoothly. I diplomatically went off for a chat with her on my own, as I genuinely empathised with her; I, too, feel daunted by being photographed. Even Nigel, with his reputation as a charmer, was hard pushed to coax Gina out of her misery. Miraculously, against all the odds, I achieved a few lovely relaxed shots in about four locations, which would be just enough for a spread. Then, quite suddenly, Gina left, and that was that, no warning, just gone like a startled gazelle.

The following day my doorbell rang, and there was Gina clutching a huge bunch of flowers. She was effusively apologetic and said that it was not personal; on the contrary, she said, she had liked me. I tripped over my words, telling her how much I liked her and that I admired her work as an actress. I was touched by her gesture, and when Gina left, I thought about how alien it was for me to cajole unwilling actors into photo shoots.

As it turned out, another positive to come out of that assignment was my association with *OK!*'s art director, Andy Becker. I had gone to the National Portrait Gallery in London to see an exhibition of the work of

Nigel Havers, Cliveden, Buckinghamshire, UK, late 1990s

Nigel Havers and Gina McKee, Cliveden, Buckinghamshire, UK, late 1990s

Annie Leibovitz. The grand dame was there in person; it was not difficult to spot the star of the show, for Annie's height and distinctive features made her stand out in the crowd. She looked awesome with her lion's mane of unruly blonde hair contrasting sharply with her clean-cut black trouser suit.

I felt small and insignificant beside this Amazon, who immediately asked me if I was a photographer and promptly bombarded me with questions. I said that I had been photographing six-page spreads for *OK!* magazine.

"Oh, you have arrived. I have met your art editor, Andy Becker, and I would really love to work with him."

It did not stop there. People were queuing to meet the star of the show but still the questions poured forth. What other publications had I worked for? Where had I trained? I mentioned my meetings over the years with Eve Arnold. "I'd love to have someone do a shoot with the three of us discussing our views as women photographers," she said. Enthusiasm is empowering, and I left the gallery feeling energised.

Every once in a while in the life of a photographer, a shoot becomes a hilarious and highly original episode. One such experience was my encounter with Christian, the lion cub. In the late 60s, Harrods boasted that you could buy anything, and so it came to pass that two Australian men about town, John Rendall and Anthony "Ace" Bourke, bought Christian. The happy trio lived for some time in their shared flat in Chelsea, and the actress Barbara Kelly invited me to take pictures of her with Christian. All did not go quite as planned. Christian liked to play but was unaware of his own strength, plus he was used to playing with two muscular young men. So first he bowled Barbara over playfully, then it was not long before I became the focus of his attention. Crouching down, buttocks wiggling, we locked eyes through my camera lens and after one giant leap I was flat on my back with two big bruises on my arms. Hardly the usual type of scene in the smartest area of London.

Barbara Kelly with Christian the lion cub, bought at Harrods, London, 1969

All good things came to an end when, inevitably, Christian grew too big. The boys were introduced to Virginia McKenna, who together with her husband Bill Travers founded the Born Free Foundation after starring as Joy and George Adamson in *Born Free,* about the couple raising a lion. Christian was returned to Africa, and a year later, when the young men flew into the wild, miraculously Christian appeared and welcomed them with open paws wrapped around John and Anthony's necks and kisses on their cheeks: two lionesses calmly watched.

I first crossed paths with Lord Patrick Lichfield in the 80s at a drinks party. Roll on to the 90s, when I was well established in my career and one of my best friends, Sir Geoffrey Shakerley, was also a photographer and Patrick's brother-in-law. Patrick and two other fellow photographer friends had all been to Harrow and were part of Lichfield Studios in Notting Hill. I joined the team from time to time. We all had one thing in common: we knew how to work hard and to have fun. On one occasion when I met Patrick at the studio, I had just been commissioned to photograph the wedding of the ex-queen of Italy's grandson. He was mildly put out that I was to photograph the royal wedding as he felt that was his territory, but he never appeared to take himself too seriously. Now Patrick was seeing me in a new light. He was one of several well-known people who had agreed to be photographed in rooms designed by different decorators in a show house created for a charity in Chelsea Wharf in London. I was asked to take the pictures as I knew many of the celebrities. With his usual sense of fun Patrick was happy to be a sport and to pose in a ludicrous room with a dog he did not know. I had asked Geoffrey Shakerley if he would assist me: a brilliant idea, as it it turned out. Patrick arrived late, looked surprised to see his brother-in-law, and asked, "What are you doing here, Geoffrey?"

I stepped forward and said, tongue in cheek, "Excuse me, Lord Lichfield, could you please not speak to my assistant?"

From then on, the day was filled with lots of laughter and a lengthy, boozy lunch generously paid for by Patrick. Thereafter, I often saw him in the Portobello market with his black bike. The stall holders loved him and he captured all the great characters there through his lens to add to his vast collection of the most famous of every profession in that era.

Lord Patrick Lichfield, Canary Wharf, London, 1990s

CHAPTER 10

WRITERS IN THE LENS

"*The pen is mightier than the sword.*" – Edward Bulwer-Lytton

For several months, a veil of grief had dulled my senses. I felt like a sailing boat becalmed at sea in a mist; I was in the doldrums. My father had just died, leaving my mother alone after sixty-five years of marriage. Daddy was quite simply my best friend. All my life he had been there for me through joys and sorrows, loving me unconditionally. My mother now had a carer living in permanently, which meant we could no longer share the close relationship we had once had. The family home had been broken up. Then the picture became broader with a cluster of deaths of people close to me, of all ages and different causes. Life seemed more transitory than ever and Shakespeare's words had never rung so true: "When sorrows come, they come not single spies, but in battalions."

My routine of visiting my mother, caring for my home and doing my portraits kept me going as I learned to "adjust". The butcher remarked that I always made him laugh and that helped me, too. This was, after all, just part of the experience of living, the wheel of life referred to in all the religions. I had to make sense of it all, and to that end I felt a subtle shift in consciousness, an acute awareness of my mortality, a wake-up call to make the most of every day. My father's words rang in my head: "Keep creating."

Then a call came from David Wilkinson, who was editing Ronald Harwood's book *Adaptations* on the art of screen writing. Would I photograph Ronnie for the cover? I was delighted, for I was fully aware of his prolific body of work and awards. Plays, books and screenplays for television and film had earned him huge respect in the business. Ronnie won an Oscar nomination for Best Adapted Screenplay for his work on *The Pianist*, directed by Roman Polanski, and now he had been nominated for his adaptation of the French screenplay of *The Diving Bell and the Butterfly*.

I felt ready for a challenge, to feel again that twinge of excitement when brushing up against great talent. The timing could not have been better, and at this moment Ronnie's words from *Adaptations* rang true to me, jumping off the page like a cork popping out of a bottle: "Writing is psychoanalysis, writing is the therapy. If you're in trouble, write it out." The same, of course, could be said of photography, and what better subject for me than writers, with their quest to make sense of the world through their creativity?

As I looked at the tall, red-brick Edwardian block of flats dwarfing the pretty Victorian houses on a side street off the Fulham Road, I felt I was the right person in the right place on this grey day in November. My assistant appeared to be more interested in his new Blackberry. I searched

Ronald Harwood, London, 2007

with difficulty through the maze of numbers on the highly polished brass plate and pressed the tiny button. Before I had time to wonder if I had got the right bell, a distant voice echoed through the entry phone. The Harwoods were on the ground floor and doors flew open and out popped Ronnie, closely followed by his wife, Natasha. "Can I help with the bags?" Ronnie oozed bonhomie and charm but I could tell that he did not want to waste time. His mask of good manners did nothing to conceal his dislike of being photographed, whether through shyness, boredom or simply having too much work on his mind.

The day was not auspicious. A fine mizzle fell as we unloaded cameras and lighting, letting the damp air seep into the cosy flat. This was not the sort of place you would want to disrupt; I sensed the atmosphere of a home where the two of them spent a great deal of time, and I felt like a clumsy invader. The good manners prevailed and coffee arrived, served in dainty cups and saucers, and I suggested that Ronnie and Natasha should leave us to set up the lights in their sitting room. Preparation is what takes the time, the placing of lights, rearranging the furniture, light readings. As the sitting room, which led into the dining room, was extremely small and cosily cluttered, it cried out not to be disturbed. By the time we had unpacked the lights and cameras there was hardly room to swing a cat, and if you did you would break a precious object.

"All ready," I called, and Ronald was there in seconds like a tightly coiled spring, all charm with a cigarette permanently on the go, but I sensed he could be short fused. Ronnie's breakthrough had come with his stage play *The Dresser*, which he then adapted for the screen. "I saw *The Dresser*," I said, "my father had a dresser called Zip who could have been your character's double." A look of disbelief came over his face. I did not know whether the surprise was due to the dresser's name or the fact that my father had a similar character in his life. "No, honestly, that was his name," I said, plumping for the former. Ronnie wanted to know my father's name. "Hugh Latimer!" At that split second, everything changed. "I saw every one of the plays he was in. What a wonderful actor," said Ronnie, and I knew then that I was no longer just another visiting photographer. It would be trite to say that this meeting was meant to be, but really it was a strange coincidence to have been in that situation at that time in my life.

Our conversation then flowed with ease, but just as I started to relax and enjoy our shoot a shrill ringing tone broke the magic. My insensitive assistant answered his mobile. I could say nothing, because it would have caused more of an atmosphere, but I was quite ready for Ronnie to explode; he did not, but the magic was broken.

Luckily, I had taken my best shot before my assistant's mobile rang. Sitting in an antique armchair by a window draped in generous golden wild silk curtains, Ronnie looked at ease; he had, after all, trained as an actor. Paradoxically, his eyes were more expressive than most of the actors I have photographed. In a split second they could change from humorous, to hard, to pensive, to full of compassion. A cluster of books, his well-used ash tray and a haze of smoke as he puffed his way through the session completed the scene. I was captivated by my animated subject as I snapped away, but deep down I was haunted by the feeling that we had invaded the Harwoods' home at the wrong time and, indeed, we had, for Natasha was newly discharged from hospital. She looked pale, frail and positively ethereal wrapped up in her dressing gown.

The job was well done, but there was no relaxed lunch or an extra cup of coffee. It was not the time or place, and I sensed the need to depart as quickly as possible in order to leave a good impression. I was reminded of the importance of using time well if you have a goal, to remain focused and not become a busy fool. Passion in what you do is the driving force, and as Ronnie had pointed out in his writings, creative artists have to find their own way and the rest is sheer hard work.

I have to admit that I was not particularly looking forward to my meeting with the author William Golding and his wife at their charming home near Bowerchalke in Wiltshire. I had jumped to the conclusion that the man who had written *Lord of the Flies* must be a dark soul. While I was truly in awe of Sir William, who had won the Nobel Prize for his large body of work and had once famously said that he had lost his belief that humans are inherently good and even children were inherently cruel when left to their own devices.

I immediately revised my preconceived opinions when a charming couple welcomed me into their cosy home and I noticed immediately that Sir William had an aura of inner peace and contentment about him. The idea was to photograph him with his collection of orchids, which had

William Golding, Bowerchalke, Wiltshire, UK, 1980s

become his passion in later years. I snapped away, observing the well-known stern expression characteristic of the man; he was polite but gave nothing away.

With the pictures finished, Sir William's wife, Ann, asked me back into the sitting room with its low-beamed ceiling, the late afternoon sunlight gently shafting through the windows. The conversation became warm and animated as the pair talked of their long and happy life together, and I felt that all the cynicism in this man had been exorcised through his writing, leaving a purity of heart. Sir William was quoted at some point as saying, "I think that good will overcome evil in the end, I don't know quite how, but I have that simple faith."

Quite suddenly, I saw the picture I wanted. My camera with black-and-white film was resting in my lap, so I glanced down, guessed the exposure, and without a pause raised it to my eye and clicked; at that split second, Sir William was smiling at his wife. And now, recorded for posterity, is a picture of true love. All the other pictures faded into the background as I took this simple black-and-white portrait. This was my gift to this charming couple and myself. Ann was delighted with the print I sent and remarked that it was her favourite picture of Bill.

Sir William died in 1993 and was closely followed by Ann two years later. I rather suspect that she felt ready to join her soulmate of fifty-four years. Their shared gravestone at Bowerchalke reads, "Remember with Love, It is the star to every wandering barque".

I also had reservations about meeting Doris Lessing. Given the huge following of readers and devotees of her avant-garde political and feminist views, there must have been many photographers who would have given their eye teeth to meet this great lady who had written the feminist classic *The Golden Notebook*. She was not someone I would naturally have been drawn towards.

Steps led up to the front door of the grey-brick Victorian house in West Hampstead. The entry was a chaotic clutter of coats and shoes, old worn carpets and tired painted walls. This was the kind of hallway that is not exactly welcoming but lived in, like a student's rented accommodation. However, this was the home of one of our most awarded writers and where she worked all day. The living room was full of comfortable, well-worn chairs, threadbare rugs and throws that were not fashionable but colourful and mainly crocheted. Doris's clothes and hairstyle sent out a different message: a tight greying bun with a centre parting framed her unmade-up, plump, round face; the impression was unflattering, but her wise old eyes became surprisingly youthful when she smiled. The suit with a boxy jacket and mid-calf skirt teamed with thick stockings and ankle-strap button shoes with kitten heels was pure Bloomsbury set. I was reminded of pictures of Virginia Woolf in similar attire but the likeness ended there, for this was not a lady with a melancholy air. Strong, positive and aware of her superior intellect, she was happily opinionated and totally unconcerned about my reactions.

Doris was friendly and offered me tea, which we made together in a cluttered kitchen that had an air of permanent work in progress. Vegetable peelings had yet to be cleaned away, unwashed plates were piled up with knives and forks between each one. The tea mugs were mainly chipped, with tannin stains etched into their rims. Yum Yum,

Doris Lessing, Hampstead, London, 1980s

the black cat, wove her way on the kitchen table, silently affirming her position in the household by pointing her bottom at me as cats are wont to do. A small conservatory off the living room was overflowing with straggling plants, many with brown leaves; the garden was in much the same condition, with a host of fading, rotting vegetables suggesting good intentions unfulfilled. We later wandered around the garden and I realised that Doris really cared about each plant; she was just letting them die back naturally and had probably feasted on a number of homegrown vegetables throughout the summer.

Doris was relaxed and happy to be photographed wherever I chose. I witnessed none of her acid views on society, probably because I was not carrying out an interview, nor was I an English man. When the time came to leave, I was surprised at how much I had enjoyed my visit with such a famously controversial writer.

Arianna Huffington (née Stassinopoulos) had already taken London by storm when I met her. She and Bernard Levin, whom she referred to as the big love of her life, mentor and role model, were a literary powerhouse, and she had just completed her biography of the singer Maria Callas. She needed a good picture for the cover and the accompanying publicity, and

Arianna Huffington, London, 1980s

one photo session in my studio led to my becoming a regular visitor to the mansion block in Chelsea where Arianna lived with her sister Agapi and her mother Elli. Large and always dressed in a kaftan, Elli would trundle around bare-footed, smoking cigars; she had style. Although every visit held a new surprise, the one constant was that Elli would prepare a small feast in the kitchen for Emily, my King Charles Spaniel. Meanwhile, Arianna introduced me to a form of Iyengar yoga that I have adhered to ever since, and numerous other new experiences, and the teachers and gurus who passed through the portals of the mansion block and the experiences they brought with them were fascinating.

Then, abruptly, the Greek household abandoned me and moved to New York. I felt sad, as their absence left a sizeable gap in my life; I had become accustomed to the constant stimulation of new experiences with them. Several years later, Arianna kindly invited me to stay for a week. Once inside the door of the brownstone, everything felt the same as it had in London. Agapi was in residence and there was one other guest, the actress Gayle Hunnicutt.

At this point in her life, Arianna was seriously looking for a suitable husband. She was not waiting for destiny, and she was taking the matter into her own hands. She would go out with a different man every night, dressed in stunning designer clothes. Her success on the social scene in New York ensured she was lent all the dresses she could dream of wearing. With her height and stunning thick, dark hair, she looked positively regal and did justice to all the borrowed clothing. Each evening, we three handmaidens would be asked to join Arianna and the various beaux for a drink before they left; the following morning we were expected to give our opinions.

Although ambitious for herself, Arianna was generous to a fault with her friends and would do what she could to promote their work. One night as I lay in bed reading there was a tap at my door. Arianna appeared, as immaculate as when she had departed for dinner. Accompanied by her date. He too looked too groomed to be true. I was to show this gentleman my portfolio. "Oh, how sweet," exclaimed Arianna, "you look like a little girl with your pink cheeks." I felt utterly ridiculous as I clambered out of bed in my borrowed white night-dress, three sizes too big, like a child being visited by her parents at bedtime. This was the strangest way to be showing my portfolio, but life with Arianna was never ordinary.

After Arianna married Michael Huffington, the entourage moved to a fairy-tale Spanish-style house in Santa Barbara. There, Arianna and her husband became involved in politics and raised two daughters, Isabella and Christina. Their ambitions were for Michael to become President of the United States. The next incarnation of Arianna, now divorced, was to relocate again to New York, where she founded the *Huffington Post*. Arianna has published numerous books over the years since I photographed her for the cover of *Callas*. Amongst them is the bestseller *Thrive*, my favourite, as it is dedicated to her visionary mother, Elli, who had meant so much to me.

Barbara Taylor Bradford also established herself in Britain before moving to the United States, where I met her in her white clapboard home in Litchfield, Connecticut. She had just published *Dangerous to Know*, one of her forty bestselling novels, and she confided to me that her great friend Pamela Harriman had just called to say that she could not put the book down.

The purpose of this trip was to take publicity shots for the novel, and my assistant Robert Simpson and I had travelled to New York and settled into an unappealing hotel in Manhattan, with the subway thundering

Barbara Taylor Bradford, Litchfield, Connecticut, 1990s

beneath us and views of an air extractor. I kept affirming to Rob that the contrast between here and Connecticut would be enriching. Then the morning to make the trip north came and there was our driver, Manny, beaming from ear to ear and ushering us into an air-conditioned limousine that was as welcome as Cinderella's coach. We were swept out of the hot, humid concrete jungle to the lush greenery of Connecticut and were soon sitting in rocking chairs on the porch of the lakeside Boulders Inn, with tea and a bowl of fresh fruit. I picked up the phone to call Barbara. She sounded so enthusiastic and welcoming that I was eager to get going, even though our bags were still packed.

Barbara and her husband Bob gave us the warmest welcome and immediately offered us food and drink, but I wanted to get the first set-up

Barbara and Bob Taylor Bradford, Litchfield, Connecticut, 1990s

Barbara Taylor Bradford, Litchfield, Connecticut, 1990s

under my belt before a break. The air was so humid that Barbara could not go outside in case her hair collapsed like a damp mop. So we ran around choosing interior set-ups and pulling out garments to enhance each room while Rob set up the lights. Clothes galore flew out of cupboards. Nothing was too much trouble. Somehow, the forces were against us; one of the lighting heads became faulty, then the Polaroids came out alarmingly bleached; all was chaos. Taking a deep breath, I realised we had to trust the meter and only the meter.

Suddenly an enormous storm broke, blacking out the windows. The electricity in the air had affected the power in our lighting. I shot with a sinking heart; we would not see the results until we returned to England. During lunch, I hid my anxiety and prayed that we would have a smooth run from then on.

Just before sunset the sky cleared and we dashed out to catch a few shots in a swarm of no-see-ums, tiny little insects which bite and, needless to say, are quite invisible. Luckily, Barbara and I had bonded, despite the pitfalls. She mentioned that my voice reminded her of Deborah Kerr's and I took this as a tremendous compliment. So, after a hard day of shooting, all was well, nothing broken, no scratches on the soft wooden floors, no faux pas made, and Robert and I were heading back to the Boulders Inn.

CHAPTER 11
AT HOME

> *"My home shall be open for the sun and wind and the voices of the sea – like a Greek temple and light, light everywhere."* – Axel Munthe

I believe all photographers are voyeurs at heart, spending our time watching others through our lenses. I admit to taking particular pleasure in photographing my subjects in their homes, where I can get a glimpse of their tastes, their habits, and how they relax and find peace of mind.

Certainly one of the most pleasurable as well as memorable home shoots I have ever undertaken was an afternoon with Dirk Bogarde in Opio near Grasse in the South of France during one of the several times I attended the Cannes Film Festival. I was given time off from my hard work to visit Dirk, whom I had got to know whilst working as a publicist on the film *Justine*. Quentin Falk, one of the few writers I could trust, joined me on the visit. He is much revered in the film industry as a serious writer, so his reputation would have been noted by Dirk. Later, Quentin went on to write the official biographies of Anthony Hopkins and Albert Finney, among others.

I had been warned that the drive through pine-covered hills was arduous, along unpaved roads that became no more than confusing tracks leading up to Dirk's house, Le Pigeonnier. The honey-coloured house, on a small plateau protected on three sides by mountains and facing a sliver of the sea, eventually came into view. I stopped just before we approached to check my appearance in the driving mirror and attempted to disguise my tiredness with fresh make-up. I was hardly a treat to behold. My linen clothes resembled a scrunched-up candy wrapper, the humid heat had done its worst, and to cap it all I still looked tired. I adored Dirk, but the shyness welled up as flashbacks of the besotted school girl with her cuttings books recurred like a cow ruminating its feed. On top of this, I remembered only too well that Dirk did not suffer fools gladly. I recalled

Dirk Bogarde, Opio, South of France, 1980s

the story of the woman who greeted him on the Croisette in Cannes. "You must come to dinner," she gushed. "Must I?" came Dirk's acid reply.

As we pulled onto the perfectly raked gravel cul-de-sac in front of the house, I slowed to a snail's pace so that I would not leave unsightly ridges. Dirk appeared, reed slim as ever, and elegant in white slacks, an immaculate blue-checked shirt and flip-flops. He could make anything look elegant. He greeted me warmly, delighted that I had found the way. Kisses on both cheeks in the French style were followed by a gasp of concern. "Oh, how tired you look, that wretched festival, so draining. You must have a cup of tea at once." At that moment Tony Forwood, Dirk's dear friend with whom he shared the house, greeted me equally effusively and disappeared to organise the refreshments. Dirk continued to make sympathetic noises and to vent his spleen on the festival; he seemed delighted to have an excuse to be rude about the whole event. The tea was Earl Grey accompanied by the largest, richest brown sugar lumps I had ever seen. Dirk was keen to tell me they were from Harrods, the only place to get good brown sugar lumps.

I sat back contentedly as I watched Quentin conducting what was certainly the most professional interview on the film business that I have ever witnessed. Two grand masters of their careers. I casually took a few pictures of Dirk right there on the terrace, relaxing with his beloved dogs while he poured out his disappointment with the industry. Dirk remarked he was thrilled that *The Night Porter*, with Charlotte Rampling and directed by Liliana Cavani, who Dirk adored, had opened in Rome, where the Americans were unable to censor any part of it because of a law passed in the time of Michelangelo decreeing that a piece of art cannot be doctored. Now he had just completed filming on Visconti's *Death in Venice*, an artistic triumph but controversial with the Americans, who questioned why anyone would want to see "a film about a dying old man lusting after a young boy".

The minute the interview with Quentin was over I packed my camera away, albeit reluctantly, and metamorphosed into the friend from several years before. Everybody relaxed and Dirk whisked me off on a private tour of his home. This lovely man, renowned for being an obsessive recluse, was treating me as he would his most trusted of friends and soon we were bonding over the pleasure we had in restoring our homes.

At just fifty, Dirk had stepped off the merry-go-round to take control of his life. He now had plenty of time to build a wall, do his paintings and sketches, write, and read in the winter by the huge log fire; these were halcyon days. Tony, at first, had to reluctantly be cajoled into leaving the UK for the depths of rural France. Now he had thrown his weight behind the whole project and was the steadying hand on the ship. While Dirk had his faithful rescue dog, Labo, found in Rome, Tony had his own dog, Daisy. Dirk had created a fish pond and a vegetable garden in spite of every hiatus known to man, from flooding under his precious olive trees to a leak in the septic tank that created what he thought was a natural spring. In blissful ignorance he had taken a drink before he noticed a piece of lavatory paper floating past him.

Now we were up and running, with Dirk enthusing about the long living room created by knocking down walls between three rooms, everything the antithesis of a film star's home. I loved it all. Then to my surprise we were heading up to the bedrooms and Dirk revealed that the bedroom with the huge double bed was Tony's, while his was tiny, with

Dirk Bogarde, Opio, South of France, 1980s

a narrow bed. He explained that he felt desperately insecure in big beds and preferred to be wrapped tightly like a mummy.

When we returned to the terrace, Kathleen Tynan seemed to have appeared out of thin air; it transpired that she was a house guest. Kathleen had to be one of the most talented and beautiful women I had ever met. She was married to the brilliant but controversial theatre critic Kenneth Tynan, who, in spite of his impressive body of work, is most well-known for being the first man to use the "f" word on television. The three hours passed too quickly and before I realised it I was saying my farewells. I felt sad as I watched Dirk and Tony through my driving mirror, waving until they were out of sight. "Let me know when you are here again," wrote Dirk on a postcard, reminding me that the visit had gone well.

I never returned to Le Pigeonnier, as Dirk and Tony had to sell their beloved home soon after my visit and move back to England. Tony fell sick and had several grim years in and out of hospital. It was only when Tony sadly died too young that Dirk had to learn everything to do with running his life, even how to write a cheque.

I later read that in the 90s, when Dirk lived alone in London, he had written in his flat behind Sloane Square about the loneliness of weekends and how he would go out to buy a newspaper just so he could speak to someone. I visualised his pain at losing almost everything dear to him when Tony fell sick: the dream home with all its privacy, the dogs, gardening, trips to the market, and above all peace of mind. I have few regrets in life but one is that I did not offer to visit Dirk in those lonely times. I stupidly presumed that, being so famous, he would have too many visitors, especially since he had been knighted. I wrote to Dirk later when I learned he had been in hospital, and soon after I received a letter thanking me for my "very sweet and encouraging card … Much love Dirk".

LE HAUT CLERMONT
06740 CHATEAUNEUF DE GRASSE
FRANCE

22nd. October. '79.

Dear Carole.

Think nothing of it ! I did get a script from someone who said that you had suggested he send it. Which surprised me a little since it was not at all the sort of subject I would consider,and not awfully well written to boot ! However that has been dealt with now,I hope.
Glad that you enjoyed "Snakes" and understand very well about "Postillion"...childhood is far less distressing usually..and more fun too.
Very wise of you to seek advice from Eve Arnold. She really is one of the very finest photographers in the world,apart from being a perfectly splendid lady and a wonderful counciller.
Loved the picture of Theo.
Not thr sort of thing to hang in the parlour...but alright for the rogues gallery up here in my Studio...thank you very much for sending it.

With very best wishes,

as ever,

Dirk

Dirk Bogarde.

9.9.92

Dear Carol -

What a pleasant surprise! I'd almost forgotten that Tunis interlude..and Opio seems such a very long time ago. And no Theo.A great loss indeed,but what a way to go!

I am delighted that you have 'taken off' so spectacularly as a photographer:it must give you infinate pleasure.As for me I dont give interviews OR have photographs taken any more.
The interviewing (in this country anyway) is so viscious and ugly that I refuse all of them.To the consternation of Viking at first who were,with reason,worried a bit! But now that I flog my books and their wares doing my 'Show' all over the UK they are much more understanding...we do well without interviews.Except I do do provincial Press.Gentler and,usually,less venal and envious.
I have had my picture took for this session..so I dont think I'll have to do it again for a year or so..until the next book.
But I'll remember Latimer if,and when,that time comes.
At my age,honey,it is better to forget photographs unless you long to look like a very aged lizard.
And I dont.
The new book,in October,is more Winnie The Bloody Pooh Than you can believe.It's the full 'frow up' department.Sorry!
Good to hear from you again..much love..

I had never met Peter Cook before I made my way to his home in Hampstead to photograph him, but of course I knew the comedian by reputation. I had spent a huge chunk of my youth watching the brilliant sketches of Peter and Dudley Moore, and of course the ground-breaking production *Beyond the Fringe*, to which they gave birth along with Jonathan Miller and Alan Bennett. Every so often, a university spawns a group of stars who shine brighter than the rest, and such was this quartet from Cambridge.

So I was delighted when Cynthia Young, the picture editor of *TV Guide*, assigned me to do a shoot with Peter. Even so, I was well aware of his brilliant mind and sharp acidic wit, and I would have felt more at ease without a picture editor breathing down my neck. We spent the afternoon huddled indoors with the rain pouring down outside as Peter sipped gin and Campari. He talked animatedly, only interrupting himself occasionally to check the cricket on television. I was relieved to find that we had friends in common and shared interests, as this made the day pass smoothly. This brilliant man seemed perfectly happy to let me spend the entire afternoon fiddling around with my lighting and snapping in all corners of his home.

When I finally packed up my black bags and piled them into the car, Peter said that although he had no idea how the pictures would turn out, he had thoroughly enjoyed the afternoon. I felt greatly relieved, as this was not a man who suffered fools gladly. The pictures, unfortunately, were not well received, because every shot featured a pile of cigarette butts. Peter's habit was still acceptable in England, but was considered a cardinal sin in the States. So, as this shoot was exclusively for *TV Guide* all the best interior shots were killed and an outdoor picture of Peter looking fairly self-conscious as he leaned against a sundial was all that was left.

Peter Cook, at home in Hampstead, London, 1981

At-home shoots can often provide a glimpse into admirably contented domesticity. This was certainly the case with actor, mountaineer and adventure enthusiast Brian Blessed. I received a chaotic and warm welcome as I arrived at the low-built, greenery-covered house. This came from my host as well as his equally friendly dogs, ducks and Shetland pony. It was a treat for me, an animal lover, to be here in Chobham, Surrey, on a sunny day, enjoying my assignment. Brian and his wife, Hildegard Neil, lived in a house that was simple and functional by design, open to the many animals that trooped in and out and where even the Shetland pony believed he was welcome in the kitchen.

Brian is a home lover who is passionate about gardening, and he showed off a lush garden and pond. Even so, his childhood dreams of being an explorer have never diminished, and when the call of the wild beckons, he is off. His ultimate goal is to travel into space, as befits an actor who has appeared in *Flash Gordon* and *Star Wars*. "We are all children of stardust," he once remarked.

Brian Blessed, Surrey, UK, 1990s

Brian Blessed and his daughter Ann, Surrey, UK, 1990s

I had a similarly favourable impression of contentment when I visited the red-brick home of actor Herbert Lom in Primrose Hill. I arrived to find a mature man, well preserved, immaculately dressed and surrounded by several small grandchildren. All was calm and orderly and the children were duly respectful, with no screaming or rushing around. Herbert was lovingly in control. These youngsters were, I soon discovered, the all-consuming love of his life.

Having snapped away indoors and out, we settled down to cups of tea and biscuits as Herbert shuffled through endless photographs, cuttings, letters from fans and stills from productions, all evidence of an extremely full life. Herbert was born in Prague but arrived in London before the Second World War, where he started his career in radio then moved on to appear in more than a hundred films. There was the handsome, smooth young man, appearing as Napoleon, then as the king of Siam; he had a wonderful chameleon quality that made him suitable for so many parts. Among the memorabilia and awards was a soft toy version of a pink

Herbert Lom, Primrose Hill, London, 1993

panther, for Herbert will always be remembered for his role as Chief Inspector Charles Dreyfus, boss of Inspector Jacques Clouseau (played by Peter Sellers) in *The Pink Panther* films.

There was nothing particularly theatrical about this talented man, and no evidence of that hungry need to please that I see in actors desperate for the next job; he had closed that chapter and now his role was that of a grandfather. As the children played quietly around us, I was conscious that this was no fantasy role, but genuine reality. I had enjoyed my afternoon with Herbert. No dramas, no great excitement, just the feeling that I was in the presence of a man who was comfortable in his own skin.

Things did not go as smoothly with Ken Russell. The prolific and talented film director was perhaps best known for his film adaptation of *Women in Love*, from the novel by D.H. Lawrence, which gained five Oscar nominations. With a ground-breaking nude wrestling scene between Oliver Reed and Alan Bates, this was but one of Russell's films with heavy sexual undertones. His earlier works about composers, including *Song of Summer*, about the life of Delius, had a particularly profound effect upon me.

I was keen to make a good impression as I knew that the director could be difficult. Ken's disturbing film *The Devils* was strongly on my mind as we arrived under stygian skies, drenched to the bone and dripping like water spaniels straight out of a pond. Puddles appeared all over the pristine parquet floor in the hallway of the bijou mews house off the Marylebone Road. Our camera cases and lighting equipment filled the tiny space, merely emphasising the chaos we were about to inflict on the household. I was mortified and so was my assistant. Ken was not amused.

As we set up the lights, Ken fussed around us, worrying that our lighting stands would scratch his floor. The living room had such an ordinary feel about it that it could have belonged to a conventional nine-to-five couple; only the bookshelves full of dark literature told a different story. The morning did not improve, as Ken decided to be negative about

Ken Russell and Hetty Baynes, London, late 1990s

every set-up I planned, and I just knew that he was getting a vicarious pleasure out of trying to unnerve me. Meanwhile, his wife, the actress Hetty Baynes, was trying her hardest to keep everyone happy as she clutched her stomach, bulging with soon-to-be Rex.

These were not my best pictures, however hard I tried, and I felt like a mouse being played with by a cat. When Ken finally left the room, I took better shots of Hetty on her own, as I could now relax and choose my set-up. I left feeling depressed but also glad that this was not how I usually felt at the end of an assignment. The day did not diminish my admiration for Ken's work.

I photographed Lady Diana Cooper for personal reasons. My beloved paternal grandmother had just died at the age of ninety-four, and trying to cling on to a part of her era before letting go and moving on, I picked up a copy of Lady Diana's biography. I was struck by the similarity of the two women. They were close in age, both considered great beauties in their youth, and they moved in the same social circles.

Lady Diana was known for her great sense of humour and she became famous for her looks when she starred as the Madonna in Max Reinhardt's 1912 *The Miracle*, the first Technicolor film made in Great Britain. She married the writer, statesman and diplomat Duff Cooper and the couple spent a great deal of their official life in Singapore, Algiers and France. Duff Cooper was devoted to his wife in spite of his frequent infidelities, which she tolerated, perhaps because she was extremely popular with the opposite sex herself. The couple lived in Chantilly, outside Paris, for many years, and after Duff Cooper died, Diana moved back to London and settled on Warwick Avenue.

Since my grandmother and Lady Diana moved in the same circles, I was not surprised to see Granny's cousin, "Scatters" Wilson, leap off the page of the biography I was reading. Lady Diana considered him a great friend; I had always heard gossip that he was a "bounder". I was intrigued to meet Lady Diana, so I called her number and discussed our mutual connection with "Scatters". She sounded delighted and volunteered a meeting. As I arrived at her pretty Regency-style house, with Greek columns topped with Corinthian capitals, I pushed open the wrought-iron gate and was greeted by a front garden festooned with nothing but yellow roses. I was clutching a small bunch of yellow roses freshly picked from my garden in Notting Hill and wondered if Lady Diana would think that I had picked her roses, or would she be surprised at the coincidence?

A maid showed me into a small, darkened drawing room, gloomy yet peaceful, deeply entrenched in the past and full of memorabilia, family portraits and precious china ornaments; books lined the walls. The maid reappeared so quietly that I nearly jumped out of my skin. We climbed the stairs and I was shown into Lady Diana's bedroom. She spent the mornings in bed making her telephone calls and arranging her social life, then dressed to go out to lunch. She would drive forth in her small brown Mini and park it wherever she wanted, much to the consternation of the traffic wardens.

The room was cool and dimly lit, filled with plenty of lace and soft yellow furnishings. Lady Diana languished in her grand double bed, holding a brown Chihuahua that was entirely bald. Doggi spent so many hours under the bed-linen that he no longer grew fur; he was an alarming sight as he gobbled what seemed like an endless supply of choc-drops. I mentioned "Scatters" again as I wanted to know more about this ladies man, and Lady Diana's face lit up. "I was most fond of 'Scatters'," she enthused, without elaborating on the friendship any further. She seemed to prefer to talk almost entirely about the past.

I noticed that downstairs and in her bedroom there were numerous unicorns in all shapes, sizes and fabrics, as well as pictures of them and even a stone one at the front door. When I mentioned that I also collected unicorns, Lady Diana said, "Then you must see a manuscript about unicorns written especially for me by my dear friend, Sir Colin Anderson." I could hardly believe it. Yet another coincidence; Sir Colin was a cousin of my step-grandfather, Sir Alexander Anderson. She too was delighted by these coincidences and pointed agitatedly at a chest of drawers where the tome was kept. I tried in vain to find it. Then suddenly the languishing Lady Diana leapt from her bed, reaching the chest of drawers in record time, and pulled out the precious bundle. I was amazed when she offered to lend it to me, and I returned it to her intact, but not before I had photocopied the entire story.

Two days later, I returned to take pictures. This time Lady Diana was a picture of beauty, hair immaculate, make-up perfect, a pretty lace bed jacket in soft yellow and cream to match the room, and a string of

Lady Diana Cooper, Maida Vale, London, UK, 1980s

pearls around her neck. Having greeted me warmly, she said capriciously, "Did I really say you could photograph me?" She had forgotten that her appearance was a total give away, and she had obviously gone to great trouble to make herself as beautiful as possible. As I raised the lens, her head came up, smoothing out her neck, and there before me was the legendary pose she had struck many moons ago as the Madonna, her beauty still shining through.

CHAPTER 12

SINGERS

"To do what you like is freedom. To like what you do is happiness." – Frank Tyger

The 70s ushered in early punk and heavy metal, and one of the first groups on the scene was the New York Dolls. The drug-fuelled crazy performances of these weird androgynous boys in high heels influenced such groups as Guns N' Roses and the Sex Pistols, and the Dolls also happened to be the first of many musician subjects I would photograph over the years. I met them at Blakes Hotel after the release of their album *Too Much, Too Soon*. I engaged with them well on the surface but they made me feel wary and I was in no way seduced into their world. They were very much a product of their time, and part of a world I kept well away from.

New York Dolls, Blakes Hotel, London, 1973 (left to right, Sylvain Sylvain, Johnny Thunder, David Johansen, Jerry Nolan and Arthur Kane)

I had a much warmer reaction to another one of my early subjects, Liberace. Female photographers were still sparse in the 1970s, so it was hardly a surprise that my colleagues at a press reception for the entertainer were all male. I found a suitable place in the front, thinking that as I am small, the men could see over the top of me, and nobody would object. Then, as Liberace walked in, I felt a sharp pain in my back followed by a bash on the head: a rude awakening to the world of paparazzi.

Liberace was gracious, turning in each direction to please all concerned. Even so, unprepared, I had not achieved one single shot by the time that he announced the shoot was over. He thanked the men, then added without a pause, "Now, the lady can stay so that she can get her pictures." My predicament had not gone unnoticed. I could have wept with joy, as now I had the great man exclusively gazing into my lens. I shall never forget his kindness.

Liberace, London launch of his autobiography, 1973

Barry Manilow, during rehearsals for
Copacabana, Los Angeles, 1980s

I landed the opportunity to photograph another icon, Barry Manilow, when I was making the rounds in Los Angeles and told PR agent Dick Guttman that I used natural light for my portraits. I was astounded by his reaction: he leapt into the corridor and shouted to all the other agents, "Hey guys, come see this girl's work, she photographs by natural light!" And so my reputation spread like wildfire and the die was cast. Of course, the expertise needed to control the light in order to flatter was another major factor. So, when I was invited to photograph Barry rehearsing in a vast theatre, I had to live up to my reputation. This was the best I could do, as there was no daylight backstage. Funnily enough, the mean light captured the atmosphere and mood of the theatre during the rehearsal, and Barry's tired eyes speak volumes.

I was given the exclusive rights to photograph Eddie Fisher on his visit to London to record his valedictory album with the London Philharmonic Orchestra. According to the arranger/conductor Vincent Falcone, it was his best and yet it was never released.

A legend in his time, Eddie was in the autumn of his life but singing still defined him, as it always had. We met at the Athenaeum Hotel in London. My first impression, at a distance, was of a slim boy-like man who still had the pretty-boy looks of the Eddie Fisher I had seen in pictures taken in the 50s. As he walked towards me, the veils slipped away, revealing the telltale signs of cosmetic surgery. Jet-black hair, an eye job and a facelift, along with a dubious tan, gave him the appearance of a French-polished piece of furniture and disguised a life of excess. His petite wiry body was dressed immaculately in the old-fashioned Hollywood style: perfectly pressed white shirt, tie, jacket accessorised with bejewelled rings, and well-cut trousers. The whole ensemble tapered down to a pair of red, tan and black snake-skin loafers. We went into full flirtatious acting mode, as after all, here was the man who had seduced all the most beautiful women in Hollywood, along with hundreds of lesser mortals, and the

Eddie Fisher, London, 1995

myth of his allure had to be maintained. I got the overall impression of a man both hugely talented and deeply insecure.

Eddie was one of the most successful pop singles artists of the first half of the 50s. In 1953, at age twenty-five, he was earning ten times the salary of President Eisenhower. By the time I met Eddie, he had been through numerous marriages, and his ex-wives famously included Debbie Reynolds, Elizabeth Taylor and Connie Stevens. He had gone bankrupt, been treated for addiction and bounced back. Just two years before I met him he had married the businesswoman Betty Lin, who helped to keep him on the straight and narrow. He was such a strange mixture of weakness and strength.

I followed Eddie for several days in London to watch him record his album with the London Philharmonic. I could photograph Eddie at work and get a chance to hear the still-legendary voice. Although he was old enough to be my father, I felt slightly uncomfortable while alone with him in the studio, and I did not get the impression that he had changed his ways. But this sexual energy is good for achieving the best portraits, so these pictures have life.

I did like Eddie and, as he had been attractive to almost everybody in his youth, it was even more important not to dent his ego. I was grateful that he still had the enthusiasm to give his best; his life story poured forth, and it was riveting to hear his tales of excess involving drugs and women coupled with a childlike quality in his honesty about his faults and his total lack of fear of spending every dollar on his family, friends and women. There was something so touching in seeing him fired up in the studio, fully in the moment, patiently working on each song as if it were his first. Here was the boy who opened his mouth at the age of three and out came this beautiful sound; now around seventy, with a life behind him that would have killed most people ten times over, the songbird still sang.

Although we had a great deal of time to talk over the three days I spent with him, when Eddie asked me about my life, it paled into insignificance compared to his, but for fun, I mentioned I was named after Carole Lombard. This was not forgotten, and when we came to say goodbye, he gave me a copy of his autobiography. The inscription read, "Lombard, you're the only one who can get me in bed!! at any time, love Eddie Baby".

Eddie Fisher, recording his valedictory album with the London Philharmonic Orchestra in 1995. Although the album has never been released, it was considered to feature some of his best vocals.

Photographing the opera singer Lesley Garrett was a thrill in itself but came with another perk: the chance to work with David Emanuel, her favourite fashion designer. How I loved working with David. We lent my handsome assistant a hand to hump the heavy lighting equipment from one location in the Lansdowne Hotel in London to another: the ballroom, the staircase and finally, a suite at the top. "No egos!" David would cry as we tackled a new location with all its pitfalls; he and I knew that this was the best approach with any creative project, and along with laughter was as good as any drug in fuelling our weary limbs. When Lesley arrived, she announced that she only had one hour; in the end she was there for three and, happily, suggested champagne all round at the end. What a star ...

I had often daydreamed about doing an album cover and would thumb through the racks of CDs in Tower Records gathering ideas in case the opportunity arose. I had no contacts in the music industry and no idea how to begin making this happen. The chance finally came through the tenant in the basement flat of my house, a publicist who specialised in the

Lesley Garrett, wearing a David Emanuel dress and everything else that the designer brought to the shoot at the Lansdowne Hotel, London, 1990s

David Emanuel, dress designer, and Lesley Garrett, opera singer, Lansdowne Hotel, London, 1990s

music industry. Graeme and I got along like a house on fire, but even so, I was extremely surprised when he offered me the chance to photograph one of his clients; after all, I was his "landlady".

Jane McDonald had risen to dizzying heights as a result of her appearance on a BBC fly-on-the-wall programme *The Cruise*. This sudden fame was a dream come true for the talented singer, who had long been practising her craft on the club circuit and was no ingenue. If we got along at a preliminary meeting, I was to shoot her first album cover.

We met for coffee on a wet, windy morning in a dreary hotel opposite the BBC in Portland Square. Everyone except Jane was dressed in fashionable grey and black to match the thunderous sky. Her manager, PR rep, assistant PR rep, hairdresser and myself all blended into the view from the hotel's vast windows. Dressed in cream which enhanced her olive skin and glossy dark hair, Jane reminded me of a star on a darkened stage lit by a single shaft of silver light, mesmerising her audience. We immediately bonded. Her down-to-earth, warm, no-nonsense approach to life was typical of a true Yorkshire lass. I presented her with my one and only idea for the shoot, and luckily she was enthusiastic. As I walked out of the hotel, I was so lost in my dream of purple and silver that I barely noticed the stormy skies.

I was excited. Simplicity had to be the theme, partly out of choice and partly because the budget was limited: purple background, silver dress, theatrical gobo for a sea effect. Would the producer Don Reedman agree? Yes, he would. How wonderful! My favourite studio was available, my chosen stylist was free, and Bill Burlington, an experienced photographer, was happy to assist. Then came the bombshell that the BBC would be filming us filming Jane. What a nightmare. All the possibilities loomed: faulty Polaroids, temper tantrums, dresses that did not fit, bad language, anything could happen and it would all be captured on film. I was horrified.

The day of the shoot, I arrived with my car bulging with black bags full of lights and cameras. I was dressed in black from head to toe like a stick of liquorice. I think I wanted to camouflage myself, to blend with the bundles of photographic equipment. The studio, with my name emblazoned on the door, was already buzzing with activity and full of people: the hairdresser, stylist, my assistant, Jane's publicist, the publicist's sidekick, Jane's manager, the manager's friend. Then, of course, there was the BBC film crew, three not two as promised, and now they were no longer staying for the first hour as agreed but for the entire day. We had only catered for half that number for lunch.

I loved the large Battersea studio, which held wonderful memories. This is where I had photographed my actress friend Fiona Mollison for a Martini commercial and Maureen Lipman with the cast of the television show *Agony Again.* Both shoots had gone exceptionally well, so this studio had brought me good fortune and I intended to keep it that way.

Having greeted Jane's courtiers I made my way up to the dressing room, where I knew she would be in the raw having her make-up applied. To my horror, I found her face bare to the cold north light with the film crew filming up her nose. Graeme, her publicist, was performing like a warm-up artist on a live comedy show, desperately trying to keep her spirits high. Jane was bearing up well, but I saw no reason why she should have been exposed in such a way when we were about to capture her at her most glamorous; it seemed counterproductive. Sadly, stars no longer have any privacy or mystique; I think this is our loss as much as theirs.

The make-up was completed, or was it? Her beautiful olive complexion had been masked in a layer of lighter make-up which, to my horror, made her face look like a powdered bap (highlighting a sprinkling of fine black hairs which otherwise disappeared). I made a mental note that there would be no close-up shots, just long shots. Her hair was so dark that I knew it would be hard to light, and it was crying out for a few subtle highlights to enhance her crowning glory. And so the day went on, and half the dresses did not fit, and if they did, they made Jane look like a suburban housewife. Only one dress, a long silver tube of sequins, looked fabulous, so I was determined to use it, although any attempt to fasten the zipper was fruitless. The stylist had not bothered to check the size, so we cheated with huge safety pins.

I now had a beautiful woman who could not be photographed close-up or from the side or back, and all the time we were being filmed for television, warts and all. Mobile phones exploded every few minutes and the endless irritation of too many superfluous people in the studio was driving me crazy. As the day wore on, I became so resigned to the situation that I no longer cared that I was wired for sound. Boom and camera lurked around me all day long like wasps around the sandwiches at a picnic. Jane looked like a beautiful wild animal caught in a snare but still we pushed on, making good eye contact, knowing we trusted each other. I am never happy to stop for lunch until we have the key shot in the bag, so by the time we had overcome all the pitfalls we were starving.

The silver dress was stunning when photographed straight on against the rich purple backdrop and one further shot of Jane lying down on the safety pins, like Houdini on a bed of nails, looked sensational. Jane hugged me and said she knew we had cracked it. I apologised for being a little ratty about the mobiles ringing and she burst out laughing; they had driven her equally crazy. My team had behaved well. Bill, who was assisting me, was nicknamed "Silent Bill" because he kept his phone switched off and only talked when he had something intelligent to say. His public school upbringing and his destiny as the Duke of Devonshire had taught him good manners from an early age. Sarah, the stylist, was equally polite.

To my relief the footage shot by the film crew was never shown on television, as they had too much other material on Jane. The album went to number one in the UK charts and the cover was well received; I had fulfilled my dream.

Jane McDonald, album cover, 1998

CHAPTER 13

STARS' GARDENS, BRITISH STYLE

"Sympathy with nature is a known evidence of perfect health. You cannot perceive of beauty but with a serene mind." – Henry David Thoreau

Gardens have always been a part of my life. In my childhood in Hampstead in London, we had gardens back and front and the Heath was across the road. My mother loved her garden and above all the roses; an oval-shaped bed in the front garden took pride of place with a fine selection of hybrid tea roses. "Peace" was the first word she ever taught me to recognise, probably because the last thing you could call our household was peaceful; loving, yes, but all the world was a stage and there was always plenty of drama. I discovered early on that buddleias attracted butterflies and that snails had to be stamped on if you did not want the garden to be desecrated. I used to justify my attacks by saying that they were scrambled eggs for the birds, even though the birds never touched this ready-made food. At the back, a small conservatory with a giant coral geranium led on to a garden, which stretched around the side of the house to a modest vegetable patch.

There was hardly a creature that I did not nurture when given the chance. Each spring I would fill an old stone sink with water into which I would empty jam jars full of tadpoles from the Hampstead ponds. The tiny black creatures, looking like wiggling apostrophes, would grow little legs and finally turn into mini frogs. Then came the frantic struggle to survive as they hopped onto the edges, exposing themselves to predators. I was always saddened by all the little deaths, as so few survived. Outside the basement bathroom, beneath an iron grid, was a dark, damp, concrete area carpeted with piles of damp rotting leaves; a perfect location for toads. What was horrific to my mother was a joy to her daughter. I would stand in the bath under the window and dip my hands into the moist leaves like a child eager to land the best prize in a tombola. In this way

I would retrieve whole families of toads and drop them into the bath. There, to my great joy, they would leap around until it was time for them to be returned to their leafy home.

Then there was my grandmother's garden in Surrey. Every year, we dug up the dahlias, picked the blackberries, bottled the fruit, picked the yellow tomatoes from under the frames, climbed inside the nets to pick the raspberries, and made lavender bags for the people in the village. I was surprised to see that celery thrived in the dark, growing tall beneath chimney pots. My private memories were watching the chickens laying their eggs through the gap in the wooden slats around their houses and crawling through the undergrowth to see how many cheroots my step-grandfather had thrown out of his smoking-room window; his secret was kept forever by the giant rhododendrons. In the summer heat, lizards basked in the rockery amongst the wild strawberries and my grandmother and I would sit on the porch eating Victoria plums for "elevenses". Water boatmen rowed rhythmically across the pond while dragonflies hovered above them like helicopters, and a leaden lion's head spewed out an endless waterfall. After tea, there was always a game of croquet. In the autumn my grandparents made jam and planted out the chrysanthemums; nothing changed except the seasons.

One of my most cherished memories of being amongst beautiful greenery is from a stay with Sarah Miles, the actress turned author, who now runs a healing centre. Her husband Robert Bolt was still alive at the time. He was one of our finest screenwriters, with such memorable films as *Lawrence of Arabia*, *Doctor Zhivago* and *Ryan's Daughter* to his name. Robert also wrote *Lady Caroline Lamb* especially for Sarah. I have a wonderful picture fixed in my mind of Robert resting on his antique wooden chaise-longue under his apple tree. It was one of those rare warm summer evenings and the sun was setting. Sarah was sitting there oozing undoubted love and devotion. I thought back to a conversation I'd had with Robert at breakfast time. A severe stroke had left his speech hesitant, but he explained that it had been his own fault. "I was working too hard and I did not listen to my body," he said with a certain amount of difficulty, and added, "but God has given me a second chance and for that I am eternally grateful." He was a remarkable man to have remained so positive in the face of adversity and to be grateful for life

Sarah Miles, West Sussex, England, early 2000s

itself. His lack of bitterness was a lesson to us all, along with his steadfast belief in the meaning of life and the thereafter. Robert now rests in peace at the end of the garden under a simple but beautiful headstone that reads "A Man for All Seasons", the title of one of his most famous screenplays. When Sarah goes for walks with her dogs she greets him on their return. His spirit lives on.

It only seemed natural to undertake a project in which I would capture other celebrities in their gardens as extensions of their personalities. Why not, I decided, begin with two of the greatest British actors, Sir John Mills and Sir John Gielgud?

Lady Mills greeted me warmly at the door of the picturesque cottage where she and Sir John lived in Denham village, an oasis near London where time has stood still. Rows of tiny cottages covered in climbing plants nestle around the village church that watches over them, its traditional yew tree shading the graveyard. It all seemed like a film set, perhaps an appropriate home for one of the most prolific film actors of his time, who appeared in more than 120 films, was acclaimed for his outstanding

Sir John and Lady Mills, Denham, UK, 1990s

performance as Pip in *Great Expectations*, and won an Oscar for Best Supporting Actor in *Ryan's Daughter*.

We bustled into the back garden, not wasting a second in case the light should change, as Lady Mills muttered all the while about the lily beetle which had never bothered her before. "They not only destroy the plants but also leave their unsightly doodoos all over the leaves," she remarked. We found Sir John pottering around with a wheelbarrow: he greeted me in a gentlemanly manner with praise for my father, "Such a charming man and such a good actor." When he voiced his concern that their gardener was looking older, Lady Mills pointed out that he did not have his false teeth in that day and the subject was closed. "Both John and I are mad about gardens," she continued, as the three of us strolled around in the warm afternoon sunshine. "We find them therapeutic although I had to teach him everything. He was apt to say, 'Look at that marvellous tulip' and I would have to say, 'No, no, that's a rose.'" Sir John was happy to let his wife take centre stage. He added that he found gardening a wonderful form of relaxation where he could get totally immersed.

A winding path framed on either side by herbaceous borders bursting with colour was a perfect spot to photograph the couple. They were charmingly enthusiastic as I shouted directions from afar but they needed no direction, for they naturally formed a perfect picture. Almost the same height and obviously as close as ever, they talked animatedly to each other while gazing into each other's eyes. By now the sun was sinking and the warm glow was cooling to a blueish light. Before I had time to wonder whether I should leave they were insisting that I stay and I was soon being offered a glass of chilled Chablis in their cosy drawing room, where the last of the sunlight shafted through the low windows.

Sir John Gielgud was not only one of our finest theatre actors, but his sense of humour and razor-sharp wit were legendary. I felt I knew him already after two telephone calls to arrange a shoot, having been introduced by my friend, Milton Goldman, the legendary American agent who referred to the actor as "Johnnie G". So I should have known what to expect when, soon after my arrival at Sir John's estate in Buckinghamshire, I noted he was growing a particular lily that I had seen in the Mills's garden. I remarked that Lady Mills had complained bitterly about her lily beetle and without a pause Sir John retorted, "Lady Mills complains bitterly most of the time."

The huge, immaculate garden ("the gardener has cleared every fallen leaf," Sir John assured me) was divided into several rooms, each one

Sir John Gielgud, Buckinghamshire, UK, 1990s

framed by ancient walls. Strategically placed statues and beautifully manicured lawns neatly edged with herbaceous borders gave me a wealth of photographic opportunities. Sir John said that though the garden brought him great joy, his work in the performing arts meant there was little time to pursue the things that others do in their spare time. So he left the planning to his friend Martin.

As we went into lunch through a wooden door heavily camouflaged by climbing roses and greenery, two enormous dogs leapt out to greet me. Feeling the warmth of Sir John and Martin, I felt comfortable enough to comment, "The dogs seem to be glad to see me!" as they revealed they were definitely male. This was greeted with more laughter and the day continued in the same vein. I loved Sir John's grand manner.

Drinks before lunch were taken in the drawing room, grandly decorated in the French style of the eighteenth century, with a great deal of brocade, gilt and ormolu. I was amused by the way Martin rather quaintly referred to his friend as "Gielgud", while I enjoyed the correct form of address, which was Sir John. We then adjourned to the dining room, where Sir John and I sat at either end of a long table decorated by two vast candelabras lit to warm the day. I remarked that this scene reminded me of *Brideshead Revisited*, in which Sir John played Jeremy Irons's austere father. "The last person to dine here, my dear, was the Queen Mother."

"Well, Sir John," I replied, "I am truly honoured."

It was quite hard enough hoping that I was being entertaining throughout our long lunch without the added worry that the candles were becoming brighter and brighter as storm clouds gathered; all I could do was pray. Miraculously, by the time the coffee was poured, the candles had faded and I sprang into the garden to photograph as much as possible while the light held. I then summoned Sir John to the perfect location, which I had gently lit to enhance the great man. By chance, I had chosen his favourite spot, a gate which he had brought back from Italy. We laughed and joked as he posed; I was delighted, as I had seldom seen Sir John smiling in photographs. Tea was then offered, and as I left he remarked that he hoped we could meet again soon. The day had clearly gone well, so I sent him a print of my favourite image.

"Dear Carole," came the reply in teeny-weeny, spidery writing. "Thank you so much for your kind good wishes and the photographs of the garden, which are indeed splendid. More power to your clever elbow. Yours ever, John Gielgud."

Sir Hardy Amies was not an actor but a legend in his own right, one of Britain's great twentieth-century fashion designers and Couturier to the Queen. Sir Hardy once famously referred to Her Majesty as "frumpish" and, probably amused by such indiscretions, she later awarded him the Royal Warrant, which he admitted was a terrific "money-spinner". According to hearsay, Sir Hardy was arrogant, patronising, selfish, indiscreet and a terrific snob, but he could also be charming, amusing and extremely agreeable. There was yet another side to this dandy, who had been a key figure in Churchill's Special Operations Executive. He spoke fluent French and German, parachuted behind enemy lines five times, and recruited a lethal trio of agents codenamed the Rat, the Goat and the Vole. Working in the fashion world long after the official retirement age, he would joke that he spent his old-age pension on sex.

So I hardly knew what to expect as I arrived in a charming Cotswold village of weathered grey stone to photograph Sir Hardy in his garden. His cottage, an old schoolhouse that he had bought and converted to be close to his sister Peg, was festooned with climbing roses. I opened the wooden gate and was welcomed by the man himself, who was polite but brisk.

As we walked out into the back garden Hardy thawed, waving his elegant stick. Although nearly ninety years old, he was soon scuttling around like a beetle, the love of the garden spurring him on. "When I first came here, there was no garden, as it must have been the playground." The square back garden had beds bursting with harmonising shades of mainly pink and white, neatly framed by low box hedges. "I think, as did the Elizabethans, that it is only with symmetry that you can create a sense of calm. Your central path must go somewhere, so that is why I built the summerhouse, which is in the Cotswold style, my first piece of architecture." I wondered if we could call this the heart of the garden, but he didn't seem to hear me. "I also learnt from a gardener in London that you must have paths that are wide enough. It's maddening to go along narrow paths like a geisha girl."

I felt that my contribution to the day was minimal, as I barely got a chance to ask a question. Hardy was passionate about his garden, and his retentive memory meant total recall of the Latin name of every plant.

Sir Hardy Amies, dress designer, Cotswolds, UK, 1990s

His favourite was the rose, so I decided to photograph him by a standard white shrub rose, secateurs in hand and a trug at his feet. "Then there are lots of aromatics. The smell I remember most vividly as a child was 'Old Man' Artemis, whatever it's called; it's over there, the rather raggedy one." The elegant cane flew up in the direction of the Cotswold stone wall. Hardy had barely stopped to draw breath.

I met Sir Hardy on two other occasions. One was at his annual summer party, on this occasion to celebrate his ninetieth birthday. It was an enormous affair, and people came from far and wide. Dressed in an immaculate off-white suit, with his legs elegantly crossed, Hardy looked positively regal. Peg, his devoted sister, sat sturdily with legs akimbo and her stick firmly planted between them.

On the next visit I stayed the whole weekend, to take a photograph of an ancient bridge for Hardy's Christmas card. Hardy was like a spring chick, and the excitement of the project made the years roll back. We set off for the river early, as it was unbearably hot. The reeds were so high along the banks that there was no way I could get a view of the bridge. There was only one thing to do. The local publican found a chap with a boat whose wife was a huge fan of Hardy's, so he offered to take us out.

I now had the daunting task of helping a man who was unsteady on land on to an even less steady boat. We reached the middle of the river, where we rocked about under the roasting midday sun as I tried to steady my tripod and camera. All the time I had to keep one beady eye on Hardy as he rose, precariously, to join in the shoot. Time disappeared as we became lost in our project, while the poor boatman waited endlessly for us to finish. Eventually, we thanked him profusely and Hardy took the man's address; I assume the resulting letter brought joy to his wife.

When I heard that Hardy had passed away peacefully in his bed, gazing out at his garden, I remembered how we had walked, arm in arm, to the village church and he had pointed out where he intended to be buried. At the time his directness had taken me aback, but his acceptance of the transition into the spirit world was reassuring. It was an honour to have known Sir Hardy.

One of the oddest gardens I've photographed in England, or anywhere for that matter, was also in the Cotswolds, and also the creation of a performer: the actor and singer Edward Woodward. Edward, his wife Michele and their daughter Emily lived in an ancient, dark, womb-like house near Stratford-upon-Avon, with thick walls and tiny lead windows that admitted shafts of sunlight that dramatically highlighted the beautiful antiques and treasured ornaments.

Edward, Michele and Emily Woodward, Cotswolds, UK, 1990s

The house brought back bittersweet memories, as I had spent a charmed weekend there a few years earlier. Edward and Michele hadn't been there, but I was the guest of Michele's sister, the actress Karen Dotrice. Those days of swimming, tennis and weather to match left us burnt, as the English are apt to become when faced with a rare heatwave. Among the guests was Steven Martindale, a charismatic man who brought a little magic into everyone's lives. A brilliant raconteur with phenomenal energy, he was a lawyer and government affairs councillor based in Washington, D.C., with an awesome range of friends in show business and politics.

One night not long after that weekend, when I had not heard from Steven for a bit, I suddenly sat bolt upright in bed as something told me to call him. "I cannot believe you are doing this", he said. "I am just trying to compose the hardest fax to tell you that I have AIDS." Typical of his positive nature, he added, "but you are not to worry because I know I shall beat it". Wishful thinking, and I miss him hugely to this day.

On this occasion, Edward pointed out that they lived at the end of the Cotswold chain of hills, which had once been underwater for billions of years. "We have fossils in practically every stone in the garden," he remarked. "The fossils are of sea creatures, including an actual flaccid sea horse which had been trapped in a rock pool."

Lunch over, we wandered out into the heat of the summer's day. "I'm not a very good gardener", Edward apologised, "as I don't have much patience with plants, but I do have endless patience with inanimate objects. I love to build rockeries, as I find the use of stone fascinating.

"You have to be very hardy to survive as a flower in this garden. We get winds seemingly coming from Siberia nonstop and we quite often get snowed in, which is great fun." Antirrhinums and poppies were obvious favourites, and they grew in profusion around us. Intriguingly, a Jacob's ladder zigzagged up one side of the garden, ingeniously designed to handle the steep slope. The rope which ran from top to toe was one of the old fly ropes from the Strand Theatre in London. Well-manicured terraces, cut out of bare rock, seemed to cuddle us in an atmosphere of peacefulness. "I think greenery is necessary for one's sanity," Edward explained. "As every Englishman who has ever travelled knows, there is no green like the English green."

CHAPTER 14

STARS' GARDENS, CALIFORNIAN STYLE

On my many trips to Los Angeles, I have always missed the garden at the back of my house in Notting Hill. I created this refuge from a rubble heap and after years of nurturing, the cherry tree celebrates the arrival of spring with a mass of white blossoms, and flowerbeds burgeon with all shades of pink, purple, dark red, white and grey. Fish glide around in a pond full of water lilies, and the flowers attract a variety of birds, insects and spiders, with their magical webs. Most welcome of all were Bob and Ethel, a pair of splendid toads who appeared each summer and sat by the pond.

I had been photographing celebrity gardens in England, so it only made sense to continue the project on my trips to Los Angeles. This would keep me connected to my gardening roots, and I knew that many of the gardens that flourish in the southern California climate are spectacular. Also, I was eager to discover the feelings and the memories of an eclectic mix of well-known people as they wandered in their creations.

I felt a wave of homesickness when I visited Phyllis Diller and sat in what she called "my very English garden". Phyllis was a true anglophile, so it was no surprise when she told me, "The rose is the prettiest bloom in the world. The name of my house is 'Rosewall West' and over the years, I have probably received more roses on my opening nights than the Queen Mother."

When I photographed Phyllis in her garden in the 1990s, she had lived there for twenty-five years. When she bought the property, the garden had been mainly lawn and hedges, but soon she was adding flowers. Then came the battle with herself whether to install a swimming pool or a tennis court, but she resisted the temptation. "I kept thinking of that lawn, the biggest area of lawn in Los Angeles; how could I desecrate it in such a way? It's so peaceful to sit and watch the sunset over the greenery and bursts of colour."

Phyllis was brought up on a fifty-five-acre farm in Ohio, where there were no flowers, just animals, wheat and corn. "I always adored flowers

Phyllis Diller, Los Angeles, 1990s

and grass, so as soon as I had my own garden, I filled it with flora. My cut flowers last for ages because I trim their stems every day; the weaker they get, the shorter I cut them; it is the caring that counts."

Phyllis, famous for her longevity as a working comedienne and beloved among her friends for her old-fashioned good manners, passed away in her home at ninety-five with a smile on her face. I still keep a packet of Sweet'N Low on which she'd printed "Sweet Thoughts From Phyllis Diller".

The actress Cloris Leachman was another Hollywood legend who lived to be ninety-five and also stayed close to her rural roots. She spent her childhood in Iowa on an acreage with a few scattered houses and a pasture with cows at the back of the house. She reminisced, "I have a vivid childhood memory of opening the door after heavy rain and almost falling backwards with the freshness of the air mixed with the lilacs."

When Cloris found her wooden house hidden away amongst the trees in Mandeville Canyon, she realised that this was to be her home for life. A winding driveway with magnolia trees (considered to bring good luck to a home) led to a hidden, sprawling bungalow, surrounded by pink, scented jasmine on one side and hanging vines and trees on the other.

Cloris Leachman, Mandeville Canyon, Los Angeles, 1990s

"I used to drive home from the studios at lunchtime, even if it was only for an hour," she told me. "It was like a blood transfusion for me. My wellbeing comes from the smell of earth, trees, azaleas, night-smelling jasmine and the bougainvillea bursting over my walls. I would diminish if I couldn't be with these living, breathing things."

It was hardly surprising that the British actress Samantha Eggar attempted to surround her cosy, ultra-feminine little house in the Hollywood Hills with an English garden. "Lavender, dahlias, roses; my whole garden was an homage to England," she said. That plan, unfortunately, was not a success. As she explained, "Well, of course, everything died, even the willow trees and rhododendrons. I soon realised that this was a semi-desert and I better get myself into the right vein of what nature wanted me to plant in this part of the world."

When I visited, Samantha pulled on a pair of Wellington boots, very British and at odds with her cotton skirt and the sunny day, and we wandered around her garden with her devoted Dalmatian following closely. "I started with plants that don't need water," she explained. "I planted an entire area with cactus, all the succulents which do well here

Samantha Eggar, Mulholland Drive, Los Angeles, 1990s

and have the brightest, prettiest flowers but do not need water. Nature taught me to move practically every plant until I find its right spot."

Samantha, who appeared in more than ninety films alongside such leading men as Terence Stamp, Sean Connery, Rex Harrison and Roger Moore, was only too happy to talk about her early memories of gardens. She was born during the first year of the Second World War and her parents disappeared to serve their country. Consequently, she was brought up by a painter friend, Gwen Barnes. "During the first five years of my life, Gwen injected nature into us and that stayed with me for my entire life. Everything I feel about nature, life, art, painting is due to that woman." They lived in Bledlow, in Buckinghamshire, England, and Samantha recounted walks in the country, eating cobnuts and rose hips, and "looking down into the faces of the biggest sunflowers I have ever seen, and we thought these were very odd with their big, black centres and floppy petals like fishes hanging down".

Samantha referred to living in her L.A. house as being in a perpetual theatre. "I not only have the view of my garden but fifty miles of valley and then the mountains to look at every single day from every single window." She added, "The moon comes right over my house and I can see the full moon and the new moon from every window." I was impressed that in this sprawling, polluted city Samantha had found a way to stay in touch with the natural world that called out to her soul.

The talented actress Linda Purl also attempted to recapture her childhood connection with nature. For her house in Beverly Hills, set well back from the road up Coldwater Canyon, a landscape designer created a garden with strong Japanese influences, bringing back Linda's memories of the home in which she grew up in Tokyo. "We had about an acre of land, which was unheard of in Tokyo, and it was like a park with lots of trees," she reminisced. "I spent a great deal of time out there away from Nanny's scolding."

Linda Purl, Coldwater Canyon, Los Angeles, 1990s

I stayed with Linda briefly after a serious falling-out with a host, and she was most sympathetic. "You must come and stay immediately," she said with conviction. Soon I was safely cocooned in her cosy guest bedroom, with swathes of tulip tree framing the window, bringing back joyful memories of my grandmother's home in England. Walking through the gate into the garden was like stepping a million miles away from the frenetic atmosphere of L.A. Eastern figurines were dotted around, peeping out from behind plants or just sitting on the ground. "My mother and I collected many of these little figures from Indonesia," Linda told me, pointing to a rather grotesque little couple. "They were absolutely filthy and when we washed them, they turned out to be wonderful." I particularly loved Linda's chimes, known as Fu Ling, which broke the silence from time to time when I was alone. The Chinese believe that chimes capture the spirit of the wind, and that the wind is an invisible element until it touches a physical entity.

My photographs of Linda in her beautiful home and garden have always brought back fond memories of a magical time and a good friendship. So a few years ago I was delighted to learn that Linda was coming to the UK to tour in the play *Catch Me if You Can*. She would be starring with Patrick Duffy, who made his fame as Bobby in the

Linda Purl, Coldwater Canyon, Los Angeles, 1990s

legendary television show *Dallas*, and the couple also enjoyed a relationship offstage. When they came to London, I was able to see them briefly at a theatre in Richmond between performances. It was heartwarming to meet again after over twenty years, all the more so knowing that Linda and Patrick were so happy together.

I have no such warm feelings for the Canadian fashion tycoon Peter Nygård. I met him in the 1990s, when I drove south along the coast from Los Angeles to Marina del Rey one late afternoon to photograph this man who claimed to have the best private indoor gardens in the world. So far, every subject had confirmed my theory that people who love nature are gentle folk with healthy values. I was about to be proved spectacularly wrong.

Nygård's condo stood in an unremarkable row close to the road, backing onto the vast expanse of Venice Beach and the rolling surf that The Beach Boys made famous in the 60s. There was no sign of an outside garden. Beckoned into the house by a voice on the intercom, I stood uneasily in a vast hall, a large pond full of huge koi carp framed by lush greenery, all stunningly beautiful, but for some reason a chill ran through my veins. I climbed the stairs to be transported into a maze of grottoes that on closer inspection turned out to be rooms in disguise. At this point, as I was feeling somewhat intimidated by my surroundings, two Great Danes and three parrots appeared, followed by Peter Nygård himself, still bottle blond, his gaping white shirt flaunting a generous brown belly. He greeted me with charm and offered to show me around. No refreshments were on offer, but the natural light was sinking fast, so I was happy for a quick tour of the venue I had come to photograph.

"I enjoy nature," he said. "I like to take a shower under a natural rock waterfall. Every house I build has a roof garden." At this point we had reached his master bedroom, with a California king-sized bed looking out over the rolling surf and the setting sun. I wished I did not feel so uneasy as I realised that this was the perfect room in which to photograph him in the equally perfect light.

Peter lounged across the bed as the two Great Danes jumped up on either side. He insisted that I join him to watch the sun setting, a miracle he paid homage to every day. I had the excuse that I needed to start photographing swiftly before the golden hour was gone. He breezed on, lounging back, shirt now gaping ominously low, but I had my camera firmly positioned between us like a crucifix before a vampire.

By the time the light had faded to a dramatic chiaroscuro effect, Nygård was stroking his stomach. I could no longer see how low his hand had travelled, but I still felt that my imagination might be running away with me. So I continued to concentrate on the settings on my light meter and translating them onto my camera while gauging the camera shake at such a low speed and the limited depth of field with such a wide-open aperture. At this point Nygård added, "I have the biggest fish, the biggest dogs, the biggest parrots and the biggest women. I like everything big. I'm a Leo." I was not too concerned, but it was still unsettling, especially as we were alone in the house and the Great Danes would probably take his side if there were a struggle.

Peter Nygård, Los Angeles, 1990s

Peter Nygård's waterfall bath and shower, Los Angeles, 1990s

Having got my photos, I left hastily in the dark, breathing a sigh of relief as, unmolested, I reached the safety of my car and a bottle of water that has never tasted so good. Only now, over twenty-five years later, has my instinct been confirmed. An unforgettable moon-shaped face, framed by straggly blond hair, sometimes stares out of the newspapers at me: it is Nygård, charged with sex trafficking, transportation of a minor for prostitution, and racketeering conspiracy, along with other offences.

Another garden I photographed belonged to a man who many also once considered a villain, others a hero, and who was undeniably one of the greatest legends of the 60s. Timothy Leary, the Harvard professor, as charismatic as the Pied Piper of Hamelin, had encouraged a whole generation to open their minds with the help of LSD.

Now Leary was living with his third wife, Barbara, whom I had met through a mutual friend in Los Angeles. I soon discovered that she was not just a beautiful face, but had her own career as a movie producer, editor and financial consultant. I was delighted to be invited by Barbara to photograph the couple at their home, and soon I was driving into the quiet and verdant heights of Bel Air. Their garden was fairly pedestrian, a well-manicured lawn with few flowers apart from the all too common birds of paradise nestling under a formidable palm tree. Barbara, who was a good twenty years younger than Leary, greeted me warmly. She looked stunning, dressed in a well-cut white trouser suit, with her short brown hair swept straight back off her face. We drank tea on the patio, discussed the mutual friend who had introduced us, and planned the shoot. Only when everything was in place, the golden retriever at the ready, Polaroid shot, and a space left for Leary, did she summon the great man. He smiled and made small talk, then snap. The shots were taken and he politely took his leave. Barbara offered her apologies for his hasty retreat, as apparently he was working to a deadline.

Leary remained a complete enigma; I have never been so puzzled by anyone I have photographed. The smile had remained fixed throughout

Timothy and Barbara Leary, Bel Air, Los Angeles,1990s

our entire meeting, which reminded me that a smile can be misleading. I have come to realise that while many smiles can come from joy, some are mere masks. Barbara had so successfully guarded her man that I had no idea where his mind was.

I did know from Leary's writings that he considered his post-prison years, living with Barbara and her son Zachary after serving a two-and-a-half-year prison sentence, as the most loving, tranquil and productive time of his life. I also learned a little about his relationship to his garden: "Our plants are very talkative," he wrote. "At our quietest and best moments we eavesdrop on their gossip, their spring-time sighs, their summer laziness, their murmured requests for water and attention."

Another mystery to me, at least at first, was Don Bachardy, the artist who was the longtime companion of the writer Christopher Isherwood. Don seemed quite distant when he greeted me on a wrought-iron balcony bursting with shaded terracotta pots of vibrant plants, open on one side to a vast expanse of the Pacific Ocean. I understood, as it was just a few years after Christopher's death, a time that can still be so raw, and I had the sense that Don wanted time to see whether he would want my presence in his home. He had shared his life with Isherwood for thirty-three years, and they called this house the "Basket". The real basket was their bed, where they would sleep entwined in the belief that they communicated with each other in the night. Don and Christopher had met on Valentine's Day when Don was just eighteen and Isherwood a mature man of forty-eight. Don had been drawing prolifically since childhood but with Christopher, doors opened to another world. They made a unique creative team in the 60s and 70s in southern California. During the last six months of Christopher's life, he was Don's only subject; these pictures are, in Don's opinion, his finest works.

Don's fine white hair, artfully cut, slightly long with a fringe, along with his lithe body, still gave the overall appearance of youthfulness. Dressed in spotless white, right down to soft cream suede shoes, he looked cool and chic with a pair of bright red socks, suggesting just a hint of the free spirit. We set off immediately to view the garden, climbing up and down the vertiginous slope, a challenge for anyone suffering from vertigo and certainly not for the frail. The plants appeared to be indigenous, and the star of the show was a large fan-shaped date palm, under which stood a splendid pair of bright pink plastic flamingos, a humorous touch, almost matching Don's splash of red with his cream outfit. As I snapped away, I favoured Don's beautiful, classic profile.

Don then offered me a cup of tea in his living room, where each large window was framed with greenery and flowers and looked out to the

Don Bachardy, Santa Monica, Los Angeles, 1990s

Don Bachardy, Santa Monica, Los Angeles, 1990s

ocean. I could well understand why he never liked to leave home. Looking at Don's face, I realised I wanted to capture the luminous quality in those humorous, hazel eyes that were observing me just as keenly as I was observing him; "One artist to another," as he generously referred to me as he signed a copy of his book for me on my return to show him the pictures.

The ultimate affirmation of acceptance was when Don asked me if I would take a picture of him with his current boyfriend. At that point, I saw a reversal of roles where Don had become the older man and perhaps the teacher. Though his life had been defined by Christopher, Don has come into his own. His talent has blossomed and his prolific body of work is awesome, with many of his pictures exhibited in the Metropolitan Museum of Art in New York and the National Gallery in London.

One of the most striking settings for a garden shoot was Calabasas, in the Santa Monica Mountains just north-west of Los Angeles. Unlike the perpetually sunny weather of southern California, there is a definite feeling of seasons up here in the foothills, where the actor Tom Skerritt lives in a low, dark wooden house camouflaged by fir trees in a barren landscape of boulders.

As we walked around the property, I found the silence eerie after the hurly-burly of Los Angeles. I was also struck by the gentle simplicity of this man, who expended no nervous energy trying to impress. He seldom made eye contact and came across as shy. "Here, in the wilds," he said, resting on a boulder and stroking his dog, "I hide away from Hollywood and its parties with my child, girlfriend and best friend, my dog." Tom radiated an inner calm and seemed at peace in this rural setting.

The garden was natural and included a tiny orchard with four little apple trees, just about the height of a man. A small fence surrounded the vegetable garden, although it was barely recognisable as such, blanketed with weeds and couch grass. Tom noticed my expression as I viewed the patch, wondering about my photographs, and said, "Sadly, my vegetables have been neglected to the point of extinction this year because I have been away filming all summer."

We went down to the bottom of the property, where I was reminded of the film in which Tom stars as Brad Pitt's father, *A River Runs Through It*. That is exactly what was happening in this scene, all just as natural and tranquil as could be.

Tom Skerritt, Calabasas, California, 1990s

CHAPTER 15

KATHARINE HEPBURN

"Here we go again, the old juice. Ah, guaranteed heart melter. A few female tears ... Stronger than any acid."

– Spencer Tracy to Katharine Hepburn in the film *Adam's Rib*

All my best opportunities have come about by chance, and this was one of them, happening out of the blue. Having the opportunity to photograph Katharine Hepburn was also proof that patience is one of the most important qualities in freelance work. The actress Fairuza Balk and her mother were staying in my house in London and gave me the Los Angeles number of the director George Schaefer. While visiting George in his office in Culver City to present my portfolio, I mentioned my garden project and he asked if I would be interested in Katharine Hepburn, knowing she loved her gardens. He had been working with her on her last three television projects. I was surprised both by the suggestion and that she was working so hard in her late eighties. I had many big names on my wish list, but this was beyond my wildest dreams. So I wrote a long letter to Miss Hepburn, explaining my project and mentioning that I had worked with John Huston for four months. I felt that a shared experience might further my chances, but unbeknownst to me she was writing a book about her experiences working with Huston on *The African Queen*. The answer came back, polite and not disinterested, but she said that she would be out of town and unavailable.

Two years later I was in New York and decided to drop the shortest line to Miss Hepburn's East Forty-Ninth Street address, just to give destiny one more chance. A few days later I returned to the Fifth Avenue apartment where I was staying, drenched to the skin from a New York downpour, when Susan, my elderly, extremely deaf hostess, screamed, "Carole, Miss Hepburn called." "Did she leave a number?" I shouted, not daring to stray farther than her doormat. "I don't have time to take numbers," she snapped, "I've got far too much to do." My heart sank.

Why had I agreed to stay with a woman I did not know? Here I was, so close to talking to the woman voted the most famous actress of the twentieth century, the brightest star in the firmament, and yet my hostess had not had the time to take down her number.

I called my friend Milton Goldman, one of the best-known theatrical agents in New York, begging him for his help, and he instantly reassured me that he would be able to find the number and, once again, saved the day. "Could I please speak to Miss Hepburn?" I hazarded on the phone. "This is she," came the retort. Why was I even asking? With that unique, rich-toned, trembling voice it was so obvious to whom I was talking. I continued nervously as I tried to explain, in brief, my idea of photographing people in their gardens, to show the garden as an extension of their personalities. "I don't know a great deal about gardening," came the brisk retort. I knew full well that she loved her gardens and was often seen planting bulbs for the spring. So I pointed out that many of the people I had talked to did not know a great deal technically, but that I wanted their feelings and memories of gardens. "What are you trying to do, make jackasses of us all?" came another challenging response. My heart sank, I felt desperately shy and wondered why George Schaefer had ever thought that this was a good idea. I tried again, stumbling over my words as I blurted out that my intentions were quite the reverse. I said that I wanted to know whether greenery was important to her wellbeing. "We'll all be under it soon enough ..." A long pause. I had no intention of being used as a verbal punchbag any longer, and anyway, I was almost ready to admit defeat. Why should such a great lady be interested in me or my wretched project? I simply let the pause continue, let her take the lead. And lead on she did. "Call me at nine tomorrow," came the capricious reply and down went the receiver.

I wondered why fate had brought me in contact with this star who I, like millions of others, had worshipped from afar. Katharine Hepburn was a legend in her field, whose wit, beauty, brains and burning ambition had all conspired to seal her place in history. In spite of her extreme fame, she remained loyal to her close friends, and as for her family, she had, as she put it, never left her home in Fenwick, Connecticut. A short marriage to Ludlow Smith at too young an age was followed by several affairs, including one with the eccentric and mega-rich Howard Hughes.

Then came her long and enduring love affair with Spencer Tracy. As she wrote in her autobiography *Me*, "Love has nothing to do with what you are expecting to get – only with what you are expecting to give – which is everything." True to her word, she devoted herself to Tracy at his weakest times, when the demon drink took over. They took over two cottages on the Los Angeles estate of George Cukor, with whom Hepburn had a working relationship and friendship for fifty years. One of her first collaborations with Cukor was on *Little Women*, which secured Hepburn her first Oscar in 1933. The first of several Cukor films in which Tracy and Hepburn starred together was *Keeper of the Flame*. The last film in which the pair starred was *Guess Who's Coming to Dinner*, at which point Tracy was a dying man. Hepburn nursed him with devotion during the filming and after, until he died of a massive heart attack.

The star kept everyone on their toes, for she was also known for her unpredictable outbursts. Her friend James Prideaux said that even in her eighties she could reduce him to tears. She could be ruthless and opinionated about other leading actors, and on meeting Jane Fonda she simply stated unequivocally, "I don't like you." Soon after this, however, when they started filming *On Golden Pond* together, Hepburn saw that Jane had a genuine problem with her father, Henry Fonda, and she retracted this statement. In her opinion Hank, as she called him, was "cold, cold, cold!" and she never felt that she got to know him.

So how, I wondered, would I fare with Miss Hepburn if we did meet? I gazed at the oppressive fake leopard-skin walls in my room. I felt fragile in New York, and my confidence was draining away like sand in an egg timer. However, I decided to stay come hell or high water, because now that I had Miss Hepburn's number, I did not want to jettison the admittedly slim chance of meeting my heroine. And so I started a period of patience and endurance.

It was November and the stunning golden treetops that heralded me from my bedroom window were rapidly turning into grey skeletons. As instructed, I phoned Miss Hepburn every morning at nine, and was ready to go: squeaky-clean hair, pristine clothes, cameras packed, lights charged up. Day after day, for three weeks, Miss Hepburn answered the phone on about the third ring and each day she said, "Call me at nine tomorrow," with no explanation, no "goodbye," just down went the receiver. Each time I had butterflies in the stomach that today might be the day, then the anti-climax. My days had to be filled with visits to museums and walks in Central Park; I was so alone and Thanksgiving was looming. Every day the east wind chased me down the streets like Mary Poppins in full flight, followed by intermittent torrential downpours of driving rain. What on earth did I expect to do in Miss Hepburn's garden if I did firm up a date? In the now rapidly approaching winter, in what kind of bedraggled state would I find the garden? I waited and I waited.

I was so near and yet so far, just killing time waiting, until one day Milton called and invited me to a tribute to Josh Logan, followed by a reception. Suddenly, after walking lonely as a cloud, day after day, I was meeting some of the greatest legends in the film industry. Mary Martin, who had starred in the film version of *Annie Get Your Gun*, seemed particularly interested in me. I was overjoyed, as she had made a lasting impression on me when I had seen the film as a child. It was only when Milton pointed out that she "fancied" me that my ego deflated like a pricked balloon. Then, walking down the staircase towards me was Jimmy Stewart with his wife of many years. I was so overcome that I barely heard Milton by my side saying, "Hi, Jimmy. This is Carole Latimer, my talented photographer friend from London." The star's handsome face, most familiar to me in black and white, smiled down at me. He shook my hand and, in that split second, he made me feel as though I was the most interesting person in the whole wide world. The gentle charm translated perfectly from screen to reality. The four of us chatted briefly as though shooting the breeze with the neighbours. The magic of meeting legendary stars is that their longevity is already guaranteed; the younger ones may be gone tomorrow.

After spending an idyllic Thanksgiving with my English friends, the Drummond-Hayes, in Rowayton, Connecticut, I returned to New York. Two days later the summons from Hepburn finally came: "I shall be back at three," and down went the phone.

The day was grey, with that flat, colourless, wintry light which saps the energy. But my nervous excitement caused the adrenalin to kick in like a booster rocket. The cab driver was helpful and annoyingly talkative, as he was all too aware of who lived in the tall brownstone at 244 East Forty-Ninth Street. As I stood on the doorstep, surrounded by a mound of camera equipment and lights, I had time to wonder what I would do if no one answered. I could hardly leave all my worldly goods to go and find another taxi. Suppose Miss Hepburn was just stating a fact when she put down the phone? At that moment her chauffeur of many years opened the door. “Is Miss Hepburn expecting me?” I blurted out. I need not have bothered to ask, for there behind him was a petite figure walking down the narrow hallway, unavoidably recognisable with the familiar twist of hair on the top of her slightly nodding head. The tremor generally thought to be Parkinson’s disease was a rare hereditary condition that had cruelly caused the shaking, although she had kept the condition at bay for well over twenty years. The clothes were familiar, a black turtle-neck sweater, khaki jacket and pants, so fashionable only because she had started the trend more than fifty years before. But oh dear, no make-up and the famous redhead’s skin, shiny as a waxed apple with huge red patches caused by sun damage. I would simply have to do a big touch-up job, as this was before the digital era.

We headed straight for the garden, where the light was sinking fast; the surrounding skyscrapers would steal it by four. The outlook, as I had feared, was now bleak; damp brown leaves and unsightly black plastic bags spoke of autumnal work in progress as the plants drooped in harmony with the dreariness of the day. “I hope this won’t take long,” Miss Hepburn muttered ominously. I tried to reassure her that it would not, and it couldn’t, as it would soon be dark. I told her to stay in the warm house while I prepared myself. All too soon she was back. “Hurry up!” came the discouraging voice from the French windows as I frantically put up a couple of lights to balance what was left of the natural light. Too impatient to wait any longer she now joined me in the chill east wind and off we went, Miss Hepburn rushing around like a meerkat on speed searching the garden for ideas while I hastily dragged my lights around, as most of the natural light was coming from above. A little statue of a monkey was fiddled with but her head shook and the light was so low that the shutter speed would fail me. And then she volunteered, “I can be carrying logs!” As she lifted the logs her head stopped nodding for a split second on the upward movement. Click! I knew I had got the shot and that it had been worth the wait; two years to reach my goal and five minutes to shoot the picture. I hastily packed my bags in the freezing wind just as darkness fell.

Katharine Hepburn, New York, 1990s

I walked through the French windows from the gathering dusk into the warm golden glow of the cosily cluttered dining room. Phyllis Wilbourn, Miss Hepburn's faithful friend and assistant for more than thirty years, sat at the large round table festooned with paperwork. By this stage the dining room had become an office where work was being carried out on Miss Hepburn's book, but at the same time it was so homely; the antithesis of what I had envisaged in any house in New York. Presuming that my finest hour was over after weathering a fair battering of Miss Hepburn's feistiness, I was taken totally by surprise when a choice of tea was offered; Earl Grey was mutually agreed upon and we withdrew to the back living room on the first floor. My first impression was of a real home, a room well lived in. Beautiful dark antique furniture was cluttered with pre-Columbian sculptures and a pair of candlesticks which Miss Hepburn's father had given her, along with sculptures of birds crafted by the actress herself. Suspended above her favourite chair, in which she now sat, was Spencer Tracy's carved-wood flying goose. On the table next to her a black telephone sat under a chubby lamp along with a jumble of crystals, carvings, a clock and a magnifying mirror. The open fireplace which no one else was allowed to tend was loaded with logs and ash which spoke of regular use, and Persian rugs and throws over the chairs gave an air of genuine, shabby chic.

This was a room seeped in memories of sparkling conversation and wit from many of the greatest artistic talents of her time. Miss Hepburn poured the tea and offered me a cookie. I declined, nervous now as to which way the wind would blow. Another scolding for not taking a cookie and a diatribe about the foolishness of young people slimming. I hastily grabbed two cookies, pointing out that I have never slimmed nor felt the need to. The atmosphere thawed, so we continued to talk about gardens as that was the reason I was here, or so I thought. "I have always had a lifestyle of a great deal of exercise which has always included gardening." Miss Hepburn was now in full flow, and the ice was melting. "I would say that I am a fair gardener even though I am ignorant about the technicalities. I love gardens. I work hard. So that's why I don't have to diet. I eat lots of everything, red meat, sugar; I don't think fish and chicken give you enough strength. I know that eating as much chocolate as I do can't be good for me; I can eat a pound without noticing. I'm a very heavy eater of fattening things.

"I'm lucky. Weight is all about how much you sweat and how much exercise you get; I find people who diet madly exhausting." I had time to relax as there was no pause in the flow of conversation, so I devoured the entire plate of cookies, which closed the subject of diet once and for all.

Miss Hepburn reminisced, for a while, about being brought up in the country, in a charming old house built in 1840 which backed onto a golf course. "This is why I don't like living in apartments. However, I must add that I was lucky to have had the choice." One of six children, she grew up with a father who was mad about gardening; he was particularly keen on trees and bushes. Her mother was more interested in people, but they were both reformers. One of her sisters was a great gardener: vegetables, flowers, everything, she had a gift. Miss Hepburn particularly loved herbs and spring flowers. "I plant all the bulbs here and I planted that mountain ash myself, and you can see it is now up to the second storey," she said, pointing out of the window into the dusk.

"My favourite flowers depend more upon colour; I like a lemon yellow, not a deep one. Delphiniums are a favourite in most of their colours, I love all the pinks and blues. I like lilies but not necessarily calla lilies. I like white so much that my entire house is painted white inside. Sadly, I can't have a favourite perfume, as I have lost my sense of smell. The cause is terrible, I used to dip Q-tips in 4711 to refresh my nose when I was working on dusty film sets. It almost totally destroyed my sense of smell and I never noticed at the time. If I do smell something I know it must be terribly strong; luckily I can smell smoke."

The subject of smoke led Miss Hepburn to divulge that she used to enjoy smoking, until she played Shakespeare on stage and found she had "no lungs", as she put it. I remarked that my father, too, had given up smoking while playing the lead in a play in London. He had a cold and when it lingered on he realised that the smoking was causing most of the misery. As Miss Hepburn refilled our dainty little flowered cups with saucers to match, she returned to gardens. "I think greenery is very important in my life." (No more sharp remarks about being under it soon enough.) "Luckily, I have advertised the fact, so I get sent a great many flowers which is heaven as I am horrified by the price of cut flowers. Only the other day, I asked the price of carnations because they last so long and the man replied one seventy-five a stem. I dropped them in horror."

Suddenly, Miss Hepburn looked at me intently, as though truly acknowledging my presence for the first time. "You're wearing exactly the same clothes as I am," she commented. "Well, isn't that the funniest thing." And then there was no stopping her. Where was I brought up? Hampstead in London, I replied. Yes, she knew Hampstead well. She always stayed in Frognal with friends called the Stewarts whenever she went to London. We both had cocker spaniels in our childhoods.

We both had bought our first and only homes in our respective cities in our early twenties. As Miss Hepburn seemed highly amused by the coincidences, I went on to tell her that I, too, had five floors, with my dining room on the ground floor, French windows onto the garden, and an L-shaped sitting room on the first floor. We were two women talking as if we had known each other forever. At last I felt truly relaxed and I was enjoying myself. I thanked her for starting the fashion for wearing trousers, which has given women so much freedom. She laughed and added that she dressed in rags. "I can't bear wearing stockings, that's what started me." However, I noticed that she was not that unconcerned about her looks when she remarked that she was pleased she had retained a good deal of her natural hair colour, unlike her sisters, who went grey quite young. I mentioned that I liked her hairstyle, so timeless and always the same. "My hair is a goof of feathers," she retorted, and added, "just like yours." Yes, I suppose she was right: fine and unruly, unless popped on the top of the head in a little bun.

We had so many tastes in common I forgot there was a huge age gap between us, until she mentioned that she had never believed in looking ahead to when she could not do so much, as it might never happen. But now that she was older, she had a weak back and could not carry so much. "I have never had a back ache like this; it's just stiff all the time and it really affects my character. It's boring, that's all, boring to talk about and boring to happen to you," she added briskly. "I don't know why the good Lord didn't arrange things more sensibly. My family were encouraged not to grumble, rather British, don't grump and don't tell people how you feel, it's not interesting. However, I can't help wondering what I shall do when I can't carry my breakfast tray upstairs. I shall just have to crawl up and come down on my seat." I wondered how long she would cope with her three floors and discovered later that she had managed to stay there until the last seven years of her life, when she was just on ninety. She finally moved to Fenwick, with fewer stairs and a wonderful view of the sea, and where she could watch the sun rising in the east and setting in the west.

Quite suddenly, out of the blue, Miss Hepburn asked me if I had slept with John Huston. Taken aback and thrown from my reverie, I said, "No," and countered by asking her the same question. "No, certainly not, a very dangerous man." She laughed. Just two women discussing that same old subject. I suppose at this point she had given me an entrée to be more inquisitive about her love life, but I had read about it many times and, anyway, Miss Hepburn was known for her brilliance at self-publicity in promoting her career. I thought of her obsession with the lyrebird, which she came across when filming in Australia. Here is a bird that mates for life and if its mate dies, spends the rest of its life in the company of its own sex. I could not help but see a parallel with her great love for Spencer Tracy, after whose death she chose to surround herself with women.

Never one to moan because it is boring, Miss Hepburn had lost none of her vim with the passing of the years. That day she seemed ageless to me, so much her own person that you could never put her in a group or a category; there would never be a sweet little old lady. There was only Katharine Hepburn.

It was time for me to take my leave; after all we had been chatting like old friends for more than two hours and I wanted to leave on a high. As she rose to see me out, she concluded, "The older I get, the more I laugh at everything; I've only to walk down the street to see so many things that amuse me." I got the distinct feeling that this attitude had a great deal to do with her longevity. I had had a memorable afternoon. The next day it all felt like a dream, and only the pictures proved otherwise.

CHAPTER 16

CANDLELIGHT EXHIBITIONS, LONDON AND L.A.

"Photography for me is not looking, it is feeling. If you can't feel what you are looking at, then you are never going to get others to feel anything when they look at your pictures." – Don McCullin

Sleepless nights … sometimes I find these the most creative times, when original ideas can manifest without the incessant interruptions of the day. The streets are quiet; only a lone fox searches the dustbins. The sky can help to still my mind and once I have gazed at the stars and tried to work out their constellations (big bear, little bear, the plough and Venus, brightest of all) there is little to do but think my thoughts. And so I gave birth to the idea of photographing men, semi-nude, bathed in candlelight. By using daylight-balanced film and candlelight, I would transform their pale bodies to a deep shade of gold. The strong shadows created by the candles would emulate the chiaroscuro effect much loved by the painters Rembrandt, Caravaggio and Joseph Wright of Derby, amongst others.

For years I had wanted to attempt the nude. This was before the male form had been romanticised by photographers; images were either ludicrous or pornographic. Since I held my exhibitions I have seen a far wider variety of male nudes and, in fact, they have become so fashionable that the market is saturated, but to this day I have never seen anything remotely similar to my pictures by candlelight.

I did not start to shoot my idea for a year after its conception: how would I overcome my inhibitions? The thought of making a man sit stark naked on a sheet of paper surrounded by candles brought a blush to my cheeks. What if the pictures did not work? Nude photography can either be beautiful or, paradoxically, ugly or ridiculous; it is a fine line.

The next challenge was to find my first model. He would have to be physically perfect and also artistic so that he could help me create the

Dan McBride, London and L.A., 1980s

pictures. Once I had an example of what I had in my mind, I would be able to approach other subjects with something tangible to show. Dan McBride was the answer to my prayers. Born and bred in Arizona he had, in his early twenties, the perfect physique. An artist and songwriter by profession, Dan was modelling around Europe in order to see the world.

I bought a bundle of church candles, and my father made wooden stands with long nails on which to secure them. Dan brought a bottle of tequila, so only my trusted Hasselblad was needed to complete the project. When the printer admired the shots and asked me how I had achieved them, I had to admit that a large amount of tequila had been consumed, so I had failed to make notes on the exposures and shutter speeds. Soon I had so many beautiful men offering to pose for me that I was spoilt for choice.

As none of my models were professional nudes, there was always a moment of embarrassment as clothing fell to the floor. This was swiftly overcome as soon as we were busy concentrating on the work. All the obstacles of low light, including smoke from the candles, made each picture a real challenge. The men came in all colours and ages, ranging from thirteen to forty; the theme was romantic. The collection grew and it was not long before a designer friend, Kenneth Partridge, suggested that I should try to exhibit the photos in a gallery.

Once I had the gallery in Knightsbridge, the machinery was in motion. Jill Thornton and Peter Lumley were to do the publicity, prints had to be framed, interviews set up; there was never a dull moment. We were too close to the launch date to approach the magazines, but the attention the exhibition received from the dailies, Sundays, television and radio was beyond my wildest dreams. The publicity ranged from a half page in the *Sunday Telegraph*, in praise of the pictures, to a tacky heading of "Carole's beefcake" in the *Mail.* The highest accolade was a vote of approval from broadcaster and art expert Dr Roy Strong. Television was by far the hardest. I was terrified talking about why I wanted to photograph men in the nude; I would like to have been witty, but that would have been counterproductive. The laughter had to be saved for behind the scenes, and with Jill and Peter there was never a shortage of wit. Janet Street-Porter's all-women chat show was the toughest ordeal. All the women except for me chose to slag off men, but I had the last laugh in the bar afterwards: all the crew were eager to buy me drinks and to voice their fears of the "Sisterhood".

Several years passed before I tried again to exhibit these pictures. On one of my many trips to Los Angeles I decided to knock on a few doors to see if there was any interest. It only took three knocks and I was offered an exhibition at the Tribeca restaurant in Beverly Hills, owned by Robert De Niro. I returned to England, but it was not long before I was packing my bags and all my cameras and setting off, yet again, to Los Angeles, clutching a large tube of glorious prints ready for the exhibition.

"Season of mists and mellow fruitfulness" (Keats) was ringing in my ears as we swept down the M4 bound for Heathrow, with the sun shining softly over fields that were partially cloaked in diaphanous patches of mist like fluffy white duvets. Cows grazed peacefully in the glistening, dewy fields, the river shone like pewter in its stillness, and I was deeply conscious of the beauty of the English countryside in the autumn.

As I scuffled around in my handbag checking that I had my ticket, passport and money, a flock of police bikes flew past, followed by a plum-coloured Bentley state limousine with a small flag fluttering on the bonnet. Inside, sitting bolt upright, was a majestic couple: the Queen and Prince Philip. I was overwhelmed by a surge of patriotism as the Bentley swept off the motorway homeward bound for Windsor.

As I struggled through the entrance of Terminal 4 reality hit hard. I was not prepared for the new and stricter measures taken at customs of ferreting through my suitcases and photographic equipment, piled head high on my trolley. I was a sitting duck with my mountain of equipment pulsating with healthy batteries. "I'm sorry," said the lady in uniform, "but we do have to be careful these days." On the X-ray machine the contents of my suitcase did look remarkably like a dismantled bomb. I had this eccentric idea that I could use my clothes as cushioning for my photographic equipment: tripods swaddled in T-shirts, knickers and socks protecting delicate items like my exposure meter and a travelling clock. I looked on with horror as every item was unravelled and checked. I prayed that this was not a bad omen for the trip.

Eventually I sank back into my large seat in Clipper Class, unable to believe my luck that I had been upgraded. This was my chance to make the most of every second of the ten-hour journey, but no sooner was I truly nested than I heard a voice above me: "Excuse me, Ma'am," said the air hostess, "would you mind changing your seat? This couple would like to sit next to each other." I suddenly felt deeply uncharitable and muttered that I wanted to be left in peace but then, feeling mildly guilty at my unfriendliness, I glanced around and noticed a single girl behind me with one seat beside her, so I suggested that she move forward next to me, leaving two seats clear for the "lovebirds". She readily agreed and soon we were comparing notes on the joys of travelling Clipper Class.

Seat belts fastened, we coasted down the runway. As the engines went into full throttle, we fell into silence to enjoy a moment which never ceases to astound me: the mighty force of the take-off. The pedestrian behaviour of people on planes is such a contrast to what is outside, everyone packed tight like sardines busily fiddling with their televisions, sound systems, possessions, exploring the magazines, undoing little packets of food; few of them sit quietly and gaze at the wonders of the universe. As I had blanked out the general restlessness, it took a few seconds for me to realise that a waterfall of the foulest-smelling liquid was cascading down between the two of us. There was no escape, as the seat-belt sign was still on. I could not believe my nose, it could not be, it was not possible … "Pee," I suggested, as daintily as I could. "Urine," replied my companion. Yes, we were right, and it was streaming from the luggage hold above us. The air hostesses rushed forward with soda syphons and champagne. "This is the first time this has ever happened to me," she wailed. I had not realised that this could be a regular occurrence. It transpired that an unfortunate passenger in a wheelchair with a urostomy bag had been parted from both these items for the duration of the flight, and the paraphernalia had been stowed above us with the hand luggage. My new companion and I sat drenched in soda water, bonding merrily through our mutual experience. So much for my efforts to arrive in Los Angeles looking cool, calm and collected. I had accepted the fact that all my well-ironed clothes were crumpled, having been hurled back into my suitcases at customs, and now my clean travelling clothes, doused in soda water, were scrunched up like squeezed-out rags. Yes, I would have a glass of champagne. I do not normally drink alcohol when flying, but this was not an ordinary flight.

LAX is a hostile airport: every time I arrive, my large amount of camera equipment attracts attention and I am treated like an illegal immigrant, even though I have proved, time and time again, that I am merely visiting and not settling for good. So I always feel a sense of relief when I clear the airport. My taxi driver was from India, and he did not like Los Angeles much, but he was making good money and sending it home to his family. Even though he had come from halfway round the world, he was amazed that I had come from London.

Although I know L.A. well, I did not know the apartment where I was going to stay. Before I had time to dwell on the matter, Elizabeth Mason flew down the steps, bubbling over with warmth and greetings. I was bombarded with questions, leaving no time for the answers, blonde curls bouncing around her face, which she joked was the shape of a peanut with huge pleading eyes. At that time, Elizabeth was just another talented actress in Hollywood seeking stardom. All her apartments looked the same once she had moved in; all became transformed into a second-hand dress shop. Instead of pictures, the walls were covered in hats, the bathroom walls were festooned with necklaces like stalactites in a cave, drawers overflowed with underwear and make-up, and cupboards bulged with clothes. Little did I know that Elizabeth would turn this chaos into a successful vintage clothes business called The Paper Bag Princess, with showrooms in Los Angeles and Toronto. Elizabeth had written to me frequently, inviting me to come to stay. So there I was in the embryonic stage of her business, with a futon mattress to rest my weary head and surrounded by clothes racks and a cupboard stuffed to the gills. I was going to have to spend six weeks literally living out of my suitcase. I did not have time to dwell on this fact as Elizabeth's kindness in inviting me was what mattered and she was already organising my life. "We", which included her boyfriend, were going straight out to dinner. I did not have the energy to protest and, anyway, I was glad to be with friends after a long journey.

We talked at double speed in the Thai restaurant, ate at the same pace and ordered the check as though we had a train to catch. Having returned to the apartment I sank into the armchair and watched Elizabeth doing her kitten act on her new boyfriend, Hank. It took me a little time to adjust to Hank, as Fernando, her Spanish boyfriend, who was so handsome, had been in residence on my last trip. But he was history, and here was Hank, curly haired, rather plain, with a voice that sounded like a bear growling. But he was kind and malleable, whereas Fernando had turned out to be a monster. I wondered idly whether her new man would be spending the night but, no, Elizabeth was going to leave me on my own and disappear into the night down to the beach with Hank; they hoped I would be okay.

Bemused and rather lonely, I battled with double locks and light switches and cleared a small space for myself in my bedroom. The futon looked inviting and I love firm beds, but lying on this one proved to be more like sleeping on Brighton Beach, with its shoals of grey pebbles. This was going to be a real thrill for six weeks. Ah well, I thought as I tried to sleep, I can handle it.

I awoke early, expecting the glorious sunrise of sunny southern California. I drew back the curtains to discover an ugly view of tin rooftops and a slate-grey sky. My back was aching badly and I felt distinctly low, but I reassured myself that I was just over-tired and that everything would look different after a shower and a cup of coffee. I called my friend Dale, who had been using my vintage car in my absence. Dale loved my beautiful Chevrolet with her spoke wheels, whitewall tyres and ten-year-old tomato-red body glistening with good maintenance. He had named her Belinda, and the name stuck.

Once I had wheels I could start to put together my exhibition at Tribeca. I could not wait to get to the framers and to organise the cocktail party, a generous gift from Daisy Hall, a beautiful young actress who came from an old monied family on the East Coast. We had struck up a friendship after I photographed her and we went on to do several shoots each time I came to Los Angeles. Daisy was divinely photogenic and could reinvent herself from nymph, with long chestnut hair, to a blonde Marilyn Monroe lookalike, and then back to brunette.

My first phone call was to Donna, the manager of the restaurant, who had been so wildly excited about the idea of exhibiting my candlelit pictures. I was mildly taken aback that she could not see me for four days, as we only had two weeks to prepare before the party; the PR for the restaurant was equally busy. I did not like waiting around with so much to organise, so one morning, I decided to lie on the floor and relax totally for ten minutes before calling "the world" to invite them to the party. Just as every muscle had softened and melted into the floorboards, the whole place shook; hats flew off their hooks and small objects leapt from the shelves. My God, I thought, could this be the big one? But no, yet again a false alarm, but big enough to get the media buzzing followed by warnings of precautions to be taken and a few more earthquake kits sold before all was forgotten until the next rumble.

At last the day of the meeting with Donna came. I dressed smartly in a silk shirt and trousers, my hair and make-up done impeccably in a way that is only possible in desert conditions where the air is dry.

Donna arrived late and took her time stopping to greet members of the staff. On spotting me in a darkly lit corner, her arms flew wide in an embrace, then came a deluge of compliments. "You look great, so slim, I love your shirt, great colour with your eyes," she gushed. She was sorry to be late but she had been for a run on the beach. She looked faultlessly fit, hair still wet from the shower, skin glowing from the salt air, and all without a trace of make-up. I doubted the sincerity of all the compliments she paid me.

"Down to business," said Donna, baring a perfect set of bleached teeth. "Just a slight change of plan. The pictures are to be hung upstairs for only one night and then re-hung downstairs the following day," she continued, without a hint of an apology. It transpired that there was an important football match that night so the television had to be on downstairs. She added that, on both occasions, I would have to hang the pictures myself; as each picture was over 100 centimetres square, this was no easy task.

I was horrified. I had travelled halfway round the world, spent a small fortune on prints and first-class frames, and now I was getting second-class treatment. I muttered that I could not hang the pictures on my own as they were too heavy. "I'm afraid that's your problem," she snapped. The practised smile quickly returned to disguise her impatience and gushing affirmations flowed again. "It's going to be just great." I was inwardly furious. I had turned down several assignments in London, and I expected more cooperation.

The following day, I drove to the airport to meet my agent, David Maskrey, who was arriving from London. Never had I been so glad to see him. David only had one week in L.A. and other business to attend to apart from my exhibition. Although he had never been to California, he instantly offered to help to hang the pictures. "We'll have it done in a trice," he assured me; he did not know L.A.

Sunday morning came and I was determined to cast aside all bad feelings. Did they have a hammer, some nails? No, but we could go to a hardware shop, and where would that be? we wondered. A measuring tape? No luck there either. Maybe one of the waiters hanging around could help. No way. We were on our own. I kept thinking of David's precious free day being swallowed up in the restaurant. At last we stood back to admire our work: the prints looked fabulous. Only the large picture dominating the staircase was a problem, as the reflection of a light bulb fell smack in the middle, obliterating most of the image. "Easy," said David, "we'll just get them to remove the light bulb." No, not so easy. They did not "have the authority to change a light bulb". David has never forgotten this bit of nonsense. Smoothing my ruffled feathers for the millionth time I decided that, against all odds, we would have a great party. This was a wonderful opportunity to bring together all the clients and friends I had made over many years of visiting California.

I had to invite everyone by telephone, as there was not enough time to send out invitations, and emails and texts had yet to be invented. I got the full range of reactions, from "Of course we'll be there for you," to "I'm afraid Sunday is the day I always spend with my child." A couple of friends, unhelpfully, confirmed that Sunday was not a good night for a party in L.A. Magically, the restaurant was full. My friend Marian Neuman was quick to tell me that her husband Jack (who had just produced *Inside the Third Reich*) hardly ever went to parties and I should be truly honoured. I was equally glad to see those who enjoyed a good party. It is nerve-racking enough giving a party, but when you put your work up for judgement the experience is petrifying. Once under way I had a ball, and compliments flowed along with the champagne and canapés, which included the restaurant's famous crab cakes. I was quietly amused

Jake Kennedy, London, 1980s

to see that Donna had not bothered to dress up, presupposing that my guests would not be worth the effort; I could see that she regretted her decision. The guests stayed late and, Cinderella-like, I was whisked off by my friends Andy and Bronya Galef, the coach replaced by a gleaming Bentley. We drove east on Sunset until we came to an exclusive new club, so new that it was not even open to the public. We ate more food, drank more wine and held a post-mortem on the evening's events. We were a select group comprising Daisy, who had generously hosted the party; David, my agent; and Andy and Bronya.

The next morning I awoke to the ringing of the telephone. Bronya was bursting with excitement. The curator of the Museum of Modern Art liked the pictures so much that he wanted to send his second in command for a second opinion.

Springing from my bed of pebbles, I showered swiftly and called the restaurant to warn them of their imminent guest. Too late. The pictures had already been taken down and piled up in a storeroom next to the kitchen. All the joy of the evening dissolved; I was apoplectic with rage, my biggest opportunity thrown to the wind, and I knew there would be no second chance with someone of such calibre. When I arrived at the restaurant I lost my cool with Donna, thereby signing my death warrant with Tribeca. I would get no further cooperation. The pictures were soon back on the walls, but with no promised spotlights installed. Dimly lit, the pictures blended into the walls as though they were fixtures, and nobody knew they were for sale. When customers would comment on a photo or inquire about a price, Donna failed to pass the messages on to me. The curator had come and gone while the pictures were buried in the basement. Bronya was furious and I was left feeling partly to blame for allowing this to happen. It did not seem fair, but life is not fair.

It was another new best friend who came to my rescue, Jane Gottlieb, a fellow photographer who had had a similar experience with another restaurant. Jane was highly successful with many of her pictures on permanent display in museums. She was quick to offer her help, so we laid our plans.

Two days before Thanksgiving we arrived smartly dressed at cocktail hour. Looking like a couple of spare ladies hoping to pick up a date, we ordered our drinks, strong, knocked them back and paid the check.

Suddenly, Jane looked me straight in the eye and whispered, "Let's go!" Speedily but dexterously we removed the huge frames and whisked them out to her Mercedes estate. The barman gaped in disbelief and only a pair of painted ladies muttered that it was an understanding management who would let us remove the pictures during cocktail hour. "Very," I assured one of them, without pausing to explain. We stacked all the pictures neatly in the car, wrapped in towels; only the one at the top of the stairs remained. Quick as a pair of cheetahs, we shot up the stairs, took a side each of the picture and bolted out of the door onto Beverly Drive. I glanced back in time to see Donna appearing at the top of the stairs; she had just spotted us, but she was too late: the walls were empty. Closure ... Happy Thanksgiving.

The Tribeca episode was in keeping with other experiences in L.A., where I have had some of the best and some of the worst times of my life. Everything is extreme in Hollywood: the highs are higher and the lows are lower. As a visitor from across the pond I have been picked up, spoilt and treated like a celebrity, while at other times I have been horrendously let down.

The day before I left California I interviewed Ned Beatty, who had just finished working on a film about the Pan Am Lockerbie disaster. Although I was travelling with the same airline, I did not give the matter more than a passing thought. As we taxied down the runway the captain suddenly announced that we had to return to base as there was an unidentified suitcase on board. I sweated for the next ten minutes, until the bag was safely removed and I was homeward bound, but the drama did not stop there. Due to thick fog we were diverted to Manchester, where we sat on the tarmac for an hour or so until we flew back over London, where we stacked for two hours until the captain announced that we would make "a try for it". We came to a screeching halt in a thick blanket of fog. At least the trip had been consistent: the best of times and the worst of times.

By the time I drove into London the sun was shining hazily and the soft mist hung like fluffy blankets over the countryside. I was back to that magical English light: I wondered if it had been like that all the time I was away.

CHAPTER 17

HOUSE GUEST

"Every person, all the events in your life are there because you have drawn them there. What you choose to do with them is up to you." – Richard Bach, *Illusions*

As a photographer, I am well accustomed to spending time in the homes of others, trying not to intrude as I set up my equipment, adjust the lighting and go about the shoot. On many wonderful occasions, I have been more than a professional on assignment but a guest made to feel at home.

Gyearbuor (Christopher) Asante

One of the most exotic of these occasions was with Christopher Asante, the actor who later became known as Gyearbuor Asante. When Christopher first arrived at my door to have his portrait taken, he already had a long list of credits in film and television. He was best known for his role as Matthew, the eternal student, in the popular television comedy series *Desmond's*. The show ran for seventy-one episodes from 1989 to 1994, and Christopher appeared in sixty-nine of them. This show was masterminded by the television producer Humphrey Barclay, with whom Christopher shared a house for twenty-eight years. The show was groundbreaking, as it opened the floodgates to Black actors in the UK.

I was pleasantly surprised at how dark Christopher's skin was, even by African standards. He was a beautiful shade of deep aubergine, and he was a royal of the Akan tribe in Ghana. Soon we were laughing at my predicament about how to light him, but the results were wonderful, and our close friendship lasted until the end of his life.

Over the years, I went to many memorable parties that Humphrey and Christopher hosted. There was always a cluster of well-known show business personalities mixed in with close friends. On one occasion, I felt disinclined to go, as I was licking my wounds from a broken relationship and my self-esteem was at an all-time low. It was a dark, wet, autumnal night and the pavements were strewn with damp leaves, so the contrast of the brightly lit living room was startling. The first person I saw instantly made eye contact with me and never left my side for the rest of the evening. A film star in his ascendancy, Tom Berenger had the laid-back manner at which Americans can excel, a kind of unsophisticated, boy-next-door demeanour which instantly put me at my ease. Dressed simply in jeans and a checked shirt, his deep suntan made him stand out from the crowd.

Tom's presence was soon explained. Christopher had been working with him on the film *Dogs of War* and they had struck up a friendship. All evening I sparkled with joy as I sat close to this attractive man, necking a bottle of beer amidst a room full of thespians. Every so often our bodies touched, sending little electric shocks through me, and by the time Tom held my hand my heart was pumping like a steam engine. My hosts were observing the whole episode with quiet amusement coupled with surprise, and for the rest of his life Christopher would playfully tease me about this dalliance. The end of the evening came too soon and I prepared myself to say goodbye, but this was not to be. The invitation came: would I come back to his hotel? We awoke the following morning with no regrets, and Tom soon returned to the States to work on another movie. We hoped to keep in touch but we never saw each other again. No regrets, and as a friend commented, "one hour was sunlit".

When Christopher had amassed what, by Ghanaian standards, was a great deal of money, he built a home outside the city of Accra. I felt honoured to be, apart from Humphrey, his first British friend to visit. Here he was known by his Ghanaian name Gyearbuor, which he eventually adopted professionally for his appearance in *Desmond's* and other shows.

I flew from London to Accra two days before Christmas and arrived after dark in a dimly lit terminal. The ride from the airport introduced me to the hazards of driving in Ghana, where I soon became accustomed to saying my prayers every time we got into a car. Our driver sped through Accra like a maniac, scattering chickens and children in a cloud of dust.

Christopher's house, in the smart suburb of McCarthy Hill, was vast, modern and surrounded by a walled garden with large iron gates. Every morning I would walk onto the terrace, letting the warmth of the sun melt all the tension in my body created by the damp English winter. The boy servant would appear with freshly squeezed pineapple that I would sip while I gazed out across the garden full of bananas, flamboyants, frangipani and many other indigenous trees. Under a couple of trees were tethered a goat and a handsome white ram with curly horns; pets, I presumed.

Most days, a leisurely shower was followed by an even more leisurely breakfast on the balcony served by the servants or the students living in the house. Christopher housed his mother, Kofi, an orphaned nephew and two cousins, brothers Kwabena and Yali; the three young men were all studying at the university in Accra, and Christopher paid their fees. It is customary for the younger generation to wait on the older members of the family, a role reversal from the practice in the UK, where mothers often slave away for their full-grown children. Apart from the students, there were two servants, one of them a girl who was responsible for my washing. I thought it strange that so many young men were named either Kwamena or Kwabena until I discovered that every Ghanaian has a "day-name": Kwamena means born on Saturday, and Kwabena born on Tuesday.

Ghanaians are not great meat eaters, and I soon became used to a diet of yams, fried and boiled, served in various sauces and sometimes appearing twice in one day. Otherwise, there were rice or plantains and occasionally fish, which Christopher would devour with his huge white gnashers, heads and all, a custom I could not imitate with my tiny set of teeth. On Christmas Day, I awoke early to the sound of one bleat and went out to the terrace. The goat no longer grazed under the tree. Soon I smelled roasting flesh. Now I understood. The goat was Christmas lunch. I had arrived in time to see a full store cupboard, a sheep, a goat and several chickens, and by the end of the festive week the garden was empty.

I thought that I had grown accustomed to Ghanaian culinary tastes until one day, as we were driving along, Christopher shouted gleefully to the driver to stop. I looked out to see a young boy holding a long tail attached to a giant, bloated rat. "Ah," said Christopher with great joy, "that's a grasscutter. It's a delicacy." That night, a stew appeared and in the middle was a long, squiggly tail; the grasscutter had made it to the pot.

A great part of each day was taken up with visiting cousins, uncles, aunts; the well appeared to be bottomless. Trips had to be made on the off chance of finding the relatives in, as few had telephones. A beer was offered, a few snippets of conversation exchanged (never personal, as that was considered rude), lots of laughter. Then we were off again, careering down the dusty roads sending chickens and dogs fleeing for their lives. Women came to the house to meet me, bearing gifts of pineapples. They enveloped me in their arms and my tiny frame would disappear from sight into their massive bosoms. They dressed up to visit in bright cotton frocks with large leg-of-mutton sleeves and lots of gold jewellery, their heads covered in colourful scarves as protection against the Harmattan. This wind was blowing during most of my visit, drying the air and leaving a fine brown dust over everything. One morning I awoke to find my bedroom looking like the Sahara Desert, though nets on the windows saved me from being bitten to death by mosquitoes, and a blue lamp fried any unfortunate bug attracted to the brightness. Sometimes a giant moth would get caught; after a terrible sizzling sound, the smell of roasting meat would fill the room.

In the late afternoons, we would often go to one of the long beaches to swim in the warm turbulent waves. The countryside was remarkably fertile with rich chestnut-brown soil and the dark green of the mango trees contrasting with the lighter shades of other trees. In the towns, vendors sat under giant straw hats selling their wares. Chickens and goats ran free and children were bathed outdoors in tin basins. When we visited the markets, young men would rush up to ask for my hand in marriage, begging me to take them to England; all white people were presumed to be rich. I thought of Christopher coming to England in the 70s with that same optimism; how overwhelming the culture shock must have been, despite his privileged background.

I was pleasantly surprised to find that most people were happy to be photographed, so I gradually became bolder and ceased to bother to ask permission. Then early one morning, Christopher and I went to a fish market on the coast. While walking along the beach, I spotted groups of fishermen, looked through my camera, and before I had time to click they were racing into my lens. I fled, eyes bulging with fright, like a gazelle running for its life. I had not realised that the coastal Ga tribe

Gyearbuor (Christopher) Asante, Accra, Ghana, late 1990s

felt differently about the camera. Christopher was laughing his head off at my predicament.

I accompanied Christopher to several official events, almost standing in as his wife. On one occasion, a military display. Christopher was dressed in flowing robes and I sat amongst the dignitaries in the elite stand. A procession of soldiers, sailors and airmen flowed by and an awesome air display unfolded in front of our eyes. That night, I saw my happy face on the Ghanaian national news, all too noticeable in the exotic crowd of chieftains, their wives and other dignitaries.

One grim tourist site I needed to see was Cape Coast Castle, infamous for its involvement in the transatlantic slave trade. I had read of how the slaves were kept in the dungeons beneath the castle awaiting shipment, but it was not until I stood in the cramped, low, damp, dark cells that I felt the full impact of the horrifying conditions. I was quite literally overwhelmed by the atmosphere and a feeling of guilt for the sins of my forefathers. Christopher squeezed my hand and gently informed me that some African tribes sold their own people as slaves, something I had never known.

By sheer chance, thirty years later, I learned that an ancestor of mine had died at Cape Coast Castle. Lady Elizabeth "Letitia" Landon was considered not only a great beauty, but was one of the most celebrated poets of her day, known as "L.E.L.". The tide turned when Letitia hit her thirties, when her reputation was destroyed as her free-spirited ways came up against the strict moral values of the Victorians. Finding a husband was now difficult, but a solution was found in George Maclean, Governor of Cape Coast Castle, who was about to return to the distant Gold Coast. Letitia was unhappy with the match, but had little choice. After just two years of marriage, at the age of thirty-six she was found dead in her room in the castle, clutching an empty bottle of prussic acid. This knowledge has certainly enhanced my ominous impressions of the castle.

When the time came to leave Ghana, I felt extremely emotional; I had embraced my new way of life to such a degree that the thought of returning to my world filled me with apprehension. I had grown to admire Christopher even more for his courage in coming to England. Groundbreakers like Christopher have led the way to a more harmonious blending of nations, and long may it continue. His spirit lives on through Humphrey Barclay, who, after Christopher's death, accepted the honour of being made a development chief in Christopher's ancestral town of Kwahu Tafo, where he is buried. I miss him madly but his spirit lives on and so do my pictures of him. It is customary in Ghana to display a large picture of the deceased at their funeral, and my picture of the young actor in his prime was chosen.

Julian and Emma Fellowes

One of the happiest, and I must say, quintessentially English country-house parties I have ever enjoyed was at the estate of Julian and Emma Fellowes. I spent a memorable early May weekend there, enjoying hospitality that brought back memories of my visits to my Auntie Mamie in her grand house in Scotland. The weekend was all one would expect from my dear friend who won an Oscar for Best Original Screenplay for *Gosford Park* and went on to create *Downton Abbey*. As if that is not enough, Julian sits in the House of Lords as Lord Fellowes of West

Stafford. However, he has not forgotten his friends of old and loyally invites them to his stately pile in Dorset.

I arrived on a Friday afternoon when an icy wind was at variance with the warmth of the spring sunshine. After crossing a tiny bridge, there the manor stood: pale-grey hamstone walls bathed in the late afternoon sunshine in a sea of emerald green; a veritable Thomas Hardy landscape. The bright, clear spring light with a brisk north wind, which invariably brings with it crystal-clear blue skies, shafted across the fields, turning the grass to silver. The sheer beauty of it all brought tears of joy to my eyes; the young, light-green leaves on the trees, the cow parsley in full bloom, the lambs with their mothers, the cows and their calves grazing in the lush fields by a clear, fast-running river.

Stafford House is not just an impressive Jacobean manor but a hamlet, built in about 1460 during the troubled reign of Henry VI. The Elizabethans altered the house, as did the Caroleans, and more changes came during the Regency period before the Victorians turned the whole place back to front. So the house is a proper English mishmash. One lone couple occupies the house farthest from the central cluster; they are much loved and their lights in the evening can be a reassuring presence on the isolated estate.

As I drove across the vast sweeping gravel forecourt, Emma, Julian's wife, appeared at her casement bedroom window and greeted me warmly with cries of welcome. Arms akimbo, she was downstairs in seconds, tall, slim, in chocolate brown with a large cluster of pink stones around her neck. Her dramatic mane of dark hair, wrapped in her customary wide scarf, flowed thick and shining down her back to her waist. The bandeau was a light pink to tone with the necklace. With a larger-than-life personality, my hostess did not seem out of place in these opulent surroundings; on the contrary, she was to the manner born. Descended from Lord Kitchener, Emma proudly kept the spirit of her family alive on marrying, and the couple's family name is Kitchener-Fellowes. On his part, Julian is, as he observes, "both upper class and nouveau riche".

Slowly throughout the early evening, the other house guests arrived, each one heralded by a cacophony of barks, echoing off the stone floor in the hallway. Humbug the dachshund loved the sound of her own bark, and Meg the blue collie was happy to join in. The dining room, arguably

Julian Fellowes, 2008

the most impressive room in the house, was where we spent a great deal of time over the weekend and was the setting for all of our meals. More a banqueting hall, this magnificent room has large casement windows on two sides, through which sunlight streamed most of the day. At night, the vast windows were hidden by huge oak shutters, creating an extraordinarily warm, cosy feeling for such a large room, especially with logs crackling robustly in the grate beneath huge fire dogs. Various ancestors stared down at us from the walls, with a modern oil painting of Emma placed above the marble mantelpiece; at meal times, a row of silver candlesticks with soft pink shades lit the long table. There was never a dead grate throughout the entire weekend. Fires were ablaze in whichever room we used, whether it was the library, drawing room, morning room or dining room.

Mornings dawned with bright spring sunshine that woke me at six as it streamed through the cracks in my shutters. I opened them and gazed out at a fairy-tale view of lawns brushed with a silvery layer of frost and dew. The misty sunlight cast a soft hue, like looking through a piece of chiffon. Following the smell of fresh coffee I would descend to the breakfast room, though I was ready like a horse at the stable door to get out in the fresh air onto the long strip of grass that ran straight down to a beautiful, crystal-clear, shallow river. Reeds flowing like green tresses reminded me of the drowning Ophelia painted by Millais.

The grass had been mown along the banks to create paths, with intermittent small wooden bridges crisscrossing from side to side. One larger bridge led on to an island with a charming summer house. There was evidence of many new and exotic plants, while many of the old tree stumps, covered in lichen, ferns and moss, reminiscent of illustrations in children's books, had been left as natural habitats for wildlife.

Saturday luncheon was especially memorable, as Emma's mother, Ursula, sat on my right. Halfway through lunch I turned to converse with her. I had been told that she is very outspoken these days and is apt to mention that she does not like the colour you are wearing. Armed and forewarned, I was pleasantly surprised to have a delightful conversation with her. I was at a huge advantage in that she had loved my father's work in the theatre and remarked on his wonderful sense of timing. A staunch Conservative, Ursula was well informed on her subject and up to date as to who might take over from Michael Howard. "David Cameron," she whispered conspiratorially, "watch out for that young man." She sat back with the look of confidence of someone in the know.

What followed was as good a dinner party as I have ever attended, proof that a good house party, like this one, is the richest way to get to know new people, a chance to go beyond the social façades. Logs crackled robustly in the hearth as if to match the sparkling conversation, much of it about politics, rich in content with none of the usual parochial chat about people I did not know.

The third act of the weekend was Sunday luncheon. To my surprise, in walked a couple who I knew from years before, Jonathan and Penny Marland (now Lord and Lady Marland). I could still picture them sitting side by side in my studio on a square corduroy seat for their engagement shots. Jonathan had teased Penny, who had gone to the trouble of having her make-up done professionally. She was endearingly shy about the whole process, and I had jokingly chastised Jonathan for his callous behaviour. I had known Penny in her single days when she worked for an extremely wealthy Israeli with the bizarre name of Ronald Fuhrer. She had recommended me to take the photographs of his wedding, which proved to be one of my more difficult assignments, as my assistants had to wear white or cream silk clothing and the party had to be covered until five in the morning.

Other than the Marlands, who greeted me warmly, the guests included a chap on my right who was intent on telling me how I should take pictures. It does seem strange that there are still men out there who think they know best just because they are men. A lighter conversation was to be had on the other side with a man whose tie I recognised, as it came from a shop on the street where I live in Notting Hill. Then, no sooner had the last guest departed than our host was taking his leave. I and the other house guests were in the hallway with our bags and in a blink the party was over.

All too soon I was ready to go on to a quiet two days with more friends who live in deepest Dorset at the end of a long road to nowhere surrounded only by working farms. There I would have the time to digest the joys of the weekend before returning to the hurly-burly of London. I took my leave from Stafford House and I shall not forget the image of Emma, the wind tossing her hair like a horse's mane, holding a parasol shaped to match the curves of the house. "We love you, darling Carole," she cried out as she waved me off down the drive, and I could still see her waving as I reached the gate and turned into the leafy road, its banks loaded with wild garlic and lords and ladies. I took pleasure in knowing that this was not a house where people had the time or inclination to pick their guests to pieces the minute they were out of sight.

Colin Stone

A far less traditional host was a unique, colourful personality named Colin Stone. Colin came into my life through a chance meeting at Morton's, the ultra-chic restaurant in Berkeley Square. I was on time but my date was late. Before I had time to feel self-conscious, a large, flamboyant figure with steel-rimmed spectacles and a spiky beard rescued me. "Colin Stone," he announced loudly. He too was waiting for a friend, so he filled the void with his charm, wit and apparent ability to conquer the world. A far-fetched dream of bringing dolphins into England in a huge boat with large pools on board was followed by a more practical scheme, creating garden centres and reinventing garden gnomes, a project that made him into a millionaire by the age of twenty-one; this was the 80s, when millionaires were as rare as hen's teeth. Colin regaled me with the terms of his trade: "gnomology"; "There's no place like gnome"; "Nostronomus"; "the Stately Gnomes of England". Colin, of course, was King Gnome. I found it all hilarious, and we promised to meet again.

One Friday afternoon of a glorious early autumn day I invited Colin to meet my friend Caroline, PA to film director John Schlesinger. Unexpectedly, Caroline turned up with her house guest, Dan McBride, one of the best-looking young men I had ever seen. He was using his physical assets to gain modelling work as a means to travel, putting his life as a singer/songwriter and artist on hold. Colin was in full flow when, quite suddenly, he announced that we should all pack hastily as we were going to drive to Herefordshire to see a house he intended to buy.

One hour later, we were in the car pounding west down the M4 motorway amid the Friday evening traffic with the setting sun smack in our eyes. It grew dark and we started to wonder how much longer the journey would take. West and still farther west we travelled, over the suspension bridge and into Herefordshire, then through verdant valleys with rivers rushing through them. At last we swung through large wrought-iron gates, with darkness all around apart from our headlights as we drove along a lengthy driveway. A field of sheep appeared to be curtsying as they were lit up; they were peeing with fright. Finally, a huge, grey-stone Victorian mansion loomed up, and as we slowed down, my dog Emily started jumping around the car with excitement. We came to a stop on a large, circular gravel forecourt. A shaft of yellow light flooded the entrance as the heavy front door opened to reveal a vast stone hallway with burgundy carpets and a wide staircase with chunky, highly polished dark wood banisters which branched into two flights at the first landing. In the centre hung a huge chandelier.

The owner greeted us and ushered us into the library on the left, and only now did we realise we were in a country-house hotel. Sandwiches and drinks were offered as it was now too late for dinner, and while we guzzled hungrily, we bombarded the owner with questions about the house. Our host was keen to tell us that this house had been the family seat of the Baskervilles and the inspiration for Arthur Conan Doyle's famous Sherlock Holmes story *The Hound of the Baskervilles*. We were promised a visit to the dogs' graves, marked by small but beautiful Victorian gravestones tucked away in the undergrowth to the side of the house. We climbed two floors to our rooms, which, to our surprise, we were told we would have to share. We presumed that the hotel was full and, as we had booked at the last minute, we felt lucky to be there. Colin and Dan doubled up and Caroline and I did the same. An adjoining door between our two rooms created an apartment, and each room had an ensuite bathroom.

The next morning we were up bright and early, and we could not wait to explore. Breakfast was served next to a beautiful indoor pool with a mural depicting a garden scene along the far wall. To the other side was a spectacular view of the lawns and a cedar tree, branches akimbo, and the Black Mountains in the distance. The dining area was empty, but we thought nothing of it and remarked that the other guests must have all eaten earlier. After a large breakfast, we went into the garden, where Colin spouted the names of every plant and tree. It was only when we sat down to lunch that we realised the hotel was completely empty; we were the only guests. Just as we felt as though the place belonged only to us, a member of staff would appear unexpectedly from nowhere and then we would be alone again for several hours.

All weekend we played like children, swam in the pool, steamed in the sauna, gorged ourselves on good food. We climbed the Black Mountains, visited the home of the famous local diarist, Reverend Francis Kilvert, and explored the bookshops of Hay-on-Wye. By Sunday night, we were utterly exhausted. That night just happened to be the airing of the last episode of the American television series *Roots*. Caroline and I decided we would get into our nightclothes and watch the programme in our huge double bed. Soon there was a knock at the door and there were the boys. Could they get in with us to watch the programme? So there we were, looking like characters in a scene from the 1969 film *Bob and Carol and Ted and Alice*. Caroline had smothered her face with cream and sported a large bath cap on her head. I wore a vast T-shirt, and the chaps were in their underwear; a bizarre group, totally unaware of our appearance and lost in the tragedy and drama of *Roots*.

We heard a knock at the door and the owner asking to come in to discuss our departure. Colin panicked, leapt from the bed just as the owner slowly opened the door, caught his foot in the sheet and fell flat on the floor, lying like an enormous whale caught in a fishing net. Caroline was jumping around the room in her bath cap, and Dan and I sat calmly in the bed. We tried to explain, but it was impossible. We were weak with laughter, and the more we tried to explain, the worse it got, until the owner left, extremely red in the face. Only then did we realise Colin was in terrible pain. He had twisted his ankle severely, and it was blowing up like a barrage balloon. Now we had to get a doctor, and again we had to try to explain, and again we simply could not get the story out straight. We did nothing for the reputation of city dwellers in these country people's eyes.

In spite of our disastrous reputation at Clyro, Colin bought the hotel and filled it with friends. The house parties on the weekends were memorable. One day, in a mischievous mood, a group of us decided it would be fun to do a photo shoot of Colin in a foam bath in the round pink bath of the quarters we called the "Greece suite"; we knew he would agree, being an extrovert and a total eccentric. Somebody had the idea that as we did not have any bubble bath, we could use Fairy Liquid, which worked a treat as a naked Colin frolicked in the foam. Shortly afterwards we greeted smart guests to lunch and to our horror, clouds of billowing foam were flying down the drive. The Fairy Liquid had escaped from the outdoor lead pipes and was flying around like celestial entities; we feigned ignorance.

I shall always be grateful to Colin for expanding my awareness and knowledge of our planet. He was a fountain of knowledge about the countryside and intensely aware of the damage that we are causing to nature. Colin was also immensely generous when friends needed help. Among them was Lionel Bart, the composer of the musical *Oliver!*, who became depressive and drank himself into oblivion, spending all day in

Colin Stone, Clyro Court, Herefordshire, England, 1980s

bed unless coaxed out by a friend. One day Colin and I went to his mews house in Kensington to get him out of bed at lunchtime. Colin dressed him as he would a small child and we took him to Kensington Gardens, where he responded with huge appreciation to our nannying. Sadly, we could not be there all the time and Lionel passed away in 1999.

Clyro soon became my place of solitude, and I often went there when I needed a break from London. I was never more content and my dog Emily was in heaven. She was the only dog ever in residence, so as far as she was concerned, this was her domain. I often wondered what drew me so strongly to this part of the world. Was it the house or the countryside? Or was there a connection between Ledbury Road, my home in London, and the town called Ledbury near Clyro? Whatever it was, I could not explain the feeling of otherworldliness I enjoyed here.

When I first went to Clyro, I heard mutterings about all the "ley lines" converging in the area. As I investigated further, I learned that these "leys" or "leas" are said to exert a subterranean magnetic force that comes to the fore at Stonehenge and other sacred sites.

I witnessed a couple of strange experiences in Clyro. Once, I was sitting alone reading in the study when Emily, curled up asleep in the other winged armchair, quite suddenly sat bolt upright and her soft, silky hair stood up on her back. This is a fact I shall never forget, as I did not know till then that this type of fur could stand up. Emily stared into the corner of the room for several minutes as I kept quite still, then she lay down and went back to sleep. I mentioned this to various people who knew the house; they did not laugh and knew the ghost well.

Another incident was far more alarming and sounds absurdly like a sci-fi movie. I was driving through the Herefordshire countryside on a full moon when a bright light, like a huge star, appeared in my rear window. I thought the object must be an aeroplane and yet I knew it was not. As I drove faster and faster along the winding country lanes, the bright light followed me and I panicked. Luckily I was near to Clyro, and when I arrived I thought I would not mention the matter, as they might think me certifiable. I walked into the study and there to my amazement was Colin, white faced and shaking, along with a policeman taking notes. He too had seen this same phenomenon and what was more, the policeman told us that several other people had made similar sightings in the area. All these years later, I am fascinated that the US government is prepared to investigate UFO sightings, giving some kind of validation to my own experience.

My visits to Clyro were magical, but all good things come to an end and, in this case, they came dramatically. Reality was not for Colin, and the magic mushrooms which grew around the Herefordshire countryside had enhanced his fantasy world until, unbeknownst to any of us, he had run up huge debts. Soon this charismatic Walter Mitty character was behind bars. The shock of prison was too much for Colin and he died young; like Elton John's song, he was "a candle in the wind".

Lionel Bart, Kensington Gardens, London, early 1990s

CHAPTER 18

NEW MEXICO AND LONDON, SUMMING UP

"But in contentment still I feel the need of some imperishable bliss." – Wallace Stevens

Two places on earth speak to me with special force, and they are forever linked. The first is London, my home since childhood, where my roots run deep and where I continue to thrive. The other is New Mexico, with mountains, mesas, vast open spaces and endless skies that reach out to me even when I am thousands of miles away and, I can honestly say, is the one place on God's earth where I have experienced a state of bliss. New Mexico is also about Dan McBride, who introduced me to this wild terrain and is a soulmate who has been ever in my heart since the 70s.

In London, thousands of talented artistic people have passed through my door and climbed to the studio that has been home base for a career in

Dan McBride, Santa Fe, New Mexico, late 1990s

which I have photographed hundreds of actors on film sets, in plays and on location. The 80s were an especially prolific time in my career, when I was pushing forward, striving to carve myself a niche in an over-crowded profession. Dan, meanwhile, was working as a singer, songwriter and artist in Los Angeles, and he was always a presence on my many trips there.

Our paths crossed again after an especially difficult time for me. My grandmother was slowly fading gently into the good night. My sister and I took it in turns to drive backwards and forwards to Surrey, a pattern which continued over a period of three months. During the last six weeks the trips became daily, and I worked in the mornings then sat with Granny in the afternoons, never knowing if it would be the last day. The only thing that mattered to us was that she could remain in her own home. Granny lay in her wonderful four-poster bed gazing out at the garden, which was her great passion, with my Cavalier King Charles spaniel, Emily, nestled by her side. When the time came, my sister collected our parents from the airport and I stayed with Granny until they arrived to say goodbye. It was early evening, and the bedroom was softly lit by one small bedside lamp throwing a warm glow around her in her soft pink crocheted bed jacket. My parents arrived just in time to say goodbye and I held my father's hand as he came to terms with her farewell. I stayed by Granny's side until she slept peacefully, then took a short break, and by the time I returned I sensed that there was nobody there. I walked to the end of the garden in the rain and stood by the tiny wrought-iron gate and let the tears flow; solitude was all I craved so that I could be close to her spirit.

Within days, I was on an aeroplane to Palm Springs, where Dan was then living. The sizzling desert heat kept us indoors as much as the cold can do in the winter. Only the early morning and evenings were pleasant, but I did not mind, for I was with the one person who truly understood and instinctively knew how to help. This gentle man's presence by my side night and day restored me and soon we were laughing like old times.

A few more years slipped by before I saw Dan again. Now he was living in New Mexico, way up in Dixon, a tiny artists' community. Dan had now thrown himself wholeheartedly into painting, with a strong Native American influence. I fell so passionately for an enormous acrylic painting of a dog under a night sky that I brought it back to England. Another painting by Dan, which has haunted me to this day, is of Crow

Mother rising from a kiva, the subterranean ceremonial chamber of the Pueblo Indians. Her face was shrouded and barely visible, her hands half male half female, as she represents both sexes in one being.

By day, we swam naked in freezing mountain rivers and basked like lizards on hot rocks. Dan took a landscape shot of me which I love, as I look, at a distance, like the statue of the mermaid in Copenhagen or a white seal. At night, we relaxed in natural rock hot tubs under the stars and tried to figure out their constellations. I felt liberated, at one with nature, in a way that I had never experienced since childhood. I embraced the Native American culture, attending the burning of old father gloom on a new moon and attending a deer dance followed by dinner in their homes.

One day Dan and I walked way up into the mountains along a narrow trail worn into the rocks by the Native Americans' moccasins. I picked crystals from the rocks and I have them to this day. As we started back down the mountain, the heavens opened and we were caught in a monumental thunderstorm; I felt exhilarated and was reminded of the carefree days of childhood. Another day we attended a Native American deer dance and picnicked at sunset to watch the sun disappear behind the mountains, to be instantly replaced by the thinnest silver thread, the new moon. I almost wept at the beauty of the crystal-clear night sky.

Another few years slipped by before I returned to New Mexico, when Dan was undergoing surgery to replace a smashed vertebra. Bone had been removed from his hip to replace the vertebra and two metal rods inserted to support his back. It was winter, and thick snow blanketed the majestic Sangre de Cristo mountain range. Only the dark green of the pinyon bushes, and the soft brown sugar colour of the adobe houses broke up the monotonous white landscape. Dan had his own small studio with kitchen, bathroom, music room and bedroom, all bathed in magical light; everything he could want.

I awoke that first morning feeling rejuvenated and free, no buildings imprisoning me, no cars swishing endlessly past my door, just huge vistas of majestic mountains changing colour from dawn to dusk. Now I understood why artists flock to New Mexico. The air is so dry that it makes the landscape crystal clear; none of those misty scenes so familiar in England. No wonder Ansel Adams spent years working in this photogenic paradise. I could not wait to tackle this new opportunity, as landscapes under snow have always thrilled me; I have many shots taken on Hampstead Heath where I was brought up, but this would be superlative. Dan painted all day long as I waited patiently for the shadows to lengthen. Then I became like a dog waiting for its walk, pacing restlessly, anxious not to miss the rich afternoon light. As an artist, Dan was the perfect person to take me to the best spots, as an artist's eye works equally well for a photographer.

I also wanted to capture Black Mesa, a huge tabletop mountain. It was not easy, for no sooner had we found the best spot than a group of Native Americans appeared out of the blue and demanded that we move on. They do not want the white man to photograph land which belongs to them; the feeling of hostility is still strong. How were they to know that we respected their way of life? Sadly we left but, just in time, we found another spot and the mountain filled my lens just as a storm was building up in the background.

Another outing took us to a kiva high up in a mountain. Snow had fallen again in the night and even for New Mexico the snowpack was exceptionally deep. We arrived at the end of a road, where we parked and started off along a long, winding path through a pine forest. The mountains rose up beside us as we followed a small frozen stream. Then the sign appeared, "Do not walk, danger due to snow". We ignored the warning. I was already tired with the weight of the tripod and Hasselblad but it is amazing what extra Herculean strength I could summon up when the will was there. Just when we thought we would never find the kiva, I saw a cave way up near the top of the mountain. Just visible to the naked eye was a mound with a pole sticking out. The climb was long and steep, so steep that the last stage could only be reached by narrow wooden ladders. The sun was sinking fast as the sky changed like a kaleidoscope, from blue to various shades of pink, the exposure dropping by the second. I raced to set up my tripod, attached the Hasselblad, loaded the film and shot as fast as I could; I had to be so quick off my feet finding the best composition.

In a blink, the light had gone and the dusk enveloped us as we started down the mountain. The imminent darkness chased us as we hurried back along the valley. Our limbs were aching from carrying the equipment in the deep snow, a flicker of fear overtook me at the thought of becoming lost in the dark as I quoted Bunyan's *Pilgrim's Progress*: "Yea, though I walk through the valley of death, my rod and staff, they comfort me still ..." Trying hard to remember all the words took my mind off our predicament. Just as night descended, the car loomed up before us.

Darkness loomed on another front as well. Dan was to undergo a further operation on his back, this time to remove the metal rods which kept him in constant pain. The operation was life threatening, and I made my choice to return after a short stay in Los Angeles to support him all the way. The skies truly darkened. The operation was horrendous. No painkillers could override the degree of agony caused by the damage to nerves and muscles when removing the rods. All day I sat by Dan's bedside, and only the view of the Sangre de Cristo mountains outside the hospital window kept me sane. At the end of each day, I would return to the apartment drained. All I wanted was solitude, the company of the two birds in a cage, and the long-lasting candle which I kept alight throughout Dan's stay in hospital. Meditation, prayer and walks in the mountains were my strength. Dan's friends kindly invited me to their homes but, for the first time in my life, I really discovered how solitude in adversity can provide more strength than company. This story has a happy ending, as Dan made a miraculous recovery. Before my departure he requested that I take pictures of the scars on his back, an artistic challenge, and I rose to the occasion to gain something positive out of darkness.

Dan McBride, Santa Fe, New Mexico, late 1990s

Part of my soul will always be in New Mexico, and those landscapes and skies still seem close to me after many years and even here in faraway England, and especially in my studio, with its huge windows looking out over the rooftops of West London. The light continually changes with the seasons as I watch the sun setting behind a distant stream of silhouetted aeroplanes that descend in a wide curve like arrows from a bow towards Heathrow. The full moon bathes the studio with its silvery light, which is sometimes so bright that I am fooled into believing that I have left the lights on. The new moon quietly offers its eternal promise of new beginnings.

Home for me is not just London, but Notting Hill, a village unto itself that has long been home to an eclectic mix of artists and artisans. The Portobello Market is the heartbeat of the area, with market stalls displaying fresh fruit and vegetables. A true village feel extends to the ritual that takes place in the market when one of the stallholders dies. Wreaths are placed on the stalls and the coffin is born on a Victorian black carriage with an ornate awning; four black horses sporting black plumes on their heads clip clop along the narrow streets.

My time away, including long stints in Los Angeles, has made me truly appreciate Notting Hill and I understand why fans of the film *Notting Hill* still invade the locations. One day in 2009, a more interesting-looking production popped up. I watched as shops and sections of roads were transformed back to the late 60s for a film called *Hippie Hippie Shake*, based on the memoirs of Richard Neville, who was the editor of the notorious Australian magazine *Oz*. The scenes being shot in the Portobello Road had been meticulously set-dressed to mimic the King's Road in Chelsea in 1967. This was the year when, still living with my parents, I walked onto a film set for the first time as an assistant unit publicist on *Joanna*, a cult movie which epitomised that era where everything hip revolved around the King's Road or Carnaby Street.

I felt as though I was experiencing time travel as I came upon the set. I blinked in disbelief; the camera crew looked exactly like the men I had known on feature films many moons ago, in their fifties and sixties with chiselled, weather-beaten faces, shoulder-length grey hair, anoraks and heavy boots. There is no glamour in working at night in the freezing cold or getting up at five in the morning in the winter months. These were outdoor guys, hardened by many years of exposure to the elements, who conjured up images of explorers battling against the elements.

A crowd dotted with paparazzi gazed up the road trying to get a glimpse of Sienna Miller, and I was able to take a photograph of Cillian Murphy, making a mental note that it would be worth keeping. I felt drawn towards the familiar group gathered around the Arriflex camera, which looked exactly like those used in the 70s. No digital cameras in sight; they were shooting on 35mm film. The low autumn sunshine was warm but when dusk fell and the temperature dropped, I was only too happy to return home to the warmth.

Cillian Murphy, on location, *Hippie Hippie Shake*, Notting Hill, London, 2009

As the sun set and sent a kaleidoscope of colours across the sky, I saw from my studio a huge crane silhouetted against the skyline. Later, as the darkness rolled in all I could see was a bank of lights suspended in mid-air, reminding me that the crew was still huddled around the camera in the cold night air.

Bumping into the film set was like an affirmation that I had been right all those years ago when I closed a chapter of my life in which I travelled all over working on films. I had been enchanted by the stardust, but now I saw with crystal clarity that I had been right to concentrate on the portraits that kept on rolling in. Then, after many years as a portrait photographer, I once again felt I needed change. Fate stepped in when my beloved Daddy passed away and all my energy was with him. That night a moon lily that I had grown from a leaf brought back from our house in Mallorca miraculously flowered for the first time, after six years. What compounded this mysterious event was that the moon lily only flowers for one night and wilts by dawn: the flower became the star of my next venture: an exhibition.

My imagination took flight, as the flower looked uncannily like a creature in the deep ocean where there is no daylight. I then picked up a single bract from a handkerchief tree that was so beautiful that I was drawn to photograph it against a black background. One after another, I photographed more flowers that seemed to resemble their soulmates in the deep. Then chance favoured me again as I met Karine Giannamore, an established artists' agent, and we put together an exhibition at the 20th Century Theatre in Notting Hill, *Floating Flowers: Mirroring the Deep*. The evening was a wonderful opportunity to bring together many of my talented clients and friends, along with potential buyers.

Sadly, my darling mother passed away soon after the exhibition, and I took a break from photography. But with a huge body of work behind me, I had the opportunity to donate pictures to charities that I support.

I have broadened my horizons again and begun photographing a wider variety of subjects. Having heard that Annie Leibovitz was holding an exhibition of her digital iPhone pictures, I embraced the versatility of my iPhone Pro. I was excited, as this felt so fresh. I could walk so far without worrying about the weight of cameras and I took on a new lease of life. When the pandemic struck, everyone was in the doldrums, a position

The moon lily (kawana)

with which I have long been familiar ever since I first went freelance. So in a way, this gave me an advantage in finding a new sense of direction that saved me from despair in the challenging times of the pandemic. I have always known the importance of living in the moment, and with the encouragement of several friends I realised that was the perfect time to complete this memoir that has been marinating for so long.

INDEX

ill refers to an illustration; *port* to a portrait